Generalized Functions, Convergence Structures, and Their Applications

Generalized Functions, Convergence Structures, and Their Applications

Edited by

Bogoljub Stanković, Endre Pap, and Stevan Pilipović

Institute of Mathematics
Novi Sad, Yugoslavia

and

Vasilij S. Vladimirov

Steklov Institute of Mathematics
Moscow, USSR

PLENUM PRESS • NEW YORK AND LONDON

Library of Congress Cataloging in Publication Data

Generalized functions, convergence structures, and their applications / edited by Bogoljub Stanković . . . [et al.].
  p.   cm.
"A collection of papers presented at the International Conference, 'Generalized Functions, Convergence Structures, and Their Applications,' held from June 23-27, 1987, in Dubrovnik, Yugoslavia" — Pref.
Includes bibliographies and index.
ISBN-13: 978-1-4612-8312-6   e-ISBN-13: 978-1-4613-1055-6
DOI: 10.1007/ 978-1-4613-1055-6
  1. Distributions, Theory of (Functional analysis) — Congresses. 2. Convergence — Congresses. I. Stanković, Bogoljub II. International Conference, "Generalized Functions, Convergence Structures, and Their Applications" (1987: Dubrovnik, Croatia)
QA324.G46   1988   88-22547
515.7'82 — dc19   CIP

Proceedings of an international conference on Generalized Functions, Convergence Structures, and Their Applications, held June 23-27, 1987, in Dubrovnik, Yugoslavia

© 1988 Plenum Press, New York
Softcover reprint of the hardcover 1st edition 1988
A Division of Plenum Publishing Corporation
233 Spring Street, New York, N.Y. 10013

All rights reserved

No part of this book may be reproduced, stored in a retrieval system, or transmitted in any form or by any means, electronic, mechanical, photocopying, microfilming, recording, or otherwise, without written permission from the Publisher

PREFACE

This Proceedings consists of a collection of papers presented at the International Conference "Generalized functions, convergence structures and their applications" held from June 23-27, 1987 in Dubrovnik, Yugoslavia (GFCA-87). 71 participants from 21 countries from all over the world took part in the Conference.

Proceedings reflects the work of the Conference. Plenary lectures of J. Burzyk, J. F. Colombeau, W. Gähler, H. Keiter, H. Komatsu, B. Stanković, H. G. Tillman, V. S. Vladimirov provide an up-to-date account of the current state of the subject. All these lectures, except H. G. Tillman's, are published in this volume. The published communications give the contemporary problems and achievements in the theory of generalized functions, in the theory of convergence structures and in their applications, specially in the theory of partial differential equations and in the mathematical physics. New approaches to the theory of generalized functions are presented, motivated by concrete problems of applications. The presence of articles of experts in mathematical physics contributed to this aim.

At the end of the volume one can find presented open problems which also point to further course of development in the theory of generalized functions and convergence structures.

We are very grateful to Mr. Milan Manojlović who typed these Proceedings with extreme skill and diligence and with inexhaustible patience.

THE CONFERENCE GFCA-87 WAS SUPPORTED BY THE FOLLOWING ORGANIZATIONS

YUGOSLAV SUPPORTING ORGANIZATIONS

    Institut za matematiku PMF u Novom Sadu
    Vojvođanska akademija nauka i umetnosti
    Samoupravna interesna zajednica za naučni rad SAP Vojvodine
    Savez republičko-pokrajinskih samoupravnih interesnih zajednica za naučne delatnosti u SFRJ
    Pokrajinski zavod za međunarodnu naučnu, kulturnu, prosvetnu i tehničku saradnju

INTERNATIONAL SUPPORTING ORGANIZATION

    International Centre for Theoretical Physics, Trieste (Italy)

The Conference is organized under the supervision of the Union of Mathematicians, Physicists and Astronomers of Yugoslavia

CONTENTS

SECTION I. PLENARY LECTURES

Nonharmonic solutions of the Laplace equation 3
    J. Burzyk

Generalized functions; multiplication of distributions; applications
    to elasticity, elastoplasticity, fluid dynamics and acoustics . 13
    J.F. Colombeau

Monads and convergence 29
    W. Gähler

Simple applications of generalized functions in theoretical physics:
    the case of many-body perturbation expansions 47
    H.F.G. Keiter

Laplace transforms of hyperfunctions: another foundation of the
    Heaviside operational calculus 57
    H. Komatsu

S-asymptotic of distributions 71
    B. Stanković

The Wiener-Hopf equation in the Nevanlinna and Smirnov algebras and
    ultradistributions 83
    V.S. Vladimirov

SECTION II. GENERALIZED FUNCTIONS

On nonlinear systems of ordinary differential equations 99
    L. Berg

A new construction of continuous endomorphisms of the operator field 113
    A. Bleyer

Some comments on the Burzyk-Paley-Wiener theorem for regular
    operators 121
    T.K. Boehme

Two theorems on the differentiation of regular convolution quotients 125
    T.K. Boehme

Values on the topological boundary of tubes 131
    R.D. Carmichael

Abelian theorem for the distributional Stieltjes transformation . . 139
  D. Nikolić-Despotović, S. Pilipović

Some results on the neutrix convolution product of distributions . 147
  B. Fisher

On generalized transcedental functions and distributional
    transforms 157
  A.N. Goyal, V.K. Chaturvedi

An algebraic approach to distribution theories 171
  J. de Graaf, A.F.M. ter Elst

Products of Wiener functionals on an abstract Wiener space 179
  Sh. Ishikawa

Convolution in $K'\{M_p\}$-spaces 187
  A. Kaminski, J. Uryga

The problem of the jump and the Sokhotski formulas in the space of
    generalized functions on a segment of the real axis 197
  L.V. Kartashova, V.S. Rogozhin

A generalized fractional calculus and integral transforms 205
  V. Kiryakova

On the generalized Meijer transformation 219
  E.L. Koh, E.Y. Deeba, M.A. Ali

The construction of regular spaces and hyperspaces with respect to
    a particular operator 227
  G. Liu

Operational calculus with derivative $\hat{S} = S^2$ 235
  E. Mieloszyk

Solvability of nonlinear operator equations with applications to
    hyperbolic equations 245
  P.S. Milojević

Some important results of distribution theory 251
  O.P. Misra

Hyperbolic systems with discontinuous coefficients: examples . . . 257
  M. Oberguggenberger

Estimations for the solutions of operator linear differential
    equations 267
  E. Pap, Đ. Takači

Invariance of the Cauchy problem for distribution differential
    equations 279
  J. Persson

On the space $D'^{(M_p)}_{L^q}$, $q \in [1,\infty]$ 285
  S. Pilipović

Peetre's theorem and generalized functions 297
  J.W. de Roever

Infinite dimensional Fock spaces and an associated generalized
    Laplacian operator 305
  J. Schmeelk

The n-dimensional Stieltjes transformation 317
  A. Takači

Colombeau's generalized functions and non-standard analysis 327
  T.D. Todorov

One product of distributions 341
  M. Tomić

Abel summability for a distribution sampling theorem 349
  G.G. Walter

On the value of a distribution at a point 359
  R. Wawak

SECTION III. CONVERGENCE STRUCTURES

On interchange of limits 367
  P. Antosik

Countability, completeness and the closed graph theorem 375
  R. Beattie, H.-P. Butzmann

Inductive limits of Riesz spaces 383
  W. Filter

Convergence completion of partially ordered groups 393
  I. Fleischer

Some results from nonlinear analysis in limit vector spaces 399
  O. Hadžić

Completions of Cauchy vector spaces 409
  D.C. Kent, G.D. Richardson

Regular inductive limits 415
  J. Kucera

Weak convergence in a K-space 419
  E. Pap

The Banach-Steinhaus theorem for ordered spaces 425
  C. Swartz

SECTION IV. OPEN PROBLEMS

Open problems 435

Participants 443

Index 449

SECTION I. PLENARY LECTURES

NONHARMONIC SOLUTIONS OF THE LAPLACE EQUATION

Jozef Burzyk

Institute of Mathematics
Polish Academy of Sciences
Katowice, Poland

1. It is known that solutions of the Laplace equation

$$\frac{\partial^2 u}{\partial x^2} + \frac{\partial^2 u}{\partial y^2} = 0$$

considered in the space of distributions (hyperfunctions) are always classical solutions called harmonic functions. In this paper we shall consider the Laplace equation in the space of so-called boehmians and show that there may appear solutions which are not classical. The boehmians, we are dealing with, are particular cases of the more general concept of generalized functions introduced in [1], p. 120. Here, they are defined by using delta-sequences.

By a delta-sequence in $\mathbb{R}^q$ we mean a sequence $\{\delta_n\}$ of functions in $C^\infty(\mathbb{R}^q)$, such that the following conditions hold:

1° supp $\delta_n \subset K(0,\varepsilon_n)$ for $n \in \mathbb{N}$ with $\varepsilon_n \to 0$, where $K(0,\varepsilon_n)$ is a ball in $\mathbb{R}^q$;

2° $\int_{\mathbb{R}^q} \delta_n(t)dt = 1$ for $n \in \mathbb{N}$;

3° $\int_{\mathbb{R}^q} |\delta_n(t)|dt < M < \infty$ for $n \in \mathbb{N}$.

A sequence (f_n,δ_n) of ordered pairs (f_n,δ_n) where $f_n \in C^\infty(\mathbb{R}^q)$ and $\{\delta_n\}$ is a delta-sequence is said to be fundamental if, for every $m,n \in \mathbb{N}$,

$$f_m * \delta_n = f_n * \delta_m,$$

where the convolution is understood in the usual sense.

We say that two fundamental sequences $\{(f_n,\delta_n)\}$ and $\{(g_n,\rho_n)\}$ are equivalent and we write

$$\{(f_n,\delta_n)\} \sim \{(g_n,\rho_n)\},$$

if and only if the interlaced sequence

$$(f_1,\delta_1),(g_1,\rho_1),(f_2,\delta_2),\ldots$$

is fundamental. It can be shown that this is an equivalence relation which partitions the set of all the fundamental sequences into equivalence classes. These equivalence classes are called boehmians. The boehmian determined by the fundamental sequence $\{(f_n,\delta_n)\}$ will be denoted by the symbol $[\{(f_n,\delta_n)\}]$.

To every distribution f in $\mathbb{R}^q$ we assign a boehmian $[\{(f * \delta_n,\delta_n)\}]$, where $\{\delta_n\}$ is a delta-sequence. From the results of this paper it will follow that there are boehmians which are not distributions. We assume the following definitions of operations on boehmians:

(a) $[\{(f_n,\delta_n)\}] \pm [\{(g_n,\rho_n)\}] = [\{(f_n * \rho_n \pm g_n * \delta_n, \rho_n * \delta_n)\}];$

(b) $\lambda [\{(f_n,\delta_n)\}] = [\{(\lambda f_n,\delta_n)\}]$ where $\lambda \in \mathbb{R}$;

(c) $[\{(f_n,\delta_n)\}] * \phi = [\{(f_n * \phi,\delta_n)\}]$ if $\phi \in \mathcal{D}$,

i.e., ϕ is a smooth function of a bounded support;

(d) $\dfrac{\partial^{|\alpha|}}{\partial t_1^{\alpha_1} \cdots \partial t_q^{\alpha_q}} [\{(f_n,\delta_n)\}] = [\{(\dfrac{\partial^{|\alpha|} f_n}{\partial t_1^{\alpha_1} \cdots \partial t_q^{\alpha_q}},\delta_n)\}]$

where α is an arbitrary multiindex.

It can be checked that all the above operations are well defined, they coincide with operations on distributions and they have the usual basic properties. The set of all the boehmians with the above operations is denoted by $B(\mathbb{R}^q)$, or shortly, by B. Thus, we see that $B(\mathbb{R}^q)$ is a linear space which is closed under the derivation and convolution with the smooth functions of a bounded support. In the sequel of this paper, the elements of $B(\mathbb{R}^q)$ will be denoted by x,y,z,... .

We say that a sequence $\{x_n\}$ of boehmians x_n converges to a boehmian x and we write $x_n \to x$ in $B(\mathbb{R}^q)$, if there is a delta-sequence $\{\delta_n\}$, such that $x_n * \delta_k \to x * \delta_k$ in $C(\mathbb{R}^q)$, i.e., $x_n * \delta_k$ and $x * \delta_k$ are continuous for n,k $\in \mathbb{N}$ and the sequence $\{x_n * \delta_k\}$ converges to $x * \delta_k$ almost uniformly in $\mathbb{R}^q$ as $n \to \infty$ for $k \in \mathbb{N}$.

The algebraic operations, the convolution and derivation defined in

$B(\mathbb{R}^q)$ are continuous with respect to the defined convergence in $B(\mathbb{R}^q)$.

A series
$$\sum_{n=1}^{\infty} x_n$$
is convergent in $B(\mathbb{R}^q)$, if the sequence of its partial sums
$$\sum_{k=1}^{n} x_k$$
converges in $B(\mathbb{R}^q)$. Obviously, if the series converges in $B(\mathbb{R}^q)$, then $x_n \to 0$ in $B(\mathbb{R}^q)$.

2. The following theorem is the main result in this paper.

Theorem Assume that $f_n \in C(\mathbb{R}^q)$ and $f_n \to 0$ in $B(\mathbb{R}^q)$. Then, there exists a subsequence $\{g_n\}$ of $\{f_n\}$ such that the series
$$\sum_{n=1}^{\infty} g_n$$
converges in $B(\mathbb{R}^q)$ and if $\phi \in \mathcal{D}$, then
$$\left(\sum_{n=1}^{\infty} g_n \right) * \phi \in C(\mathbb{R}^q),$$
iff the series
$$\sum_{n=1}^{\infty} g_n * \phi$$
converges in $C(\mathbb{R}^q)$, i.e., converges almost uniformly.

Proof Assume that $\{\rho_k\}$ is a delta-sequence such that $\text{supp}\rho_k \subset K(0,1)$, $f_n * \rho_k \to 0$ as $n \to \infty$ for $k \in \mathbb{N}$. We note that $f_n * \rho_k \to f_n$ as $k \to \infty$ for $n \in \mathbb{N}$. Let $\alpha_1 = 1$. We find β_1 such that
$$\| f_{\alpha_1} - f_{\alpha_1} * \rho_m \|_1 < \frac{1}{2} \text{ for } m \geq \beta_1.$$
Next, we find the indices $\alpha_2 > \alpha_1$ and $\beta_2 > \beta_1$ such that
$$\| f_{\alpha_2} * \rho_{\beta_1} \|_2 < 2^{-2} \text{ and } \| f_{\alpha_2} - f_{\alpha_2} * \rho_m \|_2 < 2^{-2} \text{ for } m \geq \beta_2.$$
Here, and in the sequal of this proof, we mean that if $f \in C(\mathbb{R}^q)$, then for

every positive number T

$$\|f\|_T = \max\{|f(t)| : |t| \le T\}.$$

By induction, we find two increasing sequences $\{\alpha_n\}$ and $\{\beta_n\}$, such that

$$\|f_{\alpha_n} * \rho_{\beta_k}\|_n < 2^{-n} \text{ for } k < n$$

and

$$\|f_{\alpha_n} - f_{\alpha_n} * \rho_{\beta_k}\|_n < 2^{-n} \text{ for } k \ge n.$$

Assume that $g_n = f_{\alpha_n}$ and $\delta_n = \rho_{\beta_n}$ for $n \in \mathbb{N}$. Then, we have

$$\|g_n * \delta_k\|_n < 2^{-n} \text{ for } k < n \tag{1}$$

and

$$\|g_n - g_n * \delta_k\|_n < 2^{-n} \text{ for } k \ge n. \tag{2}$$

From (1) it follows that for every $k \in \mathbb{N}$, the series

$$\sum_{n=1}^{\infty} g_n * \delta_k$$

converges almost uniformly. This means that the series

$$\sum_{n=1}^{\infty} g_n$$

converges in $B(\mathbb{R}^q)$, and

$$\sum_{n=1}^{\infty} g_n = x$$

for some $x \in B(\mathbb{R}^q)$. To prove the second part of the Theorem we put

$$E = \{\phi \in \mathcal{D}: x * \phi \in C(\mathbb{R}^q)\}.$$

We remark that if series

$$\sum_{n=1}^{\infty} g_n * \phi \tag{3}$$

converges almost uniformly, then $\phi \in E$. To complete the proof, we should show the converse, i.e., if $\phi \in E$, then (3) converges almost uniformly in $C(\mathbb{R}^q)$. To this end we shall introduce the following notations.

For every $\phi \in E$ and for every $n \in \mathbb{N}$, we put

$$p_n(\phi) = \|x * \phi\|_n + \int |\phi| + \sum_{i=1}^{n} \|g_i * \phi\|_n,$$

and note that p_n is a norm on E for $n \in \mathbb{N}$. For every $r > 0$, we put

$$\mathcal{D}_r = \{\phi \in \mathcal{D} : \operatorname{supp} \phi \subset K(0,r)\} \text{ and } E_r = \mathcal{D}_r \cap E.$$

Assume that $\phi \in E$ and

$$S_m(\phi) = \sum_{i=1}^{m} g_i * \phi$$

for $m \in \mathbb{N}$. We are going to show that the sequence $\{S_m(\phi)\}$ converges almost uniformly in $C(\mathbb{R}^q)$, or equivalently, series (3) converges almost uniformly in $C(\mathbb{R}^q)$. Let r be a positive number such that $\phi \in E_r$, let T be a fixed positive number and let $n_0 = E(T + r) + 1$. We may assume that $r \geq 1$ and $m > n_0$. Then, we have

$$\|S_m(\phi)\|_T \leq \sum_{n=1}^{n_0} \|g_n * \phi\|_T + \|\sum_{n=n_0+1}^{m} g_n * \phi\|_T$$

$$\leq p_{n_0}(\phi) + \sum_{n=n_0+1}^{m} \|(g_n - g_n * \delta_m) * \phi\|_T + \|\sum_{n=n_0+1}^{m} g_n * \delta_m * \phi\|_T. \tag{4}$$

We note that $\|f * \phi\|_T \leq \|f\|_{T+r} \int |\phi|$, if $\phi \in E_r$. Hence, by (2), we get

$$\sum_{n=n_0+1}^{m} \|(g_n - g_n * \delta_m) * \delta\|_T \leq \sum_{n=n_0+1}^{m} \|g_n - g_n * \delta_n\|_{T+r} \int |\phi|$$

$$\leq \sum_{n=n_0+1}^{m} \|g_n - g_n * \delta_m\|_n \int |\phi| \leq p_{n_0}(\phi). \tag{5}$$

We recall that $\sup \delta_m \in K(0,1)$ for $m \in \mathbb{N}$, $1 < r$, $n_0 > T+r > T+1$. Hence, we get

$$\|\sum_{n=n_0+1}^{m} g_n * \phi * \delta_m\|_T \leq \|x * \phi * \delta_m\|_T + \sum_{n=1}^{n_0} \|g_n * \phi * \delta_m\|_T$$

$$+ \sum_{n=m+1}^{\infty} \|g_n * \phi * \delta_m\|_T \leq \|x * \phi\|_{T+1} \int |\delta_m| + \sum_{n=1}^{n_0} \|g_n * \phi\|_{T+r} \int |\delta_m|$$

7

$$\sum_{n=m+1}^{\infty} \|g_n * \delta_m\|_{T+r} \int |\phi| \leq M(\|x * \phi\|_{n_0} + \sum_{n=m+1}^{n_0} \|g_n * \delta_m\|_{n_0} \int |\phi|)$$

$$+ \sum_{n=m+1}^{\infty} \|g_n * \delta_m\|_{n_0} \int |\phi| \leq 2Mp_{n_0}(\phi) + \sum_{n=m+1}^{\infty} \frac{1}{2^n} \int |\phi| \leq 3Mp_{n_0}(\phi) \text{ if } \int |\phi| < M.$$

(6)

By (1), we have

$$\|g_n * \delta_m\|_{n_0} \leq \|g_n * \delta_m\|_n < 2^{-n},$$

if $n > m$. From (4), (5) and (6) we conclude that

$$\|S_m(\phi)\|_T \leq Cp_{n_0}(\phi),$$

whenever $n_0 = E(T+r)+1$ and $m \geq n_0$, $r \geq 1$ and $\phi \in E_r$. Now, we are going to show that $\{S_m(\phi)\}$ is a Cauchy sequence in $C(\mathbb{R}^q)$. To this end we note that

$$\|S_n(\phi) - S_m(\phi)\|_T \leq \|S_n(\phi) - S_n(\phi * \delta_k)\|_T +$$

$$+ \|S_n(\phi * \delta_k) - S_m(\phi * \delta_k)\|_T + \|S_m(\phi) - S_m(\phi * \delta_k)\|_T.$$

Let ε be a positive number. We have

$$\|S_m(\phi) - S_m(\phi * \delta_k)\|_T = \|S_m(\phi - \phi * \delta_k)\|_T.$$

We note that $\phi - \phi * \delta_k \in E_{r+1}$. Let $n_1 = E(T+r+1)+1$. By what have been proved, we can write

$$\|S_m(\phi - \phi * \delta_k)\|_T \leq C_1 p_{n_1}(\phi - \phi * \delta_k)$$

for every $m > n_1$ and $k \in \mathbb{N}$. Since

$$p_{n_1}(\phi - \phi * \delta_k) = \|x * (\phi - \phi * \delta_k)\|_T + \int |\phi - \phi * \delta_k| +$$

$$+ \sum_{i=1}^{n_1} \|g_i * (\phi - \phi * \delta_k)\|_T$$

and $\{\phi - \phi * \delta_k\}$ converges to zero almost uniformly, we see that $p_{n_1}(\phi - \phi * \delta_k) \to 0$ as $k \to \infty$. Thus, there is an index k_0 such that $p_{n_1}(\phi - \phi * \delta_{k_0}) < \varepsilon/3$. Hence, we get

$$\|S_m(\phi) - S_m(\phi * \delta_{k_0})\|_T \leq \varepsilon/3$$

and

$$\|S_n(\phi) - S_n(\phi * \delta_{k_0})\|_T < \varepsilon/3$$

for $n,m \geq n_1$. By (3) and (1), the sequence $\{S_n(\phi * \delta_{k_0})\}$ converges almost uniformly. Therefore, there exists an index n_2, such that

$$\|S_n(\phi * \delta_{k_0}) - S_m(\phi * \delta_{k_0})\|_T < \varepsilon/3$$

for $m,n > n_2$. Consequently, we get

$$\|S_n(\phi) - S_m(\phi)\|_T < \varepsilon,$$

whenever $n,m > \max(n_1, n_2)$. This shows that $\{S_n(\phi)\}$ is a Cauchy sequence in $C(\mathbb{R}^q)$. Therefore, the sequence converges almost uniformly or, equivalently, series (3) converges almost uniformly, which completes the proof of the Theorem. □

Corollary If $f_n \in C(\mathbb{R}^q)$ for $n \in \mathbb{N}$, $f_n \to 0$ in $B(\mathbb{R}^q)$ and $f_n \not\to 0$ in $\mathcal{D}'$, then there are a subsequence $\{g_n\}$ of $\{f_n\}$ and an $x \in B(\mathbb{R}^q)$ such that

$$\sum_{n=1}^{\infty} g_n = x$$

in $B(\mathbb{R}^q)$, and x is not a distribution.

Proof Since $f_n \not\to 0$ in $\mathcal{D}'$, there is a subsequence $\{h_n\}$ of $\{f_n\}$ such that for every subsequence $\{g_n\}$ of $\{h_n\}$ we have $g_n \not\to 0$ in $\mathcal{D}'$. By the Theorem, there is a subsequence $\{g_n\}$ of $\{h_n\}$ and an $x \in B(\mathbb{R}^q)$, such that

$$\sum_{n=1}^{\infty} g_n = x$$

in $B(\mathbb{R}^q)$ and if $\phi \in \mathcal{D}$, then $x * \phi \in C(\mathbb{R}^q)$, iff the series

$$\sum_{n=1}^{\infty} g_n * \phi \qquad (7)$$

converges almost uniformly in $C(\mathbb{R}^q)$. We claim that x is not a distribution. Otherwise, for every $\phi \in \mathcal{D}$, $x * \phi \in C(\mathbb{R}^q)$. Consequently, series (7) converges almost uniformly for every $\phi \in \mathcal{D}$. In particular, the series

$$\sum_{n=1}^{\infty} (g_n, \tilde{\phi}),$$

where $\tilde{\phi}(x) = \phi(-x)$ converges for every $\phi \in \mathcal{D}$, or, equivalently, the series

$$\sum_{n=1}^{\infty} g_n$$

converges in $\mathcal{D}'$. Therefore $g_n \to 0$ in $\mathcal{D}'$. On the other hand, $g_n \not\to 0$ in $\mathcal{D}'$. This contradiction shows that x is not a distribution. □

3. In this section we shall construct a nonharmonic boehmian which is a solution of the Laplace equation. We precede the construction with two propositions.

Proposition 1 If $\alpha_n > 0$ and $\sum_{n=1}^{\infty} 1/\alpha_n < \infty$, then the series

$$\sum_{n=1}^{\infty} \lambda_n e^{\alpha_n x} \cos \alpha_n y$$

converges in $B(\mathbb{R}^q)$ for every numerical sequence $\{\lambda_n\}$.

Proof Assume $\varepsilon_k = 1/\alpha_k$ and assume that ℓ_k is a function such that $\ell_k(y) = (2\varepsilon_k \pi)^{-1}$, if $y \in [-\varepsilon_k \pi, \varepsilon_k \pi]$ and $\ell_k(y) = 0$, if $y \notin [-\varepsilon_k \pi, \varepsilon_k \pi]$ for $k = 1, 2, \ldots$. Next, we put

$$\rho_k = \ell_k * \ell_{k+1} * \cdots .$$

By routine calculations, one can check that $\{\rho_k\}$ is a delta-sequnece and

$$(\cos \alpha_n y) * \rho_k(y) = 0$$

for $n \geq k$ and $n, k \in \mathbb{N}$. Now, let

$$\delta_k(x,y) = \rho_k(x) \rho_k(y)$$

for $k \in \mathbb{N}$. We note that

$$(e^{\alpha_n x} \cos \alpha_n y) * \delta_k(x,y) = [e^{\alpha_n x} * \rho_k(x)][(\cos \alpha_n y) * \rho_k(y)] = 0$$

for $n \geq k$. Hence, for every $k \in \mathbb{N}$,

$$\sum_{i=1}^{n} (\lambda_i e^{\alpha_i x} \cos \alpha_i y) * \delta_k(x,y) = \sum_{i=1}^{k-1} (\lambda_i e^{\alpha_i x} \cos \alpha_i y) * \delta_k(x,y)$$

for $n \geq k$. This shows that series (8) converges in $B(\mathbb{R}^q)$ for every numerical sequence $\{\lambda_n\}$. □

Proposition 2 There are two numerical sequences $\{\alpha_n\}$ and $\{\lambda_n\}$, such that $\alpha_n > 0$ for $n \in \mathbb{N}$, $\sum_{n=1}^{\infty} 1/\alpha_n < \infty$ and $\lambda_n e^{\alpha_n x} \cos \alpha_n y \not\to 0$ in $\mathcal{D}'$.

Proof Let $\{\alpha_n\}$ be a sequence such that $\alpha_n > 0$ and $\sum 1/\alpha_n < 0$ and let ϕ be a function in $\mathcal{D}$ such that $\phi \geq 0$ and $\phi(0) = 1$. Then,

$$< e^{\alpha_n x} \cos \alpha_n y, \phi(x)\phi(y) > \neq 0$$

for a sufficiently large n. Let $\{\lambda_n\}$ be a numerical sequence such that

$$\lambda_n < e^{\alpha_n x} \cos \alpha_n y, \phi(x)\phi(y) > \to \infty.$$

This means that $\lambda_n e^{\alpha_n x} \cos \alpha_n y \not\to 0$ in $\mathcal{D}'$, which was to be proved. □

Now, we can construct an example of a boehmian u, such that $\Delta u = 0$ and u is not a distribution. Let $\{\alpha_n\}$ and $\{\lambda_n\}$ be numerical sequences satisfying the conditions of Proposition 2. By Proposition 1, $\lambda_n e^{\alpha_n x} \cos \alpha_n y \to 0$ in $B(\mathbb{R}^q)$ and at the same time $\lambda_n e^{\alpha_n x} \cos \alpha_n y \not\to 0$ in $\mathcal{D}'$. Hence, by the Corollary of the Theorem there are a subsequence $\{\lambda_{m_n} e^{\alpha_{m_n} x} \cos \alpha_{m_n} y\}$ and $u \in B(\mathbb{R}^q) \setminus \mathcal{D}'(\mathbb{R}^q)$ such that

$$\sum_{n=1}^{\infty} \lambda_{m_n} e^{\alpha_{m_n} x} \cos \alpha_{m_n} y = u$$

in $B(\mathbb{R}^q)$. Hence, and from the continuity of derivation, we get

$$\Delta u = \Delta \sum_{n=1}^{\infty} \lambda_{m_n} e^{\alpha_{m_n} x} \cos \alpha_{m_n} y = \sum_{n=1}^{\infty} \Delta \lambda_{m_n} e^{\alpha_{m_n} x} \cos \alpha_{m_n} y = 0.$$

Consequently, u is a boehmian which is not a harmonic function but $\Delta u = 0$.

REFERENCES

1. J. Mikusiński, T. Boehme, "Operational Calculus", Vol. II, PWN-Warszawa, Pergamon Press, (1987).

GENERALIZED FUNCTIONS; MULTIPLICATION OF DISTRIBUTIONS; APPLICATIONS TO

ELASTICITY, ELASTOPLASTICITY, FLUID DYNAMICS AND ACOUSTICS

J. F. Colombeau

U.E.R. de Mathematiques et Informatique
Université de Bordeaux I
33405 Talence, France

1. RECALLS

If Ω denotes any open set in $\mathbb{R}^n$, I have defined an algebra $G(\Omega)$ of "generalized functions" on Ω. One has the set of inclusions

$$C^\infty(\Omega) \subset \mathcal{D}'(\Omega) \subset G(\Omega)$$

where $C^\infty(\Omega)$ (respectively $\mathcal{D}'(\Omega)$) denotes the set of all C^∞ functions (resp. all distributions) on Ω. Two basic points have to be stressed:

- $C^\infty(\Omega)$, with its usual pointwise multiplication, is a subalgebra of $G(\Omega)$

- any element of $G(\Omega)$ admits partial derivatives of any order which generalize exactly those in $\mathcal{D}'(\Omega)$.

Let $C(\Omega)$ denote the algebra of all the continuous functions on Ω. Then, the famous Schwartz "impossibility result" (1954) [53] implies that $C(\Omega)$ cannot be a subalgebra of $G(\Omega)$.

In connection with this, I have recently noticed that elastoplastic shock waves impose the use of "several different Heaviside functions" (to be explained in part IV below). The same conclusion was obtained for shock waves in viscous media and acoustics in a medium made of several immiscible fluids.

Let us present the following mathematical computation. Let Y denote the Heaviside function and let $n = 2,3,\ldots$. In the algebra $C_f(\mathbb{R})$ of all piecewise C^∞ functions on $\mathbb{R}$, one has

$$Y^n = Y.$$

Now, let us derive and multiply freely (as in the case of C_f^∞ functions).

Derivation gives

$$nY^{n-1}Y' = Y'. \qquad (1)$$

Let us multiply (1) by Y:

$$nY^n Y' = YY'.$$

Applying (1) to both members yields

$$\frac{n}{n+1} Y' = \frac{1}{2} Y',$$

which is absurd since $n \neq 1$ and $Y' \neq 0$. From the above remark on the need of "several different Heaviside functions", an interpretation of this absurdity is

$$Y^n \neq Y \text{ if } n \neq 1.$$

Y^n and Y should be two different Heaviside functions! This can be illustrated in the figure below in which one considers that Y is indeed a continuous function, whose jump takes place on an interval $[-\varepsilon, +\varepsilon]$ with $\varepsilon > 0$ "very small". Y^n is represented by a dotted line and one "sees" a difference between Y^n and Y. In $G(\mathbb{R})$ one checks that $Y^n \neq Y$, but, as naturally for every $\psi \in \mathcal{D}(\mathbb{R})$ (i.e. ψ is a C^∞ function on $\mathbb{R}$ with a compact support), one has in the natural sense of a limit when $\varepsilon \to 0$

$$\int Y^n(x)\psi(x) \, dx = \int Y(x)\psi(x) \, dx = \int_0^{+\infty} \psi(x) \, dx,$$

i.e., when viewed "as distributions" Y^n and Y look equal. This concept (an aspect of a generalized function through mean values with test functions $\psi \in \mathcal{D}(\Omega)$) can be naturally introduced in $G(\Omega)$: two elements G_1, G_2 of $G(\Omega)$ are said to be associated (notation $G_1 \approx G_2$) iff

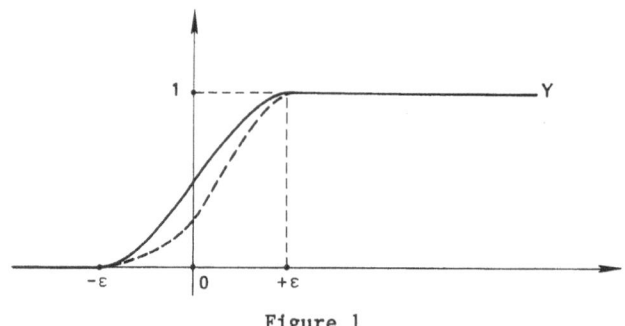

Figure 1

$$\forall \psi \in \mathcal{D}(\Omega) \quad \int G_1(x)\psi(x)\, dx = \int G_2(x)\psi(x)\, dx$$

(in which these integrals are defined in a natural sense). Of course, $G_1 = G_2 \Rightarrow G_1 \approx G_2$, but the converse is false: one has

$$Y^n \neq Y \quad \text{and} \quad Y^n \approx Y.$$

From the definitions one proves easily

Proposition 1 In $\mathcal{D}'(\Omega)$ the concepts = and $\approx$ coincide.

Proposition 2 Let $f, g \in C(\Omega)$. Then their new product in $G(\Omega)$ is associated with their classical product. The same result holds for the product of a C^∞ function and a distribution.

This splitting of the classical equality into the two concepts = and $\approx$ (in $G(\Omega)$) is a basic feature of our theory.

2. THE SEARCH FOR THE "BEST DEFINITION" OF $G(\Omega)$; AN ILLUSTRATION: THE PROBLEM OF "SAFETY BARRIERS"

There is no precise definition of $G(\Omega)$ which would be obviously the "best one" (as least up to now and as far as I can understand) but a "cloud" of very similar definitions. For the sake of maximum simplicity some of them are accepted with defects, provided they are some simplified aspect of a better definition. One of them is used in [17], App. 5, [3], [40 - 41], and it has the defect of not being invariant under a linear change of coordinates; this is a simplification of an invariant definition elaborated in detail in [9].

I am afraid that, from this state of affairs, several mathematicians have got the impression that the theory has serious defects. As far as I know up to now, this impression would be false since all the defects could be repaired, at once, by minor changes in the definition of $G(\Omega)$. Here is a recent venture. In the more elementary definition [17] chap. 1, there is $G \in G(\mathbb{R})$ solution of

$$G^2 - G = 0$$

which is not identical to the constant function 0 or the constant function 1 (such a G is an object which appears to be both 0 and 1: intuitively think for instance of an object that would have some probability p to be

0 and a probability 1-p to be 1). This fact was considered as a basic defect. An immediate and minor change in the definition gives

Proposition 3   Let I be an open interval of $\mathbb{R}$ and P a nonzero polynomial in one variable. Then,
$$\left.\begin{array}{l} P(G) = 0 \\ \text{in } G(I) \end{array}\right\} \Leftrightarrow G \text{ is identical to a (classical) root of } P.$$

More generally, the following has been proved by B. Perrot.

Theorem 1   Let P be a nonzero polynomial in two variables. Then,
$$\left.\begin{array}{l} P(x,G) = 0 \\ \text{in } G(\mathbb{R}) \end{array}\right\} \Leftrightarrow G \text{ is identical to a } C^{\infty} \text{ solution}.$$

Of course, the classical continuous solutions are recovered with the association:

Proposition 4   Let $G \in C(I)$. Then

$P(x,G) \approx 0 \Leftrightarrow G$ is a classical solution.

One recovers, also, the distribution solutions ($x\delta \approx 0$) and a few other solutions ($x^n \delta^2 \approx 0$ if $n = 2,3,\ldots$). The difference between = and $\approx$ in $G(I)$ appears thus as the difference between the C^{∞} solutions and the other (in particular the continuous) solutions.

Corollary of Theorem 1   Let $m \in \mathbb{N}$. Then,
$$\left.\begin{array}{l} P(x,G^{(m)}) = 0 \\ \exists x_0 \in I \text{ such that } G(x_0), G'(x_0), \ldots, G^{(m-1)}(x_0) \\ \text{are classical complex numbers} \end{array}\right\} \Leftrightarrow \begin{array}{l} G \text{ is a classical} \\ C^{\infty} \text{ solution}. \end{array}$$

Now, what about the case when P is a polynomial in (m+2) variables? Of course,

Proposition 5   If $G \in C^m(I)$, then

$P(x,G,G',\ldots,G^{(m)}) \approx 0 \Leftrightarrow G$ is a classical solution (of class $C^m(I)$).

For systems of a finite number of equations $Y' = F(x,Y)$ with F a vector valued C^{∞} function and Y a vector, one has easily from the C^{∞} dependence of the solution of a parameter and of the initial data:

Propositon 6

$$Y' = F(x,Y) \text{ in } G(I)$$
$$\exists x_0 \in I \text{ such that } X(x_0) \text{ is a classical vector} \Bigg\} \Leftrightarrow Y \text{ is a classical } C^\infty \text{ solution.}$$

But the general case of algebraic differential equations (ADE) is more complicated: the ADE

$$2xy + (x^2 - 1)^2 y' = 0$$

has a solution $y \in G(\mathbb{R})$ which is null on $(-\infty, -1[\cup]1, +\infty)$ and which is (in some precise sense) infinite on $]-1, +1[$!

Anyway, one has, in many cases, very strong "safety barriers" i.e. results showing that for equations which have enough classical solutions, the "generalized solutions" are automatically classical solutions.

Unfortunately, the fact that an "optimal definition" of $G(\Omega)$ is not yet fixed is at the origin of misunderstandings and from them the theory faces problems for its divulgation.

3. SHOCK WAVES IN VISCOUS MEDIA [7]

A one dimensional system of equations of fluid dynamics in a viscous medium is

$$\left. \begin{array}{ll} \rho_t + (\rho u)_x = 0 & \text{mass conservation} \\ (\rho u)_t + (\rho u^2 + p)_x = (\nu(\rho) u_x)_x & \text{momentum conservation} \\ (\rho e)_t + (\rho e u + p u)_x = (\nu(\rho) u \, u_x)_x & \text{energy conservation} \\ p = \phi(\rho, I) & \text{equation of state} \end{array} \right\} \quad (1)$$

in which ρ = density, u = velocity, p = pressure, e = total specific energy, $I = e - \frac{u^2}{2}$ = internal specific energy, ν = coefficient of viscosity. Shock waves do exist in viscous media (for instance, supersonic bangs in air). They are represented by a set of discontinuous functions that should be a solution of (1). The second members $\nu(\rho) u_x$ and $\nu(\rho) u \, u_x$ in (1) appear under the form $Y \cdot \delta$ of the product of a discontinuous function with the derivative of a function discontinuous at the same point. The product $Y \cdot \delta$ does not make sense within distribution theory.

Let us consider a theory of multiplication of distributions in which there would be only one Heaviside function Y and one Dirac "function" δ, thus, with a precise formula for the product $Y \cdot \delta$ (usually one adopts the formula $Y\delta = \frac{1}{2}\delta$). Then, immediate computations show that in such a theory (1) cannot have shock waves solutions. In the setting of $G(\mathbb{R}^2)$, one finds shock waves solutions. By definition, steady shocks are those which propagate without deformation and with a constant velocity.

Theorem 2 If system (1) is stated in $G(\mathbb{R}^2)$ with the equalities in (1) replaced by association symbols $\approx$, then (1) has steady shock solutions. The steady shocks in the form of step functions satisfy exactly the jump conditions of the nonviscous case $\nu = 0$.

This result is known to mathematical physicists ([48], p. 375). It does not hold any longer for unsteady shocks, for which viscosity influences the jump conditions. In this last case, as far as we know, there is a lack of precise experimental results.

4. SHOCK WAVES IN ELASTICITY AND ELASTOPLASTICITY

In order to design the armor of modern battle tanks, engineers need numerical simulations of collisions. These are done from systems of elasticity, elastoplasticity and fluid dynamics. A one dimensional system of elasticity used by engineers is the following (in a simplified model)

$$\left. \begin{array}{ll} \rho_t + (\rho u)_x = 0 & \text{mass conservation} \\ (\rho u)_t + (\rho u^2)_x = \sigma_x & \text{momentum conservation} \\ \sigma_t + u\sigma_t = k^2 u_x & \text{Hooke's law} \end{array} \right\} \quad (2)$$

in which ρ = density, u = velocity, σ = stress and $k^2 > 0$ is a constant depending on the material. The term $u\sigma_x$ is usually dropped for simplification in literature on elasticity, but in the case of numerical simulations used in the design of armor, it is not at all negligible. The projectiles produce shock waves in the armor and then the term $u\sigma_x$ appears in the form $Y \cdot \delta$.

Theorem 3 System (2) when written with three equalities in $G(\mathbb{R}^2)$ has no steady shock solutions.

System (2) when written with three associations has an infinite number of steady shock solutions, which have different jump formulas.

Both results are unsatisfactory from the physical point of view. But

Theorem 3' If system (2) is written in the form

$$\begin{aligned}
\rho_t + (\rho u)_x &= 0 \\
(\rho u)_t + (\rho u^2)_x &= \sigma_x \\
\sigma_t + u\sigma_x &= k^2 u_x
\end{aligned}$$

then it has shock waves solutions with uniquely defined jump formulas.

This is satisfactory and the formulas thus obtained are in agreement with the results of experiments. In this particular case, a theory with one Heaviside function Y, one Dirac function δ and the formula $Y\delta = \frac{1}{2}\delta$ would give exactly the correct result. But, this is an accident, as will now be made clear from the following study in elastoplasticity. A one dimensional model of elastoplasticity is as follows:

$$\begin{aligned}
&\rho_t + (\rho u)_x = 0 &&\text{mass conservation} \\
&(\rho u)_t + (\rho u^2)_x + (p - S)_x = 0 &&\text{momentum conservation} \\
&(\rho e)_t + [\rho eu + (p - S)u]_x = 0 &&\text{energy conservation} \\
&S_t + uS_x - \frac{4}{3}\mu(S)u_x = 0 &&\text{Hooke's law} \\
&p = \phi(\rho, I) &&\text{equation of state}
\end{aligned} \quad (3)$$

S is a component of the stress deviation tensor; μ is a function of S depending of the medium: usually

- if $|S| < S_0$, then $\mu(S)$ remains equal to a constant μ: the material is in the elastic state,
- if $|S| = S_0$, then $\mu(S) = 0$: the material is in the plastic state, i.e. it is a fluid.

Of course, one starts with a value $|S| < S_0$, and then $|S|$ cannot become larger than S_0.

Shock waves solutions of (3) put in evidence multiplications of distributions in the terms uS_x and $\mu(S)u_x$. It is well known to engineers that there are shock waves in which the material passes from the elastic state into the plastic state. For $w = \rho, u, e, p, S, \ldots$, let us seek shock waves solutions of the form (steady shocks)

$$w(x,t) = \Delta w\, Y_w(x - ct) + w_\ell$$

with Y_w a Heaviside function (depending on w). In a shock with an elastic-plastic phase transition, the function Y_S is quite different from the functions Y_w, $w \neq S$:

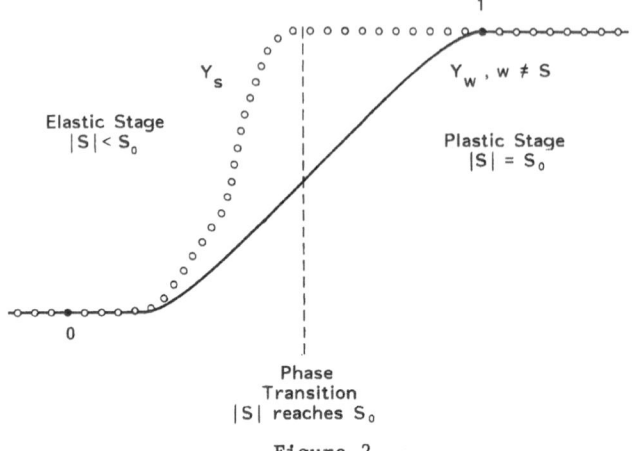

Figure 2

Since $\mu(S) = 0$ as soon as $|S|$ reaches S_0, then $|S|$ ceases to increase as soon as it reaches S_0, while the other functions $\rho, u, p, \ldots$ still vary in the plastic stage. Thus, Y_S is quite different from Y_w if $w \neq S$. The term uS_x in (3) contains a product $Y_u(Y_S)'$ of the form $Y \cdot \delta$. The value of this quantity is far different from $\frac{1}{2}\delta$ and depends on the shock: when the phase transition takes place early within the shock, then $Y_u(Y_S)'$ is approximately 0, which is quite clear from the figure. This is evidence of the need to consider several different Heaviside functions.

5. LINEAR ACOUSTICS IN A NONHOMOGENEOUS MEDIUM WITH DISCONTINUOUS CHARACTERISTICS (FOR INSTANCE A MEDIUM WATER/AIR)

In one dimension and Eulerian coordinates a system of equations is ([45 - 47])

$$\rho_t + (\rho_0 u)_x = 0 \quad \text{mass conservation}$$

$$\rho_0 u_t + p_x = 0 \quad \text{momentum conservation}$$

$$s_t + u(s_0)_x = 0 \quad \text{entropy conservation}$$

equation of state

(4)

with ρ = density, u = velocity, p = pressure, s = entropy; ρ_0, u_0 and s_0 are the given functions discontinuous at $x = 0$, C^∞ outside $x = 0$ with left-

and right-hand side derivatives. The term $u \cdot (s_0)_x$ has the form $Y \cdot \delta$. In Lagrangian coordinates one has the system (Poirée [45 - 47])

$$\left. \begin{array}{l} \rho_t + \rho_0 u_x = 0 \\ \rho_0 u_t + p_x = 0 \\ p = c_0^2 \rho \end{array} \right\} \tag{5}$$

with c_0^2 given discontinuous at $x = 0$, C^∞ outside $x = 0$ with left-and right-hand side derivatives. Then, the term $\rho_0 u_x$ appears in the form $Y \cdot \delta$. Problem (5) has been studied by Oberguggenberger ([42]) and Lafon ([54]), as a particular case of a semilinear hyperbolic system with discontinuous coefficients. One obtains the physically expected result of continuity of u and p at $x = 0$. In two space dimension, an important problem is to find the junction conditions at singular points of the curve separating two immiscible media.

6. OTHER APPLICATIONS

There are lots of other applications which have not been developed here: for instance, for the existence-uniqueness results for solutions of various hyperbolic and parabolic equations (see Colombeau [17], chap 8, Biagoni [9], chap 3, Rosinger [52], Oberguggenberger [40 - 42], Lafon [54], Cauret-Colombeau-Le Roux [14 - 15], Colombeau-Langlais [29 - 30]). There are also new numerical schemes for systems of PDEs in engineering (see Colombeau-Le Roux [31 - 32], Cauret [13], Noussair [36]).

7. PROBLEMS

1. Physical premises

The equations are written "formally"; the problem is: write them rigorously from adequate physical premises expressed in this setting. This would be perhaps the key to getting rid of ambiguities which occur in some cases: such ambiguities could disappear with a more precise formulation of the physical equations, which would be made possible by our setting.

2. Mathematical premises

Contribute to fix the "best definition" of $G(\Omega)$.

3. Clarification of the connections with nonstandard analysis

Such work is being done by T. D. Todorov.

4. "Wild constants"

In order to reproduce the formalism of C^∞ functions, I have constructed $G(\Omega)$ so that

$\forall G \in G(\Omega)$, $\forall x \in \Omega$, then the pointvalue $G(x)$ is defined and can be considered a constant function: example $\delta(0)$. This construction introduces nonclassical constants that some mathematicians called "wild constants". Problem: are there such nonclassical constants in the subalgebra of $G(\Omega)$ spanned by the distributions?

5. Algebraic differential equations

Let I be an open interval of $\mathbb{R}$ and let P be a nonzero polynomial in $m+2$ variables. We known that

$$\left. \begin{array}{l} P(x,G,G',\ldots,G^{(m)}) = 0 \\ \\ G|_\omega \text{ is a classical constant for} \\ \text{some nonvoid open set } \omega \subset I \end{array} \right\} \text{ does not imply that G is a } C^\infty \text{ function.}$$

Under what condition is this statement true?

6. Cauchy-Riemann and Laplace equations

Let Ω be a connected open set in $\mathbb{C}$. Let $G \in G(\Omega)$ be a solution of $\bar{\partial} G = 0$ (then we call G a holomorphic generalized function). Let us assume there is a nonvoid open subset ω of Ω in which the restriction $G|_\omega$ is a distribution (and thus $G|_\omega$ is a classical holomorphic function, from a classical result of distribution theory). If $G|_\omega$ can be extended to Ω as a classical holomorphic function F, then F and G coincide: this follows at once from the result of the uniqueness of the holomorphic continuation for generalized functions (see Colombeau-Gále [28]). Thus G is identical to a holomorphic function on Ω.

<u>Problem</u>: Let $G \in G(\Omega)$, which satisfies $\bar{\partial} G = 0$; assume there is a nonvoid open subset ω of Ω in which G is a distribution. Is G a classical holomorphic function on Ω? The same problem for the Laplace equation.

7. Find a convenient convergence structure on the space $G(\Omega)$ and use it. (a nonvector topology is elaborated in Biagoni-Colombeau [12] and Biagoni [9]).

Of course, each reader can easily find many other problems, such as problems of division, of solutions of PDEs... .

8. Physics: Acoustics

Solve the system of linear acoustics in two space dimensions in a medium made of two immiscible fluids and obtain the junction conditions at a singular point.

9. Quantum Field Theory

Using this setting obtain a correct mathematical understanding (see problem 1).

10. Applications to many other domains of physics in which there appear multiplications of distributions

11. Numerical analysis

Even in classical fluid dynamics one gets, from this theory, new numerical schemes, (see Colombeau-Le Roux [31 - 32], Noussair [36]). Numerical tests in one dimension show that in certain circumstances they can be more efficient than the classical ones, (see Noussair [36]), for instance. Go on with these schemes and tests in several space dimensions.

8. REFERENCES

We have tried to give a complete set of references on this new theory of generalized functions.

1. M. Adamczewski, Vectorisation, analyse et optimisation d'un code bidimensionnel eulérien, These, Bordeaux (1986).
2. J. Aragona, Théorèmes d'existence pour l'opérateur $\bar{\partial}$ sur les formes différentielles généralisées, Computes Rendus. Acad. Sci. Paris, 300, 239-242 (1985).
3. J. Aragona, On existence theorem for $\bar{\partial}$ operator on generalized differential forms, Proc. London Math. Soc. (3), 53, 474-488 (1986).
4. J. Aragona, H. A. Biagioni, Restriction and composition for Colombeau's generalized functions, (in press).
5. J. Aragona, J. F. Colombeau, The $\bar{\partial}$ equation for generalized functions, J. of Math. Ana. Appl., 110, 1, 179-199 (1985).

6. J. Aragona, J. F. Colombeau, The interpolation theorem for generalized functions, Annales Polonici Mathematici, (1988).

7. J. Aragona, J. F. Colombeau, C. K. Raju, Multiplication of distributions and shock waves in viscous media, (in press).

8. H. A. Biagioni, The Cauchy problem for hyperbolic systems with generalized functions as initial data, Resultate des Math., (1988).

9. H. A. Biagioni, Colombeau's nonlinear theory of generalized functions: an introduction, Book, preprint series Notes des Matematica, IMECC UNICAMP, 13100, Campinas, SP, Brazil.

10. H. A. Biagioni, J. F. Colombeau, Borel's theorem for generalized functions, Studia Math. 81, 179-183 (1985).

11. H. A. Biagioni, J. F. Colombeau, Whitney's extension theorem for generalized functions, J. of Math. Ana. Appl., 114, 374-383 (1986).

12. H. A. Biagioni, J. F. Colombeau, New generalized functions and C^{∞} functions valued in generalized complex numbers, J. of the London Math. Soc. (2), 33, 169-179 (1986)

13. J. J. Cauret, Analyse et développement d'un code bidimensionel élastoplastique, Thèse, Bordeaux (1986).

14. J. J. Cauret, J. F. Colombeau, A. Y. Le Roux, Solutions généralisées discontinues de problèmes hyperboliques non conservatifs, Comptes Rendus Acad. Sci. Paris, 302, 436-437 (1986).

15. J. J. Cauret, J. F. Colombeau, A. Y. Le Roux, Discontinuous generalized solutions of nonlinear nonconservative hyperbolic equations, J. of Math. Anal. Appl., (in press).

16. J. F. Colombeau, "New Generalized Functions and Multiplication of Distributions", North Holland, (1984).

17. J. F. Colombeau, "Elementary Introduction to New Generalized Functions", North Holland, (1985).

18. J. F. Colombeau, New General Existence Results for Partial Differential Equations in the C^{∞} case, (in press).

19. J. F. Colombeau, A Mathematical Analysis Adapted to the Multiplication of Distributions, (in press).

20. J. F. Colombeau, A Multiplication of Distributions, J. of Math. Ana. and Appl., 94, 1, 96-115 (1983).

21. J. F. Colombeau, New Generalized Functions, Multiplications of Distributions; Physical Applications, Portugaliae Math. vol. 41, Fasc. 1-4, 57-69 (1982).
22. J. F. Colombeau, Une Multiplication Generale des Distributions, Comptes Rendus Acad. Sci. Paris I, 296, 357-360 (1983).
23. J. F. Colombeau, Some Aspects of Infinite Dimensional Holomorphy in Mathematical Physics. In "Aspects of Mathematics and its Applications", editor J. A. Barroso, North Holland, 253-263 (1986).
24. J. F. Colombeau, A New Theory of Generalized Functions. In "Complex Analysis, Functional Analysis and Approximation Theory", editor J. Mujica, North Holland, 57-66 (1986).
25. J. F. Colombeau, "Differential Calculus and Holomorphy. Real and Complex Analysis in Locally Convex Spaces", North Holland, (1982).
26. J. F. Colombeau, Nouvelles Solutions d'Equations aux Dérivées Partielles, Comptes Rendus Acad. Sci. Paris, 301, 281-283 (1985).
27. J. F. Colombeau, J. E. Gale, Holomorphic Generalized Functions, J. of Math. Ana. Appl., 103, 1, 117-133 (1984).
28. J. F. Colombeau, J. E. Gale, The analytic continuation for generalized holomorphic functions, Acta Math. Hungarica, (1988).
29. J. F. Colombeau, M. Langlais, Existence et unicité des solutions d'équations paraboliques non linéaires avec conditions initiales distributions, Comptes Rendus Acad, Sci. Paris, 302, 379-382 (1986).
30. J. F. Colombeau, M. Langlais, Generalized solutions of nonlinear parabolic equations with distributions as initial conditions, (1988).
31. J. F. Colombeau, A. Y. Le Roux, Numerical Tehniques in Elastodynamics. In "Nonlinear Hyperbolic Problems", Lecture Notes in Mathematics 1270, 103-114 (1987).
32. J. F. Colombeau, A. Y. Le Roux, Numerical methods for hyperbolic systems in nonconservation form using products of distributions. In "Advances in Computer Methods for Partial Differential Equations VI", edited by R. Vichnevetsky and R. S. Stepleman, Publ. IMACS, 28-37 (1987).
33. J. F. Colombeau, A. Y. Le Roux, B. Perrot, Multiplication de distributions et ondes de choc elastiques ou hydrodynamiques en dimension un, Comptes Rendus Acad. Sci. Paris, 305, 453-456, (1987).

34. J. Jelinek, Characterization of the Colombeau product of distributions, Commentationes Math. Univ. Carolinae, 27, 2, 377-394 (1986).
35. C. O. Kiselman, Sur la définition de l'opérateur de Monge Ampère complexe, Lecture Notes in Math. 1904, Springer, 139-150 (1984).
36. A. Noussair, "Etudes sur les systemes non conservatifs", Memoire de DEA, Bordeaux, (1987).
37. M. Oberguggenberger, Weak limits of solutions to semilinear hyperbolic systems, Math. Annalen, 274, 599-607 (1986).
38. M. Oberguggenberger, Distributions associated with multiplications of distributions in the Colombeau algebra $G(\Omega)$, Boll. Unione Mat. Ital., (6), 423-429 (1986).
39. M. Oberguggenberger, Products of distributions, J. für die reine ang. Math., 365, 1-11 (1986).
40. M. Oberguggenberger, Generalized solutions to semi linear hyperbolic systems, Monatshefte für Mathematik, 103, 133-144, (1987).
41. M. Oberguggenberger, Solutions généralisées de problèmes hyperboliques non linéaires, Comptes rendus Acad. Sci. Paris, t. 305, I, 17-18 (1987).
42. M. Oberguggenberger, Hyperbolic systems with discontinuous coefficients: generalized solutions and a transmision problem in acoustics.
43. B. Perrot, Topologie mixte, équations différentielles en dimension infinie et équation d'evolution ..., Thèse, Bordeaux, (1984).
44. B. Perrot, Personal communication.
45. B. Poiree, Les équations de l'acoustique linéaire et nonlinéaire, Acustica, 57, 525 (1985).
46. B. Poiree, Thèse, Paris VI, (1982).
47. B. Poiree, Equations de perturbation et équations de passage associées, Revue du Cethedec, 69, 1-49 (1981).
48. R. D. Richtmyer, "Principles of Advanced Mathematical Physics I", Springer Verlag, (1978).
49. E. E. Rosinger, Distributions and nonlinear PDE, Lectures Notes in Mathematics 684, Springer Verlag (1978).
50. E. E. Rosinger, "Nonlinear Partial Differential Equations. Sequential and Weak Solutions", North Holland, (1980).
51. E. E. Rosinger, "Nonlinear Equivalence, Reduction of PDE to ODE and fast convergent numerical schemes", Pitman, (1982).

52. E. E. Rosinger, "Generalized Solutions of Nonlinear Partial Differential Equations", North Holland, (1987).
53. L. Schwartz, Sur l'Impossibilité de la Multiplication des Distributions, <u>Comptes Rendus Acad. Sci. Paris</u>, 239, 847-848 (1954).
54. F. Lafon, Generalized solutions of a one dimensional semi-linear hyperbolic system with discontinuous coefficients; application to an electron transport problem.

MONADS AND CONVERGENCE

Werner Gähler

Karl-Weierstraß-Institut für Mathematik
Akademie der Wissenschaften der DDR
Mohrenstraße 39, 1086 Berlin, DDR

The paper investigates algebraical and topological structures from a common point of view. The main idea consists in weakening the axioms of an Eilenberg-Moore algebra such that they become natural conditions of a topological structure. By some of these conditions continuous lattices can be characterized.

1. THE NOTIONS OF CONVERGENCE SPACE AND EILENBERG-MOORE ALGEBRA

In this first three sections some notions and results are collected, most of them are well-known.

Let ϕ be a covariant set functor, that is, a covariant functor ϕ : SET $\to$ SET where SET is the category of sets and mappings (of sets). ϕ assings to each set X and mapping $f : X \to Y$ a set ϕX and mapping $\phi f : \phi X \to \phi Y$, respectively. A ϕ-convergence space is a pair (X,s) consisting of a set X and a relation $s : \phi X \leftrightarrow X$, that is , a subset of $\phi X \times X$. Instead of $(\mathfrak{m},x) \in s$ we also write $\mathfrak{m} \xrightarrow{s} x$ and say that $\mathfrak{m}$ converges to x (with respect to s) or that $\mathfrak{m}$ s-convergences to x. A ϕ-continuous mapping $f : (X,s) \to (Y,t)$ between ϕ-convergence spaces is a mapping f of X into Y such that one of the following equivalent conditions is fulfilled: (i) $\mathfrak{m} \xrightarrow{s} x$ implies $\phi f(\mathfrak{m}) \xrightarrow{t} f(x)$, (ii) $f \circ s \subseteq t \circ \phi f$, (iii) $f \circ s \circ (\phi f)^{-1} \subseteq t$. Let ϕ-CON denote the category of ϕ-convergence spaces and ϕ-continuous mappings. By ϕ-CON$_{map}$ we mean the full subcategory of ϕ-CON of all ϕ-convergence spaces (X,s) with s a mapping of ϕX into X.

A monad over SET is a triple (ϕ,η,μ) where ϕ is a covariant set functor and $\eta : \text{id} \to \phi$ and $\mu : \phi \circ \phi \to \phi$ are natural transformations such that

$$1_\phi = \mu \circ \phi\eta, \quad 1_\phi = \mu \circ \eta\phi \text{ and } \mu \circ \phi\mu = \mu \circ \mu\phi. \tag{1}$$

id means the identity set functor: $idX = X$ and $idf = f$. By definition, η and μ are resp. families of mappings $\eta_X : X \to \phi X$ and $\mu_X : \phi\phi X \to \phi X$, $X \in$ obSET. $1_\phi, \phi\eta, \eta\phi, \ldots$ are defined by $(1_\phi)_X = 1_{\phi X}$ (identity mapping of ϕX), $(\phi\eta)_X = \phi(\eta_X)$, $(\eta\phi)_X = \eta_{\phi X}, \ldots$. Axioms (1) can be read argumentwise: $1_{\phi X} = \mu_X \cdot \phi\eta_X, \ldots$. There is a well-known analog between (1) and the axioms of a monoid which are suggested to call (ϕ, ν, μ) a monad and η and μ its <u>unit</u> and <u>multiplication</u>, respectively.

Let $\Phi = (\phi, \eta, \mu)$ be a monad over SET. An <u>Eilenberg-Moore algebra</u> with respect to Φ or a <u>Φ-algebra</u> is an object (X, s) of $\phi\text{-CON}_{\text{map}}$ such that

$$1_X = s \circ \phi_X \tag{2}$$

and

$$s \circ \phi s = s \circ \mu_X. \tag{3}$$

In the case of Φ-algebras, ϕ-continuity means $f \quad s = t \quad \phi f$. Let Φ-ALG denote the full subcategory of $\phi\text{-CON}_{\text{map}}$ of all Φ-algebras.

2. THE EXAMPLE OF EQUATIONALLY DETERMINED ALGEBRAS

The classical example of Eilenberg-Moore algebras gives a characterization of equationally determined algebras, also called varietes (cf [6] and [7]). The following part will deal with these well-known results.

By an <u>operator domain</u> we mean a disjoint union $\overset{\bullet}{\underset{n \le k}{\Sigma}} \Omega_n = \underset{n \le k}{\cup} \{n\} \times \Omega_n$ of sets Ω_n where n runs over all cardinals less than or equal to some cardinal k. Each operator domain $\overset{\bullet}{\underset{n \le k}{\Sigma}} \Omega_n$ is associated with a covariant set functor, namely $\underset{n \le k}{\Sigma} \Omega_n id^n$ which assigns to each set X the disjoint union $\overset{\bullet}{\underset{n \le k}{\Sigma}} (\Omega_n \times X^n) = \underset{n \le k}{\cup} \{n\} \times \Omega_n \times X^n$ and to each mapping $f : X \to Y$ the mapping $(n, \omega, (x_i)_{i \in n}) \mapsto (n, \omega, (f(x_i))_{i \in n})$ $(n \le k, \omega \in \Omega_n, (x_i)_{i \in n} \in X^n)$.

Let ϕ and ψ be covariant set fuctors. ϕ is a <u>subfunctor</u> of ψ provided $\phi X \subseteq \psi X$ for each set X and ϕf is a domain-codomain restriction of ψf for each mapping f. ϕ is a <u>quotient functor</u> $\psi/\sim$ of ψ, if for each set X, ϕX is a quotient set $\psi X/\sim$ and for each mapping $f : X \to Y$, ψf preserves the equivalence relation $(m \sim n \Longrightarrow \psi f(m) \sim \psi f(n))$ and ψf is the mapping $\tilde{m} \mapsto \psi f(m)^\sim$ $(m \in \psi X)$.

If I is a set and ϕ_i, $i \in I$, are covariant set functors such that for each mapping $f : X \to Y$, from $m \in \phi_i X \cap \phi_j X$ it follows $\phi_i f(m) = \phi_j f(m)$, then the <u>union functor</u> $\underset{i \in I}{\cup} \phi_i$ can be defined: it assigns $\underset{i \in I}{\cup} \phi_i X$ to each set X and $m \mapsto \phi_i f(m)$ $(\exists i \in I \ m \in \phi_i X)$ to each mapping $f : X \to Y$.

Let ϕ be a covariant set functor and X a set. An $m \in \phi X$ has an <u>order</u>

$\le k$ provided there is a subset M of X of cardinality $\le k$ and a $\bar{k} \in \phi M$ with $\phi\iota(\bar{k}) = m$ where ι is the identical embedding of M into X. ord m is the least one of these $\bar{k}$'s. If sup{ord m | X $\in$ obSET, m $\subset \phi X$} exists, then it is called the <u>rank</u> of ϕ. A covariant set functor need not have a rank. $\sum_{n \le k} \Omega_n \text{id}^n$ has a rank $\le k$.

Let $\Omega = \dot{\sum}_{n \le k} \Omega_n$ be a fixed operator domain. An Ω-<u>algebra</u> is a pair $(X, (s_{n\omega})_{(n,\omega) \in \Omega})$ consisting of a set X and a family of mappings $s_{n\omega} : X^n \to X$. A mapping $f : (X, (s_{n\omega})_{(n,\omega) \in \Omega}) \to (Y, (t_{n\omega})_{(n,\omega) \in \Omega})$ between Ω-algebras is called an Ω-<u>homomorphism</u> provided $f(s_{n\omega}((x_i)_{i \in n})) = t_{n\omega}((fx_i)_{i \in n})$ for each $(n,\omega) \in \Omega$ and $(x_i)_{i \in n} \in X^n$. Let Ω-ALG denote the category of Ω-algebras and Ω-homomorphisms. An Ω-<u>equation</u> $\{\kappa, \lambda\}$ is an unordered pair of natural transformations $\kappa : F^n \to F$ and $\lambda : F^n \to F$, where F is the forgetful functor of Ω-ALG to SET, n is a cardinal and $F^n = \text{id}^n \circ F$ with id^n the n-th power of id: $\text{id}^n X = X^n$, $\text{id}^n f : (x_i)_{i \in n} \mapsto (f(x_i))_{i \in n}$ $((x_i)_{i \in n} \in X^n)$. κ and λ are families of mappings $\kappa_A : X^n \to X$ and $\lambda_A : X^n \to X$ where A runs over all Ω-algebras and FA = X.

Let E be a fixed set-indexed family of Ω-equations $\{\kappa_i, \lambda_i\}$, $i \in I$. An Ω/E-<u>algebra</u> is an Ω-algebra A which <u>fulfills the Ω-equations</u> of E, that is, for which $\kappa_{iA} = \lambda_{iA}$ for all $i \in I$. Let Ω/E-ALG denote the full subcategory of Ω-ALG of all Ω/E-algebras.

We are going to define a monad Φ over SET by means of Ω and E. Let ψ be the covariant set functor given by transfinite induction as follows:
$\psi_0 = \text{id}$, $\psi_\alpha = (\sum_{n \le k} \Omega_n \text{id}^n) \circ \cup_{\beta < \alpha} \psi_\beta$ for each positive ordinal α and $\psi = \cup_{\alpha < k^+} \psi_\alpha$ where k^+ is the least cardinal greater than k and $\aleph_0$.

<u>Proposition 2.1</u> ψ is the least covariant set functor which has id and $(\sum_{n \le k} \Omega_n \text{id}^n) \circ \psi$ as subfunctors.

ψ has the following properties: For each set X, $(\psi X, (\sigma_{n\omega})_{(n,\omega) \in \Omega})$ with $\sigma_{n\omega}((m_i)_{i \in n}) = (n, \omega, (m_i)_{i \in n})$ $((n,\omega) \in \Omega$ and $m_i \in \psi X$ for each $i \in n)$ is an Ω-algebra. Given an Ω-algebra $B = (Y, (t_{n\omega})_{(n,\omega) \in \Omega})$, each mapping $f : X \to Y$ can be extended by transfinite induction to the Ω-homomorphism $f^+ :$ $(\psi X, (\sigma_{n\omega})_{(n,\omega) \in \Omega}) \to B$ as follows: $f^+ | \psi_0 X = f$ and $f^+(n, \omega, (m_i)_{i \in n}) = t_{n\omega}((f^+(m_i))_{i \in n})$ for each positive ordinal $\alpha < k^+$ and $(n, \omega, (m_i)_{i \in n}) \in \psi_\alpha X$. f^+ is called the Ω-<u>extension</u> of f <u>associated to</u> B.

Let $\phi = \psi/\sim$ be the quotient functor where for each set X the equivalence relation on ψX is defined by:

$m \sim n \Leftrightarrow f^+(m) = f^+(n)$ for each Ω/E-algebra B and mapping $f : X \to FB$

where f^+ is the Ω-extension of f associated to B.

Both functors ϕ and ψ have a rank $\le \max\{k, \aleph_0\}$. ψ has the following properties: for each set X, $(\phi X, (\bar{\sigma}_{n\omega})_{(n,\omega) \in \Omega})$ with $\bar{\sigma}_{n\omega}((\tilde{m_i})_{i \in n}) = (n,\omega,(m_i)_{i \in n})\tilde{}$ is an Ω/E-algebra. Given an Ω/E-algebra B, each mapping f : X → FB can be extended to the Ω-homomorphism $f^\# : (\phi X, (\bar{\sigma}_{n\omega})_{(n,\omega) \in \Omega})$ → B by $f^\#(\tilde{m}) = f^+(m)$. $f^\#$ is called the $\underline{\Omega/\text{E-extension}}$ of f $\underline{\text{associated to}}$ B.

Let $\eta : \text{id} \to \phi$ and $\mu : \phi \circ \phi \to \phi$ be the natural transformations defined for each set X as follows: $\eta_X(x) = x\tilde{}$ for each $x \in X$. μ_X is the Ω/E-extension of the identity mapping $1_{\phi X}$ associated to $(\phi X, (\bar{\sigma}_{n\omega})_{(n,\omega) \in \Omega}) : \mu_X = 1_{\phi X}^\#$.

$\Phi = (\phi, \eta, \mu)$ is a monad. If $(X, (s_{n\omega})_{(n,\omega) \in \Omega})$ is an Ω/E-algebra, then one gets an Φ-algebra (X,s) defining by transfinite induction:

$$s(x\tilde{}) = x \text{ for each } x \in X \text{ and } s(\tilde{m}) = s_{n\omega}((s(\tilde{m_i}))_{i \in n}) \text{ for each positive ordinal } \alpha < k^+ \text{ and } (n,\omega,(m_i)_{i \in n}) \in \psi_\alpha X. \quad (4)$$

Now, we are ready to formulate one of the main results of categorical algebra (cf. e. g. [6]):

Theorem 2.2 Ω/E-ALG is concretely isomorphic to Φ-ALG. A concrete isomorphism is given assigning each Ω/E-algebra $(X, (s_{n\omega})_{(n,\omega) \in \Omega})$ the Φ-algebra (X,s) with s defined by (4).

This theorem can be completed as follows: If $\Phi = (\phi, \eta, \mu)$ is a monad over SET whose underlying set functor ϕ has a rank, say k, then there is a set-indexed family E of Ω-equations with $\Omega = \overset{\bullet}{\underset{n \leq k}{\Sigma}} \phi n$ such that Ω/E-ALG is concretely isomorphic to Φ-ALG (for details cf. e. g. [6], 1.5.40).

Remark If $\Phi = (\phi, \eta, \mu)$ is a monad over SET such that ϕ has not a rank, then we proceed as follows: For each positive cardinal k we get a monad $\Phi_k = (\phi_k, \eta_k, \mu_k)$ in taking as ϕ_k the subfunctor of ϕ of rank $\leq k$ for which $\phi_k X = \{m \in \phi X \mid \text{ord } m \leq k\}$ for each set X and in choosing η_k and μ_k in such a way that $\eta = e_k \circ \eta_k$ and $e_k \circ \mu_k = \mu \circ \phi e_k \circ e_k \phi_k$, where $e_k : \phi_k \to \phi$ is the natural transformation consisting of identical embeddings. Using the notion of power series of set functors given in [5], each Φ-algebra (X,s) can be proved to be the limit (in the sense of [5]) of the Φ_k-algebras $(X, s \mid \phi_k X)$ and therefore a limit of equationally determined algebras.

3. SOME FURTHER EXAMPLES OF EILENBERG-MOORE ALGEBRAS

In the following examples the underlying functors of the monads have not a rank. As ϕ will be considered the powerset functor, proper powerset functor, filter functor, proper filter functor, and ultrafilter functor, denoted P, P_0, F, F_0, and U, respectively. For each set X, PX is the set of all subsets of X, $P_0 X = PX \setminus \{\emptyset\}$, FX consists of all filters on X, $F_0 X = FX \setminus \{PX\}$, and UX is the set of all ultrafilters on X. Each mapping $f : X \to Y$ are assigned by P and F the mappings $M \mapsto f[M]$ ($M \subseteq X$) and $m \mapsto \{N \subseteq Y \mid f[M] \subseteq N$ for some $M \in m\}$ ($m \in FX$), respectively. P_0 is a subfunctor of P and F_0 and U are subfunctors of F. Each of these functors defines in a natural way a monad over SET, called the <u>powerset monad</u>, <u>proper powerset monad</u>, <u>filter monad</u>, <u>proper filter monad</u>, and <u>ultrafilter monad</u>, respectively. In the cases of P and P_0 the unit and multiplication consist of the mappings $\eta_X : x \mapsto \{x\}$ and $\mu_X : m \mapsto \bigcup_{M \in m} M$, respectively, whereas in the cases of F, F_0 and U, these natural transformations consist of the mappings $\eta_X : x \mapsto \dot{x}$ (= $\{N \subseteq X \mid x \in N\}$) and $\mu_X : m \mapsto \bigcup_{M \in m} \bigwedge_{n \in M} n$, respectively. Here $\bigwedge_{n \in M} n$ means $\bigcap_{n \in M} n$, if $M \neq \emptyset$ and PX otherwise.

Let INF denote the category of complete semilattices and infima preserving mappings where the structure of each complete semilattice is given as an infimum mapping $\inf : M \mapsto \inf M$ ($M \subseteq X$). inf M is indeed the infimum of M, namely with respect to the partial ordering defined by: $x \leq y \Longleftrightarrow x = \inf \{x,y\}$. Interpreting also the structure of each powerset monad algebra as an infimum mapping, we get the following well-known result.

<u>Proposition 3.1</u>   INF coincides with the category of all powerset monad algebras.

Let each set X be extended by an element $1_X \notin X$ (e. g. $1_X = X$) to $X \cup \{1_X\}$ and define a covariant functor $S : P_0 - CON_{map} \to P - CON_{map}$ in assigning to each object (X,s) and morphism $f : (X,s) \to (Y,t)$ of $P_0 - CON_{map}$ resp. the object $(X \cup \{1_X\}, \bar{s})$ of $P - CON_{map}$ with

$$\bar{s}(M) = \begin{cases} s(M \cap X) & \text{if } M \cap X \neq \emptyset \\ 1_X & \text{otherwise} \end{cases}$$

and the mapping $\bar{f} : (X \cup \{1_X\}, \bar{s}) \to (Y \cup \{1_Y\}, \bar{t})$ with $\bar{f} \mid X = f$ and $\bar{f}(1_X) = 1_Y$. An object (X,s) of $P_0 - CON_{map}$ will be called a <u>quasi-complete semilattice</u> provided $S(X,s)$ is a complete semilattice. We shall also interpret the structure of each quasi-complete semilattice (X,s) as an infimum mapping however only over the set of all non-empty subsets of X. P_0-continuous mappings between quasi-complete semilattices are then precisely the mappings which preserve infima of non-empty sets. Let INF_0 denote the full subcategory of $P_0 - CON_{map}$ of all quasi-complete semilattices. Analogously as in 3.1, one easily shows:

33

Proposition 3.2 INF_0 coincides with the category of all proper powerset monad algebras.

Let (X,inf) be a complete semilattice. The <u>way below relation</u> $\ll$ on X is given by: $u \ll x \Longleftrightarrow$ for each directed subset M of X with $x \leq \sup M$ there is a $v \in M$ with $u \leq v$. (X,inf) is called a <u>continuous semilattice</u> (or a <u>continuous lattice</u>) provided $x = \sup\{u \in X \mid u \ll x\}$ for each $x \in X$. Let CSLAT denote the category of continuous semilattices whose morphisms are the mappings between continuous semilattices which preserve arbitrary infima as well as suprema of directed subsets. CSLAT is a subcategory of INF.

There is the following important characterization of filter monad algebras given by A. Day [3] (for further characterizations see [9]).

Theorem 3.3 The category of filter monad algebras is concretely isomorphic to CSLAT. A concrete isomorphism is given in assigning to each filter monad algebra (X,s) the continuous semilattice (X,inf) with $\inf M = s([M])$ for each $M \in PX$. [M] means the principal filter $\{N \subseteq X \mid M \subseteq N\}$ on X.

The objects (X,s) and morphisms $f : (X,s) \to (Y,t)$ of INF_0 for which $S(X,s)$ and $\bar{f} : S(X,s) \to S(Y,t)$ are resp. objects and morphisms of CSLAT define a subcategory of INF_0, which will be denoted $CSLAT_0$.

Theorem 3.3 has the following analogue.

Theorem 3.4 The category of proper filter monad algebra is concretely isomorphic to $CSLAT_0$. A concrete isomorphism is given in assigning to each proper filter monad algebra (X,s) the quasi-complete semilattice (X,inf) with $\inf M = s([M])$ for each $M \in P_0 X$.

Let $T : F_0 - CON_{map} \to F - CON_{map}$ denote the covariant functor that assigns to each object (X,s) and morphism $f : (X,s) \to (Y,t)$ of $F_0 - CON_{map}$ resp. the object $(X \cup \{1_X\}, \bar{s})$ of $F - CON_{map}$ with

$$\bar{s}(\ ) = \begin{cases} s(\{M \cap X \mid M \in m\}) & \text{if } \{M \cap X \mid M \in m\} \neq PX \\ 1_X & \text{otherwise} \end{cases}$$

and the mapping $\bar{f} : (X \cup \{1_X\}, \bar{s}) \to (Y \cup \{1_Y\}, \bar{t})$ with $\bar{f} \mid X = f$ and $\bar{f}(1_X) = 1_Y$.

Proposition 3.5 The category of proper filter monad algebras is the full subcategory of F_0 - CON_{map} of all objects (X,s) for which T(X,s) are filter monad algebras.

The category of ultrafilter monad algebras has a well-known characterization ([6]) : It is concretely isomorphic to the category of all compact Hausdorff spaces. A concrete isomorphism is given in assigning to each ultrafilter monad algebra (X,s) the Hausdorff space with underlying set X and s its ultrafilter convergence.

4. STABILITY, INTERATEDNESS AND REGULARITY

Let $\Phi = (\phi, \eta, \mu)$ be a monad over SET. In this section the axioms (2) and (3) will be extended to ϕ-convergence spaces (X,s), where s is any relation between ϕX and X.

There is not any problem to extend axiom (2). A ϕ-convergence space (X,s) is said to be η-stable provided $1_X \subseteq s \circ \eta_X$ or, equivalently, $\eta_X(x) \underset{s}{\to} x$ for each $x \in X$. Clearly, if s is a mapping of ϕX into X, then η-stability consides with axiom (2).

As will be shown in the sequel, η-stability together with the Fréchet axiom has an interesting consequence. A ϕ-convergence space (X,s) is said to satisfy the Fréchet axiom with respect to a preordering $\leq$ on ϕX provided $m \underset{s}{\to} x$ and $m \leq n$ imply $n \underset{s}{\to} x$. If $\leq$ is down-directed then η-stability together with this "topological" axiom is incompatible with the "algebraic" property of s to be a mapping of ϕX into X. We namely have:

Proposition 4.1 Let (X,s) be an η-stable ϕ-convergence space which satisfies the Fréchet axiom with respect to a down-directed preordering $\leq$ on ϕX. Suppose that s is a mapping of ϕX into X. Then X consists of at most one element.

Proof Let x and y be two elements of X. There is an $m \in \phi X$ with $m \leq \eta_X(x)$ and $m \leq \eta_X(y)$. It follows that $s(m) = s(\eta_X(x)) = x$ and $s(m) = s(\eta_X(y)) = y$. □

In the cases of $\phi = F$ and F_0, the inclusion $\subseteq$ on ϕX is down-directed. Consequently, neither any filter monad algebra nor any proper filter monad algebra of at least two elements is (resp. as F- and F_0-convergence space) a pseudo-topological space (cf. [4]) or even a topological space. In the cases of $\phi = P$ and P_0, the inverse of the inclusion on ϕX is down-directed.

In order to extend axiom (3), we have at first to attend to the problem that ϕf is not defined for any relation f between sets. There is the following well-known method of extension of ϕ (cf. [1]): Let $f : X \mapsto Y$ be a relation. To point out if f is meant as a subset of $X \times Y$, we shall write then gr f instead of f. $f = f_2 \cdot f_1^{-1}$ where $f_1 : \text{gr } f \to X$ and $f_2 : \text{gr } f \to Y$ are resp. the first and second projection $(x,y) \mapsto x$ and $(x,y) \mapsto y$ of gr f. Define $\bar{\phi}f = \phi f_2 \circ (\phi f_1)^{-1}$. Then $\bar{\phi}f^{-1} = (\bar{\phi}f)^{-1}$. Moreover, if f is a mapping of X into Y, then $\bar{\phi}f = \phi f$.

Lemma 4.2 Let $f : X \mapsto Y$ be a relation. In the cases of $\phi = P$ and P_0

$$(M,N) \in \bar{\phi}f \iff f[M] \supseteq N \text{ and } f^{-1}[N] \supseteq M$$

for all $M \in \phi X$ and $N \in \phi Y$. In the cases of $\phi = F$, F_0 and U

$$(\mathfrak{m},\mathfrak{n}) \in \bar{\phi}f \iff f(\mathfrak{m}) \subseteq \mathfrak{n} \text{ and } f^{-1}(\mathfrak{n}) \subseteq \mathfrak{m}$$

for all $\mathfrak{m} \in \phi X$ and $\mathfrak{n} \in \phi Y$. Here $f(\mathfrak{m}) = Ff(\mathfrak{m})$ and $f^{-1}(\mathfrak{n})$ means the filter on X with the base $\{f^{-1}[N] \mid N \in \mathfrak{n}\}$.

Proof 1) Let $\phi = F$. Suppose at first that $(\mathfrak{m},\mathfrak{n}) \in \bar{F}f$. Then there exists a filter $\mathfrak{l} \in F\text{gr } f$ with $Ff_1(\mathfrak{l}) = \mathfrak{m}$ and $Ff_2(\mathfrak{l}) = \mathfrak{n}$. For each $M \in \mathfrak{m}$ there is an $L \in \mathfrak{l}$ with $M \supseteq f_1[L]$ and therefore with $f[M] \supseteq f_2[L]$. Hence $f(\mathfrak{m}) \subseteq \mathfrak{n}$. Analogously, $f^{-1}(\mathfrak{n}) \subseteq \mathfrak{m}$ follows.

Suppose now that $f(\mathfrak{m}) \subseteq \mathfrak{n}$ and $f^{-1}(\mathfrak{n}) \subseteq \mathfrak{m}$. $\mathfrak{m} = PX$ if and only if $\mathfrak{n} = PY$. Let, in the following, $\mathfrak{m}$ and $\mathfrak{n}$ be proper filters. Then $\{(M \times N) \cap \text{gr } f \mid M \in \mathfrak{m}, N \in \mathfrak{n}\}$ is the base of a proper filter $\mathfrak{k}$ on gr f such that $Ff_1(\mathfrak{k}) = \mathfrak{m}$ and $Ff_2(\mathfrak{k}) = \mathfrak{n}$.

2) In the case of $\phi = F_0$ and also in the cases of $\phi = P$ and P_0, the proof follows easily by part 1.

3) Let $\phi = U$. Analogously, as in the case of $\phi = F$, $(\mathfrak{m},\mathfrak{n}) \in \bar{U}f$ implies $f(\mathfrak{m}) \subseteq \mathfrak{n}$ and $f^{-1}(\mathfrak{n}) \subseteq \mathfrak{m}$. If $f(\mathfrak{m}) \subseteq \mathfrak{n}$ and $f^{-1}(\mathfrak{n}) \subseteq \mathfrak{m}$, then define a proper filter $\mathfrak{k}$ on gr f as in part 1. Let then $\mathfrak{l}$ be an ultrafilter with $\mathfrak{k} \subseteq \mathfrak{l}$. It follows $Uf_1(\mathfrak{l}) = \mathfrak{m}$ and $Uf_2(\mathfrak{l}) = \mathfrak{n}$. □

If $f : X \mapsto Y$ and $g : Y \mapsto Z$ are relations then $\bar{\phi}(g \circ f) \subseteq \bar{\phi}g \circ \bar{\phi}f$. In the case that $\bar{\phi}(g \circ f) = \bar{\phi}g \circ \bar{\phi}f$ for all relations $f : X \mapsto Y$ and $g : Y \mapsto Z$, ϕ has an obvious extension to a covariant functor $\bar{\phi} : \text{REL} \to \text{REL}$ where REL is the category of all sets with all relations between sets as morphisms. By means of Lemma 4.1 we see that in each of the cases $\phi = P, P_0, F, F_0$, and U this extension exists.

In the following part of this section it may be open whether $\bar{\phi}$ exists or not.

A ϕ-convergence space (X,s) is said to be <u>μ-iterated</u>, provided

$s \circ \bar{\phi}s \subseteq s \circ \mu_X$ holds. If s is a mapping of ϕX into X, then μ-iteratedness coincides with axiom (3).

Proposition 4.3 A ϕ-convergence space (X,s) is μ-iterated, if and only if $s \circ \phi s_2 \subseteq s \circ \mu_X \circ \phi s_1$ holds where s_1 and s_2 are the first and second projection of gr s, respectively.

Proof $s \circ \bar{\phi}s \subseteq s \circ \mu_X$ implies $s \circ \phi s_2 \subseteq s \circ \bar{\phi}s \circ \phi s_1 \subseteq s \circ \mu_X \circ \phi s_1$
From $s \circ \phi s_2 \subseteq s \circ \mu_X \circ \phi s_1$ it follows that $s \circ \bar{\phi}s \subseteq s \circ \mu_X \circ \phi s_1 \circ (\phi s_1)^{-1} \subseteq s \circ \mu_X$. □

In [1] η-stable, μ-iterated ϕ-convergence spaces are investigated denoted there <u>relational ϕ-algebras</u>.

There is an interesting second way of extending axiom (3). A ϕ-convergence space (X,s) is said to be <u>μ-regular</u> provided $s \circ \mu_X \circ (\bar{\phi}s)^{-1} \subseteq s$ holds. If s is a mapping of ϕX into X, then also μ-regularity coincides with axiom (3).

Proposition 4.4 A ϕ-convergence space (X,s) is μ-regular if and only if $s \circ \mu_X \circ \phi s_1 \subseteq s \circ \phi s_2$ holds.

Proof $s \circ \mu_X \circ (\bar{\phi}s)^{-1} \subseteq s$ implies $s \circ \mu_X \circ \phi s_1 \subseteq s \circ \mu_X \circ (\bar{\phi}s)^{-1} \circ s_2 \subseteq s \circ \phi s_2$. From $s \circ \mu_X \circ \phi s_1 \subseteq s \circ \phi s_2$ one gets $s \circ \mu_X \circ (\bar{\phi}s)^{-1} \subseteq s \circ \phi s_2 \circ (\phi s_2)^{-1} \subseteq s$. □

The following example in the case of the proper filter monad (F_0,η,μ) shows that μ-regularity need not imply $s \circ \mu_X \subseteq s \circ \bar{\phi}s$. Let $\mathfrak{s}$ be the usual topology of $\mathbb{R}$. $(\mathbb{R},s)$ is μ-regular (see 5.4 below). Let $\mathfrak{m}$ be the filter on $F_0\mathbb{R}$ with the base $\{\{[I_m] \mid m \geq n\} \mid n = 1,2,\ldots\}$ where $[I_m]$ is the principal filter with I_m the open interval $]-1/m, 1/m[$. $\mu_X(\mathfrak{m}) = \bigcup_{n=1,2,\ldots} [I_n]$ is the zero-neighbourhood filter on $\mathbb{R}$. Thus, $\mu_X(\mathfrak{m}) \xrightarrow{s} 0$, i.e. $(\mathfrak{m},0) \in s \cdot \mu_X$. Since none of the filters $[I_m]$ converges, $s(\mathfrak{m})$ is the improper filter. Hence, there is not any filter $\mathfrak{n}$ on X with $(\mathfrak{m},\mathfrak{n}) \in \bar{F}_0 s$.

5. THE CASE THE PROPER FILTER MONAD

Let η and μ be the unit and multiplication of the proper filter monad. η-stability of an F_0-convergence space (X,s) means $\dot{x} \xrightarrow{s} x$ for each $x \in X$, which also will be called <u>stability</u> only.

Assume that if $(I_\ell)_{\ell \in L}$ is a set-indexed family of directed sets, then $\prod_{\ell \in L} I_\ell$ is directed by: $(i_\ell)_{\ell \in L} \leq (i'_\ell)_{\ell \in L} \iff i_\ell \leq i'_\ell$ for each $\ell \in L$. A net

$(x_i)_{i \in I}$ is called __non-empty__ provided I is non-empty. A non-empty net $(x_i)_{i \in I}$ on an F_0-convergence space (X,s) __converges__ or __s-converges__ to $x \in X$, written $(x_i)_{i \in I} \xrightarrow{s} x$, provided the filter on X generated by $(x_i)_{i \in I}$ s-converges to x.

__Theorem 5.1__ Let (X,s) be an F_0-convergence space. The following are equivalent.

(i) (X,s) is μ-iterated.

(ii) (__Condition on iterated limits__). For each $x \in X$, each non-empty net $(x_i)_{i \in I} \xrightarrow{s} x$ and all non-empty nets $(x_{ij})_{j \in J_i} \xrightarrow{s} x_i$, $i \in I$, it follows that the net $(x_{ij_i})_{(i,(j_k)_{k \in I}) \in I \times \prod_{k \in I} J_k}$ s-convergences to x.

__Proof__ 1) Suppose (i) is fulfilled. Let an $x \in X$, a non-empty net $(x_i)_{i \in I} \xrightarrow{s} x$ and for each $i \in I$ a non-empty net $(x_{ij})_{j \in J_i} \xrightarrow{s} x_i$ be fixed. Denote by m_i the filter on X generated by $(x_{ij})_{j \in J_i}$. Moreover, denote by l the filter on gr s generated by the net $(m_i, x_i)_{i \in I}$. Obviously, $(x_i)_{i \in I}$ generates $F_0 s_2(l)$. Hence $F_0 s_2(l) \xrightarrow{s} x$ and because of 4.3, therefore $\mu_X(F_0 s_1(l)) = \bigcup_{i_0 \in I} \bigcap_{i \geq i_0} m_i \xrightarrow{s} x$. A subset M of X is an element of $\mu_X(F_0 s_1(l))$, if an only if there is an $i_0 \in I$ and an $j_{0i} \in J_i$ for every $i \geq i_0$ such that $M \supseteq \{x_{ij} \mid i \geq i_0, j \geq j_{0i}\}$. Taking into account that $(j_{0i})_{i \geq i_0}$ can be extended to a family $(j_{0k})_{k \in I}$, it follows that $M \in \mu_X(F_0 s_1(l))$ if and only if there is a pair $(i_0, (j_{ok})_{k \in I})$ such that $M \supseteq M \supseteq \{x_{ij_i} \mid (i,(j_k)_{k \in I}) \geq (i_0,(j_{0k})_{k \in I})\}$. Hence $\mu_X(F_0 s_1(l))$ is generated by the net $(x_{ij_i})_{(i,(j_k)_{k \in I}) \in I \times \prod_{k \in I} J_k}$, which therefore s-converges to x.

2) Suppose (ii) is fulfilled. Let l be a filter on gr s and x an element of X such that $F_0 s_2(l) \xrightarrow{s} x$. Choose a net $(m_i, x_i)_{i \in I}$ on gr s which generates l. Moreover, for each $i \in I$, choose a net $(x_{ij})_{j \in J_i}$ on X which generates m_i. It follows $(x_{ij})_{j \in J_i} \xrightarrow{s} x_i$ for every $i \in I$. Since $(x_i)_{i \in I}$ generates $F_0 s_2(l)$, we also have $(x_i)_{i \in I} \xrightarrow{s} x$. Therefore $(x_{ij_i})_{(i,(j_k)_{k \in I}) \in I \times \prod_{k \in I} J_k} \xrightarrow{s} x$. Taking into account that $(m_i)_{i \in I}$ generates $F_0 s_1(l)$, we get $\mu_X(F_0 s_1(l)) = \bigcup_{i_0 \in I} \bigcap_{i \geq i_0} m_i$. Analogously, as in part 1 of this proof, it follows that $(x_{ij_i})_{(i,(j_k)_{k \in I}) \in I \times \prod_{k \in I} J_k}$ generates $\mu_X(F_0 s_1(I))$. Hence $\mu_X(F_0 s_1(l)) \xrightarrow{s} x$. Because of 4.3, therefore (X,s) is μ-iterated. □

Theorem 5.1 leads to a characterization of proper filter monad algebras.

__Theorem 5.2__ An F_0-convergence space (X,s) is a proper filter monad

algebra if and only if the following conditions are fulfilled:
   (i)   (x,s) is Hausdorff (that is, $m \xrightarrow{s} x,y$ implies $x = y$).
   (ii) Each proper filter on X s-convergences.
   (iii) (X,s) is stable and fulfills the condition on iterated limits.

   <u>Theorem 5.3</u>   Let (X,s) be an F_0-convergence space such that for each $x \in X$ there is at least a proper filter s-converging to x.

   If (X,s) is μ-iterated, then (X,s) satisfies the following <u>neighbourhood condition</u>:

$$m \xrightarrow{s} x \text{ implies } \bigcup_{M \in m} \bigcap_{y \in M} n(y) \xrightarrow{s} x \text{ where } n(y) \text{ is the neighbourhood filter of } y.$$

   If (X,s) satisfies the Fréchet axiom (with respect to $\subseteq$) and the neighbourhood condition, then conversely (X,s) is μ-iterated.

   <u>Proof</u>   1) Suppose (X,s) is μ-iterated. Let an $x \in X$ and an $m \xrightarrow{s} x$ be fixed. Let l denote the filter on gr s with the base $\{s_2^{-1}[M] \mid M \in m\}$. Because of the assumption, $s_2[\text{gr } s] = X$ and therefore $F_0 s_2(l) = m$. Hence 4.3 implies $\mu_X(F_0 s_1(l)) \xrightarrow{s} x$. Since $\bigcap_{k \in s_2^{-1}[M]} k = \bigcap_{y \in M} n(y)$ for each $M \in m$, we have $\mu_X(F_0 s_1(l)) = \bigcup_{M \in m} \bigcap_{y \in M} n(y)$.

   2) Suppose that (X,s) satisfies the Fréchet axiom and the neighbourhood condition. Let l be a filter on gr s and x an element of X with $F_0 s_2(l) \xrightarrow{s} x$. Because of the neighbourhood condition, it follows that $\bigcup_{L \in l} \bigcap_{y \in s_2[L]} n(y) \xrightarrow{s} x$. Since $\bigcup_{L \in l} \bigcap_{y \in s_2[L]} n(y) \subseteq \mu_X(F_0 s_1(l))$ and (X,s) satisfies the Fréchet axiom, we get $\mu_X(F_0 s_1(l)) \xrightarrow{s} x$. Because of 4.3 therefore (X,s) is μ-iterated. □

   <u>Corollary</u>   Let (X,s) be an F_0-convergence space which is stable and satisfies the Fréchet axiom with respect to $\subseteq$. Then (X,s) is μ-iterated if and only if (X,s) is a topological space.

   <u>Proof</u>   Use 5.3 and note that $\dot{x} \xrightarrow{s} x$ for each $x \in X$.

   For each subset M of an F_0-convergence space (X,s), $\bar{M} = \{x \in X \mid M \in n \text{ for some } n \xrightarrow{s} x\}$ is called the <u>adherence</u> of M.

   <u>Theorem 5.4</u>   Let (X,s) be a stable F_0-convergence space.
   If (X,s) is μ-regular, then (X,s) is <u>regular</u>, that is, $m \xrightarrow{s} x$ implies $m^- \xrightarrow{s} x$ where m^- is the filter on X with the base $\{\bar{M} \mid M \in m\}$.
   If (X,s) satisfies the Fréchet axiom with respect to $\subseteq$ and is regular, then (X,s) is μ-regular.

Proof   1) Suppose that (X,s) is μ-regular. Let an $x \in X$ and $\mathfrak{m} \xrightarrow{s} x$ be fixed. Denote by 1 the filter on gr s with the base $\{L_M \mid M \in \mathfrak{m}\}$, where $L_M = \{(\mathfrak{n}, y) \in s \mid M \in \mathfrak{n}\}$. It follows $\bar{M} = s_2[L_M]$ for every $M \in \mathfrak{m}$ and therefore $\bar{\mathfrak{m}} = F_0 s_2(\mathfrak{1})$. Since $M \in \bigcap_{\mathfrak{n} \in s_1[L_M]} \mathfrak{n}$ for each $M \in \mathfrak{m}$, we get $\mathfrak{m} \subseteq \bigcup_{M \in \mathfrak{m}} \bigcap_{\mathfrak{n} \in s_1[L_M]} \mathfrak{n} = \mu_X(F_0 s_1(\mathfrak{1}))$. Let $M \in \mathfrak{m}$ and $N \in \bigcap_{\mathfrak{n} \in s_1[L_M]} \mathfrak{n}$ be fixed. Because of the stability we have $y \in s_1[L_M]$ and therefore $N \in \dot{y}$ for all $y \in M$, hence $M \subseteq N$. This implies $N \in \mathfrak{m}$ and therefore $\mathfrak{m} = \mu_X(F_0 s_1(\mathfrak{1}))$. Taking into account 4.4 and that (X,s) is μ-regular, we get $\bar{\mathfrak{m}} \xrightarrow{s} x$ and hence that (X,s) is regular.

2) Assume that (X,s) satisfies the Fréchet axiom and is regular. Choose an $x \in X$ and a filter $\mathfrak{1}$ on gr s such that $\mathfrak{m} \xrightarrow{s} x$ where $\mathfrak{m} = \mu_X(F_0 s_1(\mathfrak{1}))$. Since $\mathfrak{m} = \bigcup_{L \in \mathfrak{1}} \bigcap_{\mathfrak{n} \in s_1[L]} \mathfrak{n}$, for each $M \in \mathfrak{m}$ there is an $L \in \mathfrak{1}$ with $M \in \bigcap_{\mathfrak{n} \in s_1[L]} \mathfrak{n}$ and therefore with $s_2[L] \subseteq \bar{M}$. Thus, $\bar{\mathfrak{m}} \subseteq F_0 s_2(\mathfrak{1})$. Because of the assumption it follows $F_0 s_2(\mathfrak{1}) \xrightarrow{s} x$, because of 4.4 therefore that (X,s) is μ-regular. □

Corollary   Let (X,s) be a stable F_0-convergence space which satisfies the Fréchet axiom with respect to $\subseteq$. Then (X,s) is μ-regular if and only if (X,s) is regular.

In particular, a pseudo-toplogical space (defined as F_0-convergence space) is μ-regular if and only if it is regular (in the usual sense).

6. THE CASE OF THE FILTER MONAD

Let η and μ be the unit and multiplication of the filter monad. μ-stability of an F-convergence space (X,s) means the same as in the case of the proper filter monad: $\dot{x} \xrightarrow{s} x$ for all $x \in X$. Here, we shall also speak of stability only instead of η-stability. A net $(x_i)_{i \in I}$ on an F-convergence space (X,s) converges or s-converges to an $x \in X$, written $(x_i)_{i \in I} \xrightarrow{s} x$, provided the filter $\mathfrak{m}$ on X generated by $(x_i)_{i \in I}$ s-converges to x. If $I = \emptyset$, then $\mathfrak{m} = PX$.

Theorem 6.1   Let (X,s) be an F-convergence space. The following are equivalent:

(i) (X,s) is μ-iterated.

(ii) (Condition on iterated limits). For each $x \in X$, each non-empty net $(x_i)_{i \in I} \xrightarrow{s} x$ and all nets $(x_{ij})_{j \in J_i} \xrightarrow{s} x_i$, $i \in I$, we have: If $I_0 = \{i \in I \mid J_i \neq \emptyset\}$ is a cofinal subset of I then

$(x_{ij_i})(i,(j_k)_{k \in I_0}) \in I_0 \times \prod_{k \in I_0} J_k \xrightarrow{s} x$, otherwise $PX \xrightarrow{s} x$.

The <u>proof</u> is essentially the same as that of Theorem 5.1. Therefore, we shall only make some remarks. In the case of (i) => (ii), we define m_i, $i \in I$, and 1 analogously as in part 1 of the proof of 5.1. Now however the sets J_i may be empty. It follows $\mu_X(Fs_1(1)) = \bigcup_{i_0 \in I} \bigcap_{i \geq i_0} m_i$. If I_0 is cofinal then $\bigcup_{i_0 \in I} \bigcap_{i \geq i_0} m_i = \bigcup_{i_0 \in I_0} \bigcap_{i \geq i_0} m_i$, otherwise $\bigcup_{i \in I} \bigcap_{i \geq i_0} m_i = PX$. To show (ii) => (i) suppose that (ii) is fulfilled. Let 1 be a filter on gr s and x an element of X with $Fs_2(1) \xrightarrow{s} x$. Is 1 is improper then also $Fs_2(1)$ and $\mu_X(Fs_1(1))$ are improper Hence, then $\mu_X(Fs_1(1)) \xrightarrow{s} x$. In the following suppose that 1 is proper. Let $(m_i, x_i)_{i \in I}$ and $(x_{ij})_{j \in J_i}$, $i \in I$, be nets which are chosen analogously as in part 2 of the proof of 5.1. We have $I \neq \emptyset$, $(x_i)_{i \in I} \xrightarrow{s} x$ and $(x_{ij})_{j \in J_i} \xrightarrow{s} x_i$ for each $i \in I$. If I_0 is cofinal, then $\mu_X(Fs_1(1)) = \bigcup_{i_0 \in I_0} \bigcap_{\substack{i \geq i_0 \\ i \in I_0}} m_i \xrightarrow{s} x$, otherwise $\mu_X(Fs_1(1)) = \bigcup_{i_0 \in I} \bigcap_{i \geq i_0} m_i = PX \xrightarrow{s} x$. □

Applying Theorem 6.1 we get a characterization of filter monad algebras.

<u>Theorem 6.2</u> An F-convergence space (X,s) is a filter monad algebra if and only if the following conditions are fulfilled:

(i) (X,s) is Hausdorff (that is, $m \xrightarrow{s} x,y$ and $m \neq PX$ imply $x = y$).

(ii) Each proper filter on X s-converges.

(iii) PX s-converges to precisely one element of X.

(iv) (X,s) is stable and fulfills the condition on iterated limits.

<u>Theorem 6.3</u> Let (X,s) be an F-convergence space such that for each $x \in X$ there is at least a filter s-converging to x.

If (X,s) is μ-iterated, then (X,s) satisfies the following <u>neighbourhood condition</u>:

$m \xrightarrow{s} x$ and $m \neq PX$ imply $\bigcup_{M \in m} \bigcap_{y \in M} n(y) \xrightarrow{s} x$ where $n(y)$ is the neighbourhood filter $\bigcap_{k \xrightarrow{s} x} k$ of y.

If (X,s) satisfies the Fréchet axiom (with respect to $\subseteq$) and the neighbourhood condition, then conversely (X,s) is μ-iterated.

The <u>proof</u> is a slight modification of that of Theorem 5.3. □

Obviously, the Corollary of Theorem 5.3 remains true if F_0 is replaced by F.

If M is a subset of an F-convergence space (X,s), then $\bar{M}^a = \{x \in X \mid M \in n \text{ for some } n \xrightarrow{s} x\}$ will be called the <u>algebraic adherence</u> of M, whereas by the <u>adherence</u> of M we mean $\bar{M} = \{x \in X \mid M \in n \text{ for some proper filter}$

$n \underset{s}{\to} x$}. Note that $\bar{M}^a = \bar{M} \cup \{x \in X \mid Px \underset{s}{\to} x\}$.

Theorem 6.4 Let (X,s) be a stable F-convergence space.

If (X,s) is μ-regular, then (X,s) is <u>algebraically regular</u>, that is, $m \underset{s}{\to} x$ implies $m^{-a} \underset{s}{\to} x$ where m^{-a} is the filter on X with the base $\{\bar{M}^a \mid M \in m\}$.

If (X,s) satisfies the Fréchet axiom (with respect to $\subseteq$) and is algebraically regular, then conversely (X,s) is μ-regular.

The <u>proof</u> follows from that of Theorem 5.4 by a slight modification. □

Note that stable, algebraically regular F-convergence spaces which satisfy the Fréchet axiom with respect to $\subseteq$ are only the F-convergence spaces (X,s) with the largest structure: $s = FX \times X$ (from the conditions it follows $\dot{x} \underset{s}{\to} x$ and hence $PX \underset{s}{\to} x$ for each $x \in X$, therefore $\bar{M}^a = X$ for each $M \subseteq X$ and thus $\{X\} \underset{s}{\to} x$ for each $x \in X$).

7. SEQUENTIAL STRUCTURES AND GENERALIZED MONADS

Let seq denote the <u>sequence functor</u>. seq assigns to each set X and mapping $f : X \to Y$ the set $seq\, X = X^{\mathbb{N}}$ of all sequences $(x_n)_{n \in \mathbb{N}}$ on X and the mapping $seq\, f : (x_n)_{n \in \mathbb{N}} \mapsto (f(x_n))_{n \in \mathbb{N}} ((x_n)_{n \in \mathbb{N}} \in X^{\mathbb{N}})$. seq can be completed in a natural way to a monad over SET, namely in choosing as unit η and multiplication μ resp. the families of mappings $\eta_X : x \mapsto (x)_{n \in \mathbb{N}}$ $(x \in X)$ and $\mu_X : ((x_{mn})_{m \in \mathbb{N}})_{n \in \mathbb{N}} \mapsto (x_{nn})_{n \in \mathbb{N}} (((x_{mn})_{m \in \mathbb{N}})_{n \in \mathbb{N}} \in (X^{\mathbb{N}})^{\mathbb{N}})$. (seq, η, μ) is called the <u>sequence monad</u>.

η-stability of a seq-convergence space (X,s) is the usual <u>stability</u>: $(x)_{n \in \mathbb{N}} \underset{s}{\to} x$ for each $x \in X$. However, the notions of μ-iteratedness and μ-regularity in this case are very strong notions from the point of view of topology: μ-iteratedness means that $(x_{mn})_{m \in \mathbb{N}} \underset{s}{\to} x_n$, $n \in \mathbb{N}$, and $(x_n)_{n \in \mathbb{N}} \underset{s}{\to} x$ imply $(x_{nn})_{n \in \mathbb{N}} \underset{s}{\to} x$. μ-regularity says that from $(x_{mn})_{m \in \mathbb{N}} \underset{s}{\to} x_n$, $n \in \mathbb{N}$, and $(x_{nn})_{n \in \mathbb{N}} \underset{s}{\to} x$ it follows $(x_n)_{n \in \mathbb{N}} \underset{s}{\to} x$. As is easily seen, each Euclidian space $\mathbb{R}^n$ with $n > 0$ equipped with the usual seq-convergence structure is neither μ-iterated nor μ-regular. This reason and also some other suggest to weaken the notion of monad. One way is as follows:

Let ϕ be a covariant set functor and $\eta: id \to \phi$ a natural transformation. Moreover let $\mu : \phi \circ \phi \mapsto \phi$ be a family $(\mu_X)_{X \in obSET}$ of relations $\mu_X : \phi\phi X \mapsto \phi X$ such that

$$\mu_Y \circ \phi\phi f = \phi f \circ \mu_X \qquad (5)$$

for each mapping $f : X \to Y$. (ϕ, η, μ) is called a <u>relational monad</u> over SET if

$$1_{\phi X} = \mu_X \circ \phi \eta_{X}, \tag{6}$$

$$1_{\phi X} \in \mu_X \circ \eta_{\phi X} \tag{7}$$

and

$$\mu_X \circ \bar{\phi}\mu_X \subseteq \mu_X \circ \mu_{\phi X} \tag{8}$$

for each set X. Obviously, (8) is equivalent to $\mu_X \circ \phi(\mu_X)_2 \subseteq \mu_X \circ \mu_{\phi X} \circ \phi(\mu_X)_1$.

A relation $f : (X,s) \mapsto (Y,t)$ between ϕ-convergence spaces is called <u>I-continuous</u> (resp. <u>ϕ-cocontinuous</u>) provided $f \circ s \circ (\bar{\phi}f)^{-1} \subseteq t$ (resp. $t \circ \bar{\phi}f \subseteq f \circ s$). (8) says that $\mu_X : (\phi\phi X, \mu_{\phi X}) \mapsto (\phi X, \mu_X)$ is ϕ-cocontinuous.

Let $\Phi = (\phi, \eta, \mu)$ be a relational monad over SET. For a ϕ-convergence space (X,s), <u>η-stability</u>, <u>μ-iteratedness</u> and <u>μ-regularity</u> are defined as in the case of a monad, namely by $1_X \subseteq s \circ \eta_X$, $s \circ \bar{\phi}s \subseteq s \circ \mu_X$ and $s \circ \mu_X \circ (\bar{\phi}s)^{-1} \subseteq s$, respectively. (X,s) is μ-iterated (resp. μ-regular) if and only if $s : (\phi X, \mu_X) \mapsto (X,s)$ is ϕ-cocontinuous (resp. ϕ-continuous). $(\phi X, \mu_X)$ is an η-stable, μ-iterated ϕ-convergence space.

In the following we shall investigate the notions of μ-iteratedness and μ-regularity in a special case. Let η be the unit of the sequence monad and let $\mu : seq \circ seq \to seq$ be the family of relations $\mu_X = \{(((x_{mn})_{m \in \mathbb{N}})_{n \in \mathbb{N}}, (x_{m_n n})_{n \in \mathbb{N}}) \mid x_{mn} \in X$ for all $m, n \in \mathbb{N}$, $(m_n)_{n \in \mathbb{N}}$ any subsequence of $0,1,2,\ldots\}$ where X is an arbitrary set. As is easily seen, μ with $\phi = seq$ fulfills (5) for each mapping $f : X \to Y$.

<u>Proposition 7.1</u> (seq, η, μ) is a relational monad, called the <u>sequence relational monad</u>.

<u>Proof</u>   Let $\phi = seq$. As is easily seen, (6) is satisfied. Since $\mu_X \circ \eta_{\phi X}$ is the set of all pairs $((x_n)_{n \in \mathbb{N}}, (y_n)_{n \in \mathbb{N}})$ where $(x_n)_{n \in \mathbb{N}}$ is a sequence on X and $(y_n)_{n \in \mathbb{N}}$ a subsequence of $(x_n)_{n \in \mathbb{N}}$, it follows (7). $(\mathfrak{m}, \mathfrak{n}) \in \mu_X \circ \bar{\phi}\mu_X$ with $\mathfrak{m} = (((x_{\ell mn})_{\ell \in \mathbb{N}})_{m \in \mathbb{N}})_{n \in \mathbb{N}}$ holds if and only if there are subsequences $(\ell_m^n)_{m \in \mathbb{N}}$, $n \in \mathbb{N}$, and $(m_n)_{n \in \mathbb{N}}$ of $0,1,2,\ldots$ such that $\mathfrak{n} = (x_{\ell_{m_n}^n m_n n})_{n \in \mathbb{N}}$. On the other hand, $(\mathfrak{m}, \mathfrak{n}) \in \mu_X \circ \mu_{\phi X}$ with $\mathfrak{m} = (((x_{\ell mn})_{\ell \in \mathbb{N}})_{m \in \mathbb{N}})_{n \in \mathbb{N}}$ holds if and only if there are subsequences $(\ell_n)_{n \in \mathbb{N}}$ and $(m_n)_{n \in \mathbb{N}}$ of $0,1,2,\ldots$ such that $\mathfrak{n} = (x_{\ell_n m_n n})_{n \in \mathbb{N}}$. Hence (8) is fulfilled. □

In the case of the sequence relational monad the μ-iteratedness of a seq-convergence space means: $(x_{mn})_{m \in \mathbb{N}} \xrightarrow{s} x_n$, $n \in \mathbb{N}$, and $(x_n)_{n \in \mathbb{N}} \xrightarrow{s} x$ imply $(x_{m_n n})_{n \in \mathbb{N}} \xrightarrow{s} x$ for some subsequence $(m_n)_{n \in \mathbb{N}}$ of $0,1,2,\ldots$.

Denote the filter generated by a sequence $(x_n)_{n \in \mathbb{N}}$ on a set X by filt $(x_n)_{n \in \mathbb{N}}$.

Theorem 7.2 Let (X,s) be a *seq*-convergence space which is <u>pretopological</u> (that is, $\dot{x} \cap \bigcap_{(x_n) \vec{s} x}$ filt $(x_n)_{n \in \mathbb{N}} \subseteq$ filt $(y_n)_{n \in \mathbb{N}}$ implies $(y_n)_{n \in \mathbb{N}} \vec{s} x$).

If (X,s) is μ-iterated, then the <u>related pretopology</u> (which has $\mathfrak{n}(x) = \dot{x} \cap \bigcap_{(x_n) \vec{s} x}$ filt $(x_n)_{n \in \mathbb{N}}$ as neighbourhood filters) is a toplogy.

If the related pretopology is a first countable topology, then on the other hand (X,s) is μ-iterated.

Proof 1) Suppose (X,s) is μ-iterated. Since (X,s) satisfies the Fréchet axiom with respect to the subsequence preordering $\leq :((x_n)_{n \in \mathbb{N}} \leq (y_n)_{n \in \mathbb{N}} \Leftrightarrow (y_n)_{n \in \mathbb{N}}$ is a subsequence of $(x_n)_{n \in \mathbb{N}}$, it follows $x \in \bar{M} \Leftrightarrow X \setminus M \in \mathfrak{n}(x)$ with

$$\bar{M} = \{x \in X \mid \exists (x_n)_{n \in \mathbb{N}} \vec{s} x \; x_n \in M \text{ for each } n \in \mathbb{N}\}.$$

We have to show $\mathfrak{n}(x) \subseteq \bigcup_{M \in \mathfrak{n}} \bigcap_{y \in M} \mathfrak{n}(y)$ for each $x \in X$ which is equivalent to $\bar{\bar{M}} \subseteq \bar{M}$ for each subset M of X (cf. e.g. 3.4.4 in [4]). Let M be a subset of X and x an element of $\bar{\bar{M}}$. There is a sequence $(x_n)_{n \in \mathbb{N}} \vec{s} x$ of elements of $\bar{M}$. Hence for each $n \in \mathbb{N}$ there is a sequence $(x_{mn})_{m \in \mathbb{N}} \vec{s} x_n$ of elements of M. Since $(x_{m_n n})_{n \in \mathbb{N}} \vec{s} x$ for some subsequence $(m_n)_{n \in \mathbb{N}}$ of 0,1,2,..., this implies $x \in \bar{M}$ and therefore $\bar{\bar{M}} \subseteq \bar{M}$.

2) Suppose that the related pretopology is a first countable topology. Let an $x \in X$, a sequence $(x_n)_{n \in \mathbb{N}} \vec{s} x$ and sequences $(x_{mn})_{m \in \mathbb{N}} \vec{s} x_n$, $n \in \mathbb{N}$, be fixed. Choose a base $\{M_n \mid n \in \mathbb{N}\}$ of $\mathfrak{n}(x)$ such that $M_0 = X$, $m \leq n$ implies $M_m \supseteq M_n$ and all M_n are open. There is a subsequence $(\ell_k)_{k \in \mathbb{N}}$ of 0,1,2,... with $\ell_0 = 0$ such that $x_n \in M_k$ whenever $\ell_k \leq n < \ell_{k+1}$. Hence, there is a subsequence $(m_n)_{n \in \mathbb{N}}$ of 0,1,2,... with $x_{m_n n} \in M_k$ for all $n, k \in \mathbb{N}$ for which $\ell_k \leq n < \ell_{k+1}$. It follows $(x_{m_n n})_{n \in \mathbb{N}} \vec{s} x$ and therefore that (X,s) is μ-iterated. □

μ-regularity of a *seq*-convergence space (X,s) here means: If x is an element of X and $(x_n)_{n \in \mathbb{N}}$ and $(x_{mn})_{m \in \mathbb{N}}$, $n \in \mathbb{N}$, are sequences on X such that $(x_{mn})_{m \in \mathbb{N}} \vec{s} x_n$ for each $n \in \mathbb{N}$ and $(X_{m_n n})_{n \in \mathbb{N}} \vec{s} x$ for every subsequence $(m_n)_{n \in \mathbb{N}}$ of 0,1,2,..., then $(x_n)_{n \in \mathbb{N}} \vec{s} x$.

Theorem 7.3 Let (X,s) be a *seq*-convergence space which is pretopological.

If the related pretopology is <u>regular</u> (i.e. for each $x \in X$ and $N \in \mathfrak{n}(x)$ there is an $M \in \mathfrak{n}(x)$ with $\bar{M} \subseteq N$), then (X,s) is μ-regular.

If the related pretopology is first countable and (X,s) is μ-regular, then the related pretopology is regular.

Proof   1) Suppose that the related pretopology is regular. Let x be an element of X and $(x_n)_{n \in \mathbb{N}}$ and $(x_{mn})_{m \in \mathbb{N}}$, $n \in \mathbb{N}$, be sequences on X such that $(x_{mn})_{n \in \mathbb{N}} \xrightarrow{s} x_n$ for each $n \in \mathbb{N}$, however, not $(x_n)_{n \in \mathbb{N}} \xrightarrow{s} x$. Then, there are $M, N \in \mathfrak{n}(x)$ with $\bar{M} \subseteq N$ and $x_n \notin N$ for all elements n of an infinite subset $\mathbb{N}'$ of $\mathbb{N}$. Hence $x_{m_n n} \notin M$ for some subsequence $(m_k)_{k \in \mathbb{N}}$ of $0, 1, 2, \ldots$ and all $n \in \mathbb{N}'$. This implies that not $(x_{m_n n})_{n \in \mathbb{N}} \xrightarrow{s} x$ and that, therefore, (X, s) is μ-regular.

2) Suppose that the related pretopology is first countable and is not regular. Then, there is an $x \in X$, an $M \in \mathfrak{n}(x)$ and a base $\{M_n \mid n \in \mathbb{N}\}$ of $\mathfrak{n}(x)$ such that $\bar{M}_n \setminus M \neq \emptyset$ for all $n \in \mathbb{N}$. Choose for each $n \in \mathbb{N}$ an $x_n \in \bar{M}_n \setminus M$ and a sequence $(x_{mn})_{m \in \mathbb{N}} \xrightarrow{s} x_n$ of elements of M_n. It follows $(x_{m_n n})_{n \in \mathbb{N}} \xrightarrow{s} x$ for every subsequence $(m_n)_{n \in \mathbb{N}}$ of $0, 1, 2, \ldots$ and not $(x_n)_{n \in \mathbb{N}} \xrightarrow{s} x$. Hence (X, s) is not μ-regular. □

Let us finally attend to an arbitrary relational monad $\Phi = (\phi, \eta, \mu)$ over SET, (5) implies that for each relation $f : X \mapsto Y$,

$$\mu_Y \circ \bar{\phi}\bar{\phi}f \subseteq \bar{\phi}f \circ \mu_X \tag{9}$$

holds, that is, $\bar{\phi}f : (\phi X, \mu_X) \to (\phi Y, \mu_Y)$ is ϕ-cocontinuous. If, namely, $f : X \mapsto Y$ is any relation, then from $\mu_X \circ \phi\phi f_1 = \phi f_1 \circ \mu_{gr\ f}$, $\mu_Y \circ \phi\phi f_2 = \phi f_2 \circ \mu_{gr\ f}$ and $\bar{\phi}\bar{\phi}f = \phi\phi f_2 \circ (\phi\phi f_1)^{-1}$ it follows (9). One the other hand, if (9) holds for any relation $f : X \mapsto Y$, then, in particular we have $\mu_Y \circ \phi\phi f \subseteq \phi f \circ \mu_X$ and $\mu_X \circ (\phi\phi f)^{-1} \subseteq (\phi f)^{-1} \circ \mu_Y$ and therefore (5) for each mapping $f : X \to Y$.

Let C denote the category of all η-stable, μ-iterated ϕ-convergence spaces with all ϕ-cocontinous relations between these spaces as morphisms. Assigning to each set X the object $(\phi X, \mu_X)$ of C and to each relation $f : X \to Y$ the C-morphism $\bar{\phi}f : (\phi X, \mu_X) \mapsto (\phi Y, \mu_Y)$ leads to a lax functor $S : REL \to C$. "lax" means here that (i) $S(1_X) = 1_{SX}$, (ii) $f \subseteq g$ implies $Sf \subseteq Sg$ for all relations $f, g : X \mapsto Y$ and (iii) $S(g \circ f) \subseteq Sg \circ Sf$ for all relations $f : X \mapsto Y$ and $g : Y \mapsto Z$. Let $T : C \to REL$ be the forgetful functor. $\bar{\phi} = T \circ S : REL \to REL$ also is a lax functor. η can be considered as a lax natural transformation $\bar{\eta} : id_{REL} \to T \circ S$ where id_{REL} is the identity functor of REL. Here "lax" means that $\eta_Y \circ f \subseteq \bar{\phi}f \circ \eta_X$ for each relation $f : X \mapsto Y$. The family $(\varepsilon_{(X,s)})_{(X,s) \in obC}$ of C-morphisms $\varepsilon_{(X,s)} = s : (\phi X, \mu_X) \mapsto (X, s)$ is a lax natural transformation ε where "lax" here says that $\bar{\phi}f \circ \varepsilon_{(Y,t)} \subseteq \varepsilon_{(X,s)} \circ f$ for each C-morphism $f : (X, s) \mapsto (Y, t)$. Φ can be extended to the "lax monad" $\bar{\Phi} = (\bar{\phi}, \bar{\eta} : id_{REL} \to \bar{\phi}, \bar{\mu} : \bar{\phi} \circ \bar{\phi} \to \bar{\phi})$ over REL with $\bar{\mu} = T\varepsilon S$.

45

Because of (6) we have:

Theorem 7.4 Let X be a set. Then $(\eta_X,(\phi X,\mu_X))$ is a <u>lax T-universal pair</u> of X, that is, we have: For each object (Y,t) of C and each relation $f : X \mapsto Y$ there is a smallest morphism $g : (\phi X,\mu_X) \mapsto (Y,t)$ with $f \subseteq Tg \circ \eta_X$, namely $g = t \circ \bar{\phi}f$.

<u>Proof</u> Since (Y,t) is η-stable, we get $f \subseteq t \circ \eta_Y f \subseteq t \circ \bar{\phi}f \circ \eta_X$. If $h : (\phi X,\mu_X) \to (Y,t)$ is a C-morphism with $f \subseteq Th \circ \eta_X$, then $f \circ \bar{\phi}h \subseteq h \circ \mu_X$ and therefore $t \circ \bar{\phi}f \subseteq t \circ \bar{\phi}h \circ \phi\eta_X \subseteq h \circ \mu_X \circ \phi\eta_X = h$. □

These results can be extended and can be embedded in a more general theory of lax monads given for 2-categories in [2].

REFERENCES

1. M. Barr, Relational algebras, in: "Report of the Midwest Category Seminar IV", L. N. M., 137, Berlin - Heidelberg - New York, 39-55 (1970).
2. M. C. Bunge, Coherent extensions and relational algebras, <u>Trans. A. M. S.</u> 197, 355-390 (1974).
3. A. Day, Filter monads, continuous lattices and closure systems, Canad. J. Math. 27, 50-59 (1975).
4. W. Gähler, "Grundstrukturen der Analysis", I, II, Berlin and Basel Stuttgart (1977), (1978).
5. ***, Axioms of structures and functor power series, <u>in</u>: "Convergence Structures 1984", Berlin, 137-152 (1985).
6. E. G. Manes, "Algebraic Theories", New York - Heidelberg - Berlin (1976).
7. G. Richter, "Kategorielle Algebra", Berlin (1979).
8. O. Wyler, Algebraic theories of continuous lattices, <u>in</u>: "Continuous Lattices", L. N. M., 871, Berlin - Heidelberg - New York, 390-413, (1981).

SIMPLE APPLICATIONS OF GENERALIZED FUNCTIONS IN THEORETICAL PHYSICS:

THE CASE OF MANY-BODY PERTURBATION EXPANSIONS

H. F. G. Keiter

Institut für Physik, Universität Dortmund

D-4600 Dortmund 30, Federal Republic of Germany

ABSTRACT

Let $\hat{H} = \hat{H}_0 + \hat{V}$ be a self-adjoint operator, bounded from below and defined on a Hilbert space, representing the Hamiltonian of an interacting physical system, and $\hat{H}_0$ the one for a simpler system with known spectrum and eigenstates. Typically, physicists want to evaluate the (grand-) canonical partition function Tr $\exp(-\beta\hat{H})$, where $\beta^{-1} > 0$ is Boltzmann's constant times temperature, and Tr stands for the trace, in powers of $\hat{V}$. For a fixed power of $\hat{V}$, the expansion is unique and consists of a sum of terms, interpreted as physical processes. An individual term can be calculated only if generalized functions are introduced. This is a somewhat arbitrary procedure, however. Different schemes are presented an partial summations of individual terms through all the orders of the expansion in $\hat{V}$ are discussed.

1. THE FORMAL PERTURBATION EXPANSION

For the grand canonical partition function Z, the term $\mu\hat{N}$ (μ = chemical potential, $\hat{N}$ = operator for the number of particles) is thought to be subtracted from $\hat{H}_0$ and included in $\hat{H}_0$ in the following. The perturbation expansion starts from the contourintegral representation [1] for Z or for correlation functions involving operators $\hat{A}$ and $\hat{B}$

$$Z = \text{Tr} \exp(-\beta\hat{H}) = \text{Tr}\, \frac{1}{2\pi i} \int_C [z-\hat{H}]^{-1} \exp(-\beta z)\, dz \tag{1}$$

$$G_{\hat{A}\hat{B}}(i\omega_n) = \frac{1}{2\pi i}\, \frac{1}{Z}\, \text{Tr} \int_{C+C'} [z-\hat{H}]^{-1}\hat{A}[z+i\omega_n-\hat{H}]^{-1}\hat{B}\, \exp(-\beta z)\, dz \tag{2}$$

The contours in the complex z-plane are shown in Fig. 1. The (discrete) Matsubara-frequencies $\omega_n = \pi\beta^{-1}(2n+1)$ reflect the statistics of the particles involved: odd integers (2n+1) are used for anticommuting (Fermi-like) operators $\hat{A}$ and $\hat{B}$, even ones for commuting (Bose-like) operators.

With $\hat{H} = \hat{H}_0 + \hat{V}$, the resolvent operators in (1) and (2) are expanded in terms of $\hat{V}$ and with $Z_0 = Z|_{\hat{V}=0}$ the expansion for Z reads:

$$Z - Z_0 = \frac{1}{2\pi i} \text{Tr} \int_C \sum_{n=1}^{\infty} [z - \hat{H}_0]^{-1} (\hat{V}[z - \hat{H}_0]^{-1})^n \exp(-\beta z) dz. \tag{3}$$

As usually assumed in theoretical physics, all steps of the calculation are to be allowed from a mathematical point of view: if there are problems with the convergence of the expansion in (3), $\hat{H}$ and $\hat{H}_0$ have to be changed in such a way that convergence is granted. The expansion of eq. (2) follows along the same lines and is left out for simplicity in the following. If the contourintegral and Tr can be interchanged in eq. (3), then, because of the cyclic invariance of the trace, the resolvents may be written as derivative with respect to z, and after a partial integration one arrives at

$$Z - Z_0 = \frac{-\beta}{2\pi i} \int_C \sum_{n=1}^{\infty} \frac{1}{n} \text{Tr}([z - \hat{H}_0]^{-1}\hat{V})^n \exp(-\beta z) dz \tag{4a}$$

$$= \frac{-\beta}{2\pi i} \int_C \int_0^1 \frac{dg}{g} \text{Tr} \sum_{n=1}^{\infty} ([z - \hat{H}_0]^{-1} g\hat{V})^n \exp(-\beta z) dz. \tag{4b}$$

Next, the trace is written out as a sum on eigenstates of $\hat{H}_0$:

$$\hat{H}_0|N> = E_N|N>; \quad \sum_N |N><N| = \hat{1} \tag{5}$$

Shifting the integration on z by E_N, after integchanging Tr and z-integration, the expansion eq. (3) reads:

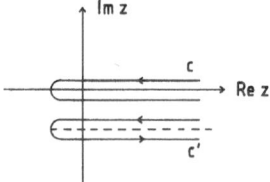

Fig. 1. The contours used in eqs. (1) and (2)

$$Z - Z_0 = \frac{1}{2\pi i} \int_C \frac{e^{-\beta z}}{z^2} \sum_N e^{-\beta E_N} \sum_{n=1}^{\infty} < N \mid \hat{V}([z + E_N - \hat{H}_0]^{-1}\hat{V})^{n-1} \mid N > dz \qquad (6)$$

This form of the expansion has the particular advantage that it contains only the excitation energies in the resolvent-matrix elements

$$< N' \mid [z + E_N - \hat{H}_0]^{-1} \mid N' > = [z + E_N - E_{N'}]^{-1}, \qquad (7)$$

while the states $\mid N >$ and $\mid N' >$ may contain an arbitrary number of particles. For the subsequent steps, $\hat{H}_0$ and $\hat{V}$ have to be specified. In view of modern applications in Solid State Physics like in "heavy Fermion systems" or in high T_c superconductors, $\hat{H}_0$ is chosen to describe strongly correlated systems. i.e. it contains many-body interactions already. In contrast to this situation, the $\hat{H}_0$ used in High Energy Physics describes asymptotically free particles without correlations. So, in the present case, the standard techniques of many-body physics, involving Feynman diagrams, cannot be applied. The mathematical problems are the same in both cases, however.

As a particularly simple example, the so-called Anderson-Hamiltonian [2] is chosen.

$$\hat{H}_0 = \sum_{\vec{k}m} \varepsilon_{\vec{k}m} d^+_{\vec{k}m} d_{\vec{k}m} + \sum_m E_m f^+_m f_m + U \sum_{\substack{m \\ m'(\neq m)}} f^+_m f_m f^+_{m'} f_{m'}, \qquad (8)$$

$$\hat{V} = \sum_{\vec{k}m} (V_{\vec{k}m} d^+_{\vec{k}m} f_m + \text{hermitean conjugate}). \qquad (9)$$

The first term in (8) describes a macroscopic number of "d"-electrons in a solid, forming an "energy-band" with energies $\varepsilon_{\vec{k}m}$ measured from the chemical potential, and characterized by a set of 4 quantum numbers, m being one out of N angular momentum ones. The second and third term in $\hat{H}_0$ stand for a small system of localized electrons on a single impurity atom, which repel each other by U. In the case $U \to \infty$, discussed later, there will be either no or one electron on this atom, yielding a nondegenerate $4f^0$ configuration or an N-fold degenerate $4f^1$. The "hybridization term" $\hat{V}$ changes this configuration by either annihilating an f- electron in $4f^1$ and emitting it into the band, whence $4f^0$ remains, or by the reverse process from $4f^0$ to $4f^1$. The elementary Fermion-operators d, d^+, f, f^+ obey

$$d_{\vec{k}m} d^+_{\vec{k}'m'} + d^+_{\vec{k}'m'} d_{\vec{k}m} = \delta_{kk'} \delta_{mm'} \qquad (10)$$

(Kronecker symbol $\delta_{\vec{k}\vec{k}'}$, a similar relation for the anti-commutator of f and f^+, and zero anti-commutators for all other combinations). The anti-commutators for the f-electrons together with the limit $U \to \infty$ lead to a considerable simplification of expansion (6). If e.g. $|N>$ is a direct product of a $4f^1$ state characterized by m and an arbitrary d-electron-state (involving an arbitrary number of d-electrons), then $\hat{V}$ acting on this state can change it only into one with $4f^0$, and the energy-difference $E_N - E_{N'}$ would be $E_m - \varepsilon_{\vec{k}m}$, say, on the other hand, with finite U, it could be changed also into $4f^2$ with $E_N - E_{N'} = E_{m'} - U + \varepsilon_{\vec{k}m'}$. Since for large U the corresponding resolvent matrix-element (7) is neigligible, for $U \to \infty$ the f-shell states alternate between $4f^1$ and $4f^0$, as shown in Fig. 2 l.s.h. Next, the partial trace on the d-electrons can be performed. For the example in Fig. 2 one easily calculates

$$\text{Tr } e^{-\beta \hat{H}_{od}} d^{+}_{\vec{k}m} d_{\vec{k}_1 m'} d^{+}_{\vec{k}_2 m'} d_{\vec{k}_3 m} / \text{Tr } e^{-\beta \hat{H}_{od}}$$

$$= f_{\vec{k}m} \delta_{\vec{k}\vec{k}_1} \delta_{mm'} \cdot f_{\vec{k}_2 m} \delta_{\vec{k}_2 \vec{k}_3} + f_{\vec{k}m} \delta_{\vec{k}\vec{k}_3} (1 - f_{\vec{k}_2 m'}) \delta_{\vec{k}_1 \vec{k}_2} \quad (11)$$

Here $\hat{H}_{od}$ is the 1st term in (8) and $f_{\vec{k}m} = (\exp(\beta \varepsilon_{\vec{k}m}) + 1)^{-1}$ denotes the Fermi distribution function. Eq. (11) is a special case of the d-electron pairing according to "Wick's theorem". It give rise to the two diagrams on the r.h.s. of Fig. 2, of which the second yileds the following contribution to (6):

$$\frac{Z^{(4)}}{Z_0} = \frac{1}{2\pi i} \sum_{\substack{mm' \\ \vec{k}\vec{k}'}} \frac{e^{-\beta E_m}}{Z_{of}} \int_C \frac{e^{-\beta z}}{z^2} \frac{|V_{\vec{k}m}|^2 |V_{\vec{k}'m'}|^2 (1 - f_{\vec{k}m}) f_{\vec{k}'m'} \, dz}{(z+E_m - \varepsilon_{\vec{k}m})^2 (z+E_m - E_{m'} - \varepsilon_{\vec{k}m} + \varepsilon_{\vec{k}'m'})} \quad (12)$$

This is an example for the general structure of the expansion: on the r.h.s. of (12) there is a sum on all the internal quantum numbers, then the occupation probability of the initial f-state,

$$P_m = e^{-\beta E_m}/Z_{of}; \quad Z_{of} = 1 + \sum_{m'} e^{-\beta E_{m'}}$$

the contour integral with a weight $e^{-\beta z} \cdot z^{-2}$ for the expansion (6), (and a

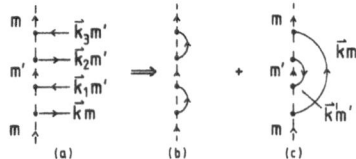

Fig. 2. A sequence of resolvent matrix-elements in 4th order in $\hat{V}$ before (a) and after (b) and (c) performing the partial trace on the band-electrons.

weight $\frac{-\beta}{z} e^{-\beta z}$ in the expansion corresponding to (4b)), multiplied by
hybridization matrix-elements and by statistical factors involving Fermi's
function and by energy denominators, which can be read off from the diagrams. By comparing (12) with the corresponding diagram in Fig. 2, rules
for calculating the diagrams are obtained, which for higher order ones
are modified only in one respect: there is a minus-sign for each crossing
of d-electron lines.

2. GENERALIZED FUNCTIONS, REGULARIZATION, ETC.

Evaluating eq. (12) further, one wants to replace the $\vec{k}$-sums (which
consist of, say, 10^{23} terms) by integrals. Approximately (and perhaps too
crude for a realistic d-band) a model density of states is introduced via

$$\sum_{\vec{k}} \rightarrow \int d\varepsilon \rho(\varepsilon). \tag{13}$$

Then, two possibilities arise: either one performs the contour-integral
first or one tries to interchange the contour-integral and the integrals on
ε and ε'. While the 2nd one immediately leads to piecewise holomorphic
functions in the complex z-plane with branch cuts along the real axis, which
may overlap and collide with the pole at $z = 0$, the first one seems to be
straightforward. But the numerical value of an <u>individual</u> diagram depends
on the type of expansion, in which it is used. Contribution (12) belongs
to expansion (6) or (3); in expansion (4a), the same diagram has a different weight (an additional factor $-\beta z/4$ under the contour integral in
(12)). Since the sum of all the contributions in a given order in $\hat{V}$ is
unique, there must be cancellations between contributions of different
diagrams, if expansion (6) or (3) is used. These cancellations are found
in a "family of diagrams with cyclically permuted vertices". An example for
such a family is given in Fig. 3. By labeling the vertices, e.g. diagram

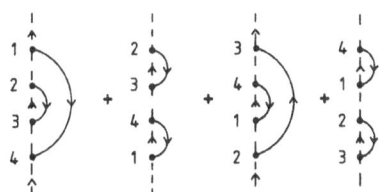

Fig. 3. A "family" of diagrams with cyclically
permuted vertices.

(2c) is duplicated, so each contribution in Fig. 3 is calculated with a factor 1/2. Also for the extensions (4a) or (4b) such a family is of interest: all its members have identical contributions, if the contour-integrals are carried out. One can then distribute the contributions in the following way: to each of the diagrams the residue at z = 0 of its contour integral is assigned. For the family in Fig. 3 one obtains

$$\frac{Z_F^{(4)}}{Z_0} = -\frac{\beta}{2} \sum_{\substack{\vec{k}m \\ \vec{k}'m'}} |V_{\vec{k}m}|^2 |V_{\vec{k}'m'}|^2 \left\{ P \sum_m \frac{f_{\vec{k}'m'}(1-f_{\vec{k}m})}{(\varepsilon_{\vec{k}m} - E_m)^2(\varepsilon_{\vec{k}m} - \varepsilon_{\vec{k}'m'} + E_{m'} - E_m)} \right.$$

$$+ \vec{k} \Leftrightarrow \vec{k}', \quad m \Leftrightarrow m' \quad \text{in the term above}$$

$$\left. + P_0 f_{\vec{k}m} f_{\vec{k}'m'} \frac{\partial}{\partial z} \left[\frac{e^{-\beta z}}{(z - E_m + \varepsilon_{\vec{k}m})(z - E_{m'} + \varepsilon_{\vec{k}'m'})} \right]_{z=0} \right\} \quad (14)$$

If z-derivative and sums are interchanged in the last term in (14), there seems to appear a square of the $\vec{k}$,m-sum. Actually, however, the contributions are still interwoven because of <u>accidentaly</u> <u>vanishing</u> <u>energy-denominators</u>. For instance, if $\varepsilon_{\vec{k}m} = E_m$ for a certain k in (14), the resulting singularity in the 1st term is compensated by a similar one in the 3rd. These denominators have to be "regularized" by introducing <u>generalized</u> <u>functions</u>, i.e. by adding small imaginary parts to any independent energy-denominator, in particular, if the continuum limit of the $\vec{k}$-sums, as in eq. (13), is used. There are several restrictions on the generalized functions: the contribution of the family must be unchanged by the regularization, they should keep the factoring features (e.g. that for the "twins" in the 2nd and 4th diagram of Fig. 3, reflected in the last term in (14)) of the diagrams intact, and the contributions of the diagrams should all be real functions.

Even with these restrictions the regularization can be realized in an infinite number of ways [3]. Applied to the family of Fig. 3, the "adiabatic" regularization furnishes [1].

$$\frac{Z_F^{(4)}}{Z_0} = \lim_{\xi_1,\xi_2 \to 0} (-\beta) \sum_{\substack{\vec{k}m \\ \vec{k}'m'}} |V_{\vec{k}m}|^2 |V_{\vec{k}'m'}|^2 \left\{ P \sum_m f_{\vec{k}m}(1 - f_{\vec{k}'m'}) \cdot \right.$$

$$\cdot \frac{1}{8} \sum_{\substack{\xi_1 \lessgtr 0 \ \xi_2 \lessgtr 0 \\ |\xi_1| \leq |\xi_2|}} \frac{1}{(\varepsilon_{\vec{k}m} - E_m + i\xi_1)^2 (\varepsilon_{\vec{k}m} - \varepsilon_{\vec{k}'m'} + E_{m'} - E_m + i\xi_1 + i\xi_2)} +$$

$$+ \frac{1}{2} P_0 f_{\vec{k}m}^{f} f_{\vec{k}'m'}^{} \frac{\partial}{\partial z} \left\{ \frac{1}{8} e^{-\beta z} \left[\frac{1}{z + \varepsilon_{\vec{k}m} - E_m + i\xi_1} + \frac{1}{z + \varepsilon_{\vec{k}m} - E_m - i\xi_1} \right] \right.$$

$$\cdot \left[\frac{1}{z + \varepsilon_{\vec{k}'m'} - E_{m'} + i\xi_2} + \frac{1}{z + \varepsilon_{\vec{k}'m'} - E_{m'} - i\xi_2} \right] \, |\xi_1| > |\xi_2|$$

$$\left. + \text{ same expression with } |\xi_1| < |\xi_2| \right\}_{z=0}. \tag{15}$$

While this is just the simplest non-trivial example for a regularization, one has to touch mathematical branches like graph-topology and-combinatorics for higher order terms. Example (15) shows, however, that the factoring property of the "twins" in Fig. 3 has been kept. Indeed, the structure of the last three lines in (15) is

$$- \beta P_0 \frac{1}{2} \frac{\partial}{\partial z} \left[e^{-\beta z} (\Gamma_0^{(2)}(2))^2 \right],$$

where $\Gamma_0^{(2)}$ is the contribution of one of the two twins. Then, introducing the concept of the "linked" parts of a diagram as those parts which cannot be cut into pieces without cutting the d-electron lines and denoting by $\Gamma_0(z)$ the sum of all the linked diagrams starting from $4f^0$, and by $\Gamma_m(z)$ the corresponding (regularized) ones starting from $4f'$, expansion (4a) can be rewritten as

$$\frac{Z}{Z_0} - 1 = - \beta \sum_{i=0,m} P_i \sum_{\ell=1}^{\infty} \frac{1}{\ell!} \frac{\partial^{\ell-1}}{\partial z^{\ell-1}} \left[e^{-\beta z} (\Gamma_i(z))^\ell \right]_{z=0}. \tag{16}$$

If $\Gamma_i(z)$ is holomorphic on and inside the unit circle, there would be a unique solution of

$$\tilde{\tilde{E}}_i = \Gamma_i(\tilde{\tilde{E}}_i), \tag{17}$$

and Langrange's expansions yileds

$$\frac{Z}{Z_0} = \sum_i P_i \exp(- \beta \tilde{\tilde{E}}_i) \tag{18}$$

This appealing form of the expansion contains the energy corrections $\tilde{\tilde{E}}_i$ to the unperturbed f-energies, yileding "Statistical quasiparticles". Provided that the expansion exists from a mathematical point of view, it may either be viewed as a justification of Landau's "Fermi-liquid-theory" at arbitrary temperature [1] or as a finite-temperature version of the Boullouin-Wigner perturbation theory [4]. From a practical point of view, already the lowest order diagrams for Γ_i have contributed greatly to the understanding of mixed-valent and heavy-fermion systems in recent years [5].

3. PARTIAL SUMMATIONS IN THE PERTURBATION EXPANSIONS

Due to the complicated regularization procedure sketched in the last section, partial summations within the diagrams for Γ_i are cumbersome. Only one simple example could be found, which is given in Ref. [1], p. 254.

In view of these difficulties, the 2nd possibility, mentioned after eq. (13), namely to interchange $\vec{k}$-sums and the contour-integral in e.g. (12), has attracted more recent attention.

The results for the family of Fig. 3 are obtained, if one slightly displaces overlapping branch cuts and poles from one another, before calculating the δ-distributions appearing under the integrals. This gives positive imaginary parts for the first three denominators in (14) and the complex conjugate of the 3rd for the 4th.

Summations of contributions under the contour integral are as easy as in standard many-body physics with Feynman-diagrams. Instead of the regularized quantities $\Gamma_i(z)$ with real z one uses unregularized ones with complex z, called self-energies $\Sigma_i(z)$. Instead of expansion (16) one obtains

$$\frac{Z}{Z_0} = \frac{1}{2\pi i} \int_C e^{-\beta z} \sum_{i=0,m} P_i(z - \Sigma_i(z))^{-1} dz$$

$$= 1 - \frac{\beta}{2\pi i} P_0 \int_0^1 \frac{dg}{g} \int_C e^{-\beta z} \Sigma_0(z,g)(z - \Sigma_0(z,g))^{-1}. \tag{19}$$

In the 2nd line of (19), the cyclic invariance of the trace was exploited to its limits. The simplest approximation for the self-energies takes into account all the diagrams without crossing d-electron lines. The resulting system of non-linear integral equations

$$\Sigma_0(z) = \sum_{\vec{k}(m)}^{(m)} |V_{\vec{k}m}|^2 \frac{f((-)\epsilon_{\vec{k}m})}{z\,(\mp)\,E_m\,(\pm)\,\epsilon_{\vec{k}m} - \Sigma_{m(0)}(z\,(\mp)\,E_m\,(\pm)\,\epsilon_{\vec{k}m})} \tag{20}$$

is the starting point for one of the most successful approaches to mixed-valent and heavy-fermion impurites.

There are many open problems in these kinds of perturbation expansions. At $\beta \to \infty$, for example, the Fermi-function $f(-\epsilon_{\vec{k}m})$ tends to a Heaviside function, and one encounters the well defined products of generalized functions. How the counter-integrals in (19) behave in this limit, however, is not completely known.

ACKNOWLEDGEMENT

The author would like to thank the organizers of the GFCA-87 for their hospitality and the mathematicians for discussions.

REFERENCES

1. H. Keiter and G. Morandi, Physics Reports, 109, 227-308, (1984).
2. P. W. Anderson, Phys. Rev., 124, 41, (1961).
3. R. Balian and C. De Dominicis, Ann. Phys. (N.Y.), 62, 229, (1971).
4. L. Brillouin, J. Phys. Paris, 3, 379, (1932); 4, 1, (1933); E. P. Wigner, Math. Naturwiss. Anz. Ungar. Akad. Wiss., 53, 477, (1935).
5. G. Czycholl, Physics Reports, 143, 277-345, (1986).

LAPLACE TRANSFORMS OF HYPERFUNCTIONS:

ANOTHER FOUNDATION OF THE HEAVISIDE OPERATIONAL CALCULUS

Hikosaburo Komatsu

Department of Mathematics

Faculty of Science, University of Tokyo, Japan

1. INTRODUCTION

The Laplace transform

$$\hat{f}(\lambda) = \int_0^\infty e^{-\lambda x} f(x) dx \tag{1}$$

is usually defined for a measurable function $f(x)$ on $[0,\infty)$ satisfying the exponential type condition

$$|f(x)| \leq C\, e^{Hx}, \quad x > 0, \tag{2}$$

with constants C and H. Then $\hat{f}(\lambda)$ is a holomorphic function on the half plane $\text{Re}\lambda > H$ and satisfies the estimates

$$|\hat{f}(\lambda)| \leq C(\text{Re}\lambda - H)^{-1}.$$

Moreover, the original function is represented by the integral

$$f(x) = \frac{1}{2\pi i} \int_{\Lambda-i\infty}^{\Lambda+i\infty} e^{\lambda x}\, \hat{f}(\lambda)\, d\lambda \tag{3}$$

almost everywhere, where Λ is an arbitrary abscissa greater than H.

The Laplace transformation was employed to justify the Heaviside operational calculus. Looking at Doetsch's book [4], we find the following solution to the initial value problem

$$\begin{cases} P(d/dx)u(x) = f(x), \\ u^{(j)}(0) = g_j, \quad j = 0,1,\ldots,m-1, \end{cases} \tag{4}$$

where

$$P(d/dx) = a_m(d/dx)^m + a_{m-1}(d/dx)^{m-1} + \ldots + a_0 \qquad (5)$$

is a linear ordinary differential operator of order m and with constant coefficients $a_i \in \mathbb{C}$.

Assume that the solution $u(x)$ and its derivateives $u^{(j)}(x)$ up to order m, all satisfy the exponential type condition (2). Then we have by integration by parts the identity

$$\hat{f}(\lambda) = P(\lambda)u(\lambda) - \{(a_m g_{m-1} + \ldots + a_1 g_0)$$

$$+ (a_m g_{m-2} + \ldots + a_2 g_0)\lambda + \ldots + a_m g_m \lambda^{m-1}\}.$$

Hence, we have

$$\hat{u}(\lambda) = P(\lambda)^{-1}\{\hat{f}(\lambda) + (a_m g_{m-1} + \ldots + a_1 g_0) + \ldots + a_m g_0 \lambda^{m-1}\}. \qquad (6)$$

Thus the solution $u(x)$ is obtained by the inversion formula (3) applied to the right-hand side.

This solution has been believed to have the following three defects:

1. In order for (6) to make sense, the datum $f(x)$ must be a global function satisfying condition (2) of exponential type.

2. No simple characterization of the Laplace image is known of the functions satisfying (2), so we do not know <u>a priori</u> whether or not $\hat{u}(\lambda)$, defined by (6), is the Laplace transform of a solution.

3. The inversion formula (3) for solution $u(x)$ does not converge absolutely.

The purpose of this paper is to show that if we extend the definition of the Laplace transforms to a class of hyperfunctions, we obtain a theory without any of these defects.

First, we shall reduce the initial value problem (4) to a simpler problem for distributions by employing the Green formula

$$(d/dx)^i(\theta(x)u(x)) = \theta(x)u^{(i)}(x) + \delta(x)u^{(i-1)}(0) + \ldots$$

$$+ \delta^{(i-1)}(x)u(0), \qquad (7)$$

where $u \in C^i(\mathbb{R})$ and $\theta(x)$ is the Heaviside function. Namely, if we consider the distributions $\theta(x)f(x)$ and $\theta(x)u(x)$ instead of the functions f and u, the problem (4) becomes equivalent to the following problem for distributions $u \in \mathcal{D}'_{[0,\infty)}$ on $\mathbb{R}$ with support in $[0,\infty)$:

Given an $f \in \mathcal{D}'_{[0,\infty)}$, find a solution $u \in \mathcal{D}'_{[0,\infty)}$ such that

$$P(d/dx)u(x) = f(x). \tag{8}$$

Actually, we consider the more general problem in which the distributions $\mathcal{D}'$ are replaced by the hyperfunctions B.

2. LAPLACE HYPERFUNCTIONS

We shall recall the definition of the hyperfunctions on $\mathbb{R}$ with support in $[a,\infty)$:

$$B_{[a,\infty)} = \mathcal{O}(\mathbb{C} \setminus [a,\infty))/\mathcal{O}(\mathbb{C}), \tag{9}$$

where $\mathcal{O}(V)$ denotes the space of all the holomorphic functions on the open set V in $\mathbb{C}$ (see Sato [15] and Komatsu [8]).

The hyperfunction $f(x)$ represented by $F(z) \in \mathcal{O}(\mathbb{C} \setminus [a,\infty))$ is denoted as

$$f(x) = F(x + io) - F(x - io). \tag{10}$$

Here, the boundary values $F(x \pm io)$ of $F(z)$ are only symbolic. However, if $f(x)$ is in a topological linear space of generalized functions such as the space of distributions, C^∞ functions or functions locally in L^p for $1 < p < \infty$, then the holomorphic function $F(x + iy)$ converges to the boundary values $f(x \pm io)$ as y tends ± 0, so that identity (10) holds topologically. The holomorphic function $F(z)$ is called a <u>defining function</u> of $f(x)$.

The integral of a hyperfunction is interpreted as the contour integral of its defining function. Therefore, the Laplace transform $\hat{f}(\lambda)$ of a hyperfunction $f(x) \in B_{[a,\infty)}$ should be defined by the integral

$$\hat{f}(\lambda) = \int_\Gamma e^{-\lambda z} F(z) dz, \tag{11}$$

where Γ is a path as follows:

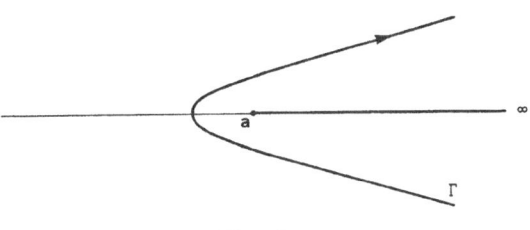

Fig. 1

However, we cannot expect that the integral converges for an arbitrary defining function F(z). Therefore, we shall introduce the following class of hyperfunctions.

Definition 1 We define the space of <u>Laplace hyperfunctions</u> with support in $[a,\infty]$, including ∞ at infinity, by

$$B^{exp}_{[a,\infty]} = O^{exp}(\mathbb{C} \setminus [a,\infty))/ O^{exp}(\mathbb{C}), \tag{12}$$

where $O^{exp}(V)$ denotes the space of holomorphic functions on V of exponential type.

To be exact, O^{exp} is a sheaf on the radial compactification

$$\mathbf{O} = \mathbb{C} \cup S^1_\infty$$

of the complex plane $\mathbb{C}$. For each open set V in $\mathbf{O}$, the section space $O^{exp}(V)$ is defined to be the space of all the holomorphic functions F(z) on $V \cap \mathbb{C}$, such that on each closed sector

$$\Sigma = \{z \in \mathbb{C}; \alpha \leq \arg(z-b) \leq \beta\} \tag{13}$$

whose closure $[\Sigma]_0$ in $\mathbf{O}$ is included in V, we have the estimates

$$|F(z)| \leq C\, e^{H|z|}, \quad z \in \Sigma, \tag{14}$$

with constants C and H.

Therefore, a more exact definition is

$$B^{exp}_{[a,\infty]} = O^{exp}(\mathbf{O} \setminus [a,\infty])/O^{exp}(\mathbf{O}). \tag{12'}$$

When F(z) is in $O^{exp}(\mathbb{C} \setminus [a,\infty))$ or $O^{exp}(\mathbf{O} \setminus [a,\infty])$, we denote its class f(x) by (10).

Theorem 1 Define the Laplace transform $\hat{f}(\lambda)$ of $f(x) = F(x + i0) - F(x - i0) \in B^{exp}_{[a,\infty]}$ by integral (11). Then $\hat{f}(\lambda)$ is a function in $O^{exp}(\tfrac{1}{2}S^1_\infty)$, independent of the defining function F, which satisfies estimates

$$\varlimsup_{r \to \infty} \frac{\log|\hat{f}(re^{i\theta})|}{r} \leq -a\cos\theta, \quad |\theta| < \pi/2, \tag{15}$$

where $O^{exp}(\tfrac{1}{2}S^1_\infty)$ denotes the inductive limit $\varinjlim O^{exp}(V)$, as the open set V tends to the half circle $\{e^{i\theta}\infty;\ |\theta| < \pi/2\}$ at infinity.

Conversely, every $\hat{f} \in O^{exp}(\tfrac{1}{2}S^1_\infty)$ satisfying (15) is the Laplace transform of a unique Laplace hyperfunction f in $B^{exp}_{[a,\infty]}$ and a defining function F of f is obtained by the absolutely convergent integral

$$F(z) = \frac{1}{2\pi i} \int_\Lambda^\infty e^{\lambda z} f(\lambda) d\lambda, \qquad (16)$$

where Λ is an arbitrary point in the domain Ω of $\hat{f}(\lambda)$ and the path of the integral is a convex curve in Ω which is eventually a ray tending to $e^{i\theta}\infty$ for a θ with $|\theta| < \pi/2$.

The proof is essentially included in Boas' book [2], as the classical works of Borel [3], Pólya [13] and Macintyre [10]. We need only a change in the order of integrations and Cauchy's integral formula. For details see Komatsu [9]. Cf. also the recent works of Ecalle [5]. □

A miracle is the following.

<u>Theorem 2</u>   The natural mapping

$$\rho : B^{\exp}_{[a,\infty]} \to B_{[a,\infty)} \qquad (17)$$

induced from the inclusion mappings $O^{\exp}(0 \setminus [a,\infty]) \to O(\mathbb{C} \setminus [a,\infty))$ and $O^{\exp}(0) \to (\mathbb{C})$ is surjective.

In other words, every hyperfunction $f(x)$ in $B_{[a,\infty)}$ has a defining function $F(z)$ in $O^{\exp}(\mathbb{C} \setminus [a,\infty))$.

An ideal proof of this theorem would be obtained if we could prove the Mittag-Leffler theorem

$$H^1(V, O^{\exp}) = 0 \qquad (18)$$

for any open set V in 0. In fact, then $B^{\exp}_{[a,\infty]}$ would be shown to be the sections on $[-\infty, \infty]$ with the support in $[a,\infty]$ of the flabby sheaf $B^{\exp} = H^1_{[-\infty,\infty]}(O^{\exp})$ as in the theory of usual hyperfunctions of one variable (see [8]). Since the restriction of $B^{\exp}$ to $\mathbb{R}$ is B, the surjectivity of ρ follows from the flabbiness.

Since we were unable to prove (18), we reduced in [9] the proof of Theorem 2 to the corresponding result

$$H^1(V, O^{\text{infexp}}) = 0$$

for the sheaf O^{infexp} of the holomorphic functions of infraexponential type due to Saburi [14], which holds for sufficiently many open sets V in 0. Here $O^{\text{infexp}}(V)$ denotes the space of all the holomorphic functions $F(z)$ on $V \cap \mathbb{C}$ such that on each closed sector $\Sigma \subset\subset V$ in $\mathbb{C}$ and each $H > 0$, estimate (14) holds with a constant C.

We could also rely on the flabbiness of the sheaf of Fourier hyperfunctions by Kawai [7]. The following direct proof is due to B. A. Taylor. We shall prove the following stronger result.

Lemma 1 Let $F(z)$ be a holomorphic function on $\mathbb{C} \setminus [0,\infty)$. Then, there is an entire function $G(z)$ such that $F(z) - G(z)$ is bounded except on $R = \{z \in \mathbb{C};\ \operatorname{Re} z > -1,\ |\operatorname{Im} z| < \pi/3\}$.

Proof Let $\chi(z) \in C^{\infty}(\mathbb{C})$ be a cut-off function which is equal to 1, except on $S = \{z \in \mathbb{C};\ \operatorname{Re} z \geq -1/2,\ |\operatorname{Im} z| \geq \pi/4\}$ and vanishes on a neighbourhood T of $[0,\infty)$. Then,

$$g = \bar{\partial}(\chi(z) z^2 F(z))$$

is a $C^{\infty}(0,1)$-form on $\mathbb{C}$ with the support in $S \setminus T$. We solve the equation

$$\bar{\partial} u = g \qquad (19)$$

by Theorem 4.4.2 of Hörmander [6]. The function

$$\psi(x + iy) = \begin{cases} e^x \cos y & |y| \leq \pi/2, \\ 0, & \text{otherwise,} \end{cases}$$

is clearly a subharmonic function on $\mathbb{C}$. If we take an increasing convex function $\gamma(t)$ of sufficiently fast growth, then

$$\phi(z) = \gamma(\psi(z))$$

is a subharmonic function for which

$$\int_{\mathbb{C}} |g|^2 e^{-\phi} dx\, dy < \infty.$$

Hence, we can find a solution $u(z)$ of (19) satisfying

$$\int_{\mathbb{C}} |u|^2 e^{-\phi} (1 + |z|^2)^{-2} dx\, dy < \infty. \qquad (20)$$

The function $\chi(z) z^2 F(z) - u(z)$ is an entire function, which we expand as $a + bz + z^2 G(z)$. Except on S, we have

$$z^2(F(z) - G(z)) = u(z) + a + bz.$$

Since $\phi(z)$ is bounded outside S, it follows from (20) that there exists a constant $C < \infty$ such that

$$|u(z)| \leq C|z|^2$$

outside R. Thus $F(z) - G(z)$ is bounded outside R. □

The mapping ρ of (17) is not injective. We denote the kernel by $B^{\exp}_{\{\infty\}}$ and call the non-zero elements Laplace hyperfunctions with support at ∞.

More generally, we say that a Laplace hyperfunction $f(x) = F(x + i0) - F(x - i0) \in B_{[a,\infty]}^{exp}$ does not have ∞ in its support, if the defining function $F(z)$ can be continued to a holomorphic function of exponential type on the set $\{z \in \mathbb{C}; |z| > C\}$ for a $C < \infty$. In this case, the support supp f of f in $B_{[a,\infty]}^{exp}$ is defined to be the support of $\rho(f)$ in $B_{[a,\infty)}$. Otherwise, it is defined to be the support of $\rho(f)$ in $B_{[a,\infty)}$ joined with $\{\infty\}$. By definition, we have

$$\text{supp } \rho(f) = \text{supp } f \cap \mathbb{R}. \tag{21}$$

When K is a closed set in $[a,\infty]$, we denote by B_K^{exp} the set of all $f \in B_{[a,\infty]}^{exp}$ with supp $f \subset K$. If $a < b < \infty$, $B_{[b,\infty]}^{exp}$, as a subspace of $B_{[a,\infty]}^{exp}$ is canonically isomorphic to $B_{[b,\infty]}^{exp}$ defined by (12), so that we identify them.

When K is a compact set in $[a,\infty)$, ρ is an isomorphism of B_K^{exp} onto $B_K = \{f \in B(\mathbb{R}); \text{supp } f \subset K\}$. In fact, ρ is injective by (21). It is surjective because the standard defining function

$$F(z) = \frac{-1}{2\pi i} \int \frac{f(x)}{z-x} dx$$

of an $f \in B_K$ is clearly of exponential type outside K. Therefore we identify B_K^{exp} with B_K.

Theorem 2 shows that

$$B_{[a,\infty)} \cong B_{[a,\infty]}^{exp} / B_{\{\infty\}}^{exp}. \tag{22}$$

Let $a < b < \infty$ and denote by $B_{[a,b)}$ the space of all the hyperfunctions on $(-\infty,b)$ with support in $[a,b)$. Every $f \in B_{[a,b)}$ can be continued to an $f_1 \in B_{[a,\infty)}$ by the flabbiness of the hyperfunctions, and then to an $f_2 \in B_{[a,\infty]}^{exp}$ by Theorem 2. The kernel of the restriction mapping $B_{[a,\infty]}^{exp} \to B_{[a,b)}$ is $B_{[b,\infty]}^{exp}$. Hence, we have the isomorphism

$$B_{[a,b)} \cong B_{[a,\infty]}^{exp} / B_{[b,\infty]}^{exp}. \tag{23}$$

In this case, the extension f_1 can also be chosen in $B_{[a,b]}$, so that we have another representation

$$B_{[a,b)} \cong B_{[a,b]} / B_{\{b\}}. \tag{24}$$

However, representation (23) is more convenient when we develop the operational calculus.

For the same reason we have the decomposition

$$B_{[a,\infty]}^{exp} = B_{[a,b]} + B_{[b,\infty]}^{exp}, \tag{25}$$

where the intersection of the two components is the hyperfunctions $B_{\{b\}}$ with support at most at b.

There seems to be no natural linear extension mapping $B_{[a,\infty)} \to B_{[a,\infty]}^{exp}$ which is a right inverse of the restriction mapping ρ of (17).

When a subclass of $B_{[a,\infty)}$ is specified, however, there can be such a mapping on the subclass. The extension by zero for the hyperfunctions with a compact support in $[a,\infty)$ is an example. A natural extension mapping is also defined for a hyperfunction f(x) which has a holomorphic extension of exponential type for sufficiently large x. In fact, if f(x) has an analytic continuation $\tilde{f}$ in $O^{exp}(S_{\alpha,\beta}^{\infty})$ with $S_{\alpha,\beta} = \{e^{i\theta}; \alpha < \theta < \beta\}$ for some $\alpha < 0 < \beta$, then the path Γ of integral (11) can be deformed for large $|z|$ into a double ray with the opposite directions tending to $e^{i\theta}\infty$ for any $\alpha < \theta < \beta$, so that the integral $\hat{f}(\lambda)$ turns out to be a function in $O^{exp}(S_{-\pi/2-\beta,\pi/2-\alpha})$. We also have the estimates of $|\hat{f}(re^{i\theta})|$.

Conversely, Macintyre [10] has proved that the inverse Laplace transform of a function in $O^{exp}(S_{-\pi/2-\beta,\pi/2-\alpha})$ is a function in $O^{exp}(S_{\alpha,\beta})$.

In particular, we have the following.

<u>Theorem 3</u> Let $a \leq b < \infty$. If $f(x) \in B_{[a,\infty)}$ is continued for $x > b$ to a holomorphic function of exponential type of the sector $\{z \in \mathbb{C}; \alpha < \arg(z-b) < \beta\}$ for some $\alpha < 0 < \beta$, then f(x) has a natural extension in $B_{[a,\infty]}^{exp}$ whose Laplace transform $\hat{f}(\lambda)$ belongs to $O^{exp}(S_{-\pi/2-\beta,\pi/2-\alpha})$ and satisfies the estimates

$$\overline{\lim_{r\to\infty}} \frac{\log|\hat{f}(re^{i\theta})|}{r} \leq \begin{cases} -a\cos\theta, & |\theta| \leq \pi/2, \\ -b\cos\theta, & \text{otherwise.} \end{cases} \quad (26)$$

Conversely, if $\hat{f}(\lambda) \in O^{exp}(S_{-\pi/2-\beta,\pi/2-\alpha})$ satisfies (26), then its inverse Laplace transform f(x) is continued analytically to a holomorphic function of exponential type of the sector $\{z \in \mathbb{C}; \alpha < \arg(z-b) < \beta\}$.

See § 4 for other cases in which the natural extension is defined.

3. OPERATIONAL CALCULUS

<u>Definition 2</u> Let

$$f(x) = F(x + i0) - F(x - i0) \quad (27)$$

be a Laplace hyperfunction in $B_{[a,\infty]}^{exp}$ and let P(d/dx) be a linear differential operator with complex coefficients. Then, its action on f(x) is defined by

$$P(d/dx)f(x) = P(d/dz)F(x + i0) - P(d/dz)F(x - i0). \tag{28}$$

Clearly, $P(d/dx)f(x)$ is a Laplace hyperfunction in $B^{exp}_{[a,\infty]}$ independent of the defining function.

Since the action of $P(d/dx)$ on a hyperfunction $f \in B_{[a,\infty)}$ is defined by the same formula, $P(d/dx)$ commutes with the restriction ρ of (17).

Theorem 4 Let $f(x) \in B^{exp}_{[a,\infty]}$ and $P(d/dx) \in \mathbb{C}[d/dx]$. Then the Laplace transform of $P(d/dx)f(x)$ is given by

$$(P(d/dx)f)^\wedge(\lambda) = P(\lambda)\hat{f}(\lambda). \tag{29}$$

Proof Consider the integral (11) with $F(z)$ replaced by $P(d/dz)F(z)$ and integrate it by parts; or differentiate the integral (16) under the integral sign. □

Now we shall consider the initial value problem (4) when $f(x)$ is in $C([0,b))$ with $0 < b \leq \infty$. By (23) we regard $f(x)$ and $\theta(x)u(x)$ as elements in the quotient space $B^{exp}_{[0,\infty]} / B^{exp}_{[b,\infty]}$. Let $\tilde{f}(x)$ and $\tilde{u}(x)$ be extensions in $B^{exp}_{[0,\infty]}$. Then, the equation we have to solve turns into the following:

$$P(d/dx)\tilde{u}(x) \equiv \tilde{f}(x) + (a_m g_{m-1} + \ldots + a_1 g_0)\delta(x) + \ldots$$
$$+ a_m g_0 \delta^{(m-1)}(x) \mod B^{exp}_{[b,\infty]}. \tag{30}$$

We denote by $\hat{u}(\lambda)$ and $\hat{f}(\lambda)$ the Laplace transforms of $\tilde{u}(x)$ and $\tilde{f}(x)$, respectively. Then it follows from Theorem 4 that

$$P(\lambda)\hat{u}(\lambda) \equiv \hat{f}(\lambda) + (a_m g_{m-1} + \ldots + a_1 g_0) + \ldots + a_m g_0 \lambda^{m-1} \mod LB^{exp}_{[b,\infty]}, \tag{30}'$$

where L denotes the Laplace transformation.

Since the multiplication by the reciprocal $p(\lambda)^{-1}$ is a mapping on $LB^{exp}_{[0,\infty]}$ into itself and on $LB^{exp}_{[b,\infty]}$ into itself, there is a unique solution

$$\hat{u}(\lambda) \equiv P(\lambda)^{-1}(\hat{f}(\lambda) + (a_m g_{m-1} + \ldots + a_1 g_0) + \ldots$$
$$+ a_m g_0 \lambda^{m-1}) \mod LB^{exp}_{[b,\infty]} \tag{31}$$

of (30)'. Hence, if we set

$$U(z) = \frac{1}{2\pi i} \int_\Lambda e^{\lambda z} P(\lambda)^{-1}(\hat{f}(\lambda) + \ldots + a_m g_0 \lambda^{m-1}) d\lambda \in \mathcal{O}^{exp}(\mathbb{C} \setminus [0,\infty)). \tag{32}$$

then

$$u(x) = U(x + i0) - U(x - i0) \tag{33}$$

is a unique solution of (4) on the interval (0,b).

Since $P(\lambda)^{-1}$ is also a multiplier on the subspace of all the elements of $O^{exp}(S_{-\pi/2-\beta}, S_{\pi/2-\alpha})$ satisfying estimates (26), the solution $u(x)$ of (4) is continued to a holomorphic function of exponential type on the sector where $f(x)$ is. In this case we have the absolutely convergent integral representation

$$u(x) = \frac{1}{2\pi i} \int_\Gamma e^{\lambda x} p(\lambda)^{-1} (f(\lambda) + \ldots + a_m g_0 \lambda^{m-1}) d\lambda \qquad (34)$$

for $x > b$, where Γ is the sum of the rays from $e^{i\theta_1}\infty$ to Λ and from Λ to $e^{i\theta_2}\infty$ for some $-\pi/2 - \beta < \theta_1 < -\pi/2$ and $\pi/2 < \theta_2 < \pi/2 - \alpha$ and for a sufficiently large Λ. Actually, (34) holds for x in the sector $\{x \in \mathbb{C}; \pi/2 - \theta_2 < \arg(x-b) < -\pi/2 - \theta_1\}$.

In particular, let $f(x)$ be an exponential polynomial $\Sigma c_{ij} x^i \exp \mu_j x$ on $(0,\infty)$. Then the Laplace transform $\hat{f}(\lambda)$ is a rational function with poles μ_j, so that $\hat{u}(\lambda)$ of (31) is a rational function with poles μ_j and the zeros of $P(\lambda)$. The integral (34) is, then, deformed into the sum of integrals of the same integrand along small circles around the poles. This representation of solution $u(x)$ as an exponential polynomial is known to be the most successful application of the operational calculus.

Thus we have seen that the method of Laplace transforms works without any of the defects we have mentioned in the introduction.

The representation (32) and (33) of the solution may be written

$$u(x) = P(d/dx)^{-1}(f(x) + \ldots + a_m g_0 \delta^{(m-1)}(x)).$$

Since the multiplication is clearly a bilinear mapping

$$LB^{exp}_{[c,\infty]} \times LB^{exp}_{[a,\infty]} \to LB^{exp}_{[a+c,\infty]}, \qquad (35)$$

we can define, more generally, the action $Q(d/dx)$ for a function $Q(\lambda)$ in $LB^{exp}_{[c,\infty]}$ as follows.

Definition 3 Let $Q(\lambda) \in LB^{exp}_{[c,\infty]}$. For an $f(x) \in B^{exp}_{[a,\infty]}$, we define $Q(d/dx)f(x) \in B^{exp}_{[a+c,\infty]}$ by

$$Q(d/dx)f(x) = G(x + i0) - G(x - i0), \qquad (36)$$

where

$$G(z) = \frac{1}{2\pi i} \int_\Lambda^\infty e^{\lambda z} Q(\lambda) \hat{f}(\lambda) d\lambda. \qquad (37)$$

Let $a < b \leq \infty$. Since $Q(d/dx)$ maps $B^{exp}_{[b,\infty]}$ into $B^{exp}_{[b+c,\infty]}$, it also induces the mapping

$$Q(d/dx) : B_{[a,b)} \to B_{[a+c,b+c)}. \tag{38}$$

Since $LB_{[a,\infty]}^{\exp}$ includes the Laplace transform $e^{-a\lambda}$ of $\delta(x-a)$, we remark that the condition $Q(\lambda) \in LB_{[c,\infty]}^{\exp}$ is also necessary in order for the multiplier $Q(\lambda)$ to map $LB_{[a,\infty]}^{\exp}$ into $LB_{[a+c,\infty]}^{\exp}$.

The action $Q(d/dx)$ may also be understood as a convolution.

Definition 4   Let $f \in B_{[a,\infty]}^{\exp}$ and $g \in B_{[b,\infty]}^{\exp}$. We define the <u>convolution</u> $f * g \in B_{[a+b,\infty]}^{\exp}$ by

$$f * g(x) = \hat{f}(d/dx)g(x) = \hat{g}(d/dx)f(x). \tag{39}$$

We note that in the finite region it concides with the usual convolution of hyperfunctions. Namely, we have

$$\rho(f * g) = \rho(f) * \rho(g). \tag{40}$$

4. COMPARISON WITH OTHER THEORIES

In order to overcome the difficulties mentioned in the introduction, many theories have been proposed. We shall compare our approach with some of them.

The most famous would be Mikusiński's algebraic foundation [11] based on the Titchmarsh theorem on convolutions of continuous functions (see also Yosida [18]). In our setting this corresponds to considering the field of all the quotients $\hat{f}(\lambda)/\hat{g}(\lambda)$ of $\hat{f}$ and $\hat{g} \in LB_{[0,\infty]}^{\exp}$ as operators. When $\hat{g}(\lambda)$ has many zeros tending to infinity, it is very difficult to give an intuitive meaning to the operator. Actually, all the operators appearing in applications have denominators in a very restricted class of functions whose Laplace transforms have few zeros.

Inspired by W. A. Ditkin's representation of Mikusiński's operator field by analytic functions, Berg [1] has established a theory employing the sequences $\{\phi_n(\lambda)\}$ of truncated Laplace transforms

$$\phi_n(\lambda) = \int_0^n e^{-\lambda x} f(x) dx$$

of $f(x) \in C([0,\infty))$ or, more generally, the sequences $\{\phi_n(\lambda)\}$ of the holomorphic function $\phi_n(\lambda)$ defined on a right half-plane and satisfying

$$\phi_m(\lambda) = \phi_n(\lambda) + O(e^{-n\lambda}), \quad m > n, \tag{41}$$

with $\phi_0(\lambda) = 0$.

Our Theorem 2 proves that for any sequence $\{\phi_n(\lambda)\}$, as above, there is a $\phi(\lambda) \in LB_{[0,\infty]}^{\exp}$ such that

$$\phi(\lambda) - \phi_n(\lambda) \in LB_{[n,\infty]}^{\exp}.$$

Therefore, each sequence is represented by a single function, though the residue classes are somewhat different.

Laplace transforms of distributions and hyperfunctions have been discussed by many authors.

Schwartz [16] and Sebastiao e Silva [17] have considered Laplace transforms of distributions of exponential type. The Laplace image is charecterized as the space of holomorphic functions $\phi(\lambda)$ defined on a half plane $\text{Re } \lambda > H$ and satisfying the estimates

$$|\phi(\lambda)| \leq C(1 + |\lambda|)^k e^{-a\text{Re}\lambda} \tag{42}$$

with constants a, k and C. Therefore, it is included in $LB_{[a,\infty]}^{\exp}$.

In Kawai's theory of Fourier hyperfunctions [7] the Laplace image is the space of holomorphic functions $\phi(\lambda)$ defined on the half-plane $\text{Re }\lambda > 0$ and satisfying, for any $\varepsilon > 0$, the estimates

$$|\phi(\lambda)| \leq C\, e^{\varepsilon|\lambda| - a\text{Re}\lambda} \tag{43}$$

with a constant C. Theorem 2 holds also in Kawai's theory, but exponential functions e^{kx} do not have natural extensions if $\text{Re } k > 0$ (the natural Laplace transform has a pole at k).

Saburi's theory of modified Foureier hyperfunctions [14] has a similar disadvantage.

Developing the ideas of Sebastião e Silva, Morimoto [12], Zharinov [19] and others have constructed the theory of Fourier ultrahyperfunctions and their Fourier and Laplace transforms. In particular, Morimoto [12] has proved that the Laplace image of the Fourier hyperfunctions of the exponential type and with support in $[a,\infty]$ is the space of holomorphic functions $\phi(\lambda)$ defined on a half-plane $\text{Re } \lambda > H$ and satisfying for any $\varepsilon > 0$ estimates (43) with a constant C. Actually, this result of Morimoto is sufficient for discussing the initial value problem (4) by Laplace transformation, but in order to treat the vector valued case, our formulation will become more important.

REFERENCES

1. L. Berg, Asymptotische Auffasung der operatorenrechnung, Studia Math., 21, 215-229 (1962).
2. R. P. Boas, Jr., "Entire Functions", Academic Press, New York, (1954).
3. E. Borel, " Leçons sur les Séries Divergentes", $2^{\acute{e}d}$., Gauthier-Villars, Paris, (1928).
4. G. Doetsch, "Theorie und Andwendung der Laplace-transformation", Springer, Berlin, (1937).
5. J. Ecalle, "Les Fonctions Résurgentes", I, II et III, Dépt. Math., Orsay (1981, 1981 et 1985).
6. L Hörmander, "An Introduction to Complex Analysis in Several Variables", Van Nostrand, Princeton, (1966).
7. T. Kawai, On the theory of Fourier hyperfunctions and its applications to partial differential equations with constant coefficients, J. Fac. Sci. Univ. Tokyo, Sec. IA, 17, 467-517 (1970)
8. H. Komatsu, An introduction to the theory of hyperfunctions, "Hyperfunctions and Pseudo-Differential Equations", Lecture Notes in Math., 287, 3-40 (1973).
9. H. Komatsu, Laplace transforms of hyperfunctions - A new foundation of the Heaviside calculus, J. Fac. Sci., Univ. Tokyo, Sec. IA, 34, 805-820, (1987).
10. A. J. Macintyre, Laplace's transformation and integral functions, Proc. London Math. Soc., (2) 45, 1-20 (1938).
11. J. Mikusiński, "Rachunek Operatorow", Warszawa, (1953).
12. M. Morimoto, Analytic functionals with non-compact carrier, Tokyo J. Math., 1, 77-103 (1978).
13. G. Pólya, Untersuchungen über Lücken und Singularitäten von Potenzreihen, Math. Z., 29, 549-640 (1929).
14. Y. Saburi, Fundamental properties of modified Fourier hyperfunctions, Tokyo J. Math., 8, 231-237, (1985).
15. M. Sato, Theory of hyperfunctions, I, J. Fac. Sci., Univ. Tokyo, Sec. I, 8, 139-193 (1959).
16. L. Schwartz, Transformation de Laplace des distributions, Medd. Lunds Math. Sem. Suppl., 196-206 (1952).
17. J. Sebastião e Silva, Les functions analytiques comme ultradistributions dans le calcul opérationnel, Math. Ann., 136, 58-96 (1958).

18. K. Yosida, "Operational Calculus-A Theory of Hyperfunctions", Springer, New York - Berlin - Heidelberg - Tokyo, (1984).
19. V. V. Zharinov, Analytic representations of a class of analytic functionals containing Fourier hyperfunctions, Math. Sb., 108, 62-77 (1979).

S-ASYMPTOTIC OF DISTRIBUTIONS

Bogoljub Stanković

Institute of Mathematics
University of Novi Sad
dr I. Đuričića 4, 21000 Novi Sad, Yugoslavia

INTRODUCTION

In the last thirty years many definitions of the asymptotic behaviour of distributions have been presented. We can roughly divide them in two sets. To the first one belong those definitions which directly use the classical definition of the asymptotic behaviour of a numerical function. The distribution T has to be equal to a numerical function f or to a derivative, in the sense of distributions, of a numerical function, $D^p f$, in a neighbourhood of infinity. The behaviour of the distribution at infinity is in reality the behaviour of the function f or corrected by p. All of these definitions are basically given in the one dimensional case.

The second set of definitions contains those in which we correspond to the distribution T a class of distributions $\{T_h\}$ depending on a parameter h. The behaviour of the distribution at infinity is, in this case, given by the behaviour of the mentioned class when the parameter h tends to infinity.

We chose as the representatives of the first set of definitions the ones given by M.J. Lightill [4] and by J. Lavoine and O.P. Misra [3].

The representatives of the second set can be: quasiasymptotic and S--asymptotic. For the quasiasymptotic we use the family $\{T_h\} = \{T(ht)\}$, $T \in S'(\Gamma)$. The most general definition of the quasiasymptotic is given in book [15]. My paper will be devoted to the S-asymptotic. All the results on S-asymptotic, we cite, are published ones except those in Part 8.

Already L. Schwartz in his book [10], in the remark on page 97 introduced: "L'order de croissance d'une distribution $T \in S'$ à l'infini" as the infinum of all the possible k, such that the set of distributions

$\{T(t+h)/(1+\|h\|^2)^{k/2}, h \in \mathbb{R}^n\}$ is bounded in D'.

The authors of book [1], on page 44 introduced "the value of a distribution T at infinity" as the limit: $\lim T(t+h)$, $h \to \infty$, if this limit exists in the sense given on page 25 of this book. They proved some properties of the introduced notion (see [1] page 45).

Yu.A. Brichkov and Yu.M. Shirokov [2] studied the "asymptotical expansion" of a distribution as a new approach to investigations of analytical properties of quantum field matrix elements. In [2] one can find references to papers from the quantum field theory which pushed forward the study of the asymptotic behaviour of distributions.

1. DEFINITION OF THE S-ASYMPTOTIC IN D' AND DIFFERENT GENERALIZATIONS [5]

Let Γ be a cone in $\mathbb{R}^n$ with the vertex at zero. By $\Sigma(\Gamma)$ we denote the set of all the real valued and positive functions $c(h)$, $h \in \Gamma$, which equal to 1 in $\Gamma \cap B(0,r)$ for a fixed r depending on the function c. $B(0,r)$ is the open ball in $\mathbb{R}^n$.

Definition 1. A distribution $T \in D'$ has the S-asymptotic in the cone Γ, related to the $c(h) \in \Sigma(\Gamma)$ and with the limit $U \in D'$, if there exists

$$\lim_{h \in \Gamma, \|h\| \to \infty} <T(t+h)/c(h), \phi(t)> = <U,\phi>, \phi \in D.$$

Then, we write $T(t+h) \stackrel{s}{\sim} c(h)U(t)$, $h \in \Gamma$.

The given definition can be changed or generalized in many directions. Let us point only at two:

We can use another limit. Let Γ_a be an accute convex cone in $\mathbb{R}^n$ with the vertex at zero. For $h_1, h_2 \in \Gamma_a$ we say that $h_1 \geq h_2$ if $h_1 \in h_2 + \Gamma_a$. Then, by definition, $g(h) \to A$, $h \to \infty$, $h \in \Gamma_a$ if for every $\varepsilon > 0$ there exists $h(\varepsilon)$ such that $g(h) \in (a-\varepsilon, A+\varepsilon)$, $h \geq h(\varepsilon)$. We can now take the limit in (1) when $h \in \Gamma_a$ and $h \to \infty$.

We can suppose also that T and U, from Definition 1, belong to a subspace of distributions A' and that ϕ belongs to the basic space A, then we have the S-asymptotic in A'.

2. RELATION OF THE S-ASYMPTOTIC WITH ASYMPTOTIC, QUASIASYMPTOTIC AND EQUIVALENCE AT INFINITY [12]

A continuous function can have the S-asymptotic, as a regular distribution, without having the usual asymptotic. Such a function g is the following one. Let $I_n = (n-e^{-2n}, n+e^{-2n})$, then $g(t) = e^{3n}(t - n) + e^n$, $t \in I_n$, $n = 1, 2, \ldots$ and $g(t) = 0$, $t \in C_R(\bigcup_{n \in N} I_n)$. Similarly, a continuous function can have the usual asymptotic without having the S-asymptotic with the limit U different from zero, as the function $g(t) = \exp t^2$ is. We can give some sufficient conditions that a regular distribution $\tilde{f}$ defined by the locally integrable function f has S-asymptotic, taking care of the asymptotical behaviour of f.

It is not easy to compare quasiasymptotic and S-asymptotic. We know that for some classes of distributions from the S-asymptotic follows the quasiasymptotic (in the case $c(h) = h^p$, $p > -1$). But in some cases the S-asymptotic gives more information about the behaviour of a distribution at infinity than the quasiasymptotic ($c(h) = h^p$, $p < -1$) [6].

The difference between the equivalence at infinity and the S-asymptotic follows from the fact that the S-asymptotic "preserves" the usual asymptotic of numerical functions, but the equivalence at infinity generalizes L'Hospital's rule.

3. RELATION BETWEEN THE S-ASYMPTOTIC AND OPERATIONS WITH DISTRIBUTIONS

From all the operations with distributions we shall restrict ourselves to: convolution, derivative, primitive of a distribution and product with a smooth function.

<u>Proposition 1.</u> [5] Let $S \in E'$ and $T \in D'$. If $T(t + h) \overset{s}{\sim} c(h)U(t)$, $h \in \Gamma$, then $(S * T)(t + h) \overset{s}{\sim} c(h)(S * U)(t)$, $h \in \Gamma$.

Proposition 1 is correct in S', as well if $S \in O_c'$ and $T \in S'$. Moreover, if we take for $S = \delta^{(k)}$, $k = (k_1, \ldots, k_n) \in N^n$, we have:

<u>Proposition 2.</u> If $T \in D'$ and $T(t + h) \overset{s}{\sim} c(h)U(t)$, $h \in \Gamma$, then $T^{(k)}(t + h) \overset{s}{\sim} c(h)U^{(k)}(t)$, $h \in \Gamma$, $k = (k_1, \ldots, k_n)$.

After this proposition arises a question: The limit U can be a constant distribution, then the derivative U' would be zero. The question is whether there exists a $c_1(h) \in \Sigma(\Gamma)$ such that T' has the S-asymptotic related to this $c_1(h)$, but with a limit different from zero. In general, the answer is negative. The function $x^2 + \sin(\exp x^2)$ shows it. Now, the problem is to find classes of distributions, large enough, for which it is possible to find the looked-for function $c_1(h)$, when $U' = 0$.

Proposition 3. Suppose that $S \in D'$, $\Gamma = \{x \in \mathbb{R}^n, x = (0,\ldots,0, x_k, 0,\ldots,0)\}$ $T = D_{x_k} S$ and $T(t+h) \stackrel{s}{\sim} c(h)U(t)$, $h \in \Gamma$. If $c(h)$ is local integrable in h_k and such that

$$c_1(h) = \int_{h_k^0}^{h_k} c(v) dv_k \to \infty \text{ as } h_k \to \infty, \; h_k^0 \geq 0.$$

Then

$$S(t+h) \stackrel{s}{\sim} c_1(h)U(t), \; h \in \Gamma.$$

Proposition 4. Let $S \in D'$ and $(D_{t_m} S)(x+h) \stackrel{s}{\sim} c(h)U(x)$, $h \in \Gamma$. If for a $V \in D'$, $D_{t_m} V = U$ for a $\phi_0 \in D(R)$, $\int_R \phi_0(t) dt = 1$ and for every $\Psi \in D$ we have

$$\lim_{h \in \Gamma, \|h\| \to \infty} <S(x+h)/c(h), \phi_0(x_m)\lambda_m(x)> = <V, \phi_0 \lambda_m>$$

where $\lambda_m(x) = \int_R \Psi(x) dx_m$, then $S(x+h) \stackrel{s}{\sim} c(h)V(x)$, $h \in \Gamma$.

Proposition 5. [5] Let $g \in E$, $c(h), c_1(h) \in \Sigma(\Gamma)$ and $g(t+h)/c_1(h)$ converges to $G(t)$ in E as $h \in \Gamma$, $\|h\| \to \infty$. If $T(t+h) \stackrel{s}{\sim} c(h)U(t)$, $h \in \Gamma$, then $g(t+h)T(t+h) \stackrel{s}{\sim} c_1(h)c(h)G(t)U(t)$, $h \in \Gamma$.

4. THE S-ASYMPTOTIC IS A LOCAL PROPERTY

Proposition 6. [5] Suppose that the distribution T_1 equals T_2 over the open set $\Omega \subset \mathbb{R}^n$, where Ω has the following property: for every $r > 0$ there exists a β_r such that the ball $B(h,r)$ is in Ω for $h \in \Gamma$ and $\|h\| \geq \beta_r$. If we have $T_1(t+h) \stackrel{s}{\sim} c(h)U(t)$, $h \in \Gamma$, then $T_2(t+h) \stackrel{s}{\sim} c(h)U(t)$, $h \in \Gamma$, as well.

Proposition 7. [14] A necessary and sufficient condition that the support of $T \in D'$ has the property: for every $r > 0$ there exists β_r such that the sets $\{\text{supp } T \cap B(h,r)\}$, $h \in \Gamma$, $\|h\| \geq \beta_r$ are empty, is that $T(t+h) \stackrel{s}{\sim} c(h) \cdot 0$, $h \in \Gamma$ for every $c(h) \in \Sigma(\Gamma)$.

The support of T in this Proposition has the property: The distance from the supp T and a point $h \in \Gamma$, $d(\text{supp } T, h)$, tends to infinity when $h \in \Gamma$, $\|h\| \to \infty$.

If we take the other limit for the S-asymptotic, then there follows a more precise result.

Proposition 8. [9] Let $T \in D'$ and Γ_a be an acute, open and convex cone with the vertex at zero. The necessary and sufficient condition that supp $T \in C_{\mathbf{R}^n}(a + \Gamma_a)$ for some $a \in \Gamma_a$ is that

$$\lim_{h \in \Gamma_a, h \to \infty} T(x + h)/c(h) = 0 \text{ in } D' \text{ for every } c(h) \in \Sigma(\Gamma).$$

In Proposition 8 the support of T can be just $C_{\mathbf{R}^n}(a + \Gamma_a)$. The question is: is it possible to obtain a similar proposition for the S-asymptotic, but with the first limit? The example

$$T(x,y) = \sum_{m \geq 1} m\delta(x-m,y) \quad \text{with} \quad \Gamma = \mathbf{R}_+^2$$

gives a negative answer.

The next example

$$T(x,y) = \sum_{m \geq 1} m\delta(x-m, y-\frac{1}{m}), \quad \Gamma = \mathbf{R}_+^2$$

shows that if

$$\lim_{h \in \gamma, \|h\| \to \infty} T(x + h)/c(h) = 0 \text{ in } D'$$

for every $c(h) \in \Sigma(\Gamma)$ and every ray $\gamma = \{pw, p > 0\}$, $w \in \Gamma_1$, this does not imply that for the cone Γ

$$\lim_{h \in \Gamma, \|h\| \to \infty} T(x + h)/c(h) = 0 \text{ in } D' \text{ for every } c(h) \in \Sigma(\Gamma).$$

5. CHARACTERIZATION OF THE NUMERICAL FUNCTION c(h) AND THE LIMIT DISTRIBUTION U [5]

Proposition 9. Let us suppose that Γ is a convex cone with its interior different from zero. If $T(x + h) \overset{s}{\sim} c(h)U(x)$, $h \in \Gamma$ and $U \neq 0$, then for every $h_0 \in \mathbf{R}^n$ there exists the limit

$$\lim_{\|h\| \to \infty} c(h + h_0)/c(h) = d(h_0), \quad h \in (h_0 + \Gamma) \cap \Gamma$$

with $d(x) = \exp(<\alpha, x>)$, $<\alpha, x> = \sum_{i=1}^{n} \alpha_i x_i$, $\alpha_i = \left[\frac{\partial}{\partial x_i} d(x)\right](0)$.

The limit distribution U is $U(x) = C \exp(<\alpha, x>)$ for some C. If we suppose Γ to be only convex, our result for c and U is less precise [5].

75

We can give an analytical expression for $c(h) \in \Sigma(R)$ if we assume that $c(h)$ satisfies some additional conditions. For this reason we shall introduce the set $\Sigma_0(R)$: $c(h) \in \Sigma_0(R)$ if: (i) $c(h)$ is positive and continuous in $[b,\infty)$ for some $b > 0$; (ii) There exists $T_c \in D'$, $w_c \in S^{n-1}$ and $U_c \neq 0$ such that $T_c(t + \beta w_c) \overset{s}{\sim} c(\beta) U_c(t)$, $\beta \in R_+$. S^{n-1} is the unit spere in $\mathbb{R}^n$.

Proposition 10. The necessary and sufficient condition that $c(h) \in \Sigma_0(R)$ is that $c(h) = \exp(\nu h) L(\exp h)$, $h \in [b,\infty)$, $\nu \in \mathbb{R}$ and L is a slowly varying function $(\lim_{x\to\infty} L(ux)/L(x) = 1, u > 0)$.

6. S-ASYMPTOTIC AND MAPPINGS OF SOME SUBSETS OF D' INTO D' [11]

If a distribution T has an S-asymptotic behaviour and if we map it by an operator L into D', the question is what can we say about the S-asymptotic of the distribution LT? In part 3 we discussed this question for some special operations. A general statement is given by the next proposition

Proposition 11. Suppose that the mapping $L : E' \to D'$ has the following property: It is linear, continuous and keeps the translation. A necessary and sufficient condition that L maps E' into the set $\{T \in D', T(t + h) \overset{s}{\sim} c(h) U_T(t), h \in \Gamma\}$ is that $(L\delta)(t + h) \overset{s}{\sim} c(h) V(t), h \in \Gamma$. In this case for $S \in E'$ $(LS)(t + h) \overset{s}{\sim} c(h)(S * V)(t), h \in \Gamma$.

7. CHARACTERIZATION OF SOME SUBSPACES OF D' BY THE S-ASYMPTOTIC [14]

In this part Γ will be $\mathbb{R}^n$ and $c(h) \in \Sigma(\mathbb{R}^n)$.

Already Proposition 7 gives the necessary and sufficient condition that $T \in E'$. We have only to take $\Gamma = \mathbb{R}^n$.

Proposition 12. The necessary and sufficient condition that $T \in E'$ is that $T(x + h) \overset{s}{\sim} c(h) \cdot 0$, $h \in \mathbb{R}^n$ for every $c(h) \in \Sigma(\mathbb{R}^n)$.

The next proposition is related to the spaces O'_c and B'.

Proposition 13. The necessary and sufficient condition that
a) $T \in O'_c$, b) $T \in B'$ is that T has S-asymptotic zero related to every $c(h)$

  a) $c(h) = \|h\|^{-\alpha}, \alpha \in R_+$ b) $c(h) \to \infty$ as $\|h\| \to \infty$.

Proposition 14. Let $T(x + h) \overset{s}{\sim} c^{-1}(h)U_c(x)$, $h \in \mathbb{R}^n$ for every $c(h) \in \Sigma(\mathbb{R}^n)$ which has a fast descent, then $T \in S'$ (U_c can be zero as well).

A similar proposition can be proved for the space K_1' when $c(h)$ is a rapidly exponentially decreasing function.

An interesting result for the elements of the space D'_{L^p}, $1 \le p < \infty$, is given by the following

Proposition 15. Every distribution $T \in D'_{L^p}$, $1 \le p < \infty$ has the S-asymptotic, related to $c(h) = 1$ just $U = 0$.

8. RELATION BETWEEN S-ASYMPTOTIC IN D' AND S-ASYMPTOTIC IN A SUBSPACE OF D'

To analyze the problem given in the title, we have first to restrain our set $\Sigma(\Gamma)$; it follows from Part 7.

Let Γ be a convex cone. By $\Sigma_p(\Gamma)$ we denote a subset of $\Sigma(\Gamma)$ such that $c(h) \in \Sigma_p(\Gamma)$ if and only if there exist positive numbers C and a positive function $p(x)$ such that

$$c(h + x) \le Cc(h)p(x), \quad h, x \in \Gamma \setminus B(0,r). \tag{1}$$

In the following G will be the set $G = \{x \in \mathbb{R}^n \setminus (\Gamma \cup B(0,r))\}$. By A we denote a barrelled vector space of smooth functions such that D lies dense in A with its topology finer than the topology induced by A. A' is the dual space of A, $A' \in D'$. We suppose also for elements ϕ of A that for every $y \in B(0,r)$ $p(x)\phi(x + y) \in L^1(\mathbb{R}^n)$.

Proposition 16. Suppose that $T \in A'$ and $c(h) \in \Sigma_p(\Gamma)$. If the sets:

$Q_1 = \{T(x + h)/c(h), h \in \Gamma\}$

$Q_2 = \{T(x + k + h)/c(h)p(k), h \in \Gamma, k \in G\}$

are weakly bounded in D', then the set Q_1 is weakly bounded in A' as well.

Proof. We supposed that Q_1 is a weakly bounded set in D', then it is bounded in D'. A set A is bounded in D' if and only if for every $\phi \in D$ the set of functions $\{T * \alpha, T \in A\}$ is bounded on every compact $K \in \mathbb{R}^n$. Now, we shall use a part of the proof of Theorem XXII from [10] T. II, p. 52. Let α be a fixed element from D_K, $K = \bar{\Omega}$, Ω be an open relative compact neighbourhood of zero, and S be from Q_1. Then $\beta \to (S * \alpha) * \beta$ is a function which maps D_k into L_B^∞, B is the ball $B(0,r)$. In the mentioned proof L. Schwartz showed that the functions $(\alpha,\beta) \to S * \alpha * \beta$, $S \in Q_1$ are equicontinuous and map $D_K \times D_K$ into L_B^∞. They can be enlarged to the equicontinuous functions which map $D_\Omega^m \times D_\Omega^m$ into L_B^∞. In such a way $\{S * \alpha * \beta; \alpha,\beta \in D_\Omega^m, S \in Q_1\}$ is the set of continuous and bounded functions over B.

Because of equicontinuity, for the ball $Z(0,\rho)$ from L_B^∞ there exists a neighbourhood of zero $V(m_1,\varepsilon_1,\bar{\omega}_1)$, $\bar{\omega}_1 \subset \Omega$, in D_Ω^m such that $|(S * \alpha * \beta)(t)| \leq \rho$, $t \in B$ and, $\alpha,\beta \in V(m_1,\varepsilon_1,\bar{\omega}_1)$, $S \in Q_1$.

In the same way we can find $V(m_2,\varepsilon_2,\bar{\omega}_2)$, such that $|(U * \alpha * \beta)(t)| \leq \rho$, $t \in B$; $\alpha,\beta \in V(m_2,\varepsilon_2,\bar{\omega}_2)$, $U \in Q_2$.

If α,β belong to the neighbourhood $V(m,\varepsilon,\bar{\omega})$, $m = \max(m_1,m_2)$, $\varepsilon = \min(\varepsilon_1,\varepsilon_2)$ and $\bar{\omega} = \bar{\omega}_1 \cap \bar{\omega}_2$, then both inequalities are satisfied.

We shall use, now, relation (VI, 6; 23) from [10] T. II, p. 47

$$\Delta^{2k} * (\gamma E * \gamma E * T) - 2\Delta^k * (\gamma E * \xi * T) + (\xi * \xi * T) = T \qquad (2)$$

where E is a solution of the iterated Laplace equation $\Delta^k E = \delta$; $\gamma \in D_\Omega$, supp $\gamma \subset \bar{\omega}$; $\xi \in D_\Omega$, supp $\xi \subset \bar{\omega}$. For large enough k $\gamma E \in D_\Omega^m$, as well.

Relation (2) shows that T is of the form

$$T = \sum_{|i| \leq 2k} D^i F_i.$$

Now, using property of the shift operator τ_{-h}, we have

$$T(x+h)/c(h) = \sum_{|i| \leq 2k} D^i F_i(x+h)/c(h), \quad h \in \Gamma \qquad (3)$$

where

$$|F_i(x+h)/c(h)| \leq C_i, \quad x \in B, \ h \in \Gamma, \ |i| \leq 2k \qquad (4)$$

and

$$|F_i(x+h+k)/c(h)p(k)| \leq C_i', \quad x \in B, \ h \in \Gamma, \ k \in G, \ |i| \leq 2k. \qquad (5)$$

Let us suppose, now, that $\phi \in A$

$$< T(x+h)/c(h), \phi(x) > = \sum_{|i| \leq 2k} (-1)^{|i|} < F_i(x+h)/c(h), \phi^{(i)}(x) >.$$

The right hand side can be divided into three parts for every $|i| \leq 2k$.

$$\int_{\mathbf{R}^n} \frac{F_i(x+h)}{c(h)} \phi^{(i)}(x)dx = \int_B + \int_{\Gamma\setminus B} + \int_{\mathbf{R}^n\setminus(\Gamma\cup B)} \frac{F_i(x+h)}{c(h)} \phi^{(i)}(x)dx.$$

The first integral is bounded because of relation (4). For the second we use relation (1) and the properties of elements from A:

$$\left| \int_{\Gamma\setminus B} \frac{F_i(x+h)}{c(h)} \phi^{(i)}(x)dx \right| \leq \int_{\Gamma\setminus B} \left| \frac{F_i(x+h)}{c(x+h)} \right| p(x) |\phi^{(i)}(x)| dx$$

$$\leq C_i \int_{\Gamma\setminus B} p(x) |\phi^{(i)}(x)| dx$$

which is bounded. For the third one we have

$$\left| \int_{\mathbf{R}^n\setminus(\Gamma\cup B)} \frac{F_i(x+h)}{c(h)} \phi^{(i)}(x)dx \right| \leq$$

$$\leq \int_{\mathbf{R}^n\setminus(\Gamma\cup B)} \left| \frac{F_i(y+k+h)}{c(h)p(k)} \right| p(k) |\phi^{(i)}(y+k)| d(y+k) \leq$$

$$\leq C_i' \int_{\mathbf{R}^n} p(x) |\phi^{(i)}(y+x)| dx.$$

We could use relation (5) because $B + G \supset \mathbf{R}^n \setminus (\Gamma \cup B)$. □

<u>Proposition 17.</u> Suppose that $T \subset A'$ and $c(h) \subset \Sigma_p(\Gamma)$. If T has the S-asymptotic in D' related to $c(h)$ with the limit U and if the set $Q_2 = \{T(x+h+k)/c(h), h \in \Gamma, k \in G\}$ is weakly bounded in D', then T has the S-asymptotic in A' related to the same $c(h)$ and U.

<u>Proof.</u> If $T \in A'$ and has the S-asymptotic in D', then the set $Q_1 = \{T(x+h)/c(h), h \in \Gamma\}$ is weakly bounded in D'. Now, all the suppositions of Proposition 16 are filled. Consequently, the set Q_1 is weakly bounded in A' as well. Since space D lies dense in A, we have only to use the Banach-Steinhaus theorem. □

At the end we shall remark that all the well known basic spaces as K_1, S, D_{L^p} ($1 \leq p < \infty$) and $\dot{B}$ satisfy our conditions for the space A. The space $B = D_{L^\infty}$ is one which does not, D is not dense in B. In such a way, our propositions 16 and 17 relate to all of these spaces and their duals.

9. S-ASYMPTOTIC AND THE FOURIER TRANSFORM [11]

As the Fourier transform suits well for the tempered distributions, we shall use the S-asymptotic in S'. By $F[T]$ we denote the Fourier transform of T, and by $F^{-1}[T]$ the inverse Fourier transform.

Proposition 18. Let $g \in S'$ and $f = F[g]$. A necessary and sufficient condition that there exists $g(t + h) \overset{s}{\sim} c(h)U(t)$, $h \in \Gamma$, in S' is the existence of the limit

$$\lim_{h \in \Gamma, \|h\| \to \infty} \frac{1}{c(h)} \exp(-i <t,h>) f(t) = V(t) \quad \text{in} \quad S'$$

and in this case $U(t) = (F^{-1}[V])(t)$.

10. APPLICATIONS OF THE S-ASYMPTOTIC TO PARTIAL DIFFERENTIAL EQUATIONS [11]

We shall give only one proposition to illustrate this application.

Proposition 19. A sufficient condition that there exists a solution X of the equation

$$L(D)X = G, \quad G \in O'_c; \quad L(D) = \sum_{|\alpha| \geq 0} a_\alpha D^\alpha, \quad a_\alpha \in R, \quad \alpha \in N_0^n$$

such that $X(t + h) \overset{s}{\sim} c(h)(G * U)(t)$, $h \in \Gamma$ in S' is that there exists the limit

$$\lim_{h \in \Gamma, \|h\| \to \infty} \frac{1}{c(h)} \exp(-i <t,h>) \operatorname{reg} \frac{1}{L(-it)} = (F[U])(t) \quad \text{in} \quad S'.$$

By $\operatorname{reg} \frac{1}{P(y)}$ we denote a solution, belonging to S', of the equation $P(y)X = 1$.

We can apply the S-asymptotic not only to linear partial differential equations with constant coefficients, but also in the case of variable coefficients using Proposition 5.

With the S-asymptotic we can prove some Abelian and Tauberian type theorems for the integral transforms as are Stieltjes, Hilbert, Weierstrass, ... [7].

11. S-ASYMPTOTIC EXPANSION [13]

We can enlarge the definition of the S-asymptotical behaviour of a distribution to the asymptotical expansion.

Definition 2. The distribution $T \in D'$ has the S-asymptotic expansion related to the asymptotic sequence $\{c_n(h)\} \subset \Sigma(\Gamma)$, we write it as

$$T(t + h) \overset{s}{\sim} \sum_{n=1}^{\infty} U_n(t,h) | \{c_n(h)\}, \quad h \in \Gamma,$$

where $U_n(t,h) \in D'$ for $n \in \mathbb{N}$ and $h \in \Gamma$, if for every $\phi \in D$

$$< T(t + h), \phi(t) > \sim \sum_{n=1}^{\infty} < U_n(t,h), \phi(t) > | \{c_n(h)\}, \quad h \in \Gamma, \|h\| \to \infty$$

in the classical sense.

The S-asymptotic expansion has similar properties as the S-asymptotic and we can prove similar propositions.

12. GENERALIZED S-ASYMPTOTIC

S. Pilipović [8] proposed a generalization of the S-asymptotic dwelling on one dimension in proving propositions for it. But all of them can be enlarged to the n-dimensional case.

Definition 3. Suppose that the function $e(x) \in C^{\infty}(\mathbb{R}^n)$ and such that $e(x) \neq 0$, $x \in \mathbb{R}^n$. The distribution $T \in D'$ has the generalized S-asymptotic in cone Γ related to $e(x)$ and with the limit $a \in \mathbb{R}$, if there exists

$$\lim_{h \in \Gamma, \|h\| \to \infty} T(x + h)/e(x + h) = \tilde{a} \quad \text{in } D'$$

where $\tilde{a}$ is the regular distribution defined by the constant function a.

If we compare Definition 1 and this one, we see that a distribution $T \in D'$ has the generalized S-asymptotic if and only if the distribution T/e has the S-asymptotic related to $c(h) = 1$ and with the limit $U = \tilde{a}$. But it has been proved in [8] that introducing the generalized S-asymptotic, we have some advantages in analyzing the behaviour of a distribution at infinity.

References

1. P. Antosik, J. Mikusiński, R. Sikorski, "Theory of Distributions, the Sequential Approach", Elsevier SPC-PWN, Amsterdam - Warszawa, (1973).
2. Yu. A. Brichkov, Yu. M. Shirokov, On the asymptotic behaviour of Fourier transform, Teor. Mat. Fiz. 4, 301 - 309 (1970) (in Russian).
3. J. Lavoine and O. P. Misra, Théorèmes abélian pour la transformation de Stieltjes des distributions, C. R. Acad. Sci. Paris, 279 Série A, 99 - 102 (1974).
4. M. J. Lightill, "Introduction to Fourier analysis and generalized functions", Cambridge Univ. Press, London (1958).
5. S. Pilipović and B. Stanković, S-asymptotic of a distribution, Pliska Stud. Math. Bulgar. (to appear).
6. S. Pilipović, The translation asymptotic and the quasiasymptotic behaviour of a distribution, Acta Mathematica Hungarica (to appear).
7. S. Pilipović, Asymptotic behaviour of the distributional Weierstrass transform, Applicable Analysis (to appear).
8. S. Pilipović, On the behaviour of distributions at infinity, Proceedings of the Conference in Szczyrk (Poland)(1983).
9. S. Pilipović, Remarks on support of distributions, Glasnik Matematički, Zagreb (to appear).
10. L. Schwartz, "Théorie des distributions", T. II, Herman, Paris (1951).
11. B. Stanković, Applications of the S-asymptotic, Review of Research, Faculty of Science, University of Novi Sad, 15,1, 1 - 9 (1985).
12. B. Stanković, S-asymptotic and other definitions of the asymptotic behaviour of distributions, Rev. Res. Sci. Univ. Novi Sad (16/1, 1-12, (1986).
13. B. Stanković, S-asymptotic expansion of distributions, Int. Journal Math. Math. Sci. Florida (to appear).
14. B. Stanković, Characterisation of some subspaces of (D') by S-asymptotic, Publ. Inst. Math. Beograd, (N. S.), 41, 55, 111-117, (1987).
15. V. S. Vladimirov, Yu. N. Drožžinov and B. I. Zavjalov, "Multidimensional Tauberian theorems for generalized functions", Nauka, Moscow, (1986) (in Russian).

THE WIENER-HOPF EQUATION IN THE NEVANLINNA AND SMIRNOV ALGEBRAS AND ULTRA-DISTRIBUTIONS

V. S. Vladimirov

Steklov Institute of Mathematics, Moscow

1. The Wiener-Hopf equation on the semi-axis

$$\phi(\xi) = \int_0^\infty k(\xi - \xi')\phi(\xi')d\xi' + f(\xi), \quad \xi \geq 0 \qquad (1.1)$$

and the associated Riemann-Hilbert problem on a real axis

$$\rho(x)\phi^+(x) = \psi^-(x) + F(x) \quad \text{a.e. on } \mathbb{R} \qquad (1.2)$$

has been investigated by many mathematicians starting from N. Wiener and E. Hopf [1] under various assumptions about kernel k and function ρ. An important contribution to their theory has been made by V. A. Fok [2], N. I. Muschelishvili [3, 4], I. N. Vekua [24], N. P. Vekua [3, 5], V. A. Ambartsumian [6], F. D. Gahov [7], S. Chandrasekhar [8], V. V. Sobolev [9], M. G. Krein [10, 11], I. I. Daniluk [26], B. V. Bojarskii [27], I. B. Simonenko [28], G. S. Litvinchuk [29], M. V. Maslennikov [12], N. B. Engibarjan [13], V. M. Kokilashvili and V. A. Paatashvili [30] and others.

In this paper we shall study the generalized Wiener-Hopf equation

$$F[P_+(\rho F^{-1}[\phi])] = f, \quad f \in F[N^+] \qquad (1.3)$$

in algebras $F[N^+]$ and $F[N_*^+]$ and analytical functionals ϕ which are the images of the Nevanlinna algebra N^+ and the Smirnov algebra N_*^+, resp., under the Fourier transform F; P_+ is a projector on the algebra N^+. We shall study also the associated Riemann-Hilbert problem (1.2) in the Nevanlinna algebras $N^\pm$: $\phi^+ \in N^+$, $\psi^- \in N^-$ and in the Smirnov algebras $N_*^\pm$: $\phi^+ \in N_*^+$, $\psi^- \in N_*^-$, resp.

We suppose that ρ is a measurable function on $\mathbb{R}$ and satisfies the condition

$$\int_{-\infty}^{\infty} \frac{|\ln|\rho(x)||}{1+x^2} dx < \infty. \tag{1.4}$$

We emphasize that the discrete analogy of the Wiener-Hopf equation was considered first in the Smirnov algebras $N_*^{\pm}$ (under some additional assumptions about ρ) by V. S. Vladimirov ans I. V. Volovich [17] in connection with the solution of some problem in statistical physics.

2. The Nevanlinna class N consits of all the functions $f(\zeta)$ which are holomorphic in the unit circle $|\zeta| < 1$ and satisfy the condition

$$\sup_{0 \leq r < 1} \int_0^{2\pi} \ln^+|f(re^{i\theta})| d\theta < \infty. \tag{2.1}$$

The Smirnov class N_* consists of such functions $f(\zeta)$ for which the set of functions

$$\{\ln^+|f(re^{i\theta})|, \ 0 \leq r < 1\} \tag{2.1'}$$

is uniformly integrable on $(0, 2\pi)$; $N_* \subset N$.

About the classes N and N_* see the books by G. M. Golusin [19], I. I. Privalov [20], N. K. Nikol'skii [21], a survey by S. V. Shwedenko [18] and a paper by A. B. Aleksandrov [16] (see also [15]).

Let $f \in N$ and

$$f(\zeta) = \sum_{k=0}^{\infty} a_k \zeta^k. \tag{2.2}$$

The set of boundary values $f^+(e^{i\theta})$ a.e. of functions $f \in N$ we shall denote by N^*:

$$f^+(e^{i\theta}) = \lim_{r \to 1-0} f(re^{i\theta}) = \lim_{r \to 1-0} \sum_{k=0}^{\infty} a_k e^{ik\theta} = \sum_{k=0}^{\infty} a_k e^{ik\theta} \quad \text{a.e.} \tag{2.3}$$

The set of boundary values $f^-(e^{i\theta})$ a.e. of functions $f(1/\zeta)$, $f \in N$ we shall denote by N^-:

$$f^-(e^{i\theta}) = \lim_{r \to 1+0} f(\frac{1}{r} e^{-i\theta}) = \lim_{r \to 1+0} \sum_{k=0}^{\infty} q_k \frac{1}{r^k} e^{-ik\theta} = \sum_{k=0}^{\infty} a_k e^{-ik\theta} \text{ a.e.} \tag{2.4}$$

We note that $\ln|f^{\pm}| \in L_1(0, 2\pi)$.

The classes $N_*^{\pm}$ are defined similarly.

The Nevanlinna class N (the Smirnov class N_*) in the upper half-plane $y > 0$, $z = x + iy$, is defined as the image of the Nevanlinna class N (the Smirnov class N_* resp.) in the unit circle $|\zeta| < 1$ under the conformal mapping

$$z = i\frac{1+\zeta}{1-\zeta}, \quad \zeta = \frac{z-i}{z+i}. \tag{2.5}$$

The classes $N^{\pm}$ and $N_*^{\pm}$ on the real axis are defined similarly. In addition, the boundary values $f^{\pm}(x)$ satisfy condition (1.4). The series (2.3) and (2.4) take the form

$$f^+(x) = \lim_{y \to +0} f(x+iy) = \lim_{y \to +0} \sum_{k=0}^{\infty} a_k \left(\frac{x+iy-i}{x+iy+i}\right)^k =$$

$$= \sum_{k=0}^{\infty} a_k \left(\frac{x-i}{x+k}\right)^k \quad \text{a.e.} \tag{2.6}$$

$$f^-(x) = \lim_{y \to -0} f(-x-iy) = \lim_{y \to -0} \sum_{k=0}^{\infty} a_k \left(\frac{x+iy+i}{x+iy-i}\right)^k =$$

$$= \sum_{k=0}^{\infty} a_k \left(\frac{x+i}{x-i}\right)^k \quad \text{a.e.} \tag{2.7}$$

Expansions (2.6) and (2.7) are unique, if $f^+ \in N^+$ and $f^- \in N^-$ resp.

Classes N, $N^{\pm}$, $N^{\circ} = N^+ \cap N^-$, $N_*^{\pm}$, N_*, $N_*^{\circ} = N_*^+ \cap N_*^-$ form algebras with a unit with respect to the usual multiplications, associative and commutative. In addition, if $f \in N$, $f \neq 0$ then $1/f \in N$.

We denote by $\tilde{N}$ an algebra which consists of all the measurable functions $\rho(x)$ on $\mathbb{R}$ satisfying condition

$$\int_{-\infty}^{\infty} \frac{\ln(1+|\rho(x)|)}{1+x^2} dx < \infty. \tag{2.8}$$

In paper [16] the following result was proved: every function $\rho \in \tilde{N}$ can be represented in the form

$$\rho = a^+ + b^-, \quad a^+ \in N_*^+, \quad b^- \in N_+^-. \tag{2.9}$$

In expansion (2.9), the pair (a^+,b^-) are defined up to an additive pair $(\rho_0, -\rho_0)$ where ρ_0 is an arbitrary function from the algebra N_*°. The components a^- and b^- are called projections of function ρ on the algebras $N_*^{\pm}$ resp. and are denoted by $a^+ = P_+(\rho)$, $b^- = P_-(\rho)$.

It follows from (2.8) and (1.4), that every function $g^- f^+$, $g^- \in N^-$,

85

$f^+ \in N^+$, belongs to the algebra $\tilde{N}$ and it can be decomposed

$$g^-f^+ = P_+(g^-f^+) + P_-(g^-f^+), \quad P_\pm(g^-f^+) \in N_*^\pm. \tag{2.10}$$

In particular, every $f^+ \in N^+$ can be represented in the form

$$f^+ = f_0 + C, \quad f_0 \in N_*^+, \quad C \in N^+ \cap N_*^-$$

where the pair (f_0, C) is defined up to an additive pair $(C_0, -C_0)$ where C_0 is an arbitrary function from N_*^o.

In paper [16], a description was given of the algebra N_*^o: N_*^o is the closure of the linear hull of the Cauchy kernels $\{1/(x' - x), x' \in \mathbb{R}\}$ in metric (2.8).

3. If $f \in N$, then it satisfies the estimate (see [15])

$$|f(z)| \leq M \exp\left[\sigma \frac{1 + |z|^2}{y}\right], \quad y > 0 \tag{3.1}$$

for some $M = M_f > 0$ and $\sigma = \sigma_f \geq 0$. Then it follows that if $f \in N$, then there exists a boundary value $f(\cdot + i0) \in \mathcal{D}_{+i0}^{(2)'}$ i.e.

$$f(x + iy) \to f(x + i0), \quad y \to +0 \text{ in } \mathcal{D}^{(2)'}. \tag{3.2}$$

Therefore, owing to (2.2) and (2.5),

$$f(x + i0) = \sum_{k=0}^\infty a_k \left(\frac{x - i}{x + i}\right)^k \tag{3.3}$$

and the series (3.3) converges in $\mathcal{D}^{(2)'}$. Here $\mathcal{D}^{(2)'}$ is the space of ultra-distributions of class (2) of the Beurling type, i.e. the space of linear continuous functionals over the space $\mathcal{D}^{(2)}$ of C^∞-functions $\phi(x)$, $x \in \mathbb{R}$, with a compact support and such that for any $h > 0$ there exists a constant $C = C(h, \phi)$, such that the following inequalities are fulfilled

$$|\phi^{(k)}(x)| \leq Ch^k(k!)^2, \quad k = 0, 1, \ldots . \tag{3.4}$$

(About the theory of ultradistributions see a survey by H. Komatsu [22, 23]; See also [15]).

The Fourier transform of a function $\phi \in \mathcal{D}^{(2)}$ is sefined by the formula

$$F[\phi](\xi) = \int_{-\infty}^{\infty} \phi(x) \, e^{-ix\xi} \, dx.$$

We shall denote the dual spaces

$$F[\mathcal{D}^{(2)'}] = H^{(2)'}, \quad F[\mathcal{D}^{(2)'}_{+i0}] = H^{(2)'}_+ \subset H^{(2)'}. \tag{3.5}$$

Here $\mathcal{D}^{(2)'}_{+i0}$ is a subspace of ultradistributions $\mathcal{D}^{(2)'}$ which consists of boundary values $f(x + i0)$ of functions $f(z)$ holomorphic in the upper-half plane and satisfying the estimate (3.1). $\mathcal{D}^{(2)'}_{+i0}$ is an algebra with a unit, associative and commutative.

Elements of the space $H^{(2)'}$ are analytical functionals, elements of the space $H^{(2)'}_+$ are analytical functionals with a support in $[0, \infty)$. The topologies of the spaces $H^{(2)'}$, $H^{(2)'}_+$ and $\mathcal{D}^{(2)'}_{+i0}$ are induced by the topology of the space $\mathcal{D}^{(2)'+}$, in accordance with the duality (3.5).

The Fourier transform of the ultradistribution (3.5) is expressed by the formula

$$F[\tilde{f}(\cdot + i0)] = \sum_{k=0}^{\infty} a_k F\left[\left(\frac{x-i}{x+i}\right)^k\right] = 2\pi \sum_{k=0}^{\infty} a_k H_k(\xi) \in H^{(2)'}_+, \tag{3.6}$$

where

$$H_0(\xi) = \delta(\xi), \quad H_k(\xi) = \frac{(-1)^k}{(2k-1)!}(1-\partial^2)^k[\theta(\xi) \, e^{-\xi} \xi^{2k-1}],$$

$$H_k = H_1 * H_{k-1} = \delta(\xi) + \theta(\xi) h_k(\xi), \quad h_k \in L_1(\mathbb{R}), \quad k = 1, 2, \ldots \tag{3.7}$$

The series in (3.6) converges in the space $H^{(2)'}$.

Let $f \in N$. Then there exists a one-to-one correspondence between boundary values $f^+(x)$ and $f(x + i0)$ (the Privalov theorem). We shall denote this correspondence by T, so that

$$f(\cdot + i0) = Tf^+, \quad f^+ \in N^+; \quad f^+ = T^{-1}f(\cdot + i0), \quad f(\cdot + i0) \in TN^+. \tag{3.8}$$

We define the Fourier transform $F[f^+]$ of a function $f^+ \in N^+$ by the formula

$$\mathcal{F}[f^+] = F[Tf^+], \quad \mathcal{F} = FT. \tag{3.9}$$

Then, its inverse Fourier transform $\mathcal{F}^{-1}[g]$ on elements $g \in \mathcal{F}[N^+]$ is defined by the formula

$$\mathcal{F}^{-1}[g] = T^{-1}F^{-1}[g], \quad \mathcal{F}^{-1} = T^{-1}F^{-1}.$$

The inversion formula

$$F^{-1}[F[f^+]] = f^+, \quad f^+ \in N^+$$

and imbedding $F[N^+] \subset H_+^{(2)'}$ hold.

We define the convolution $g_1 * g_2$ of analytical functionals g_1 and g_2 from $H_+^{(2)'}$ by the formula

$$g_1 * g_2 = 2\pi F[F^{-1}[g_1]F^{-1}[g_2]]. \tag{3.10}$$

If g_1 and g_2 belong to $F[N^+] = F[TN^+]$, then in accordance with (3.8) and (3.9), the formula (3.10) takes the form

$$g_1 * g_2 = 2\pi F[F^{-1}[g_1]F^{-1}[g_2]], \tag{3.11}$$

owing to the equalities

$$T(f_1^+ f_2^+) = Tf_1^+ Tf_2^+, \quad f_1, f_2 \in N^+;$$

$$T^{-1}(f_1(\cdot + i0)f_2(\cdot + i0)) =$$

$$T^{-1}f_1(\cdot + i0)T^{-1}f_2(\cdot + i0), \quad f_j(\cdot + i0) \in TN^+.$$

4. We say that a function ρ admits a factorization in the algebras $N^\pm$, if ρ can be represented in the form

$$\rho(x) = f^+(x)g^-(x) \quad \text{a.e. on } \mathbb{R}, \tag{4.1}$$

where $f^+ \in N^+$ and $g^- \in N^-$ and the functions $f(z)$ and $g(-z)$ have no zeros in $y > 0$.

Theorem 1 (see [15])   In order that a function ρ admits a factorization (4.1) in the algebras $N_*^\pm$, it is necessary and sufficient that it satisfies the condition (1.4). In addition,

$$f^+(x) = \pi^+(x) e^{ia^+(x)}, \quad g^-(x) = \pi^-(x) e^{ib^-(x)},$$

$$\pi^+ \in N_*^+, \quad \frac{1}{\pi^+} \in N_*^+, \quad \overline{\pi^+} = \pi^-, \quad a^+ \in H_2^+, \quad b^- \in H_2^-, \quad 0 < a^+ + b^- \le 2\pi.$$

Here H_2 is the Hardy class.

5. We shall turn to the Riemann-Hilbert problem (1.2) under the assumption that a given function F belongs to algebra $\tilde{N}$.

By theorem 1 function ρ admits a factorization (4.1), where f^+, $1/f^+$ $\in N_*^+$, g^-, $1/g^- \in N_*^-$. Substituting the expression (4.1) in (1.2), we obtain

$$g^- f^+ \phi^+ = \psi^- + F. \tag{5.1}$$

As $1/g^- \in N_*^-$, then multiplying both sides of (5.1) by $1/g^-$ we get

$$f^+ \phi^+ = \frac{\psi^-}{g^-} + \frac{F}{g^-} \tag{5.2}$$

As $\frac{F}{g^-} \in \tilde{N}$, we represent this function in the form (see (2.9))

$$\frac{F}{g^-} = P_+\left(\frac{F}{g^-}\right) + P_-\left(\frac{F}{g^-}\right), \quad P_\pm\left(\frac{F}{g^-}\right) \in N_*^\pm. \tag{5.3}$$

Substituting expression (5.3) in (5.2), we obtain the equality

$$f^+ \phi^+ - P_+\left(\frac{F}{g^-}\right) = P_-\left(\frac{F}{g^-}\right) + \frac{\psi^-}{g^-}$$

from which it follows that

$$f^+ \phi^+ = P_+\left(\frac{F}{g^-}\right) + C, \quad \frac{\psi^-}{g^-} = C - P_-\left(\frac{F}{g^-}\right),$$

where C is an arbitrary function from the algebra N^o (N_*^o resp.). Thus,

$$\phi^+ = \frac{1}{f^+} P_+\left(\frac{F}{g^-}\right) + \frac{C}{f^+}, \quad \psi^- = -g^- P_-\left(\frac{F}{g^-}\right) + g^- C \tag{5.4}$$

is a general solution of the Riemann-Hilbert problem (1.2) in the algebras $N^\pm$ ($N_*^\pm$ resp.).

So, we have just proved the following

Theorem 2 There exists a solution of the Riemann-Hilbert problem (1.2) in the algebras $N^\pm$ ($N_*^\pm$) for any given $F \in \tilde{N}$ and the general solution is given by the formulae (5.4).

Corollary For any given $F \in \tilde{N}$ there exists a solution of the problem (1.2) in the algebras $N_*^\pm$, and this solution is expressed by the formula (5.4) by $C = 0$.

The problem is now to find the function $P_+(F/g^-)$ in decomposition (5.3). We shall discuss here three cases.

1) Let $\frac{F}{g_-}(1 + x^2)^{-N}$ belong to the Wiener algebra W for some integer $N \geq 0$ i.e.

$$\frac{F}{g_-} = (1 + x^2)^N \left[\lambda + \int_{-\infty}^{\infty} \eta(\xi) e^{ix\xi} d\xi\right], \quad \eta \in L_1(\mathbb{R}).$$

Then,

$$P_+\left(\frac{F}{g_-}\right) = (1 + x^2)^N \left[\lambda_1 + \int_0^{\infty} \eta(\xi) e^{ix\xi} d\xi\right]. \tag{5.5}$$

2) Let $\frac{F}{g_-}(1 + x^2)^{-N}$ belong to $L_1(\mathbb{R})$ for some integer $N \geq 0$. Then,

$$P_+\left(\frac{F}{g_-}\right) = \frac{(1 + x^2)^N}{2\pi i} \lim_{y \to +0} \int_{-\infty}^{\infty} \frac{F(x')dx'}{g^-(x')(x' - z)(1 + x'^2)} \quad \text{a.e. on } (\mathbb{R}). \tag{5.6}$$

Formulae (5.5) and (5.6) are known (see [4, 5, 7, 10]). For instance, if ρ belongs to the Wiener algebra W and does not vanish on $\bar{\mathbb{R}}$, then the factorization (4.1) can be made precise as follows

$$\rho(x) = f^+(x) g^-(x) \left(\frac{x - i}{x + i}\right)^\kappa, \quad f^+, \frac{1}{f^+} \in W^+, \quad g^-, \frac{1}{g_-} \in W^-, \tag{5.7}$$

where κ is an integer (index)

$$\kappa = \text{ind}\rho = \frac{1}{2\pi}[\arg(+\infty) - \arg(-\infty)].$$

Let $F \in W$. A solution of the problem (1.2) for $\kappa > 0$ in algebras $W^\pm$ exists, if, and only if, the function F satisfies the condition

$$\left(\frac{x + i}{x - i}\right)^\kappa P_+\left(\frac{F}{g_-}\right) \in W^+, \tag{5.8}$$

and this solution is unique and is expressed by the formula

$$\phi^+ = \frac{1}{f^+}\left(\frac{x + i}{x - i}\right)^\kappa P_+\left(\frac{F}{g_-}\right), \quad \psi^- = - g^- P_-\left(\frac{F}{g_-}\right). \tag{5.9}$$

We note that the condition (5.8) is equivalent to $\kappa - 1$ conditions of orthogonality

$$\int_0^{\infty} \eta(\xi) \xi^k e^{-\xi} d\xi = 0, \quad k = 1, 2, \ldots, \kappa-1. \tag{5.10}$$

The last conditions follow from the formula (5.5) by $N = 0$ and

$$\lambda_1 = -\int_0^\infty \eta(\xi) e^{-\xi} d\xi \tag{5.11}$$

and also from the following lemma: a function

$$\left(\frac{x+i}{x-i}\right)^n f^+(x), \quad f^+ \in W^+, \; n > 0 \; - \text{ integer},$$

belongs to the algebra W^+ if, and only if, the function $f(z)$ has a zero at the point i of multiplicity $\geq n$.

For $\kappa \leq 0$, a solution of the problem (1.2) exists for any $F \in W$ and is expressed by the formulae

$$\phi^+ = \frac{1}{f^+} P_+ \left[\left(\frac{x-i}{x+i}\right)^{|\kappa|} \frac{F}{g^-}\right], \quad \psi^- = -\left(\frac{x+i}{x-i}\right)^{|\kappa|} g^- P_- \left[\left(\frac{x-i}{x+i}\right)^{|\kappa|} \frac{F}{g^-}\right] \tag{5.12}$$

The general solution of the homogeneous equation (1.2)

$$p\phi_0^+ = \psi_0^-, \quad \phi_0^+ \in W^+, \; \psi_0^- \in W^- \tag{1.2'}$$

has the form

$$\phi_0^+ = \frac{1}{f^+} P_{|\kappa|}\left(\frac{x-i}{x+i}\right), \quad \psi_0^- = \left(\frac{x+i}{x-i}\right)^{|\kappa|} g^- P_{|\kappa|}\left(\frac{x-i}{x+i}\right), \tag{5.13}$$

where $P_{|\kappa|}(\zeta)$ is an arbitrary polynomial of degree $\leq |\kappa|$ that is, it contains $|\kappa| + 1$ arbitrary constants.

If we assume that a solution (ϕ^+, ψ^-) vanishes at a point at infinity, $\phi^+(\infty) = \psi^-(\infty) = 0$, then for $\kappa > 0$, owing to (5.10) and (5.11), we have κ necessary and sufficient conditions of solvability (5.10) for $k = 0, 1, \ldots, \kappa-1$; for $\kappa \leq 0$ the general solution of the homogeneous equation (1.2') contains $|\kappa|$ arbitrary constants (a polynomial $P_{|\kappa|}$ has to satisfy the condition $F_{|\kappa|}(1) = 0$).

3) Let a function $F = F^+$ be "finite"

$$F^+(x) = \sum_{k=0}^N a_k \left(\frac{x-i}{x+i}\right)^k. \tag{5.14}$$

We expand the functions $\frac{1}{f^+}$ and $\frac{1}{g^-}$ in series (2.6) and (2.7), resp.

$$\frac{1}{f^+(x)} = \sum_{k=0}^\infty b_k \left(\frac{x-i}{x+i}\right)^k, \quad \frac{1}{g^-(x)} = \sum_{k=0}^\infty d_k \left(\frac{x+i}{x-i}\right)^k. \tag{5.15}$$

Then, (see [15]),

$$P_+\left(\frac{F^+}{g}\right) = \sum_{s=0}^{N}\left(\frac{x-i}{x+i}\right)^s \sum_{k=s}^{N} a_k d_{k-s} \in N_*^+,$$

$$P_-\left(\frac{F^+}{g}\right) = \sum_{s=-\infty}^{-1}\left(\frac{x-i}{x+i}\right)^s \sum_{k=0}^{N} a_k d_{k-s} \in N_*^-.$$

From theorem 2 we have

Theorem 3 A solution of the Riemann-Hilbert problem (1.2) in the algebras $N^\pm$ ($N_*^\pm$), where a function ρ satisfies the condition (1.4) and $F = F^+$ is "finite" exists, is unique up to an arbitrary function

$$C(x) = \sum_{k=0}^{\infty} c_k \left(\frac{x-i}{x+i}\right)^k = \sum_{k=-\infty}^{0} c_k \left(\frac{x-i}{x+i}\right)^k \tag{5.16}$$

from the algebra N^o (N_*^o resp.) and is expressed by the formula

$$\phi^+(x) = \sum_{k=0}^{\infty}[(Ta)_k + \varepsilon_k]\left(\frac{x-i}{x+i}\right)^k = \frac{1}{f^+}\sum_{k=0}^{N}\left(\frac{x-i}{x+i}\right)^k \sum_{j=k}^{N} a_j d_{j-k} + \frac{C}{f^+}, \tag{5.17}$$

where a is a vector $a = (a_0, a_1, \ldots, a_N, 0, \ldots)$, T is a matrix (T_{kj}),

$$T_{kj} = \sum_{s=0}^{\min(k,j)} b_{k-s} d_{j-s} = \frac{1}{b_0 d_0} \sum_{s=0}^{\min(k,j)} T_{k-s,0} T_{0,j-s}, \quad k,j = 0,1,\ldots \tag{5.18}$$

and coefficients ε_k are equal

$$\varepsilon_k = \sum_{s=0}^{k} c_j b_{k-s}, \quad k = 0,1,\ldots. \tag{5.19}$$

Remark If $\rho \geq 0$ then $d_k = \bar{b}_k$ and the matrix T is hermitian; if in addition ρ is even, then $d_k = b_k$ are real and the matrix T is symmetric.

6. We shall turn to the generalized Wiener-Hopf equation (1.3). Applying the inverse Fourier transform F^{-1} and denoting $F^{-1}[\phi] = \phi^+ \in N^+$ and $F^{-1}[f] = F^+ \in N^+$, we get the equation

$$P_+(\rho \phi^+) = F^+. \tag{6.1}$$

As $\rho \phi^+ \in \tilde{N}$, the decomposition (2.9)

$$\rho \phi^+ = P_+(\rho \phi^+) + P_-(\rho \phi^+), \quad P_\pm(\rho \phi^+) \in N_*^\pm \tag{6.2}$$

is valid. Putting $P_-(\rho\phi^+) = \psi^- \in N_*^-$ fron (6.1) and (6.2), we get the equation

$$\rho\phi^+ = F^+ + \psi^-. \tag{6.3}$$

Conversely, from (6.3) the equation (1.3) follows. Thus the generalized Wiener-Hopf equation (1.3) and the Riemann-Hilbert problem (6.3) are equivalent.

A solution of the equation (1.3) is an analytical functional $\phi = F[\phi^+] \in H_+^{(2)'}$, where ϕ^+ is a solution of the Riemann-Hilbert problem (6.3) in the algebras $N^\pm$ ($N_*^\pm$).

We denote

$$E = \frac{1}{2\pi} F\left[\frac{1}{f^+}\right] \in F[N_*^+] \subset H_+^{(2)'}. \tag{6.4}$$

$E(\xi)$ is an analytical functional with support in $[0,\infty)$. By virtue of (6.4) and (5.15), it is expanded in a series by distributions $H_k(\xi)$, $k = 0,1,\ldots$

$$E(\xi) = \sum_{k=0}^\infty b_k H_k(\xi) \tag{6.5}$$

which converges in space $H_+^{(2)'}$.

From theorem 2 and the fiormula (3.11), there follows

Theorem 4 A solution of the generalized Wiener-Hopf equation (1.3) in the algebra $F[N^+](F[N_*^+])$ exists for any $f \in F[N^+](F[N_*^+]$ resp.) and its general solution is expressed by the formula

$$\phi = E * F\left[P_+\left(\frac{F^{-1}[f]}{g_-}\right)\right] + E * F[C], \tag{6.6}$$

where C is an arbitrary function from the algebra N^o (N_*^o resp.).

We note that the solution (6.6) in the algebra $L_1^+ = [f : f \in L_1(\mathbb{R}), f(x) = 0, x < 0]$ under conditions

$$\rho(x) = 1 - F[k](x) \neq 0, \ k \in L_1(\mathbb{R}), \ \text{ind } \rho = 0$$

(the classical Wiener-Hopf equation) takes the known form (see [10]):

$$\phi(x) = f(x) + \int_0^\infty k(x,x')f(x')dx', \ f \in L_1^+, \tag{6.7}$$

where

$$k(x,x') = \varepsilon(x - x') + \eta(x' - x) + \int_0^{\min(x,x')} \varepsilon(x - x'')\eta(x' - x'')dx'', \qquad (6.8)$$

$$\frac{1}{2\pi} F\left[\frac{1}{f^+}\right] = \delta(x) + \varepsilon(x), \quad \frac{1}{2\pi} F\left[\frac{1}{g^-}\right] = \delta(x) + \eta(-x), \quad \varepsilon, \eta \in L_1^+. \qquad (6.9)$$

In fact, in this case $F = F$ and using the formulae (5.5) for $N = 0$ and (3.10) from (6.6), we get

$$\phi = E * [\theta(f * E_1)], \quad E_1 = \frac{1}{2\pi} F\left[\frac{1}{g^-}\right], \qquad (6.10)$$

where θ is the Heaviside function. The formula (6.10) is equivalent to the formula (6.7) owing to (6.8) and (6.9).

Let us suppose that f is a "finite" functional

$$f(\xi) = \sum_{k=0}^{N} a_k H_k(\xi). \qquad (6.11)$$

Applying theorems 3 and 4 we get

Theorem 5 A solution of the generalized Wiener-Hopf equation (1.3) in the class $F[N_*^+]$ under condition (1.4) and f is a "finite" functional of the form (6.11) exists and unique up to an analytical of the form

$$\sum_{k=0}^{\infty} c_k(E * H_k) = \sum_{k=0}^{\infty} \varepsilon_k H_k(\xi),$$

where c_k are coefficients of the expansion (5.16) in the algebra N_*^+ of an arbitrary function $C(x)$ from N_*^o and the coefficients ε_k are defined by the formula (5.19). A particular solution is given by the formula

$$\phi(\xi) = \sum_{k=0}^{\infty} (Ta)_k H_k(\xi), \qquad (6.12)$$

where the series converges in the space $H_+^{(2)'}$.

REFERENCES

1. N. Wiener and E. Hopf, Sitz. Berliner Akademi Wiss., 696-706, (1931).
2. V. A. Fok, Matem. sb., 14, 56, NO. 1-2, 3-50, (1944), (in Russian).

3. N. I. Muschelishvili and N. P. Vekua, Trudy Tbilis. Mathem. Inst. XII, 1-46, (1943), (in Russian).
4. N. I. Muschelishvili, "Singular Integral Equations", Moscow, (1962), (in Russian).
5. N. P. Vekua, "Systems of Singular Integral Equations", Moscow, (1970), (in Russian).
6. V. A. Ambartsumian, Nauchye trudy, v. I, Erevan, (1960), (in Russian).
7. F. D. Gahov, "Boundary Value Problems", Moscow, (1977), (in Russian).
8. S. Chandrasekhar, "Radiative transfer", Oxford, (1950).
9. V. V. Sobolev, "Radiative Transfer in Stars and Planets of Atmospheres", Moscow, (1956), (in Russian).
10. M. G. Krein, Uspehi Mathem. Nauk, v. 13, No. 5, 3-12, (1958), (in Russian).
11. I. C. Gohberg and M. G. Krein, Uspehi Mathem. Nauk, v. 13, No. 5, 3-72, (1958), (in Russian).
12. M. V. Maslennikov, Trudy Steklov Institute of Mathematics, t. 97 3-133, (1968), (in Russian).
13. L. G. Arabajan, N. B. Engibarjan, Itogi nauki i tekhniki, ser. Mathematical Analysis, t. 22, Moscow, VINITI, 174-244, (1984), (in Russian).
14. V. S. Vladimirov, Doklady AN SSSR, t. 293, NO. 2, 278-283, (1987), (in Russian).
15. V. S. Vladimirov, Izvestia AN USSR, ser. Mathematics, v. 51, No. 4. 747-784, (1987), (in Russian).
16. A. B. Aleksandrov, Lectures Notes in Mathem., 864, 1-89, (1981).
17. V. S. Vladimirov and I. V. Volovich, Theoretical and Mathematical Physics, t. 54, No. 1, 8-22, (1983), (in Russian).
18. S. V. Swedenko, Itogi nauki i tekhniki, ser. Mathematical Analysis, t. 23, Moscow, VINITI, 3-124, (1985), (in Russian).
19. G. M. Golusin, "Geometrical Theory of Functions of Complex Variable", Moscow, (1966), (in Russian).
20. I. I. Privalov, "Boundary Properies of Analitical Functions", Moscow, (1950), (in Russian).
21. N. K. Nikol'skii, "Treatise of the Shift Operator. Spectral Function Theory", Springer-Verlag, (1986).
22. H. Komatsu, J. Fac. Sci. Univ. Tokyo, Section IA, 20, 25-105, (1973).

23. H. Komatsu, J. Fac. Sci. Univ. Tokyo, Section IA, 24, 607-628, (1977).
24. I. N. Vekua, "Generalized Analitical Functions", Moscow, (1959), (in Russian).
25. I. N. Vekua, "New Method for Solution of Elliptic Equations", Moscow - Leningrad, (1948).
26. I. I. Daniluk, "Nonregular Boundary Value Problem on Plane", Moscow, 1975.
27. B. V. Bojarskii, Doklady AN USSR, t. 126, 695-698, (1959), (in Russian).
28. I. B. Simonenko, Izvestia AN USSR, ser. Mathematics, v. 28, No. 2, 277-306, (1964).
29. G. S. Litvinchuk, "Boundary Value Problems and Singular Integral Equations with Shift", Moscow, (1977), (in Russian).
30. V. M. Kokilashvili and V. A. Paatishvili, "Differential Equations", XVI, No. 9, 1650-1659, (1980), (in Russian).

SECTION II. GENERALIZED FUNCTIONS

ON NONLINEAR SYSTEMS OF ORDINARY DIFFERENTIAL EQUATIONS

Lothar Berg

Sektion Mathematik der Wilhelm-Pieck Universität Rostock

Rostock, DDR

ABSTRACT

The paper gives some analytical representations and numerical methods for the solutions of systems of ordinary differential equations with emphasis of the formal side, using the connection to the linear partial differential equations in the case first mentioned. The numerical methods are investigated concerning their stability and compared by test calculations.

1. INITIAL VALUE PROBLEMS

Let us consider the autonomous nonlinear initial value problem

$$Z'(t) = \theta(Z(t)), \quad Z(0) = z \tag{1}$$

with the vectors

$$Z(t) = (Z_1(t), \ldots, Z_n(t)), \theta(Z) = (\theta_1(Z), \ldots, \theta_n(Z)),$$
$$z = (z_1, \ldots, z_n). \tag{2}$$

We tacitly assume all appearing functions to be continuous in a certain domain and to have the necessary number of derivatives. Using the operator

$$\tilde{D} = \theta_1(Z)\frac{\partial}{\partial Z_1} + \ldots + \theta_n(Z)\frac{\partial}{\partial Z_n} \tag{3}$$

the chain rule reads

$$\frac{d}{dt} F(Z(t)) = \tilde{D} F(Z(t)), \tag{4}$$

where $Z(t)$ is the solution of (1) and $F(z)$ an arbitrary function. Introducing, further, the operator

$$D = \tilde{D}\big|_{t=0} = \theta_1(z) \frac{\partial}{\partial z_1} + \ldots + \theta_n(z) \frac{\partial}{\partial z_n}, \tag{5}$$

Taylor's formula yields the approximation

$$Z(t) = \sum_{j=0}^{m} \frac{1}{j!} t^j D^j z + O(t^{m+1}) \tag{6}$$

of the solution of (1) for small t. For holomorphic solutions equation (6) leads to the representation

$$Z(t) = e^{tD} z \tag{7}$$

of the solution, which is well known from the Lie theory (cf. W. Gröbner [3] and G. Maeß [9]).

In what follows we shall use for the Jacobi matrix of a vector $F(z) = (F_1(z),\ldots,F_n(z))$ the notation

$$\frac{\partial}{\partial z} F(z) = F'(z) = \begin{pmatrix} \frac{\partial F_1}{\partial z_1} & \cdots & \frac{\partial F_n}{\partial z_1} \\ \vdots & & \vdots \\ \frac{\partial F_1}{\partial z_n} & \cdots & \frac{\partial F_n}{\partial z_n} \end{pmatrix} \quad \text{with} \quad \frac{\partial}{\partial z} = \begin{pmatrix} \frac{\partial}{\partial z_1} \\ \vdots \\ \frac{\partial}{\partial z_n} \end{pmatrix} \tag{8}$$

or in case of $F = F(t,z)$ the notation $\frac{\partial F}{\partial z} = F_z(t,z)$. Then (5) can be written in the form $D = \theta(z) \frac{\partial}{\partial z}$. These notations are also transferred to the case of Z instead of z.

2. LINEAR PARTIAL DIFFERENTIAL EQUATIONS

The solutions $Z = Z(t,z)$ of (1) and the corresponding inverse vectors $z = z(t,Z)$ satisfy, according to the theory of characteristics (cf. E. Kamke [5]), the linear partial differential equations

$$Z_t(t,z) = \theta(z) Z_z(t,z), \tag{9}$$

$$z_t(t,Z) + \theta(Z) z_Z(t,Z) = 0. \tag{10}$$

The last equation expresses that $z(t,Z)$ is a constant vector and therefore

$\frac{d}{dt} z(t,Z) = 0$; the first is similar to (4) with $F(Z) = Z$.

To <u>prove</u> the first equation we write (1) in the form $Z_t = \theta(Z)$ and obtain by differentiation with respect to z

$$Z_{tz} = Z_z \theta'(Z). \qquad (11)$$

The initial condition $Z(0,z) = z$ implies $Z_z(0,z) = I$, the n-dimensional unit matrix, and since (11) can be considered as a linear system of ordinary differential equations for Z_z, it follows that $\det Z_z(t,z) \neq 0$ for such t, where the continuity assumptions are satisfied and the solution exists (cf. E. Kamke [5]). Hence Z_z^{-1} exists, and from $(Z_z^{-1})_t = -Z_z^{-1} Z_{zt} Z_z^{-1}$ and (11) we obtain

$$(Z_z^{-1})_t = -\theta'(Z) Z_z^{-1}$$

and after multiplication with $\theta(Z) = Z_t$

$$\theta(Z)(Z_z^{-1})_t + Z_t \theta'(Z) Z_z^{-1} = 0.$$

Since the left-hand side is equal to $(\theta(Z) Z_z^{-1})_t$, we find by integration from 0 to t in view of $Z_z(0,z) = I$

$$\theta(Z) Z_z^{-1} = \theta(z)$$

and therefore according to $Z_t = \theta(Z)$ the wanted equation (9).

Introducing the integral operator (cf. J. Mikusiński [10] and L. Berg [2])

$$T = \int_0^t \ldots d\tau$$

and using notation (5) we obtain from (9) and the initial condition in (1)

$$Z - z = TDZ,$$

i.e. $(1 - TD)Z = z$ and from this the representation

$$Z = \frac{1}{1-TD} z.$$

Expanding the operator by means of Neumann's series we obtain

$$Z = \sum_{j=0}^{m} T^j D^j z + \frac{T^{m+1} D^{m+1}}{1-TD} z,$$

which is nothing else than (6) according to $T^j c = \frac{1}{j!} t^j c$ for a constant vector c.

3. THE GROUP PROPERTY

Since (1) is an autonomous system, the solutions $Z = Z(t,z)$ possess for real s,t the group property

$$Z(t,z) = Z(t-s, Z(s,z)), \tag{12}$$

so far as they exist. According to the theory of continuous iterations (cf. M. Kuczma [6]) we expect the representations

$$Z = f(t\underline{1} + \phi(z)), \quad z = f(\phi(Z) - t\underline{1}) \tag{13}$$

with $\underline{1} = (1,1,\ldots,1)$, where the vectors $z = f(w)$ and $w = \phi(z)$ with

$$f = (f_1,\ldots,f_n), \quad \phi = (\phi_1,\ldots,\phi_n), \quad w = (w_1,\ldots,w_n)$$

are inverses of each other, i.e.

$$z = f(\phi(z)), \quad w = \phi(f(w)). \tag{14}$$

Representations (13) explain both the similarity and the difference of equations (9) and (10). In the scalar case $n = 1$ with $\theta(Z) \neq 0$ they arise immediately from (1) by separating the variables and integration.

The vectors $z = f(w)$, $w = \phi(z)$ are solutions of the partial differential equations

$$\theta(z) w'(z) = \underline{1}, \quad \underline{1} z'(w) = \theta(z), \tag{15}$$

which can be derived from (13) if we write the later equations in the form

$$\phi(Z) = t\underline{1} + \phi(z). \tag{16}$$

According to $w'(z) = (z'(z))^{-1}$ the equations (15) are equivalent to each other. However, whereas the second equation is a nonlinear system, the first consists in n independent linear equations, which are in fact one single equation

$$\theta(z) W'(z) = 1 \tag{17}$$

for all components $W(z) = w_j(z)$ with $j = 1,\ldots,n$.

Now, we shall <u>prove</u> the existence of the representations (13) for the solution of (1), under the assumption that one component of $\theta(z)$, let us say $\theta_1(z)$, is different from zero in a certain domain. Let $w(z)$ be a row vector the n components of which are solutions of (17). Then by suitable linear combinations of the rows and columns, respectively, we obtain with regard to (17)

$$\det w'(z) = \begin{vmatrix} w_{1z_1} & w_{2z_1} & \cdots & w_{nz_1} \\ w_{1z_2} & w_{2z_2} & \cdots & w_{nz_2} \\ \vdots & \vdots & & \vdots \\ w_{1z_n} & w_{2z_n} & \cdots & w_{nz_n} \end{vmatrix} = \frac{1}{\theta_1} \begin{vmatrix} 1 & 1 & \cdots & 1 \\ w_{1z_2} & w_{2z_2} & \cdots & w_{nz_2} \\ \vdots & \vdots & & \vdots \\ w_{1z_n} & w_{2z_n} & \cdots & w_{nz_n} \end{vmatrix}$$

$$= \frac{1}{\theta_1} \cdot \begin{vmatrix} 1 & 0 & \cdots & 0 \\ w_{1z_2} & (w_2-w_1)_{z_2} & \cdots & (w_n-w_1)_{z_2} \\ \vdots & \vdots & & \vdots \\ w_{1z_n} & (w_2-w_1)_{z_n} & \cdots & (w_n-w_1)_{z_n} \end{vmatrix}.$$

Since the differences $W(z) = w_j(z) - w_1(z)$ are solutions of the homogeneous equation belonging to (17), there exist such solutions so that the foregoing determinant is different from zero (cf. E. Kamke [4]). This means that the vector $w = \phi(z)$ of the solutions of the unhomogeneous equation (17) is invertible, i.e. (14) is proved. Finally, substituting (1) into $\theta(Z)\phi'(Z) = \underline{1}$ and integrating this equation, we obtain (16) and by inversion the wanted representations (13).

4. REGULARIZATION

The approximations (6) together with (12) can be used for the numerical solution of (1), however, they are unstable. Let us assume that we know an m-dimensional vector $g(t,Z) = (g_1(t,Z),\ldots,g_m(t,Z))$ with $m < n$ such that

$$g(t,Z) = c \qquad (18)$$

is a constant vector for the solution $Z = Z(t)$ of (1). This is e.g. the case if the components of $g(t,Z)$ are so-called integrals of (1). Then, we can either drop m equations of (1) and consider the remainder together with (18) as differential-algebraic equations (cf. E. Griepentrog and R. März [3]) or we can use (18) to regularize the numerical instability in the following way. Starting from an approximating vector z for $Z(s)$ with $s < t$ we use

$$\zeta = \sum_{j=0}^{m} \frac{1}{j!} h^j D^j z \qquad (19)$$

103

with $h = t-s$ according to (6) and (12) as a predictor for $Z = Z(t)$ and correct this vector by means of

$$\|Z - \zeta\|^2 = \min! \text{ subject to (18)} \tag{20}$$

using the Euclidean norm. Using the Lagrange vector $\Lambda = (\lambda_1, \ldots, \lambda_m)$, the necessary condition for (20) reads

$$Z - \zeta = \Lambda\, g_Z^T(t,Z),$$

which can be solved together with (18) by, let us say, Newton's method. A weakened version of this is the iteration method

$$Z^{(k+1)} - \zeta = \Lambda_k g_Z^T(t, Z^{(k)}),$$

$$g(t, Z^{(k)}) + (Z^{(k+1)} - Z^{(k)}) g_Z(t, Z^{(k)}) = c$$

for $k = 0, 1, 2, \ldots$ with $Z^{(0)} = \zeta$, which gives after the first step

$$Z^{(1)} = \zeta - (g(t,\zeta) - c)(g_Z^T(t,\zeta) g_Z(t,\zeta))^{-1} g_Z^T(t,\zeta). \tag{21}$$

Usually, we can terminate already after this first step, because if $(g_Z^T g_Z)^{-1} g_Z^T$ is bounded, then

$$g(t, Z^{(1)}) - c = O(\|g(t,\zeta) - c\|^2).$$

To **prove** this estimation we only have to consider Taylor's expansion

$$g(t, Z^{(1)}) = g(t,\zeta) + (Z^{(1)} - \zeta) g_Z(t,\zeta) + O(\|Z^{(1)} - \zeta\|^2)$$

$$= c + O(\|Z^{(1)} - \zeta\|^2)$$

and to use (21) once more.

Of course, in the linear case $g(t,Z) = ZC$ with an n×m-matrix $C = C(t)$ and $g_Z(t,Z) = C$ the vector $Z^{(1)}$ from (21), i.e.

$$Z = \zeta - (\zeta C - c)(C^T C)^{-1} C^T \tag{22}$$

is the exact solution of (20).

5. THE LINEAR TEST EQUATION

To investigate the influence of the integrals (18) on the method, we shall consider the case $\theta(Z) = ZA$ with a constant matrix A, i.e. we shall consider the homogeneous test equation

$$Z' = ZA. \tag{23}$$

The matrix A shall be similar to the diagonal matrix $\Lambda = \text{diag}(\lambda_1,\ldots,\lambda_n)$, i.e. we assume the representation $A = Q \Lambda Q^{-1}$, where the eigenvalues shall be ordered in such a way that $\text{Re }\lambda_j \geq \text{Re }\lambda_{j+1}$ for $j = 1,\ldots,n-1$. The solution of (23) with $Z(0) = z$ possasses the representation

$$Z = z\, e^{At} = zQe^{\Lambda t}Q^{-1}. \tag{24}$$

Now, we split up the n-dimensional unit matrix I into the matrix E consisting of the first m and F consisting of the last n-m columns, respectively, so that

$$I = EE^T + FF^T.$$

The initial vector we can choose in such a way that $zQE = 0$. Then the foregoing solution turns over into

$$Z = zQFF^T e^{\Lambda t}Q^{-1}. \tag{25}$$

This solution contains in fact only the last eigenvalues $\lambda_{m+1},\ldots,\lambda_n$, so that the usual numerical method for its calculation would be unstable in case of $\text{Re }\lambda_1 > 0 > \text{Re }\lambda_{m+1}$. However, if we choose the integral $ZQe^{-\Lambda t}E$, which, in view of $zQE = 0$, turns over into $g(t,Z) = ZQE$, we obtain $c = 0$ in (18), $C = QE$ in (22) and the first m eigenvalues fade away, which means stability.

To **prove** this we specialize (19) according to $Dz = zA$ to $\zeta = zB$ with

$$B = \sum_{j=0}^{m} \frac{1}{j!} h^j A^j = QGQ^{-1}$$

and

$$G = \sum_{j=0}^{m} \frac{1}{j!} h^j \Lambda^j,$$

and we write (22) in the form $Z = \zeta P = zBP$ with

$$P = I - QES^{-1}E^T Q^T = Q(I - ES^{-1}E^T Q^T Q)FF^T Q^{-1}$$

and $S = E^T Q^T Q E$. We remember that z and Z here are approximate values for $Z(t-h)$ and $Z(t)$, respectively, so that for $t = kh$ we can write $Z = Z_k$, $z = Z_{k-1}$ and obtain

$$Z_k = z(BP)^k \tag{26}$$

with $z = Z(0)$ in the original sense. In more detail this reads

$$Z_k = zQ(G(I - ES^{-1}E^T Q^T Q))^k Q^{-1},$$

since

$$(I - ES^{-1}E^TQ^TQ)FF^T = I - ES^{-1}E^TQ^TQ. \tag{27}$$

In view of $S = E^TQ^TQE$, we can make the splitting

$$Q^TQ = \begin{pmatrix} S & U \\ V & W \end{pmatrix}, \quad ES^{-1}E^T = \begin{pmatrix} S^{-1} & 0 \\ 0 & 0 \end{pmatrix}, \quad I = \begin{pmatrix} I & 0 \\ 0 & I \end{pmatrix}$$

with suitable block matrices U, V, W, 0, I, so that we obtain

$$I - ES^{-1}E^TQ^TQ = I - \begin{pmatrix} S^{-1} & 0 \\ 0 & 0 \end{pmatrix}\begin{pmatrix} S & U \\ V & W \end{pmatrix} = \begin{pmatrix} 0 & -S^{-1}U \\ 0 & I \end{pmatrix},$$

what underlines (27). Splitting the diagonal matrices Λ and G into the form

$$\Lambda = \begin{pmatrix} \Lambda_1 & 0 \\ 0 & \Lambda_2 \end{pmatrix}, \quad G = \begin{pmatrix} G_1 & 0 \\ 0 & G_2 \end{pmatrix}$$

and considering

$$zQG = zQ(EE^T + FF^T)G = zQFG_2F^T$$

as well as $F^TE = 0$, $F^TG = G_2F^T$, and therefore

$$zQG(I - ES^{-1}E^TQ^TQ)G = zQTG_2^2F^T,$$

and so on we find for Z_k the representation

$$Z_k = zQFG_2^kF^TQ^{-1}, \tag{28}$$

where

$$G_2 = \sum_{j=0}^{m} \frac{1}{j!} h^j \Lambda_2^j$$

contains only the last $n - m$ eigenvalues. Since G_2 is a matrix of order $n - m$, it is impossible that the first eigenvalues snack in by roundoff errors, which gives the asserted stability.

The proof becomes essentially shorter if A is symmetrical and therefore Q orthogonal, i.e. $Q^{-1} = Q^T$. Then $S = E^TE = I$ and $P = I - QEE^TQ^T = QFF^TQ^T$, so that from (26), $GFF^T = FG_2F^T$ and

$$(QFG_2F^TQ^T)^k = QFG_2^kF^TQ^T$$

we immediately obtain (28).

Since $G_2^k \to e^{\Lambda_2 t}$ for $k \to \infty$ with $hk = t$, the approximations (28) converge to the solution (24).

6. OBRESCHKOFF METHODS

Let $\Phi(x)$ be a polynomial of degree m, then from

$$\frac{d}{dx} \sum_{j=0}^{m} (-1)^j \left(\frac{h}{b-a}\right)^j \Phi^{(m-j)}(x) Z^{(j)}\left(\frac{x-a}{b-a} h\right) = (-1)^m \left(\frac{h}{b-a}\right)^{m+1} \Phi(x) Z^{(m+1)}\left(\frac{x-a}{b-a} h\right)$$

we find by integration the well known formula

$$\sum_{j=0}^{m} (-1)^j \left(\frac{h}{b-a}\right)^j \left[\Phi^{(m-j)}(b) Z^{(j)}(h) - \Phi^{(m-j)}(a) Z^{(j)}(0)\right]$$

$$= (-1)^m \left(\frac{h}{b-a}\right)^{m+1} \int_a^b \Phi(x) Z^{(m+1)}\left(\frac{x-a}{b-a} h\right) dx. \tag{29}$$

For $\Phi(x) = (b-x)^m$ it turns over into Taylor's formula (6) with $t = h$, and for $\Phi(x) = (b-x)^k (x-a)^\ell$ with $k+\ell = m$, it gives the Padé approximations of $Z(h)$ (cf. O. Perron [11] and L. Berg [1]). Replacing the right-hand side of (29) by zero we obtain the one-step Obreschkoff methods (cf. N. Obreschkoff [12] and J. D. Lambert [7]). In the case $a = -1$, $b = 1$, and $\Phi(x) = T_m(x)$, the Tschebyscheff polynomials, these are optimal methods in the sense that the factor $\Phi(x)$ in the remaindes is minimal with respect to the maximum norm. Howewer, in view of $T_m(1) = 1$ and $T_m(-1) = (-1)^m$, these methods are not convenient for stiff differential equations. Hence, we look for minimal polynomials subject to $\Phi(a) = 0$, which are for

$$a = -\cos\frac{\pi}{2m}, \quad b = 1$$

once more the Tschebyscheff polynomials $\Phi(x) = T_m(x)$. The corresponding methods in the first three cases $m = 1, 2, 3$ are

$$Z(h) - Z(0) - h Z'(h) = 0, \tag{30}$$

$$Z(h) - Z(0) - \frac{h}{1-a}(Z'(h) + \frac{1}{2}\sqrt{2}\, Z'(0)) + \frac{1}{4}\left(\frac{h}{1-a}\right)^2 Z''(h) = 0, \tag{31}$$

$$Z(h) - Z(0) - \frac{h}{1-a}(Z'(h) + \frac{1}{2}\sqrt{3}\, Z'(0)) + \frac{1}{8}\left(\frac{h}{1-a}\right)^2 (3 Z''(h) - 2Z''(0))$$
$$- \frac{1}{24}\left(\frac{h}{1-a}\right)^3 Z'''(h) = 0. \tag{32}$$

Obviously, (30) is the Euler backward method. The method (31) with $a = \frac{\sqrt{2}}{2}$ is L-stable, since for the test equation (23) with a scalar A we obtain with the abbreviation

$$\lambda = -\frac{h}{1-a} A \tag{33}$$

from (31)

$$(4 + 4\lambda + \lambda^2)z(h) = (4 - 2\sqrt{2}\,\lambda)z(0),$$

and it is easy to see that

$$|4 - 2\sqrt{2}\lambda| \leq |4 + 4\lambda + \lambda^2|$$

for Re $\lambda \geq 0$. The method (32) with $a = -\sqrt{3}/2$ goes over for (23) and (33) into

$$(24 + 24\lambda + 9\lambda^2 + \lambda^3)z(h) = (24 - 12\sqrt{3}\lambda + 6\lambda^2)z(0),$$

and according to J. Leskin [8] we have

$$|24 - 12\sqrt{3}\,\lambda + 6\lambda^2| \leq |24 + 24\lambda + 9\lambda^2 + \lambda^3| \qquad (34)$$

for $|\arg\lambda| \leq \alpha = 89.95^0$ with $\tan\alpha = 1078.19871$, so that the method is $A(\alpha)$-stable. But, according to the great value of $\tan\alpha$ it actually works like an L-stable method. The curve (34) with the equality sign and its enlargement in the neighbourhood of the imaginary axis are to be seen in the two figures also made by J. Leskien.

For the application of (31) and (32) to solve (1) numerically it is possible to calculate the derivatives in these formulas as in (6) and (19) by means of the operator D from (5).

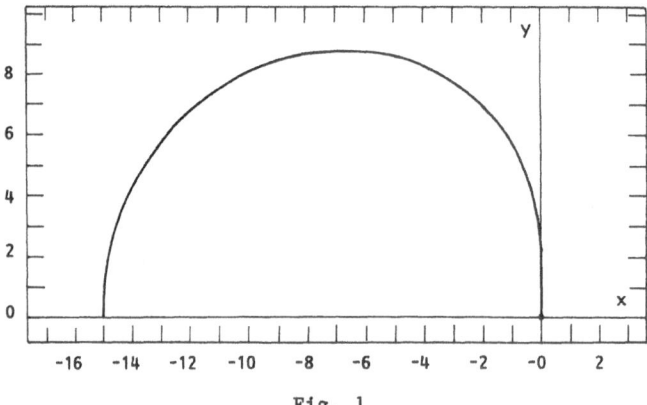

Fig. 1

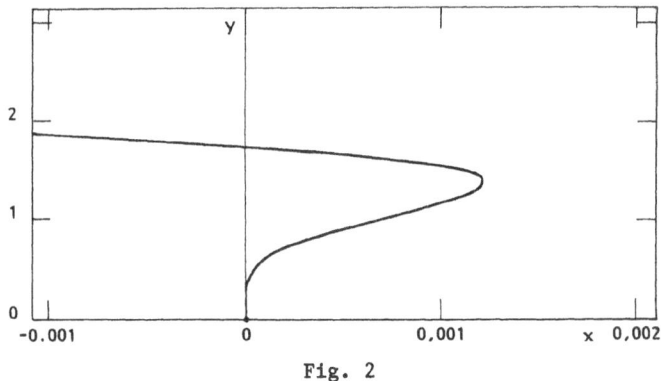

Fig. 2

7. EXAMPLES

We terminate the lecture with four examples. In the cases of $n = 2$ we use the notations

$$Z = (X,Y), \quad z = (x,y), \quad w = (u,v).$$

(1) For $X + Y > 0$ the system

$$X' = \frac{3X+Y}{\sqrt{X+Y}}, \quad Y' = \frac{X+3Y}{\sqrt{X+Y}}$$

possesses the integral $\sqrt{X+Y} - 2t = \sqrt{x+y}$, which leads to the solution

$$X = \left[t + \frac{x}{\sqrt{x+y}}\right](2t + \sqrt{x+y}), \quad Y = \left[t + \frac{y}{\sqrt{x+y}}\right](2t + \sqrt{x+y}).$$

As vectors $(x,y) = f(u,v)$ and $(u,v) = \phi(x,y)$ from (14) we find e.g. in components

$$x = u(u+v), \quad y = v(u+v), \quad u = \frac{x}{\sqrt{x+y}}, \quad v = \frac{y}{\sqrt{x+y}}.$$

(2) The linear system

$$X' = X + Y, \quad Y' = Y$$

possesses the integral $Ye^{-t} = y$ and the solution

$$X = e^t(x + ty), \quad Y = e^t y.$$

Here we have e.g.

$$u = |\ln x - y \ln|y||, \quad v = \ln|y|, \quad x = \varepsilon_1 e^u + \varepsilon_2 v e^v, \quad y = \varepsilon_2 e^v$$

with $\varepsilon_1 = \text{sign}(x - y\ln|y|)$, $\varepsilon_2 = \text{sign } y$.

(3) The solution (24) of the linear system (23) can be written in the form

$$ZQ = zQe^{\Lambda t}.$$

This means for a regular matrix Λ that the components of w read

$$w_j = \frac{1}{\lambda_j} \ln|zQ_{\cdot j}|$$

for $j = 1,\ldots,n$, where $Q_{\cdot j}$ is the vector of the jth column of Q, and we have

$$z = \left(\varepsilon_j e^{\lambda_1 w_1}, \ldots, \varepsilon_n e^{\lambda_n w_n} \right) Q^{-1}$$

with $\varepsilon_j = \text{sign}(zQ_{\cdot j})$.

(4) As an numerical example let us consider the system

$$X' = X \ln R - Y, \quad Y' = Y \ln R + X, \quad x = 1, \quad y = 0$$

with $R = \sqrt{X^2+Y^2}$ and the integral $e^{-t}\ln R = \ln\sqrt{x^2+y^2}$. For our initial values this means $R = 1$, and therefore

$$X = \cos t, \quad Y = \sin t,$$

but this circle is unstable. We have solved this problem with the three methods (19), (21) and (31), respectively, always with $m = 2$ and the steplength $h = 0.1$, and with (19) also as predictor in the last case with the constant number p of iterations of the corrector (31). The results on the minicomputer KC 85/2 for the errors $\Delta X = X - \cos t$ always after twenty steps up to about 2 periods ($14 > 4\pi = 12.6$) are contained in the table

t	(19)	(21)	p=1	p=2	p=3	cos t
2	-0.0033	-0.0030	0.0010	0.0008	0.0007	-0.4161
4	0.0008	0.0050	0.0027	-0.0009	-0.0011	-0.6536
6	0.0505	0.0027	-0.0426	-0.0047	-0.0021	0.9602
8	-0.0789	-0.0131	0.0435	0.0076	0.0047	-0.1455
10	-10.966	0.0091	0.7654	0.1693	0.0619	-0.8391
12	1.3E+07	0.0105	-0.8438	-0.6872	-0.3752	0.8439
14	owerflow	-0.0231	-0.1367	-0.1367	-0.1349	0.1367

together with the exact solution $\cos t$. The results show that in these cases (31) is better than (19), and at first also better than (21), but later on not so good than (21). After p = 3 iterations by (31) there are no changes of the first four significant decimals. This trend confirms also by Y and by other calculations in particular with smaller h and a longer interval of approximation. The general solution

$$X = R\cos(t+v), \quad Y = R\sin(t+v), \quad v = \arctan\frac{y}{x},$$

consists of spirals tending to infinity or to zero, as also do the numerical solutions, disregarding (21). Finally, let us mention that

$$x = e^{\varepsilon e^u}\cos v, \quad y = e^{\varepsilon e^u}\sin v, \quad u = \ln\left(\frac{1}{2}|\ln(x^2+y^2)|\right)$$

with $\varepsilon = \text{sign}\ln(x^2+y^2)$.

REFERENCES

1. L. Berg, Necessary stability conditions for discretization methods concerning boundary value problems of ordinary differential equations, ZAMM, 57, 342-344, (1977).

2. L. Berg, General operational calculus, Linear Algebra Appl., 84, 79-97, (1986).

3. E. Griepentrog and R. März, Differential-Algebraic Equations and Their Numerical Treatment, Leipzig: Teubner Texte zur Mathematik, 88, (1986).

4. W. Gröbner, "Die Lie-Reihen und ihre Andwendungen", Berlin, VEB DVW, (1960).

5. E. Kamke, "Differentialgleichungen I, II", Leipzig, Akad. VG Geest & Portig K.-G., (1969, 1965).

6. M. Kuczma, Functional Equations in a Single Variable, Warszawa, Pol. Sci. Publ., (1968).

7. J. D. Lambert, "Computational Methods in Ordinary Differential Equations", London, John Wiley and Sons, (1973).

8. J. Leskien, Private communication, 20 April 1987.

9. G. Maeß, Zur Bestimmung des Restgliedes von Lie-Reihen, Wiss. Z. FSU Jena, M. 14, 423-425, (1965).

10. J. Mikusiński, "Operational Calculus", London, Pergamon Press, (1959).

11. O. Perron, "Die Lehre von den Kettenbrüchen II", Stuttgart, Teubner VG, (1957).

12. N. Obreschkoff, Neue Quadraturformeln, Abh. Preuß. Akad. Wiss., Math.-nat. Kl. Nr. 4, Berlin, (1940).

A NEW CONSTRUCTION OF CONTINUOUS ENDOMORPHISMS OF THE OPERATOR FIELD

A. Bleyer

Technical University of Budapest
Electrical Engineering Faculty, Department of Mathematics
Stoczek u 3-5, 1111 Budapest, Hungary

1. INTRODUCTION AND PRELIMINARIES

In this note we shall give a method to construct continuous endomorphisms of the operator field endowed with convergence structure of type I. The problem is to find, to construct different types of endomorphisms or the linear mapping of the operator field M. In 1967 Gesztelyi published some representation theorems on linear operator transformations, nevertheless there were known only a few types of different transformations. In 1971, in Dubrovnik, R. A. Struble proposed to investigate this problem in view of finding new types of transformations.

Since our procedure will be based on the so called Ditkin-Berg model of the operator field, we shall recall its main properties. Let H be the set of all complex variable functions being holomorphic and bounded on some right half-plane. H is a linear space. The set of all the continuous Laplace transformable functions on $[0,\infty)$ will be denoted by L and Laplace transforms form a linear space H_1 being a subspace of H, $H_1 \subset H$. A sequence of functions $f_n(p)$, $f_n(p)$ from H_1, is called a ring sequence if they are holomorphic functions on a half-plane $\text{Re}(p) > a$ and satisfy the relation

$$f_n(p) = f_n(p) + O(\exp(-np)) \tag{1.1}$$

as $\text{Re}(p) \to \infty$ for each n with $f_0(p) = 0$. Here a depends on the sequence but is independent of n. Two ring sequences $\{f_n(p)\}$, $\{g_n(p)\}$ are equivalent if and only if each n

$$f_n(p) = g_n(p) + O(\exp(-np)) \tag{1.2}$$

as $\text{Re}(p) \to \infty$, in some right half-plane $\text{Re}(p) > c$, where c is independent of n. It is a classification of ring sequences and it is easy to see that these classes form a commutative C-module with addition, multiplication by numbers and product in the usual sense, that is, the operations are taken on elements of arbitrary representation of classes. Denote this C-module by R. There is an isomorphic embedding of C (the set of all the continuous functions defined on $[0,\infty)$) into R by the mapping:

$$f_n(p) = \int_0^n f(t) \exp(-tp) dt, \quad n = 0,1,2,\ldots \qquad (1.3)$$

for $f \in C$. By the Titchmarch theorem, R is an integral domain under a pointwise product which obviously corresponds to the "convolution product", in C.

Therefore the field extension Q or R consists of "sequential quotients" corresponding to convolution quotients, that is $x_n(p)$ ($n = 1,2,\ldots$) is a field sequence if

$$x_n(p) = f_n(p)/g_n(p) \qquad (1.4)$$

where $\{f_n(p)\} \in R$, $\{g_n(p)\} \in R$, moreover $\{g_n(p)\}$ is not the zero element of R, i.e.: $g_n(p) \neq 0(\exp(-np))$ as $\mathrm{Re}(p) \to \infty$. The elements of a field sequence are meromorphic functions. By the equivalence relation of ring sequences, $\{w_n(p)\}$ and $\{x_n(p)\}$ are equivalent if and only if for $n > n_0$

$$x_n(p) = w_n(p) + \exp(-np) v_n(p)/h_n(p) \qquad (1.5)$$

for $\mathrm{Re}(p) > c$, where $\{h_n(p)\} \in R$ and $h_n(p) \neq 0(\exp(-np))$, $\{v_n(p)\} \in R$, holomorphic on the half-plain $\mathrm{Re}(p) > c$ and bounded by K being independent of n.

Let (MR) be the space of the ratios of Laplace transforms, ratios of two holomorphic and bounded functions on some right half-plane. (MR) is the quotion field of H_1 and obviously (MR) is a subfield of Q.

Q and (MR) are fields, therefore, it is sufficient to find the isomorphic mappings of R into Q (or respectively of H_1 into (MR)), hence, these isomorphisms are uniquely extendable into the endomorphism of Q (respectively into (MR)). By (1.3) there is a correspondence between C and R, respectively M and Q, hence, some of the isomorphism into M can be described through the isomorphism of R into Q.

2. SUBSTITUTION MAPPINGS

Let us consider all the functions $w(p)$ from (MR) which fulfil the following conditions:

$w(p)$ is holomorphic in some half-plane $\mathrm{Re}(p) > a > 0$; (2.1)

$$\text{Re}[w(p)] > b > 0 \quad \text{if} \quad \text{Re}(p) > a \quad \text{with some a,b;} \tag{2.2}$$

$$D = \text{Int}\{z : Z = w(p), \text{Re}[w(p)] > b, \text{Re}(p) > a\} \neq 0. \tag{2.3}$$

The w(p)-substitution is defined as follows:

Let $x \in R$ and take a represent ring sequence of x; $\{x_n(p)\}$. Substitute w(p) into the argument of $x_n(p)$, if the formal ring sequence $\{x_n(w(p))\}$ represents an element of R, then we say the w(p)-substitution is carried out. The conditions (2.1) and (2.2) gurantee that the w(p) substitution maps H into H. By using the argument of analytic continuation and condition (2.3), it follows immediately that the domain of a w(p)-substitution is a subring of R and the image is in Q, and the transformation is an isomorphism of the above subrings. The domain of w(p)-substitution contains each function of (MR) which is holomorphic on the half-plane $\text{Re}(p) > c$, where $c < b < \text{Re}[w(p)]$. Therefore, we obtain:

Lemma 1 If w(p) satisfies the conditions (2.1) - (2.3) and (2.4) for each a > 0, there exists d > 0 such that

$$\text{Re}[w(p)] > a \text{ whenever } \text{Re}(p) > d$$

then the domain of w(p) substitution contains H and it can be extended to an isomorphism of Q_1, which contains (MR) as a subfield, into Q.

Let us consider an arbitrary ring sequence, i.e. an arbitrary function from C. By virtue of the definition of ring sequences

$$f_{n+1}(w(p)) = f_n(w(p)) + 0(\exp(-npw(p))) \tag{2.5}$$

holds as $\text{Re}(p) \to \infty$ for each n. Therefore, in the case of $\{f_n(w(p))\} \in Q$

$$f_n(w(p)) = a_n(p)/b_n(p) \tag{2.6}$$

for each $n > n_0$, where $\{a_n(p)\}, \{b_n(p)\} \in R$ and $b_n(p) \neq 0(\exp(-np))$, if $n > n_1$ as $\text{Re}(p) \to \infty$. Since $\{a_n(p)\}, \{b_n(p)\}$ are ring sequences by the classification relation it follows that

$$[f_n(w(p)) + 0(\exp(-nw(p)))][b_n(p) + 0(\exp(-np))] =$$

$$a_n(p) + 0(\exp(-np)) \tag{2.7}$$

for $n > n_1$. Since $b_n(p) \neq 0(\exp(-np))$ as $n > n_1$

$$0(\exp(-np)) = 0(\exp(-nw(p))) \tag{2.8}$$

as $\text{Re}(p) \to \infty$, holds for $n > n_1$.

The condition (2.8) implies that the w(p)-substitution is independent to the special choice of representation of the ring sequence. Indeed, if $\{f_n(p)\}$ and $\{g_n(p)\}$ are from the same class then

$$f_n(w(p)) - g_n(w(p)) = O(\exp(-nw(p))) = O(\exp(-np))$$

by (2.8) it holds for each $n > n_1$ as $\text{Re}(p) \to \infty$. It follows from the definition of ring sequences that a subsequence of a ring (or a field) sequence is a ring (or a field) sequence and <u>belongs to the same class</u>. Hence, a <u>refinement</u> (select subsequence with the same method) and after that an $w(p)$-substitution seems to be legal. Generally, the simple refinementless $w(p)$-substitution cannot be carried out, but it does not make any difficulties to show that a refinement and a substitution generate a transformation which is independent of the density of the refinement. The refinement substitution is not allowed whenever the condition (2.8) has not been satisfied.

Let us assume that $w(p)$ is a holomorphic function satisfying (2.1) and (2.4) but instead of satisfying (2.8) $w(p)$ fulfils only

$$O(\exp(-np)) = O(\exp(-k(n)w(p))) \qquad (2.9)$$

as $\text{Re}(p) \to \infty$ if $n > n_0$ and $k(n)$ are integers with $k(n) \to \infty$ as $n \to \infty$.

The mapping $\{f_n(p)\} \sim \{f_n(w(p))\}$ defines an operator transformation which is called a <u>refinemement-substitution</u> transformation. Using the previous argument one obtains that $\{f_n(w(p))\}$ becomes a field sequence provided $\{f_n(p)\}$ was a field sequence.

<u>Theorem</u> Let $w(p)$ be a holomorphic function satisfying (2.1), (2.4) and (2.9). Then, the $w(p)$-refinement-sibstitution defines an endomorphism of Q.

<u>Proof</u> We have to show only that the procedure is independent of the measure of refinement, i.e.: if $\{k'(n)\}$ is such a sequence of integers for which (2.9) is satisfied, then $\{f_{k'(n)}(w(p))\}$ belongs to the same class as $\{f_{k(n)}(w(p))\}$. Indeed, by the virtue of (2.9) and the classification relation:

$$f_{k(n)}(w(p)) - f_{k'(n)}(w(p)) = O(\exp(-\min(k'(n),k(n))w(p))) = O(\exp(-np)), \qquad (2.10)$$

as $\text{Re}(p) \to \infty$ if $n > n_0$. □

3. REPRESENTATION FOR CONTINUOUS OPERATOR FUNCTIONS

<u>Theorem</u> Let $w(p)$ be as in the theorem of section 2. Then, the $w(p)$-refinement-substitution transformation defines a continuous (in the sense

of Gesztelyi's theory) endomorphism of Q and the image of function f belonging to C can be represented by the formula:

$$s^2 \left\{ \frac{1}{2\pi i} \int_{c-i\infty}^{c+i\infty} \exp(pt) f_n(w(p)) p^{-2} dp \right\} \tag{3.1}$$

where the function in brackets has the same value in $[0,n]$ just as

$$\frac{1}{2\pi i} \int_{c-i\infty}^{c+i\infty} \exp(pt) f_n(w(p)) p^{-2} dp \tag{3.2}$$

where $f_n(p)$ has been defined by the ring isomorphism between C and R, and $f_n(p)$ is holomorphic on the halfplane $Re(p) > c$ for each n.

The image of the differential operator s is w(s), which is the correspondent of w(p) and w(s) is a logarithm. For all u continuous function with support bounded from left

$$F^{w(p)}(u) = \int_{-\infty}^{\infty} e^{-zw(s)} u(z) dz \tag{3.3}$$

holds, where $F^{w(p)}(u) \sim \{u_{k(n)}(w(p))\}$.

  Proof The algebraic properties of an endomorphism can be easily verified for any w(p)-refinement-substitution mapping. Let f(x) be a continuous operator function, i.e.: $f(x) = a \cdot f(x,t)$ where $a \in M$ (or $a \in Q$) and f(x,t) is a continuous function on $I * [0,\infty)$ and $F^{w(p)}(\cdot)$ denotes the w(p)-refinement-substitution.

Then,

$$F^{w(p)}(f(x)) = F^{w(p)}(a) F^{w(p)}[\{f(x,t)\}].$$

Therefore, it is sufficient to show that $F^{w(p)}\{f(x,t)\}$ is a continuous operator function. Let $*f_n(p,x)$ be defined as in isomorphic embedding of C into R; i.e.:

$$*f_n(p,x) = \int_0^n \exp(-pt) f(x,t) dt, \quad n = 0,1,2,\ldots,$$

then,

$$\frac{1}{2\pi i} \int_{c-i\infty}^{c+i\infty} \exp(pt) *f_n(w(p),x) p^{-2} dp$$

defines a continuous operator function on $[0,n]$. Taking into consideration the theorem of section 2 and Gesztelyi's representation theory, we obtain the results claimed in the theorem. □

4. EXAMPLES

1. $w(p) = p + c$, c is a number. In this case $F^{w(p)}$ coincides with the wellknown transformation $T^{-c}(\cdot)$.

2. $w(p) = p + \frac{1}{p}$. $F^{w(p)}$ is a new transformation. The operator function $\ell(\exp(-z(s+1)))$ corresponds to $p^{-1}\exp(-z(p+1/p))$ and it is the operator representation of

$$g(z,t) = \begin{cases} 0 & \text{if } 0 < t < z \\ I_0(2(z(t-z))^{0.5}) & \text{if } 0 < z < t. \end{cases}$$

Therefore, the formula for f from C is

$$F^{w(p)}(f) = \int_0^\infty e^{-z(s+1/s)} f(z) dz;$$

3. $r(p) = 0.5(p + [p^2-4]^{0.5})$ where the main branch of the root is taken. One can easily check that $r(p)$ satisfies the conditions required. $F^{r(p)}(\cdot)$ is the inverse transformation of $F^{p+1/p}$, that can be verified by the relation

$$F^{r(p)}(F^{p+1/p}(x)) = F^{p+1/p}(F^{r(p)}(x)) = x.$$

Nevertheless, $F^{r(p)}$ is a transformation of refinement-substitution with $k(n) = 2n$. For the integral representation one can use the theory of special functions.

4. $w(p) = cp$, $c > 0$. $w(p)$ generates the wellknown operator transformation $U_{1/c}(\cdot)$.

REFERENCES

1. E. Gesztelyi, Über das Stieltjes-Integral von Operatorenfunktionen, II., Publ. Math. Debrecen, 13, 313-324 (1966).
2. E. Gesztelyi, Über lineare Operatoren Transformationen, Publ. Math. Debrecen, 14, 169-206 (1967).
3. A. Bleyer, On continuous endomorphism of Mikusinski's operator field, Acta Math. Acad. Sci. Hungary, 21, 393-402 (1970).
4. A. Bleyer, On integral representation of the endomorphism of Mikusiński's operator field, Proc. Edinburgh Math. Soc., 17, Sre. II, 4, 351-367 (1971).
5. A. Bleyer, Extension of linear operator transformations, Acta Math. Sci. Hungary, 26, 3-4, 413-417 (1975).

6. A. Bleyer, A note to the construction of continuous operator transformations, Seminar report, Univ. of Florence, Facoltá d'Ingegneira, (1973).
7. A. Bleyer, A note on the general theory of operator transformations, *Comptes rendus de l'Academie Bulgare des Sciences*, 28, No. 6, 731-733 (1975).
8. A. Bleyer, On semi-groups of operator transformations, *Periodica Polytechnica*, Bp. 21, No. 2, 175-196 (1977).
9. A. Bleyer, Examples, counter-examples and applications to the theory of operator transformations, *Periodica Polytecnica*, Bp. 21, No. 2, 164-174 (1977).
10. V. A. Ditkin and A. P. Prudnikov, "Integral transforms and operational calculus", Pergamon Press, New York (1965).
11. J. Mikusiński, "Operational calculus", Pergamon Press, New York (1959).
12. A. Erdélyi, "Operational calculus and generalized functions, Holt, Rinehart and Winston, New York (1962).
13. K. Urbanik, Sur la structure non-topologique du corps des operateurs, *Studia Math.*, 14, 243-246 (1954).
14. G. Krabbe, Ratios of Laplace-Transforms, Mikusiński Operational Calculus, *Math. Annalen*, 162, 237-245 (1962).
15. T. K. Boehme, On Mikusinki operators, *Studia M.*, XXXIII, 127-140 (1969).
16. T. K. Boehme, On sequences of continuous functions and convolution, *Studia M.* XXV, 333-347 (1965).
17. T. K. Boehme, The Mikusiński operators as a toplogical space, *Am. J. Math.*, 98, 55-66 (1974).
18. J. Burzyk, On convergence in the Mikusiński operational calculus, Thesis, Polish Ac. Sc. Katowice (1981).
19. L. Berg, "Einführung in die Operatorenrechnung, VEB Berlin (1965).

SOME COMMENTS ON THE BURZYK-PALEY-WIENER THEOREM FOR REGULAR OPERATORS

Thomas K. Boehme

University of California

Santa Barbara, CA 93106, USA

0. INTRODUCTION

The proof of the Paley-Wiener type theorem of Burzyk was outlined by J. Burzyk in his talk during this conference. A full proof is to appear in print soon. The following are some comments concerning his theorem. The theorem can be posed as pertaining to regular operators or as pertaining to regular quotients on the whole line (the Boehmian's of P. and J. Mikusiński) since those with compact support coincide.

We give a brief sketch of the development of regular operators, some indications of what is used in the proof of the theorem, and point out some questions connected with regular operators which are connected to the theorem.

1. PRELIMINARIES ON REGULAR OPERATORS

A Mikusiński operator $X \in M$ is a convolution quotient, $X = f/g$, where f and g are continuous functions on $(-\infty,\infty)$, $g \neq 0$, and f and g have left bounded support. The algebra of regular operators are a subclass of the Mikusiński field, M, of convolution quotients.

By an <u>approximate identity</u> we shall mean a sequence $(\phi_n \mid n = 1,2,...)$ where for each n, $\phi_n \in C(-\infty,\infty)$,

(i) $\int \phi_n = 1$,

(ii) there is an $M > 0$ such that for all $n = 1,2,...$ $\int |\phi_n| dx < M$,

(iii) the Support of $\phi \subset [-\varepsilon_n, \varepsilon_n]$, and $\varepsilon_n \to 0$ as $n \to \infty$.

An operator is said to be <u>regular</u>, $X \in M_R$, if there is an approximate identity $(\phi_n \mid n = 1,2,\ldots)$ such that $X = f_n/\phi_n$, for each $n = 1,2,\ldots$ where the f_n are continuous functions with left bounded support.

A regular operator X is said to be zero on an open set, Ω, if $X = f_n/g_n$, $n = 1,2,\ldots$ where $(\phi_n \mid n = 1,2,\ldots)$ is an approximate identity and the functions $f_n \to 0$ uniformly on compact subsets of Ω. For each $X \in M_R$ there is a largest open set on which $X = 0$. The complement of the largest open set on which $X = 0$ is called the <u>support</u> of X. A regular operator, X, has compact support if the support of X is a compact set.

It is a theorem that a regular operator has compact support if and only if for some approximate identity $(\phi_n \mid n = 1,2,\ldots)$, $X = f_n/\phi_n$, where each f_n has compact support.

Each Distribution with left bounded support (in particular those with compact support $T \in E'$) is identified with a unique regular operator, X_T, by the identification

$$X_T = T* \phi_n/\phi_n \qquad n = 1,2,\ldots$$

where $(\phi_n \mid n = 1,2,\ldots)$ is any approximate identity which is made up of C^∞ functions. The support of T as a distribution, and the support of X_T as a regular operator are the same. There are regular operators which are not distributions.

If $X \in M_R$ has compact support one defines the Fourier-Borel transform $\hat{X}(z)$ of X bu the equation

$$\hat{\phi}_n(z)\hat{X}(z) = \hat{f}_n(z) \qquad n = 1,2,\ldots$$

where $\hat{f}_n$ and $\hat{\phi}_n$ are the Fourier transforms of f_n and ϕ_n. Since $(\phi_n \mid n = 1,2,\ldots)$ is an approximate identity, for any complex number z the numbers $\hat{\phi}_n(z) \to 1$ as $n \to \infty$. Thus $\hat{X}(z)$ is well defined, and it is clear that $X(z)$ is an entire analytic function of z.

2. THE THEOREM

<u>Theorem</u> (Paley-Wiener Theorem for Regular Operators). An entire analytic function, $F(z)$, is the Fourier transform of a Regular Operator with support on $[-a,a]$ if and only if $F(z)$ is an entire function of type less than or equal to a and

$$\int_{-\infty}^{\infty} (\ln_+|F(x)|)/(1 + x^2) \, dx < \infty.$$

(we use the notation $\ln_+ x = \text{Max}(\ln x, 0)$).

(This class of analytic functions is known as the Cartwright class of order a.)

Discussion of the proof. The primary tool used in showing that the Fourier transform of a regular operator with compact support is in the Cartwright class is Jensen's theorem from analytic function theory.

To prove the converse requires a deep theorem of Buerling and Mallivin (Acta Math Scand. 1962). Their theorem says that an entire function F is the quotient of two Fourier Transforms of measures with compact support if and only if F is in a Cartwright class and furthermore, that the support of the denominator can be taken arbitrarily small. Thus $F = \hat{f}_n/\hat{\phi}_n$, where $(\phi_n \mid n = 1,2,\ldots)$ is such that diam(Supp ϕ_n) tends to zero as $n \to \infty$. The problem is then to show that $(\phi_n \mid n = 1,2,\ldots)$ can be taken as an approximate identity. It clear that one can assume $\phi_n \in C(-\infty,\infty)$ and $\int \phi_n = 1$ for each n. The difficulty is to show that one can suppose $\int |\phi_n| dx$ is bounded as $n \to \infty$.

Burzyk first attacks the case where F is a Cartwright function of type 0, and then uses this result to show the general case.

3. PERIODIC CONVOLUTION QUOTIENT ON THE REAL LINE

It is not quiet clear what one might mean by "periodic" convolution quotients. There are two possibilities.

One possibility in the class of quotients discussed by D. Nemzer (to appear, Rocky Mtn. Journal of Math.). Nemzer considered the regular quotients on compact groups. If Γ is the unit circle, he shows that the Fourier coefficinets of such a quotient must grow no faster than $\exp(o(|n|))$, while any sequence c_n which is such that for some $A > 0$ and for some γ, $0 \le \gamma < 1$, $|c_n|/\exp(A|n|^\gamma)$ is bounded is the set of Fourier coefficients for a Boehmian on Γ.

A second possibility for "periodic" regular quotients is the sum of trigonometric series. Here the situation is not so simple. Piotr Mikusiński discussed the interesting sequence $f_n(t) = \exp(i2^n t)$, (Thesis, Katowice, 1982). If $\phi_k(t)$ is 2^k times the characteristic function of the interval $(0, 2^{-k})$ then the convolution $\phi_k * f_n(t) = \int_{-\infty}^{\infty} \phi_k(u) f_n(t-u) \, du \equiv 0$ for all $n > k$. Thus given this approximate identity $(\phi_k \mid k = 1,2,\ldots)$, let a_n be an arbitrary sequence of complex numbers. Let $S_n = \sum_{-n}^{n} a_j f_j$ be the partial sums of the series $S = \sum_{-\infty}^{\infty} a_j f_j$. Then

$$\phi_k * S = \lim_{n \to \infty} \phi_k * S_n = \lim_{n \to \infty} \phi_k * \sum_{-n}^{n} a_j f_j = \phi_k * \sum_{-k}^{k} a_j f_j = g_k \in C(-\infty,\infty)$$

for each $k = 1, 2, \ldots$. Thus the partial sums $S_n = g_{n,k}/\phi_k$ where $g_{n,k} \to g_k$ as $n \to \infty$ for each k. We have $S = g_k/\phi_k$ for each k. Therefore the trigonometric series $\Sigma\, a_n \exp(i 2^n t)$ is convergent in space of regular quotients on the real line for arbitrarily chosen coefficients a_n. Hence, there is no order of growth condition possible for the Fourier coefficients of a regular convolution quotient which is periodic in this sense.

TWO THEOREMS ON THE DIFFERENTIATION OF REGULAR CONVOLUTION QUOTIENTS

Thomas K. Boehme

University of California

Santa Barbara, CA 93106, USA

0. INTRODUCTION

We shall discuss two theorems on the derivatives of generalized functions. The class of generalized functions defined below as regular convolution quotients is a generalization of distributions and is also a generalization of the regular Mikusiński operators. Moreover, is a subclass of the quotients defined by J. and P. Mikusiński (Quotients de suites et leurs applications dans l'analyse fonctionnelle, Comptes Rendus, 239, série I (1981)). It is a subclass with some local properties, and we discuss some of these local properties. This subclass has been investigated by Piotr Mikusiński (Convergence of Boehmians, Japan. J. Math., Vol 9 (1983) and Boehmians as generalized functions, to appear Japan. J. Math.).

The proofs of many of the things we discuss for regular convolution quotients is similar to the proofs of the analogous properties for regular Mikusiński operators. These latter are given in forthcoming book by Mikusiński and Boehme, Operational Calculus, Vol. II, 2nd edition, Polish Scientific Publishers, (1987), and therefore we omit proofs here.

1. PRELIMINARIES ON REGULAR CONVOLUTION QUOTIENTS

By an <u>approximate identity</u> we shall mean a sequence $(\phi_n \mid n = 1, 2, \ldots)$ where for each n, $\phi_n \in L(-\infty, \infty)$,

  i) $\int \phi_n = 1$,

  ii) there is an $M > 0$ such that for all $n = 1, 2, \ldots$ $\int |\phi_n| dx < M$,

  iii) each function ϕ_n vanishes outside an interval $[-\varepsilon_n, \varepsilon_n]$ where

$\varepsilon_n \to 0$ as $n \to \infty$.

Let Δ = the set of all approximate identities. A defining sequence for a regular convolution quotient is a sequence of pairs $((f_n, \phi_n) \mid n = 1, 2, \ldots)$ where $f_n \in L_{loc}(-\infty, \infty)$, $(\phi_n \mid n = 1, 2, \ldots)$ is an approximate identity, and for all n,m the following convolution products are equal:

iv) $f_n * \phi_m = f_m * \phi_n$.

Two defining sequences (f_n, ϕ_n) and (g_m, ψ_m) are said to be __equivalent__ if

v) $f_n * \psi_m = g_m * \phi_n$ for all n,m.

A __regular convolution quotient__ X on $(-\infty, \infty)$ is an equivalence class of defining sequences. If X contains the defining sequence (f_n, ϕ_n) for $n = 1, 2, \ldots$ we write

$$X = f_n/\phi_n.$$

The regular convolution quotients are a vector space when the usual multiplication by scalars and addition of fractions is used. If h is a continuous function (or integrable function of distribution) with compact support and $X = f_n/\phi_n$ is a regular convolution quotient then $Y = (h * f_n)/\phi_n$ is a regular convolution quotient which we denote by

$$Y = hX.$$

In this way the regular convolution quotients are a module under convolution over the distributions with compact support. We denote the convolution by juxtaposition.

The regular convolution quotients defined in this way are the same as the class $B(L_{loc}, \Delta)$ defined by J. and P. Mikusiński and we shall stick to their notation.

The regular convolution quotients contain all continuous functions, locally integrable functions, and distributions $\mathcal{D}'$ under the isomorphism $T \in \mathcal{D}' \to (T * \phi_n)/\phi_n \in B(L_{loc}, \Delta)$ where $(\phi_n \mid n = 1, 2, \ldots)$ is any approximate identity made up of C^∞ functions. Moreover, $B(L_{loc}, \Delta)$ contains the class of all regular Mikusiński operators. Both of these containments are proper.

If $X = f_n/\phi_n$ is a regular convolution quotient we say that the approximate identity $(\phi_n \mid n = 1, 2, \ldots)$ is a __regularizing sequence__ for X. For each n we have

$$\phi_n X = f_n$$

(in the sense of the above isomorphism).

If Ω is an open set and $f \in C(\Omega)$ then a regular convolution quotient X is said to be __equal to__ f __on__ Ω if there is a representation $X = f_n/\phi_n$ such that $f_n \to f$ uniformly on compact subsets of Ω. It is a theorem that if this is true for any one representation of X then it is true for all representations of X (for a proof see Boehme, "The support of Mikusiński operators", Tran. Am. Math. Soc. (1972)).

In particular, we say $X = 0$ on Ω if there is a regularizing sequence $(\phi_n \mid n = 1,2,\ldots)$ for X such that $\phi_n X = f_n \to 0$ uniformly on compact subsets of Ω. There is a unique largest open set on which $X = 0$. The complement of this set is called the <u>support</u> of X. The complement of the largest open set on which X is a C^∞ function is called the <u>singular support</u> of X. It is a theorem that $X \in B(L_{loc}, \Delta)$ has compact support if and only if there is a regularizing sequence for X such that $\phi_n X = f_n$ has compact support for each n. Moreover, if X has compact support then this is true for <u>every</u> regularizing sequence for X.

More generally, if $X,Y \in B(L_{loc}, \Delta)$ and Ω is an open set we say that

$$X = Y \text{ on } \Omega \tag{1}$$

if $X - Y = 0$ on Ω.

If $Y \in B(L_{loc}, \Delta)$ and $Y = g_n/\psi_n$ has compact support then one can define the convolution of Y with any $X = f_n/\phi_n \in B(L_{loc}, \Delta)$ by $XY = (f_n * g_n)/(\phi_n * \psi_n) \in B(L_{loc}, \Delta)$. With this definition $B(L_{loc}, \Delta)$ is a module over the regular convolution quotients with compact support under the operation of convolution.

If $(\phi_n \mid n = 1,2,\ldots)$ is any approximate identity then $1 = \phi_n/\phi_n$ is the multiplicative identity in the module. It has support equal to $\{0\}$ since $\phi_n \to 0$ uniformly on each compact subset of $R - \{0\}$. The identity operator is identified with the Dirac delta distribution (under the above isomorphism).

We define the differentiation operator by $D = \phi_n'/\phi_n$ where ϕ_n is any continuously differentiable approximate identity. If $f \in C^1(-\infty,\infty)$ then $Df = f'$ in $B(L_{loc}, \Delta)$ where f' is the classical derivative of f. Let H be the Heaviside function, $H(t) = 1$ for $t > 0$, $H(t) = 0$ for $t < 0$. It is easy to see that the Dirac delta distribution is DH since $(DH - 1) = ((\phi_n' * H) - \phi_n)/\phi_n$ and the numerator is identicaly zero on any compact subset of $R - \{0\}$ for sufficiently large n.

2. THE TWO THEOREMS A AND B

The derivative of an $X \in B(L_{loc}, \Delta)$ is defined to be DX. Thus every regular convolution quotient has a derivative and its derivative is another regular convolution quotient. The connection between this derivative and the classical derivative is given by the following two theorems.

Theorem A Let g be a continuous function on an open set Ω. If X is a regular convolution quotient such that DX = g on Ω then there is a function f which is differentiable in the classical sense on Ω such that for all t ∈ Ω

$$g(t) = f'(t)$$

and in the above defined sense X = f on Ω.

If the open set Ω is the whole real line the result is easily seen to be true. The importance of the theorem is that it can be applied to an arbitrary open set.

In paricular, if DX = 0 on an open connected set then X = const. on that set.

Theorem B Let f be a continuously differentiable function on an open set Ω and suppose X is a regular convolution quotient. If X = f on Ω then DX = f' on Ω where f' is the derivative in the classical sense of the function f.

Rather than prove the theorem here we shall give some examples of the use of the theorems.

3. SOME EXAMPLES

1) Let n be a positive integer. Then the n-th derivative of the operator 1 is $D^n = D^{n+1}H$ where H is the Heaviside function. Since H is constant on the intervals $(-\infty,0)$ and $(0,\infty)$ it follows from theorem B that $D^n = 0$ on $R - \{0\}$. Thus the support and the singular support of D^n is the single point $\{0\}$.

2) Suppose $\beta = n + \alpha$, $0 < \alpha < 1$, and n is a positive integer. Let $H^{1-\alpha} = (\Gamma(1-\alpha)t^\alpha)^{-1}$ for $t > 0$ and $H^{1-\alpha}(t) = 0$ for $t < 0$ (here Γ denotes Euler's gamma function). Since the n+1 derivative of $t^{-\alpha}$ is $(-1)^{n+1}(\alpha)(\alpha+1)(\alpha+2) \ldots (\alpha+n)/t^{\alpha+n+1}$ we have <u>the restriction of D^β to the interval $(0,\infty)$ in the sense of regular convolution quotients is</u>

$$D^\beta(t) = 1/(\Gamma(-\beta)t^{\beta-1}) \text{ on } t > 0 \quad (\beta \neq 0,1,2,\ldots)$$

<u>and</u>

$$D^\beta(t) = 0 \text{ on } t < 0.$$

3) Finally, with an appropriate notion of convergence in $B(L_{loc},\Delta)$ one can show that if Y is defined by the following series

$$Y = H + \sum_{n=1}^{\infty} (-1)^n D^n \{H(t) \log t\}/n!(n-1)!$$

then Y is such that

$$Y(t) = \exp(1/t) \text{ on } t > 0$$

and

$$Y(t) = 0 \text{ on } t < 0.$$

VALUES ON THE TOPOLOGICAL BOUNDARY OF TUBES*

Richard D. Carmichael

Department of Mathematics
Wake Forest University
Winston-Salem, NC 27109, U.S.A.

ABSTRACT

Holomorphic functions in tube domains in $\mathbb{C}^n$ which generalize the Hardy H^p functions are shown to have boundary values on the topological boundary of the tube. The boundary values are obtained restrictedly or unrestrictedly in L^p and in the distribution space K'_1.

1. INTRODUCTION

Let B denote a proper open subset of $\mathbb{R}^n$. Let $0 < p < \infty$ and $A \geq 0$. Let $d(y)$ denote the distance from $y \in B$ to the complement of B in $\mathbb{R}^n$. The space $S_A^p(T^B)$, $T^B = \mathbb{R}^n + iB$, is the set of all functions $f(z)$, $z = x + iy \in T^B$, which are holomorphic in T^B and which satisfy

$$\|f(x + iy)\|_{L^p} \leq M (1 + (d(y))^{-r})^s \exp(2\pi A|y|), \quad y \in B, \qquad (1.1)$$

for some constants $r = r(f,p,A) \geq 0$, $s = s(f,p,A) \geq 0$, and $M = M(f,p,A,r,s)$ which are independent of $y \in B$. The $S_A^p(T^B)$ functions generalize the Hardy H^p functions [9, pp. 90-91] and those functions considered by Vladimirov in [10]. A discussion of our motivation in considering the $S_A^p(T^B)$ functions and of the special cases is contained in [1, Introduction]. In [1 - 6] we have studied the $S_A^p(T^B)$ functions; we have obtained Fourier-Laplace, Cauchy, and Poisson integral representations of functions in $S_A^p(T^B)$ for values of p and for various bases B. We have characterized H^p as a subspace of

---

* This paper is dedicated to V. S. Vladimirov.

$S_A^p(T^B)$ and have obtained holomorphic extension theorems. For $B = C$, a cone, we have proved the existence of distributional boundary values in S' on the distinguished boundary $\mathbb{R}^n + i\bar{0}$ of the tube T^C, that is as $y \to \bar{0}$, $y \in C$.

In this note we investigate boundary value properties on the topological boundary of the tube T^B of functions in a subspace of $S_A^p(T^B)$ which we have considered in [8]. All definitions and terminology concerning cones in $\mathbb{R}^n$ will be that of Vladimirov ([11], [12]); also see [7, p. 1042]. Throughout the paper $\bar{0} = (0,0,\ldots,0)$ will denote the origin in $\mathbb{R}^n$; if B is a subset of $\mathbb{R}^n$, $\bar{B}$ will denote the closure of B in $\mathbb{R}^n$.

2. VALUES ON THE TOPOLOGICAL BOUNDARY

For a base B of the tube T^B, let ∂B denote the topological boundary of B. Prior to this point in our research we have not obtained any boundary value information as $y = \text{Im}(z) \to y_0 \in \partial B$, $z = x + iy \in T^B$, for $S_A^p(T^B)$ or any of its subspaces that generalize H^p where y_0 is an arbitrary point on ∂B. In this paper we show that a subspace of $S_A^p(T^B)$ which generalises H^p does have boundary value properties on the topological boundary $\mathbb{R}^n + i\partial B$ of the tube T^B.

If either $r = 0$ or $s = 0$ in (1.1), $S_A^p(T^B)$ becomes the space $V_A^p(T^B)$ that we have considered in [8]. That is $V_A^p(T^B)$, $0 < p < \infty$, $A \geq 0$, is the set of all holomorphic functions in T^B which satisfy

$$\|f(x + iy)\|_{L^p} \leq M \exp(2\pi A|y|), \quad y \in B, \tag{2.1}$$

where $M = M(f,A)$ is independent of $y \in B$. The following theorem concerning $V_A^p(T^B)$ is basic for obtaining a well defined boundary value on $\mathbb{R}^n + i\partial B$ in our results here.

Theorem 2.1 Let B be a proper open connected subset of $\mathbb{R}^n$. Let $f(z) \in V_A^p(T^B)$, $1 < p \leq 2$, $A \geq 0$. There exists a measurable function $g(t)$, $t \in \mathbb{R}^n$, such that

$$\|e^{-2\pi \langle y,t \rangle} g(t)\|_{L^q} \leq M e^{2\pi A|y|}, \quad y \in \bar{B}, \tag{2.2}$$

$1/p + 1/q = 1$, where $M = M(f,A)$ is independent of $y \in \bar{B}$; and

$$f(z) = \int_{\mathbb{R}^n} g(t) e^{2\pi i \langle z,t \rangle} dt, \quad z \in T^B. \tag{2.3}$$

Proof Except for (2.2) holding for $y \in \partial B$ the proof is by [8, Theorem 4.1]. Let y be any point on ∂B. Choose a sequence of points $\{y_j\}_{j=1}^{\infty}$ in B such that $y_j \to y$ as $j \to \infty$. By Fatou's lemma and (2.2) for $y_j \in B$, $j = 1, 2, \ldots$, we have

$$\int_{\mathbb{R}^n} |e^{-2\pi \langle y, t \rangle} g(t)|^q \, dt \leq \liminf_{j \to \infty} \int_{\mathbb{R}^n} |\exp(-2\pi \langle y_j, t \rangle) g(t)|^q \, dt$$

$$\leq \liminf_{j \to \infty} M^q \exp(2\pi A q |y_j|) = M^q \exp(2\pi A q |y|), \tag{2.4}$$

and (2.2) is obtained for $y \in \partial B$ also. The proof is complete.

The growth (1.1) does not allow the proof of Theorem 2.1 to yield information similar to (2.2) for $y \in \partial B$ for general $S_A^P(T^B)$.

Corollary 2.1 Let the hypotheses of Theorem 2.1 be satisfied with the additional assumption that $\bar{0} \in \partial B$. The function $g(t)$ in the representation (2.3) satisfies $g(t) \in L^q$, $1/p + 1/q = 1$.

Proof Apply (2.2) with $y = \bar{0}$.

Theorem 2.1 and [8, Theorem 5.1] show that $f(z) \in V_A^2(T^B)$ if and only if there exists a measurable function $g(t)$ such that (2.2) holds for $p = q = 2$ and (2.3) holds.

Recall the distributions of exponential growth K_1' and its Fourier transform space $\mathcal{K}_1' = F[K_1']$ as defined in [7, section 4]. Using Corollary 2.1 we can show that for any $y_0 \in \partial B$, $f(x + iy) \to F[\exp(-2\pi \langle y_0, t \rangle) g(t)]$ as $y \to y_0$, $y \in B$, in the strong topology of $\mathcal{K}_1'$ for any element $f(z) \in V_A^P(T^B)$, $1 < p \leq 2$, $A \geq 0$, where B is a proper open connected subset of $\mathbb{R}^n$ such that $\bar{0} \in \partial B$. We summarize this and other distributional boundary value properties in section 3 below. In the remainder of this section we prove L^2 boundary value properties using Theorem 2.1.

Following [9, p. 97] we define an open polyhedron in $\mathbb{R}^n$ to be the interior of the convex hull of a finite subset of $\mathbb{R}^n$.

Theorem 2.2 Let B be a proper open connected subset of $\mathbb{R}^n$. Let $f(z) \in V_A^2(T^B)$, $A \geq 0$. Let $y_0 \in \partial B$ such that $y_0 \in \partial P$ where P is an open polyhedron contained in B defined by the points $y_j \in \bar{B}$, $j = 1, 2, \ldots, k$. We have

$$\lim_{\substack{y \to y_0 \\ y \in P}} f(x + iy) = F[\exp(-2\pi \langle y_0, t \rangle) g(t); x] \tag{2.5}$$

in L^2 where $g(t)$ is the function in (2.3) and the Fourier transform in (2.5) is in L^2.

Proof For each $y_0 \in \partial B$, $(\exp(-2\pi\langle y_0,t\rangle)g(t)) \in L^2$ by Theorem 2.1; the Fourier transform of the right side of (2.5) exists in L^2. From the proof of Theorem 2.1 (i.e. [8, Theorems 4.1 and 2.1]) we know that $(\exp(-2\pi\langle y,t\rangle)g(t)) \in L^1 \cap L^2$, $y \in B$, and $f(x+iy) = F[\exp(-2\pi\langle y,t\rangle)g(t);x]$, $z = x + iy \in T^B$, with the Fourier transform being interpreted in either the L^1 or L^2 sense. By the Parseval equality

$$\|f(x+iy) - F[\exp(-2\pi\langle y_0,t\rangle)g(t);x]\|_{L^2} =$$
(2.6)
$$\|(\exp(-2\pi\langle y,t\rangle)g(t)) - (\exp(-2\pi\langle y_0,t\rangle)g(t))\|_{L^2}, \quad y \in B, \; y_0 \in \partial B.$$

Now consider $y \in \bar{P}$, the convex hull of the points $y_j \in \bar{B}$, $j = 1,2,\ldots,k$. Any $y \in \bar{P}$ is of the form $y = a_1 y_1 + \ldots + a_k y_k$ where $0 \le a_j \le 1$, $j = 1,2,\ldots,k$, and $a_1 + \ldots + a_k = 1$. Now put

$$G(t) = \sum_{j=1}^{k} (\exp(-4\pi\langle y_j,t\rangle)|g(t)|^2). \tag{2.7}$$

We have $(\exp(-2\pi\langle y_j,t\rangle)g(t)) \in L^2$, $j = 1,2,\ldots,k$, by Theorem 2.1; hence $G(t) \in L^1$. $G(t)$ is independent of $y \in P$, and for any $y \in \bar{P}$ we have

$$\exp(-4\pi\langle y,t\rangle) = \prod_{j=1}^{k} (\exp(-4\pi\langle y_j,t\rangle))^{a_j} \le \sum_{j=1}^{k} a_j \exp(-4\pi\langle y_j,t\rangle). \tag{2.8}$$

Hence for any $y \in \bar{P}$ and $t \in \mathbb{R}^n$

$$\exp(-4\pi\langle y,t\rangle)|g(t)|^2 \le \sum_{j=1}^{k}(a_j \exp(-4\pi\langle y_j,t\rangle)|g(t)|^2) \le G(t). \tag{2.9}$$

For $y \in P$ and $y_0 \in \partial P$, (2.9) yields

$$|(\exp(-2\pi\langle y,t\rangle)g(t)) - (\exp(-2\pi\langle y_0,t\rangle)g(t))|^2$$
$$\le 2^2(|\exp(-2\pi\langle y,t\rangle)g(t)|^2 + |\exp(-2\pi\langle y_0,t\rangle)g(t)|^2) \tag{2.10}$$
$$\le 4(G(t) + G(t)) = 8G(t)$$

for all $t \in \mathbb{R}^n$, and $(8G(t)) \in L^1$ independently of $y \in P$. By (2.10), the Lebesgue dominated convergence theorem, and (2.6), the convergence (2.5) is proved.

Following [9, p. 99], if $f(z)$, $z = x + iy \in T^B$, attains a boundary value in a topology as $y \to y_0$, $y \in B$, $y_0 \in \partial B$, we say that the boundary value is attained unrestrictedly. If the boundary value is attained as $y \to y_0$, $y_0 \in \partial B$, inside a polyhedron in B with y_0 on the boundary of the polyhedron, we say that the boundary value is attained restrictedly. (Restricted boundary values are best possible for some bases B and some points on the boundary of the bases for some elements of $H^2(T^B) \subset V_A^2(T^B)$; see the example [9, pp. 95-97].)

Theorem 2.2 yields a direct extension of [9, Corollary 2.9, p. 97] to $V_A^2(T^B)$ functions; let P be an open polyhedron in $\mathbb{R}^n$ and $f(z) \in V_A^2(T^P)$. Extend $f(z)$ to $T^{\bar{P}}$ by putting $f(z) = F[\exp(-2\pi\langle y,t\rangle)g(t);x]$, $z = x + iy$, $x \in \mathbb{R}^n$, $y \in \partial P$, where (2.3) holds for $z \in T^P$. The mapping $y \to f(z)$ from $\bar{P}$ to $L^2(\mathbb{R}^n)$ is continuous.

If B is a proper open connected subset of $\mathbb{R}^n$, $\bar{0} \in \partial B$, and $\bar{0} \in \partial P$ for some open polyhedron P contained in B then Corollary 2.1 and Theorem 2.2 combine to prove

$$\lim_{\substack{y \to \bar{0} \\ y \in P}} f(x + iy) = F[g(t);x] \qquad (2.11)$$

in L^2 for $f(z) \in V_A^2(T^B)$.

In the remainder of this section we let the base B be an open convex cone C in $\mathbb{R}^n$. As usual the restriction of the base to a cone will give more detailed information concerning the boundary value properties of elements in $V_A^2(T^C)$. For an open convex cone C in $\mathbb{R}^n$, every $y_0 \in \partial C$ is on the boundary of some open polyhedron $P \subset C$. Thus we immediately have by Theorem 2.2 and Corollary 2.1 that $f(z) \in V_A^2(T^C)$, $A \geq 0$, satisfies (2.5) in L^2 restrictedly at every point $y_0 \in \partial C$, and the function $g(t)$ in (2.5) is in $L^2(\mathbb{R}^n)$.

The following result is a direct analog of [9, Theorem 3.1, p. 101] for the functions $V_A^2(T^C)$ with additional information on the boundary of the base C. Recall the indicatrix function $u_C(t)$ defined in [11, p. 219].

__Theorem 2.3__   Let C be an open convex cone in $\mathbb{R}^n$. $f(z) \in V_A^2(T^C)$, $A \geq 0$, if and only if

$$f(z) = \int_{\mathbb{R}^n} g(t) e^{2\pi i\langle z,t\rangle} dt, \quad z \in T^C, \qquad (2.12)$$

where $g(t) \in L^2$ and $\mathrm{supp}(g) \subseteq \{t : u_C(t) \leq A\}$ almost everywhere. In both directions we further conclude

$$\| e^{-2\pi\langle y,t\rangle} g(t) \|_{L^2} \leq M e^{2\pi A|y|}, \quad y \in \bar{C}. \qquad (2.13)$$

Proof For $f(z) \in V_A^2(T^C)$, [8, Corollary 4.1] yields a function $g(t) \in L^2$ with $\mathrm{supp}(g) \subseteq \{t : u_C(t) \leq A\}$ almost everywhere such that (2.12) holds and (2.13) holds for $y \in C$. The proof of Theorem 2.1 now yields (2.13) holding for $y \in \partial C$ also.

For the converse let $g(t) \in L^2$ with $\mathrm{supp}(g) \subseteq \{t : u_C(t) \leq A\}$ almost everywhere. By [12, Lemma 3, p. 74] we have $\{t : u_C(t) \leq A\} = C^* + \overline{N(\bar{0};A)}$ where $C^* = \{t : \langle y,t \rangle \geq 0 \text{ for all } y \in C\}$ is the dual cone of C and $\overline{N(\bar{0};A)}$ is the closure in $\mathbb{R}^n$ of the open ball $N(\bar{0};A)$ about $\bar{0}$ of radius A. Thus $t \in \{t : u_C(t) \leq A\}$ implies $t = t_1 + t_2$ where $t_1 \in C^*$ and $t_2 \in \overline{N(\bar{0};A)}$. Let $y \in C$ be arbitrary. We have

$$\int_{\mathbb{R}^n} |e^{-2\pi\langle y,t\rangle} g(t)|^2 \, dt = \int_{C^*+\overline{N(\bar{0};A)}} |e^{-2\pi\langle y,t\rangle} g(t)|^2 \, dt \qquad (2.14)$$

$$\leq K \sup_{t \in C^*+\overline{N(\bar{0};A)}} e^{-4\pi\langle y,t\rangle} \leq K \sup_{\substack{t_1 \in C^* \\ t_2 \in \overline{N(\bar{0};A)}}} \exp(-4\pi\langle y, t_1+t_2\rangle)$$

where $K = \|g\|_{L^2}^2$. For $t_2 \in \overline{N(\bar{0};A)}$ we have $|t_2| \leq A$ and

$$\exp(-4\pi\langle y,t_2\rangle) \leq \exp(4\pi|t_2||y|) \leq \exp(4\pi A|y|). \qquad (2.15)$$

For $t_1 \in C^*$ we have $\langle t_1, y\rangle \geq 0$ for all $y \in C$. Thus

$$\exp(-4\pi\langle y,t_1\rangle) \leq 1, \quad y \in C, \quad t_1 \in C^*. \qquad (2.16)$$

Using (2.15) and (2.16) in (2.14) we have

$$\int_{\mathbb{R}^n} |e^{-2\pi\langle y,t\rangle} g(t)|^2 \, dt \leq K e^{4\pi A|y|}, \quad y \in C,$$

which proves (2.13) for $y \in C$. We now form the function $f(z)$ in (2.12); by [8, Corollary 5.1], $f(z) \in V_A^2(T^C)$. The fact that (2.13) holds for $y \in C$ and the proof of Theorem 2.1 now yield that (2.13) holds also for $y \in \partial C$. The proof is complete.

By [8, Corollary 4.1] and the proof of Theorem 2.1, the sufficiency of Theorem 2.3 holds for C being an open connected cone in $\mathbb{R}^n$ for $f(z) \in V_A^p(T^C)$, $1 < p \leq 2$, $A \geq 0$, with (2.13) being replaced by (2.2) for $y \in \bar{C}$ and $1/p + 1/q = 1$.

Let C be an open convex cone in $\mathbb{R}^n$. There exists an open polyhedron

P contained in C such that $\bar{0} \in \partial P$; thus the paragraph containing (2.11) shows that $f(z) \in V_A^2(T^C)$ has a restricted limit at $\bar{0}$ in L^2. We now show as a corollary to Theorem 2.3 that this boundary limit at $\bar{0}$ in fact is attained unrestrictedly for C being an open convex cone. The Fourier transform in the following corollary is the L^2 transform.

Corollary 2.2 Let $f(z) \in V_A^2(T^C)$, $A \geq 0$, where C is an open convex cone in $\mathbb{R}^n$. For the function $g(t)$ obtained in Theorem 2.3 we have $f(x + iy) \to F[g(t);x]$ in L^2 as $y \to \bar{0}$ unrestrictedly in C.

Proof By Theorem 2.3 and Parseval's equality

$$\|f(x + iy) - F[g(t);x]\|_{L^2} = \|(e^{-2\pi\langle y,t\rangle} g(t)) - g(t)\|_{L^2}, \quad y \in C. \quad (2.17)$$

Since $\text{supp}(g) \subseteq \{t : u_C(t) \leq A\}$ almost everywhere we can use analysis obtained by Vladimirov in [11] and [12] and the Lebesgue dominated convergence theorem to show that the right side of (2.17) approaches 0 as $y \to \bar{0}$, $y \in C$. The L^2 boundary value $F[g(t);x]$ is attained independently of how $y \to \bar{0}$, $y \in C$; in particular y is not restricted to arbitrary but fixed compact subcones of C as $y \to \bar{0}$, $y \in C$.

3. DISTRIBUTIONAL BOUNDARY VALUES

Let B be a proper open connected subset of $\mathbb{R}^n$ such that $\bar{0} \in B$. Let $f(z) \in V_A^p(T^B)$, $1 < p \leq 2$, $A \geq 0$, and let $g(t)$ be the function obtained in Theorem 2.1. Let $y_0 \in \partial B$. As noted in section 2 we can show that $f(x + iy) \to F[\exp(-2\pi\langle y_0,t\rangle)g(t)]$ as $y \to y_0$, $y \in B$, unrestrictedly in the strong topology of K_1'.

We can also prove the existence of strong K_1' unrestricted boundary values on the topological boundary of the tube T^C for elements in $V_A^p(T^C)$, $2 < p < \infty$, $A \geq 0$, where C is a polygonal cone or a regular cone. (See [3, section 2] for the terminology of quadrant, quadrant cone, polygonal cone, and regular cone.) The proofs are obtained by sucessively proving the existence of the boundary value for C being a quadrant, a quadrant cone, a polygonal cone, and finally a regular cone with each successive case depending on the preceding one.

The proofs of these distributional boundary value results will appear elsewhere.

4. ACKNOWLEDGEMENT

This material is based upon work supported by the National Science Foundation under Grant No. DMS-8418435.

REFERENCES

1. R. D. Carmichael, Generalization of H^p functions in tubes, I, Complex Variables Theory Appl. 2, 79-101 (1983).
2. R. D. Carmichael, Generalization of H^p functions in tubes, II, Complex Variables Theory Appl. 2, 243-259 (1984).
3. R. D. Carmichael, Boundary values of generalizations of H^p functions in tubes, Complex Variables Theory Appl. (to appear).
4. R.D. Carmichael, Cauchy and Poisson integral representations of generalizations of H^p functions, Complex Variables Theory Appl. 6, 171-188 (1986).
5. R. D. Carmichael, Holomorphic extension of generalizations of H^p functions, Internat. J. Math. Math. Sci. 8, 417-424 (1985).
6. R. D. Carmichael, Holomorphic extension of generalizations of H^p functions, II, Internat. J. Math. Math. Sci. 10, 1-8 (1987).
7. R. D. Carmichael, Analytic functions related to the distributions of exponential growth, SIAM J. Math. Anal. 10, 1041-1068 (1979).
8. R. D. Carmichael and E. K. Hayashi, Analytic functions in tubes which are representable by Fourier-Laplace integrals, Pacific J. Math. 90, 51-61 (1980).
9. E. M. Stein and G. Weiss, "Introduction to Fourier Analysis on Euclidean Spaces", Princeton University Press, Princeton, New Jersey, (1971).
10. V. S. Vladimirov, On Cauchy-Bochner representations, Math. USSR-Izv. 6, 529-535 (1972).
11. V. S. Vladimirov, "Methods of the Theory of Functions of Many Complex Variables", M.I.T. Press, Cambridge, Massachusetts, (1966).
12. V. S. Vladimirov, "Generalized Functions in Mathematical Physics", Mir Publishers, Moscow, (1979).

ABELIAN THEOREM FOR THE DISTRIBUTIONAL STIELTJES TRANSFORMATION

Danica Nikolić-Despotović and Stevan Pilipović

Institute of Mathematics
University of Novi Sad
dr I. Đuričića 4, 21000 Novi Sad, Yugoslavia

ABSTRACT

We study the behaviour of the distributional Stieltjes transformation $(S_r f)(z)$, $z \in \mathbb{C} \setminus \mathbb{R}$, at zero of an $f \in S'$ which has the appropriate quasi-asymptotic behaviour at zero. By using the known results for the asymptotic behaviour at $\pm \infty$, we obtain a final value Abelian theorem for the distributional Stieltjes transformation at zero.

1. NOTIONS AND NOTATION

We shall denote by S the space of rapidly decreasing functions. Its dual S' is the space of tempered distributions. A real valued continuous function L defined on $(0,a)$, $a > 0$, is called slowly varying at 0^+, if for any $\lambda > 0$

$$\lim_{x \to 0^+} \frac{L(\lambda x)}{L(x)} = 1.$$

For the properties of such functions we refer to [5].
The quasiasymptotic behaviour at 0 of tempered distributions was considered in [6]. This definition was reformulated by S. Pilipović [2] in the following way.

Definition 1   Let $f \in S'$ and $c(x)$, $x \in (0,a)$, $a > 0$, be a continuous positive function. It is said that f has the quasiasymptotic at 0 with respect to $c(1/k)$ (in S') if there is $g \in S'$, $g \neq 0$, such that

$$\lim_{k\to\infty} < \frac{f(x/k)}{c(1/k)}, \phi(x) > = < g(x), \phi(x) >, \quad \phi \in S. \tag{1.1}$$

In this case, we write $f \overset{q}{\sim} g$ at 0 with respect to $c(1/k)$ (in S').

<u>Theorem I</u> [2] Let f and c satisfy the conditions of Definition 1. Then, for some $\nu \in \mathbb{R}$ and some slowly varying function L at 0^+

$$c(x) = x^\nu L(x), \quad x \in (0,a).$$

Moreover, g is homogenous with the order of homogenity ν.

Recall, $f_{\nu+1}(x) = (x^\nu / \Gamma(\nu+1))H(x)$, $x \in \mathbb{R}$, for $\nu > -1$ and for $\nu \leq -1$ $f_{\nu+1}(x) = f_{\nu+n+1}^{(n)}(x)$, where n is chosen so that $\nu + n > -1$. H is Heaviside's function.

The quasiasymptotic at 0 is a local property of a distribution. The following theorem is the so-called structural theorem:

<u>Theorem II</u> [2] Let $f \in S'$ and f have the quasiasymptotic behaviour at 0 in S' with respect to $(1/k)^\nu L(1/k)$. If $\nu > 0$ or $\nu < 0$, $\nu \neq -1,-2,\ldots$ and L is bounded in some interval $(0,a)$, $a > 0$, then there are a continuous function F defined on $(-1,1)$, an integer m and $(C_+, C_-) \neq (0,0)$, such that

$$\lim_{x\to\pm 0} \frac{F(x)}{|x|^{\nu+m} L(|x|)} = C_\pm \quad (\nu+m > 0).$$

The Stieltjes transformation of distributions was considered in [1].

<u>Definition 2</u> Let $f \in S'$. We say that $f \in J'(r)$, $r \in \mathbb{R} \setminus (-\mathbb{N})$ if there exist an $m \in \mathbb{N}_0 = \mathbb{N} \setminus \{0\}$ and a locally integrable function F such that
(a) $f = F^{(m)}$; (b) $\int_{-\infty}^{+\infty} |F(x)(x+z)^{-r-m-1}| dx < \infty$ for Im $z \neq 0$. (1.2)

We also need the definition of the space $I'(r)$, $r \in \mathbb{R} \setminus (-\mathbb{N})$.

<u>Definition 3</u> $I'(r)$ is the space of all f for which (a) holds and instead of (b), we suppose that there exist $C = C(F)$ and $\varepsilon = \varepsilon(F)$ such that

$$|F(x)| \leq C(1+|x|)^{r+m-\varepsilon}, \quad x \in \mathbb{R}. \tag{1.3}$$

The Stieltjes transformation S_r of index r, $r \in \mathbb{R} \setminus (-\mathbb{N})$ of a distribution $f \in J'(r)$ with the properties given in (1.2) is a complex valued function given by

$$(S_r f)(z) = (r+1)_m \int_{-\infty}^{+\infty} F(x)(x+z)^{-r-m-1} dx$$

$$= (r+1)_m \langle F(x), (x+z)^{-r-m-1} \rangle \text{ for Im } z \neq 0, \qquad (1.4)$$

where $(a)_n = a(a+1) \ldots (a+n-1)$, $n \in \mathbb{N}$ and $(a)_0 = 1$, $a \in \mathbb{R}$.

It is easy to see that $(S_r f)(z)$ is a holomorphic function of the complex variable z in the domain $\mathbb{C} \setminus (-\infty, +\infty)$.

We shall observe in this paper the function $(S_r f)$ in the upper half plane i.e. Im $z > 0$. This is not a restriction because one can easily show that all the assertions which are to follow hold with Im $z < 0$. We shall need the following theorem from [1].

Theorem A (i) Let $f \in S'$ and $f \overset{q}{\sim} g$ at $\pm\infty$ with respect to $k^\nu L(k)$, $\nu \in \mathbb{R} \setminus (-\mathbb{N})$ and $g(x) = C_+ f_{\nu+1}(x) + C_- f_{\nu+1}(-x)$. Let $r \in \mathbb{R} \setminus (-\mathbb{N})$ and $r > \nu$. Then for any $z \in \mathbb{C} \setminus \mathbb{R}$

$$\lim_{k \to \infty} \frac{(S_r f)(kz)}{k^{\nu-r} L(k)} = \frac{\Gamma(\nu - r)}{\Gamma(r+1)} z^{\nu-r} (C_+ + C_- e^{-\pi i(\nu - 2r - m - 1)})$$

(ii) If $L \equiv 1$, this holds uniformly in any angle of the form $\Lambda_\varepsilon = \{Re^{i\phi}, R > 0, \varepsilon \leq \phi \leq \pi - \varepsilon\}$ $(0 < \varepsilon < \pi/2)$.

We assumed in [1] that $(C_+, C_-) \neq (0,0)$ in Theorem A. This theorem also holds for $(C_+, C_-) = (0,0)$ because one can easily prove that the corresponding structural theorem for $\pm\infty$ given in [3] holds with the assumption $C_+ = C_- = 0$, as well.

2. ABELIAN-TYPE RESULTS

First, we shall prove five lemmas which will be used in the proof of the Abelian theorem.

Lemma 1 If $f \subset I'(r)$, then there is $m_0 \in \mathbb{N}_0$ such that for every $m \geq m_0$ there are locally integrable functions F_m and constants C_m such that

$$F_m^{(m)} = f \text{ and } |F_m(x)| \leq C_m (1 + |x|)^{r+m-\varepsilon}, \; x \in \mathbb{R}. \qquad (2.1)$$

141

Proof Since $f \in I'(r)$, assume that (1.2)(a) and (1.3) hold with $m = m_0$. Let

$$F_{m+1}(x) = \int_0^x F_m(t)dt, \quad x \in \mathbb{R}, \text{ for } m \geq m_0 \quad (m \in \mathbb{N}).$$

We have

$$|F_{m_0+1}(x)| \leq \int_0^x |F_{m_0}(t)|dt \leq C_{m_0}\int_0^x (1+|t|)^{m_0+r-\varepsilon} dt =$$

$$= C_{m_0+1}(1+|x|)^{m_0+1+r-\varepsilon},$$

where $C_{m_0+1} = C_{m_0}/(m_0+r-\varepsilon+1)$. The proof now follows by induction. □

Lemma 2 Let f satisfy the conditions of Theorem II, and let us put

$$F_1(x) = \int_0^x F(t)\,dt, \quad x \in \mathbb{R}.$$

Then, $F_1(x) \sim \bar{C}_{\pm}|x|^{\nu+m+1}L(|x|)$ as $x \to \pm 0$ where $(\bar{C}_+, \bar{C}_-) \neq (0,0)$.

Proof From [5] pp. 65-66, and $x > 0$ we have

$$\frac{x^{\nu+m+1}L(x)}{m+\nu+1} \sim \int_0^x t^{m+\nu}L(t)\,dt, \quad x \to 0^+.$$

By the L'Hospital rule and Theorem II we obtain

$$\lim_{x \to 0^+} \frac{\int_0^x F(t)dt}{\int_0^x t^{m+\nu}L(t)dt} = C_+.$$

For $x < 0$, the same arguments imply the assertion. □

Lemmas 1 and 2 imply that there exist $m_0 \in \mathbb{N}$ and locally integrable functions F_{m_0} and F such that

$$\left.\begin{array}{l} f = F_{m_0}^{(m_0)} \text{ and } |F_{m_0}(x)| \leq C_{m_0}(1+|x|)^{m_0+r-\varepsilon}, \quad x \in \mathbb{R}, \\ \\ f = F^{(m_0)}, \quad x \in (-1,1), \quad F(x) \sim \bar{C}_{\pm}|x|^{\nu+m_0} L(|x|), \quad x \to \pm 0. \end{array}\right\} \quad (2.2)$$

(2.2) implies that $(F_{m_0}(x) - F(x))^{(m_0)} = 0$, $x \in (-1,1)$ and so,

$$F_{m_0}(x) = f(x) + c_1 + \ldots + c_{m_0-1}x^{m_0-1}, \quad x \in (-1,1).$$

Lemma 3 If $f \in I'(r)$, $r > -1$ and f satisfy the conditions of Theorem II, then there exist $m_0 \in \mathbb{N}_0$ and $\tilde{F}$ such that

$$\tilde{F}^{(m_0)}(x) = f(x)$$

and

$$\left.\begin{array}{l} |\tilde{F}(x)| \le C_{m_0}(1 + |x|)^{m_0+r-\varepsilon}, \quad x \in \mathbb{R} \\ \\ \tilde{F}(x) = F(x), \quad x \in (-1,1) . \end{array}\right\} \quad (2.3)$$

Proof If we assume that $r > -1$, we can chose ε in (2.1) such that $m_0 + r - \varepsilon > m_0 - 1$. Let us put

$$\tilde{F}(x) = F_{m_0}(x) - c_1 - c_2 x - \ldots - c_{m_0-1} x^{m_0-1}, \quad x \in \mathbb{R}.$$

We have

$$|\tilde{F}(x)| \le \tilde{C}_{m_0}(1 + |x|)^{m_0+r-\varepsilon}, \quad x \in \mathbb{R},$$

$$\tilde{F}(x) = F(x), \quad x \in (-1,1). \quad \square$$

By the direct computation, one can prove

Lemma 4 If $\phi \in L^1$ then

$$\frac{\phi(kx)}{k^{-1}} \to C\delta, \quad k \to \infty \text{ in } (S^{t_0})' \text{ for some } t_0 \in \mathbb{N}, \text{ where } C = \int_{-\infty}^{+\infty} \phi(t)dt.$$

For a given slowly varying function at ∞ L, we put

$$L^*(x) = \int_a^x \frac{L(t)}{t} dt, \quad x > a > 0,$$

$$L^*(x) = \int_a^{-x} \frac{L(t)}{t} dt, \quad x < -a.$$

Lemma 5 (as in [4])

(i) Let $\phi \in L^1_{loc}$ and $L^*(x) < \infty$, $|x| > a$. If $\lim_{t \to -\infty} \frac{\phi(t)}{|t|^{-1} L(|t|)} = C_\pm$, then

$$\lim_{k \to \infty} \frac{\phi(kx)}{k^{-1}} = C\delta \text{ in } (S^{t_0})' \text{ for some } t_0 \in \mathbb{N}.$$

143

(ii) Let $\phi \in L^1_{loc}$, $L^*(x) \to \infty$ as $x \to \infty$ and

$$\lim_{t \to \pm\infty} \frac{\phi(t)}{|t|^{-1}L(|t|)} = C_\pm, \quad (C_+, C_-) \neq (0,0).$$

Then

$$\lim_{k \to \infty} \frac{\phi(kx)}{|k|^{-1}L(|k|)} = C\delta \quad \text{in } (S^{t_0})' \text{ for some } t_0 \in \mathbb{N}.$$

C in (i) and (ii) can be equal to 0.

Proof (i) holds trivially.

(ii) By the L'Hospital rule we have

$$\frac{\int_0^x \phi(t)\,dt}{L^*(x)} \to C_+ \quad \text{as } x \to +\infty,$$

$$\frac{\int_0^x \phi(t)\,dt}{L^*(x)} \to C_- \quad \text{as } x \to -\infty..$$

So we have

$$\lim_{k \to \infty} \frac{\int_0^{kx} \phi(t)\,dt}{L^*(k)} = C_+ H(x) + C_- H(-x) \quad \text{in } S'.$$

Now, by differentiation we obtain (ii). □

Abelian theorem Let $f \in I'(r)$ and f have (in S') the quasiasymptotic behaviour at zero with respect to $(1/k)^\nu L(1/k)$, $\nu \in \mathbb{R} \setminus (-\mathbb{N})$ and with the limit $C_+ f_{\nu+1} + C_- f_{\nu+1}(-x)$, $(C_+, C_-) \neq (0,0)$. Then

(i) for $r > \nu$, $\displaystyle\lim_{x \to \pm 0} \frac{(S_r f)(x)}{|x|^{\nu-r}L(|x|)} = \frac{\Gamma(r-\nu)}{\Gamma(r+1)}(C_+ + C_- e^{-\pi i(\nu-2r-m-1)})$.

(ii) for $r < \nu$, $\displaystyle\lim_{x \to \pm 0}(S_r f)(x) = C$.

(iii) for $r = \nu$ and $L \equiv 1$,

$$\lim_{x \to \pm 0} \frac{(S_r f)(x)}{|x|^{-1-r}\ln|x|} = C \quad \text{(C in (ii) and (iii) is different).}$$

Proof As in [4], we have

$$(z/k)^{r+m+1}(S_r f)(z/k) = (r+1)_m (S_{r+m}\phi)(k/z). \quad k > 0, \text{ Re } z > 0, \quad (2.4)$$

where

$$\phi(x) = x^{r+m-1} \tilde{F}(1/x), \quad x \neq 0, \quad (\tilde{F} \text{ is from Lemma 3}).$$

The function ϕ has the following properties

$$\phi(k) \sim \tilde{C}_\pm |k|^{r-\nu-1} L(|1/k|), \quad k \to \pm\infty, \quad \text{where } \tilde{C}_\pm = \frac{C_\pm}{\Gamma(m+r+1)}$$

or

$$\phi(k) \sim \tilde{C}_\pm |k|^{r-\nu-1} L_1(|k|), \quad k \to \pm\infty, \tag{2.5}$$

where L_1 is a slowly varying at ∞.

From Lemma 2 it follows that there exists a suitable constant C_1 such that

$$|\phi(x)| \leq C_1 |x|^{\varepsilon-1}(1+|x|)^{r+m-\varepsilon}, \quad x \neq 0. \tag{2.6}$$

(This means that ϕ is a locally integrable on $\mathbb{R}$.)

(i) Assume that $r > \nu$. Since $r-\nu-1 > -1$, (2.5) implies that $\phi \overset{q}{\sim} g$ at $\pm\infty$ with respect to $k^{r-\nu-1} L_1(k)$, where $g(x) = \tilde{\tilde{C}}_+ f_{r-\nu}(x) + \tilde{\tilde{C}}_- f_{r-\nu}(-x)$, $x \in \mathbb{R}$, $\tilde{\tilde{C}}_\pm = \tilde{C}_\pm \Gamma(r-\nu)$. Theorem A implies (i).

(ii) If $r < \nu$, then $r-\nu-1 < -1$. So $\phi \in L^1$ and by Lemma 4, it follows that ϕ has the quasiasymptotic behaviour at $\pm\infty$ with respect to k^{-1} with the limit $C\delta$ and with a suitable C.

By Lemma 4 we can find a sufficiently large m such that $(z+t)^{-r-m} \in S^{t_0}$ and

$$\int_{-\infty}^{+\infty} \phi(t)(z+t)^{-r-m} dt = \langle \phi(t), (z+t)^{-r-m} \rangle$$

in the sense of the dual pair $((S^{t_0})', S^{t_0})$. Again, we can remark that the limits in this lemma can be extended on S^{t_0} for some $t_0 \in \mathbb{N}$. Now, Theorem A completes the proof.

(iii) If $r = \nu$ then Lemma 5(ii) and Theorem A imply the proof. □

REFERENCES

1. D. Nikolić-Despotović, S. Pilipović, The quasiasymptotic of distributions and distributional Stieltjes transformation, Zbornik radova PMF Novi Sad, 16-2, 41-53, (1986).

2. S. Pilipović, On the behaviour of a distribution at 0, <u>Math. Nach.</u> (to appear).
3. S. Pilipović, Some properties of the quasiasymptotic of Schwartz distributions, Part I, Quasiasymptotic at ±∞, <u>Publ. Math. Inst. Beograd</u> (to appear).
4. S. Pilipović, On the behaviour of the distributional Stieltjes transformation at the Origin, <u>Z. Anal. Anwend.</u>, (to appear).
5. E. Seneta, "Regularly varying functions", Lecture Notes in Math., Springer-Verlag, Berlin - Heidelberg - New York (1976), Moscow (1985) (in Russian).
6. V.S. Vladimirov, Yu. N. Drožinov, B. I. Zavialov, "More dimensional Tauberian theorems for generalized functions", Nauka, Moscow (1986) (in Russian).

SOME RESULTS ON THE NEUTRIX CONVOLUTION PRODUCT OF DISTRIBUTIONS

Brian Fisher

Department of Mathematics
The University
Leicester, LE1 7RH England

1. INTRODUCTION

The convolution product of two distributions is normally defined as folows, see Gelfand and Shilov [3].

Definition 1 Let f and g be distributions satisfyng either of the following conditions:
   (a) either f or g has bounded support,
   (b) the supports of f and g are bounded of the same side.
Then the convolution product f*g is defined by

$$((f*g)(x), \phi(x)) = (g(y),(f(x),\phi(x+y)))$$

for arbitrary test function ϕ in the space K of infinitely differentiable functions with compact support.

Note that if f has bounded support then $(f(x),\phi(x+y))$ is in K and so $(g(y),(f(x),\phi(x+y)))$ is meaningful. On the other hand, if g has bounded support or the supports of f and g are bounded on the same side, then the intersection of the supports of $g(y)$ and $(f(x),\phi(x+y))$ is bounded and so $(g(y),(f(x),\phi(x+y)))$ is again meaningful.

It follows that if the convolution product f*g exists by this definition then

$$f*g = g*f, \tag{1}$$

$$(f*g)' = f*g' = f'*g. \tag{2}$$

Now let τ be an infinitely differentiable function satisfying the following conditions:

(i) $\tau(x) = \tau(-x)$,
(ii) $0 \leq \tau(x) \leq 1$,
(iii) $\tau(x) = 1$ for $|x| \leq 1/2$,
(iv) $\tau(x) = 0$ for $|x| \geq 1$.

The function τ_n is now defined by

$$\tau_n(x) = \begin{cases} 1, & |x| \leq n, \\ \tau(n^n x - n^{n+1}), & x > n, \\ \tau(n^n x + n^{n+1}), & x < -n, \end{cases}$$

for $n = 1, 2, \ldots$.

The next definition was introduced in [2] in order to extend the convolution product to a larger class of distribution. In this definition, we denote the convolution product of the distributions f and g by $f * g$ to distinguish it from the convolution product given in definition 1.

<u>Definition 2</u>   Let f and g distributions and let $f_n = f\tau_n$ for $n = 1, 2, \ldots$. Then the convolution product $f \circledast g$ is defined as the neutrix limit of the sequence $\{f_n * g\}$, providing the limit h exists in the sense that

$$\underset{n \to \infty}{N\text{-lim}} (f_n \circledast g, \phi) = (h, \phi)$$

for all test functions ϕ in K, where N is the neutrix, see van der Corput [1], having domain $N' = \{1, 2, \ldots, n, \ldots\}$ and range N'' the real numbers, with negligible functions finite linear sums of the functions

$$n^\lambda \ln^{r-1} n, \quad \ln^r n \quad (\lambda > 0, \; r = 1, 2, \ldots)$$

and all functions $E(n)$ for which $\lim_{n \to \infty} E(n) = 0$.

Note that in this definition the convolution product $f_n * g$ is in the sense of definition 1, the distribution f_n having bounded support since the support of τ_n is contained in the interval $(-n - n^{-n}, n + n^{-n})$.

The following theorem was proved in [2] and shows that definition 2 is a generalization of definition 1.

<u>Theorem 1</u>   Let f and g be distributions satisfying either condition (a) or condition (b) of definition 1. Then the convolution product $f \circledast g$ exists and

$$f \circledast g = f * g.$$

The next two theorems were also proved in [2].

Theorem 2 Let f and g be distributions and suppose that the convolution product $f \circledast g$ exists. Then the convolution product $f \circledast g'$ exists and

$$(f \circledast g)' = f \circledast g'.$$

Theorem 3 The convolution product $x_+^\lambda \circledast x^s$ exists and

$$x_+^\lambda \circledast x^s = 0$$

for $\lambda > -1$ and $s = 0, 1, 2, \ldots$.

2. PREREQUISITES

In the following, $E(n)$ denotes any function such that $\lim_{n \to \infty} E(n) = 0$ and $O(n)$ denotes any function such that $N\text{-}\lim_{n \to \infty} O(n) = 0$. We will therefore write

$$O(n) + O(n) = O(n) + E(n) = O(n).$$

We let $p(x)$ be the polynomial

$$p(x) = \sum_{i=0}^{s} a_i x^i = \sum_{i=0}^{\infty} a_i x^i$$

using the convention that

$$a_i = 0, \; i = s+1, s+2, \ldots .$$

We also note that

$$\tau_n^{(i)}(n) = \begin{cases} 1, & i = 0, \\ 0, & i > 0, \end{cases} \qquad (3)$$

$$\tau_n^{(i)}(n + n^{-n}) = 0, \; i \geq 0. \qquad (4)$$

Lemma 1
$$\int_0^n \ln x \, p(x) \, dx = O(n).$$

Proof

$$\int_0^n \ln x \, p(x) \, dx = \sum_{i=0}^{s} a_i \int_0^n x^i \ln x \, dx = \sum_{i=0}^{s} \left[\frac{n^{i+1} \ln n}{i+1} - \frac{n^{i+1}}{(i+1)^2} \right] = O(n).$$

Lemma 2
$$\int_n^{n+n^{-n}} x^\lambda (\tau_n p)(x)\, dx = E(n)$$

for all λ.

Proof
$$\left| \int_n^{n+n^{-n}} x^\lambda (\tau_n p)(x)\, dx \right|$$
$$\leq (s+1) \max\{|a_i| : 0 \leq i \leq s\} (n + n^{-n})^{s+\lambda} n^{-n} = E(n)$$

and the result of the lemma follows. □

Lemma 3
$$n^{-i}(\tau_n p)^{(r-i-1)}(n) = \frac{(r-1)!}{i!} a_{r-1} + O(n)$$

for $i = 0, 1, \ldots, r-1$ and $r = 1, 2, \ldots$.

Proof Using equation (3) we have
$$n^{-i}(\tau_n p)^{(r-i-1)}(n) = n^{-i} p^{(r-i-1)}(n) = \sum_{j=r-i-1}^{\infty} \frac{j!}{(j-r+i+1)!} a_j n^{j-r+1}$$
$$= \frac{(r-1)!}{i!} a_{r-1} + O(n). \quad \square$$

Lemma 4
$$\int_n^{n+n^{-n}} x^{-i}(\tau_n p)^{(r-1)}(x)\, dx = -(r-1)!\psi(i-1) a_{r-1}$$
$$+ (i-1)! \int_n^{n+n^{-n}} x^{-i}(\tau_n p)^{(r-i)}(x)\, dx + O(n) \tag{5}$$

for $i = 1, \ldots, r$ and $r = 1, 2, \ldots$, where
$$\psi(r) = \begin{cases} 0, & r = 0, \\ \sum_{i=1}^{r} \frac{1}{i}, & r \geq 1. \end{cases}$$

Proof Equation (5) is trivially true when $i = 1$. We will therefore assume that equation (5) holds for some $i < r$. Now
$$\int_n^{n+n^{-n}} x^{-i}(\tau_n p)^{(r-i)}(x)\, dx = \int_n^{n+n^{-n}} x^{-i} d(\tau_n p)^{(r-i-1)}(x) =$$

$$= \left[x^{-i}(\tau_n p)^{(r-i-1)}(x)\right]_n^{n+n^{-n}} + i \int_n^{n+n^{-n}} x^{-i-1}(\tau_n p)^{(r-i-1)}(x)\,dx$$

$$= -\frac{(r-1)!}{1!}\, a_{r-1} + i \int_n^{n+n^{-n}} x^{-i-1}(\tau_n p)^{(r-i-1)}(x)\,dx + O(n),$$

on using lemma 3. It follows on using our assumption that

$$\int_n^{n+n^{-n}} x^{-1}(\tau_n p)^{(r-1)}(x)\,dx = -(r-1)!\,\psi(i-1)\,a_{r-1} +$$

$$+ i! \int_n^{n+n^{-n}} x^{-i-1}(\tau_n p)^{(r-i-1)}(x)\,dx + O(n)$$

$$= -(r-1)\psi(i)\,a_{r-1} + i! \int_n^{n+n^{-n}} x^{-i-1}(\tau_n p)^{(r-i-1)}(x)\,dx + O(n)$$

and equation (5) follows by induction. □

Lemma 5
$$\int_n^{n+n^{-n}} x^{-1}(\tau_n p)^{(r-1)}(x)\,dx = -(r-1)!\,\psi(r-1)\,a_{r-1} + O(n)$$

for $r = 1, 2, \ldots$.

Proof It follows from lemma 4 with $i = r$ that

$$\int_n^{n+n^{-n}} x^{-1}(\tau_n p)^{(r-1)}(x)\,dx = -(r-1)!\,\psi(r-1)\,a_{r-1} +$$

$$+ (r-1)! \int_n^{n+n^{-n}} x^{-r}(\tau_n p)(x)\,dx + O(n) = -(r-1)!\,\psi(r-1)\,a_{r-1} + E(n) + O(n)$$

$$= -(r-1)!\,\psi(r-1)\,a_{r-1} + O(n),$$

on using lemma 2. □

3. RESULTS

We now prove the following extension of theorem 3.

Theorem 4 The convolution product $x_+^{-r} \circledast x^s$ exists and

$$x_+^{-r} \circledast x^s = \begin{cases} \binom{s}{r-1}(-1)^r \psi(r-1) x^{s-r+1}, & s \geq r-1, \\ 0, & s < r-1 \end{cases} \quad (6)$$

for $r = 1, 2, \ldots$ and $s = 0, 1, 2, \ldots$, where $\binom{s}{r}$ denotes the binomial coefficients.

Proof The locally summable $\ln x_+$ is defined by

$$\ln x_+ = \begin{cases} \ln x, & x > 0 \\ 0, & x < 0 \end{cases}$$

and the distribution x_+^{-r} is defined by

$$x_+^{-r} = \frac{(-1)^{r-1}}{(r-1)!} (\ln x_+)^{(r)},$$

for $r = 1, 2, \ldots$.

The convolution product $(x_+^{-r})_n * x^s$, where

$$(x_+^{-r})_n = x_+^{-r} \tau_n(x),$$

exists by definition 1 and so

$$((x_+^{-r})_n * x^s, \phi(x)) = ((y_+^{-r})_n, (x^s, \phi(x+y)))$$

for arbitrary test function ϕ in K. If the support of ϕ is contained in the interval (a,b), we have

$$(x^s, \phi(x+y)) = \int_{a-y}^{b-y} x^s \phi(x+y) dx = \int_a^b (t-y)^s \phi(t) dt = \sum_{i=0}^s a_i y^i = p(y),$$

where

$$a_i = \binom{s}{i}(-1)^i (t^{s-i}, \phi(t)) \quad (7)$$

for $i = 0, 1, \ldots, s$. Thus

$$((x_+^{-r})_n * x^s, \phi(x)) = ((y_+^{-r})_n, p(y)) = (y_+^{-r}, (\tau_n p)(y))$$

$$= \frac{(-1)^{r-1}}{(r-1)!} ((\ln y_+)^{(r)}, (\tau_n p)(y)) = -\frac{1}{(r-1)!} (\ln y_+, (\tau_n p)^{(r)}(y))$$

$$= -\frac{1}{(r-1)!} \int_0^n \ln y \, p^{(r)}(y) dy - \frac{1}{(r-1)!} \int_n^{n+n^{-n}} \ln y \, d(\tau_n p)^{(r-1)}(y)$$

$$= O(n) - \frac{1}{(r-1)!} \left[\ln y (\tau_n p)^{(r-1)}(y) \right]_n^{n+n^{-n}} +$$

152

$$+ \frac{1}{(r-1)!} \int_n^{n+n^{-n}} y^{-1}(\tau_n p)^{(r-1)}(y) dy = -\psi(r-1)a_{r-1} + O(n)$$

$$= \begin{cases} \binom{s}{r-1}(-1)^r \psi(r-1)(x^{s-r+1}, \phi(x)) + O(n), & s \geq r-1, \\ O(n), & s < r-1, \end{cases}$$

on using equations (4) and (7) and lemmas 1, 3 and 5. Taking the neutrix limit as n tends to infinity, the result of the lemma follows. □

Corollary   The convolution products $x_-^{-r} \circledast x^s$ and $x^{-r} \circledast x^s$ exist and

$$x_-^{-r} \circledast x^s = \begin{cases} -\binom{s}{r-1}\psi(r-1)x^{s-r+1}, & s \geq r-1, \\ 0, & s < r-1 \end{cases} \tag{8}$$

$$x^{-r} \circledast x^s = 0, \tag{9}$$

for $r = 1, 2, \ldots$ and $s = 0, 1, 2, \ldots$.

Proof   The distribution x_-^{-r} is defined by the equation

$$x_-^{-r} = (-x)_+^{-r}$$

for $r = 1, 2, \ldots$. Equation (8) follows from equation (6) on replacing x by $-x$.

The distribution x^{-r} is defined by the equation

$$x^{-r} = x_+^{-r} + (-1)^r x_-^{-r}$$

for $r = 1, 2, \ldots$. Equation (9) follows immediately from equations (6) and (8). □

Theorem 8   The convolution product $x_+^{-r} \circledast x_-^s$ exists and

$$x_+^{-r} \circledast x_-^s =$$

$$\begin{cases} \binom{s}{r-1}(-1)^{r+s}[\psi(r-1)x_+^{s-r+1} - \psi(s-r+1)x_+^{s-r+1} + x_+^{s-r+1}\ln x_+], & s \geq r-1, \\ \frac{s!(r-s-2)!}{(r-1)!} x_+^{s-r+1}, & s < r-1, \end{cases}$$

for $r = 1, 2, \ldots$ and $s = 0, 1, 2, \ldots$. In particular (10)

$$x_+^{-r} \circledast x_-^{2r-2} = \binom{2r-2}{r-1}(-1)^r x_+^{r-1} \ln x_+$$

for r = 1, 2, ...

Proof   The convolution product $\ln x_+ \circledast x_+^s$ exists by definition 1 and

$$\ln x_+ * x_+^s = \int_{-\infty}^{\infty} \ln(x-t)_+ t_+^s \, dt = x_+^{s+1} \int_0^1 [\ln x + \ln(1-u)] u^s \, du$$

where t = ux. It follows that

$$\ln x_+ * x_+^s = \frac{1}{s+1} x_+^{s+1} - \frac{\psi(s+1)}{s+1} x_+^{s+1}$$

for s = 0, 1, 2, ... Equation (2) holds and it follows that

$$x_+^{-1} * x_+^s = x_+^s \ln x_+ - \psi(s) x_+^s.$$

It now follows by induction that

$$x_+^{-r} * x_+^s = \binom{s}{r-1}(-1)^r [\psi(s-r+1) x_+^{s-r+1} - x_+^{s-r-1} \ln x_+] \qquad (11)$$

for r = 1, ..., s+1 and s = 0, 1, 2, ...

In particular we have

$$x_+^{-r} * x_+^{r-1} = -(-1)^r \ln x_+$$

and so

$$x_+^{-r} * x_+^{r-2} = -\frac{(-1)^r}{r-1} \ln x_+.$$

It now follows by induction that

$$x_+^{-r} * x_+^s = -\frac{(-1)^s s!(r-s-2)!}{(r-1)!} x_+^{s-r+1} \qquad (12)$$

for s = 0, 1, ..., r-2 and r = 2, 3, ...

Equation (10) now follows from equations (6), (11) and (12) on noting that

$$x_+^{-r} \circledast x^s = x_+^{-r} \circledast [x_+^s + (-1)^s x_-^s] = x_+^{-r} * x_+^s + (-1)^s x_+^{-r} \circledast x_-^s. \quad \square$$

The corollary follows immediately.

Corollary   The convolution products $x_-^{-r} \circledast x_-^s$, $x_-^{-r} \circledast x_+^s$ and $x_+^{-r} \circledast x_-^s$ exist and

$$x_-^{-r} \circledast x_+^s =$$

$$\begin{cases} \binom{s}{r-1}(-1)^{r+s}[(-1)^{r+s+1}\psi(r-1)x^{s-r+1} + \psi(s-r+1)x_-^{s-r+1} + x_-^{s-r+1}\ln x_-], \\ \qquad\qquad\qquad\qquad\qquad\qquad\qquad\qquad\qquad\qquad s \geq r-1, \\ \dfrac{s!(r-s-2)!}{(r-1)!} x_-^{s-r+1}, \qquad\qquad\qquad\qquad s < r-1, \end{cases}$$

$$x^{-r} \circledast x_+^s = \begin{cases} \binom{s}{r-1}(-1)^{r+1}x^{s-r+1}\ln|x|, & s \geq r-1, \\ \dfrac{(-1)^{s+1}s!(r-s-2)!}{(r-1)!} x^{s-r+1}, & s < r-1, \end{cases}$$

$$x^{-r} \circledast x_-^s = \begin{cases} \binom{s}{r-1}(-1)^s x^{s-r+1}\ln|x|, & s \geq r-1, \\ \dfrac{s!(r-s-2)!}{(r-1)!} x^{s-r+1}, & s < r-1, \end{cases}$$

for $r = 1, 2, \ldots$ and $s = 0, 1, 2, \ldots$.

REFERENCES

1. J. G. van der Corput, Introduction to the neutrix calculus, J. Analyse Math., 7, 251-398 (1959/60).
2. B. Fisher, Neutrices and the convolution of distributions, Univ. Novom Sadu, Zb. Rad. Prirod.-Mat. Fak. Ser. Mat., 17, 119-135, (1987).
3. I. M. Gelfand and G. E. Shilov, "Generalized functions", Vol. I, Academic Press (1964).

ON GENERALIZED TRANSCEDENTAL FUNCTIONS AND DISTRIBUTIONAL TRANSFORMS

A. N. Goyal and V. K. Chaturvedi

Department of Mathematics
University of Rajasthan
Jaipur, India

1. INTRODUCTION AND NOTATIONS

During the last two and half decades, Meijer C.S. [1] G-function and its generalization in one variable due to Fox C. [2] by the symbol H in two variables by Agarwal R.P. [3], Sharma B.L. [4], Mourya D.P. [5] in n-variables by Khadiya S.S. and Goyal A.N. [6] and their respective representations in H-symbol of two and n-variables have given great impetus to researches in special functions. However, J. Gopal Krishana and Muhammed Ghouse [7]; Buschman R.G. [8]; Tandon O.P. [9] have raised certain questions regarding the path of integration and existence in the case of G and H functions of two variables. During the write up of the present paper we shall use the following notations

$$A_{p,q}^{m,n}\left[x\left|\begin{matrix}((a_p,\alpha_p))\\((b_q,\beta_q))\end{matrix}\right.\right] = (2\pi\sqrt{-1})^{-1} \int_L \frac{\prod_{j=1}^{m}\Gamma(a_j-\alpha_j s)\ \prod_{j=1}^{n}\Gamma(1-b_j+\beta_j s)}{\prod_{j=1+m}^{p}\Gamma(1-a_j+\alpha_j s)\ \prod_{j=1+n}^{q}\Gamma(b_j-\beta_j s)} x^{-s}\, ds$$

(1.1)

$$A_{p+1,q+1}^{m+1,n+1}\left[x\left|\begin{matrix}(1,\lambda),((a_p,\alpha_p))\\((b_q,\beta_q)),(\nu,u)\end{matrix}\right.\right] \equiv A_{p+1,q+1}^{m+1,n+1}\left[x\mid(1,\lambda),((\ ));((\ )),(\nu,u)\right]$$

(1.2)

$$i_{c,d,\ell}^{u,v}\left[\phi(x)\right] \equiv i\left[\phi(x)\right] \quad \text{when the suffixes and indices on i are as given}$$

(1.3)

$$T_{c,d}(x)\ e^{-dx+vx}\ x^{c+u+\ell} \equiv T^*(x);\ LA\binom{u,v}{c,d} = LA\{\ \}\ \text{or}\ LA(\ )$$

(1.4)

$$((a_p, \alpha_p)) \equiv ((a_1, \alpha_1), (a_2, \alpha_2), \ldots, (a_p, \alpha_p)) \qquad (1.5)$$

2. OVERLAPPING REGIONS OF RESEARCH

In a fast developing area of resarch it is not unlikely that some over-lapping regions of resarch are created. To substantiate I reacll a few specific instance. (i) The integrals involving the product of two Meijer G-functions given by Sharma K.C. [10] and Saxena R.K [11]. (ii) A*(x,y) due to Chaturvedi K.K. and Goyal A.N. [12]; H(x,y) due to Goyal G.K. [14] [submitted for publication on 26th October 1967 containing the reference to A*(x,y) as a function defined in 1966] and their replicas by Mittal P.K. and Gupta K.C. [13], Munot P.C. and Kalla S.L. [15]; Verma R.U. [16]; Pathak R. S. [17]; Joshi C.M. [18]; Prasad Y.N. [19]; and in n-variables by Saxena R.K. [20, 21]; Srivastava H.M. and Panda R. [22]. (iii) Result due to Goyal S.P. [23] on $x^\rho y^\sigma H(x,y)$ and the one contained in the Ph. D. Thesisi of Chaturvedi K.K. [24]. Mathai A.M. and Saxena R.K. [25] as well as Srivastava H.M., Gupta K.C. and Goyal S.P. [26] added further to the confusion by assigning different and misleading priorities to the development of H(x,y).

3. UNIFIED APPROACH

The A-function due to Gautam G.P. and Goyal A.N. [27] with slight changes is

$$A(x) = \frac{1}{2\pi\sqrt{-1}} \int_L f(s) \, x^{-s} \, ds \qquad (3.1)$$

where f(s) are gamma products in the integral (1.1) and (i) empty product is interpreted as unity (ii) non-negative integers m, n, p, q satisfy $m \leq p$, $n \leq q$ (iii) $x \neq 0$, a_j, α_j, b_j, β_j are all complex (iv) poles of gamma products are simple, separable and those of $\prod_1^m \Gamma(a_j - \alpha_j s)$ lie to the right and of $\prod_1^n \Gamma(1 - b_j + \beta_j s)$ lie to the left if the contour L.

The integral on the right hand side of (3.1) converges and representes an analytic function in the domain determined by one of the following set of conditions

(i) $\quad x \neq 0, \xi = 0, \eta < 0, \ |\arg \zeta x| < \frac{\pi}{2} \eta$ \qquad (3.2)

(ii) $\quad x > 0, \xi = 0 = \eta, \nu - \sigma\lambda < -1$ \qquad (3.3)

where

$$\xi = I\left(\sum_1^p \alpha_j - \sum_1^q \beta_j\right); \quad \eta = R\left(\sum_1^m \alpha_j + \sum_1^n \beta_j - \sum_{m+1}^p \alpha_j - \sum_{n+1}^q \beta_j\right) \quad (3.4)$$

$$\zeta = \prod_1^p \alpha_j^{\alpha_j} \prod_1^q \beta_j^{-\beta_j}; \quad \lambda = R\left(+ \sum_1^p \alpha_j - \sum_1^q \beta_j\right)$$

$$\nu = R\left(+ \sum_1^p a_j - \sum_1^q b_j\right) - \frac{1}{2}(p - q) \quad (3.5)$$

and $s = \sigma + it$ on the path L when $|t| \to \infty$.

A natural extension of (3.1) by augmenting number of variables, parameters and integrals with proper conditions of convergence can be seen in Gautam G.P. and Goyal A.N. [28].

The conditions of analyticity given in (3.2) and (3.3) can be easily obtained on using the asymptotic expansion of gamma function given in Rainville E.D. [29, p. 31]. Further

$A(x) = O(|x|^\gamma)$ for large $x, \xi = 0, \eta \geq 0$; $\gamma = \max R(a_j/\alpha_j)$, $j = 1,\ldots,m$

$A(x) = O(|x|^\delta)$ for small $x, \xi = 0, \eta \geq 0$; $\delta = \min R(b_j-1)/\beta_j$,

$j = 1,\ldots,n.$ \quad (3.6)

$$\int_0^\infty \exp(-\rho t) t^{-k} A(wt^{\lambda_1}) dt = \rho^{k-1} A_{p+1,q}^{m+1,n}\left[w/\rho^\lambda \Big| \begin{matrix} (1-k,\lambda_1),((a_p,\alpha_p)) \\ ((b_q,\beta_q))\end{matrix}\right] \quad (3.7)$$

provided $0 < \rho < \lambda_1(w\zeta)^{1/\lambda}$ and $1 + R(-k) + \lambda_1 \min R((b_j-1)/\beta_j) > 0$, $j = 1,\ldots,n$. The proof is trivial and hence omitted.

4. LAPLACE A-TRANSFORM AND ITS COMPLEX INVERSION

The Laplace A-transform (LA tr) is defined as

$$L_A[f(x)] = \int_0^\infty \exp(-\rho yx) A[w(yx)^\lambda] f(x) dx. \quad (4.1)$$

The integral in (4.1) must converge absolutely and $t^{k-1} f(t) \quad L(0,\infty)$, $f(t)$ is of bounded variation in the neighbourhood $t = x$ and

$$F(k) = \int_0^\infty y^{-k} L_A[f(x)] dy \quad (4.2)$$

then

$$\tfrac{1}{2}[f(x+0) + f(x-0)] = \frac{1}{2\pi i} \lim \int_{\sigma-i\gamma}^{\sigma+i\gamma} x^{-k}\rho^{1-k} F(k)/A_{p+1,q}^{m+1,n}[w/\rho^\lambda] \, dk \qquad (4.3)$$

with the value of $A(w/\rho^\lambda)$ and conditions of convergence given in (3.7). Also the LA.tr of $|f(x)|$ exists and the integral in (4.3) converges absolutely. Symbolically

$$F(y) \triangleq LA[f(x)] \equiv \int_0^\infty \exp(-\rho y x) A[(yx)^\lambda] f(x) \, dx \qquad (4.4)$$

Proof Using (4.1) in (4.2), changing the order of integration, substituting $xy = t$ and using (3.7) as well as Mellin inversion, we obtain (4.3).

5. EXTENSION OF LA.tr TO DISTRIBUTIONAL CASE

In order to extend LA.tr to the distributional case, we require to define a testing function space over $(0,\infty)$ which contains the kernel of the transform. The complex inversion formula is also extended to the distributional case.

Spaces of the type LA{ } and their duals

$D(I)$ is the set of all smooth functions on I having compact supports where I is the open set of $0 < x < \infty$. Let d be a positive real number and c, u, v suitably fixed real numbers. Then LA{ } is the space of all smooth functions $\phi(x)$ defined on $0 < x < \infty$ such that

$$i[\phi(x)] = \sup_{0<x<\infty} |T^*(x) D_x^\ell \phi(x)| < \infty \quad \text{for} \quad \ell = 0,1,2,\ldots$$

where

$$T_{c,d}(x) = x^{vc}, \quad 1 \leq x < \infty \quad \text{and} \quad T_{c,d}(x) = x^{vd}, \quad 0 < x < 1$$

LA{ } is endowed with the topology generated by the countable multinorms $\{i_{c,d,\ell}^{u,v}\}_{\ell=0}^\infty$ as in Zemanian, A. H. [30, p. 8].

The convergence of a sequence of testing functions $\{\phi_j(x)\}_{j=1}^\infty$ is defined in LA{ } to $\phi(x)$ if for each $\ell = 0,1,2,\ldots$; $[\phi_j(x) - \phi(x)] \to 0$ as $j \to \infty$.

Theorem 1 LA{ } is a Frechet space.

Proof Trivial, hence not included. □

Countable union space

If $0 < d_1 < d_2$, then $LA(u,v;c,d_1)$ is a subspace of $LA\{\ \}$. Let $\{d_r\}_{r=1}^{\infty}$ be a monotonic increasing sequence of positive real numbers such that $d_r \to d$. The countable union space $LA\{\ \}$ is defined as the countable union $\bigcup_{r=1}^{\infty} LA(u,v;c,d_r)$. Since $LA\{\ \}$ is a Frechet space so is $\bigcup_{r=1}^{\infty} LA(u,v;c,d_r)$ and consists of all testing functions $\phi(x)$ on $0 < x < \infty$ which are smooth and

$$i[\phi(x)] = \sup_{0<x<\infty} |T^*(x)D_x^\ell \phi(x)| < \infty \text{ for } \ell = 0,1,2,\ldots.$$

Let $LA'\{\ \}$ be the dual space of $LA\{\ \}$ and contains all linear functionals in $LA\{\ \}$. Let $LA'^{u,v}_{c,d}$ be the dual of $LA^{u,v}_{c,d}$. Clearly $D(I) \subset LA^{u,v}_{c,d}$ and the topology of $D(I)$ is stronger than that induced on it by $LA^{u,v}_{c,d}$. Hence the restriction of any $f \in LA'^{u,v}_{c,d}$ to $D(I)$ is in $D'(I)$ which is the dual of $D(I)$.

Theorem 2 As a function of x, x^{k-1} is a member of $LA\{\ \}$ if $d > v$ and $R(k) \geq 1 - C(1+v) - u$.

Proof Clearly x^{k-1} is a smooth function on $0 < x < \infty$ consider for any $\ell = 0,1,2,\ldots$

$$i[x^{k-1}] = \sup_{0<x<\infty} |T^*(x)D_x^\ell(x^{k-1})| = \sup_{0<x<\infty} |T^*(x)x^{k-\ell-1} \Gamma(k)/\Gamma(k-1)|$$

which is bounded under the supremum sign. Thus $i[x^{k-1}] < \infty$ for all $\ell = 0,1,2,\ldots$ which proves that $x^{k-1} \in LA\{\ \}$ if $d > v$ and $R(k) \geq 1 - C(1+v) - u$. The theorem is proved. □

6. DISTRIBUTIONAL LA.tr

Lemma If r is a positive integer, then

$$D_x^r[A\{w(yx)^\lambda\}] = x^{-r} A_{p+1,q+1}^{m+1,n}\left[w(yx)^\lambda \Big| \begin{matrix}(1,\lambda),((a_p,\alpha_p))\\((b_q,\beta_q)),(1-k,\lambda)\end{matrix}\right].$$

Theorem

$\exp(-\rho yx) A[w(y\ x)^\lambda] \in LA\{\ \}$ for $\dfrac{d-v}{-\rho} < R(y)$ with $R(\rho) > 0$, $d > 0$

and $d > v$, $R(vd+c+u+\lambda\delta_1) \geq 0$ where $\delta_1 = \min R\{(b_j-1)/\beta_j\}$, $j = 1,\ldots,n$.

Proof It is sufficient to prove that for any $\ell = 0,1,2,\ldots$

$$i\{\exp(-\rho yx)A[w(yx)^\lambda]\} = \sup_{0 < x < \infty} |T*(x)D_x^\ell\{\exp(-\rho yx)A[w(yx)^\lambda]\}|$$

is bounded

$$= \sup_{0 < x < \infty} |T*(x) \sum_{r=0}^{\ell} \ell_{c_r} D_x^{-r} \exp(-\rho yx) D_x^r A[w(yx)^\lambda]|$$

$$\leq \sum_{r=0}^{\ell} \sup_{0 < x < \infty} \left|T*(x)x^{-\ell}(-\rho y)^{\ell-r} \exp(-\rho yx) \ell_{c_r} A_{p+1,q+1}^{m+1,n}\left[w(yx)^\lambda \bigg| \begin{matrix}(1,\lambda),((\ ))\\((\ )),(1-r,\lambda)\end{matrix}\right]\right|$$

(6.1)

consider now, for any j, $0 \leq j \leq \ell$

$$\sup_{0 < x < \ell} \left|T*(x)x^{-\ell}\ell_{c_r}(-\rho y)^{\ell-r}\exp(-\rho yx)A\left[w(yx)^\lambda \bigg| \begin{matrix}(1,\lambda),((\ ))\\((\ )),(1-r,\lambda)\end{matrix}\right]\right|$$

for $0 < x < 1$ as $x \to 0$ and using $A(x) = O(|x|^\delta)$ for small x.

$$\leq \sup_{0 < x < 1} |x^{vd} e^{-dx+vx} x^{u+c} \ell_{c_r} (-\rho y)^{\ell-r} \exp(-\rho yx) M_1 w^{\delta_1} y^{\lambda\delta_1} x^{\lambda\delta_1}|, \text{ or}$$

$$\leq \sup_{1 < x < \infty} |x^{vc} e^{-dx+vx} x^{u+c} \ell_{c_r} (-\rho y)^{\ell-r} \exp(-\rho yx) M_2 w^{\gamma_1} y^{\lambda\gamma_1} x^{\lambda\gamma_1}|$$

where M_2 is constant, $1 < x < \infty$ as $x \to \infty$, $A(x) = O(|x|^\gamma)$ for large x and $\gamma_1 = \max R(1/\lambda)R(a_j/\alpha_j)$, $j = 1,\ldots,m$ which is again bounded if $R(\rho) > 0$, $d > v$, $d > 0$ and $((d-v)/(-\rho)) < R(y)$.

7. DISTRIBUTIONAL LAPLACE A-TRANSFORM

A generalized function f shall be called Laplace A-transformable iff $f \in LA'(u,v,c,d)$ for u,v,c,d (positive) to be reals and suitably fixed. From Theorem 1 (Sec. 6) it is clear that for any generalized function f, there exists a positive real number τ_f depending on f such that $f \in LA'\{\ \}$ for all $(d-v)^f/(-\rho) < \tau_f$ where $\tau_f = \inf[d-v: f \in LA'\{\ \}]$. Since $\exp(2\rho yx)A[w(yx)^\lambda] \in LA\{\ \}$ for $(d-v)/(-\rho) < R(y)$ and $R(vd+c+u+\lambda\delta_1) \geq 0$ and $f \in LA'\{\ \}$ for $0 < R(y) < \tau_f$, then we define the Laplace A-transform in distributional sense as

$$LA[f(x)] = \langle f(x), \exp(-\rho yx)A[w(yx)^\lambda] \rangle. \quad (7.1)$$

162

Distributional complex inversion

The following lemmas will be useful in the proof of the inversion theorem

Lemma 1   If $f(t) \in LA'\{\ \}$ then

$$\int_0^\infty y^{-k} < f(t), \exp(-\rho y t) A[w(yt)^\lambda] > dy$$

$$= < f(t), \int_0^\infty y^{-k} \exp(-\rho y t) A[w(yt)^\lambda] dy > .$$

Lemma 2   Let $\phi(x) \in D(I)$ and r be a fixed positive number with $0 < r < \infty$; $f \in LA'\{\ \}$ for $0 < R(y) < \tau_f$, $Q(k) = \int_0^\infty \phi(x) x^{-k} dx$ where $k = \sigma + i\mu$ and σ is fixed with the condition $1-c(1+v)-u \le \sigma < \tau_f$. Then

$$\int_{-r}^r < f(t), t^{k-1} > Q(k) d\mu = < f(t), \int_{-r}^r t^{k-1} Q(k) d\mu > .$$

Lemma 3   If $\phi(x) \in D(I)$ and σ, r are real numbers such that $1-c(1+v)-u < 1 < \sigma$, $d > v$, $R(vd+c+u+\lambda\delta_1) \ge 0$, then

$$\frac{1}{\pi} \int_0^\infty [\phi(x)/t \, \log(t/x)](t/x)^\sigma \sin(r \, \log(t/x)) dx$$

converges to $\phi(x)$ in $LA\{\ \}$ as $r \to \infty$.

Proof   The proofs of above lemmas can be developed on similar lines as given in Zemanian [p. 121 and 66-68]. □

Lemma 4   If $f(t) \in LA'\{\ \}$ and $\sigma - R(k) > 1 > 1-c(1+v)-u$, then

$$\frac{\rho^{1-k}}{A_{p+1,q}^{m+1,n}[w/\rho^\lambda]} \int_0^\infty y^{-k} < f(t), \exp(-\rho y t) A[w(yt)^\lambda] > dy = < f(t), t^{k-1} >$$

where

$$A_{p+1,q}^{m+1,n}\left[w/\rho^\lambda \Big| \begin{matrix}(1-k,\lambda),((a_p,\alpha_p))\\((b_q,\beta_q))\end{matrix}\right] \equiv A_{p+1,q}^{m+1,n}[w/\rho^\lambda].$$

Theorem (Inversion theorem)   Let $f \in LA'\{\ \}$, $\phi \in D(I)$ and

$$F(k) = \int_0^\infty y^{-k} LA[f(x)] dy, \quad k = \sigma + i\mu \text{ and } \sigma > 1 > 1-c(1+v)-u,$$

163

then

$$\text{Lim}_{r\to\infty} < \frac{1}{2\pi i} \int_{\sigma-ir}^{\sigma+ir} [x^{-k}\rho^{1-k} F(k)]/A_{p+1,q}^{m+1,n}[w/\rho^\lambda] dk, \phi(x) > = < f, \phi >.$$

Proof The integral on k is a continuous function of x, thus the left hand side without limit may be written as

$$\frac{1}{2\pi i} \int_0^\infty \int_{\sigma-ir}^{\sigma+ir} [x^{-k}\rho^{1-k} F(k)]/A_{p+1,q}^{m+1,n}[w/\rho^\lambda] \phi(x) dx$$

since $\phi(x)$ is of bounded support and the integrand is a continuous function of (x,μ), thus order of integration may be changed, using the value of $F(k)$ and $LA[f(x)]$, we have

$$\frac{1}{2\pi} \int_{-r}^{r} \rho^{1-k}/A_{p+1,q}^{m+1,n}[w/\rho^\lambda] \int_0^\infty y^{-k} < f(t), \exp(-\rho y t) A[w(yt)^\lambda] > dy \int_0^\infty x^{-k} \phi(x) dx d\mu$$

using lemmas [4] and [3] sucessively and since $\phi(x)$ is of bounded support and the integrated is a continuous function of (x,μ) we can change the order of integration to obtain

$$< f(t), \frac{1}{2\pi} \phi(x) \int_{-r}^{r} t^{k-1} x^{-k} d\mu dx > . \qquad (7.2)$$

Now consider the integral

$$\int_{-r}^{r} t^{k-1} x^{-k} d\mu = \int_{-r}^{r} (t/x)^\sigma t^{-1} e^{-iw\log(t/x)} d\mu$$

$$= 2(t/x)^\sigma t^{-1} [\log(t/x)]^{-1} \sin\{r \log(t/x)\}. \qquad (7.3)$$

Now putting the value from (7.3) in (7.2), we have

$$< f(t), \frac{1}{\pi} \int_0^\infty \phi(x)(t/x)^\sigma \sin\{r \log(t/x)\} [t \log(t/x)]^{-1} >$$

using lemma [3], the above converges to $<f(t), \phi(t)>$ as $r \to \infty$. This completes the proof. □

8. UNIQUENESS THEOREM

The two LA-transformable generalized functions whose LA-transforms are equal must have equal restriction to $D(I)$.

Theorem If $f(x)$ and $g(x) \in D'(I)$ such that $LA[f(x)]$ and $LA[g(x)]$ are defined in $0 < R(y) < \tau_f$ and $0 < R(y) < \tau_g$ respectively and $LA[f(x)] = LA[g(x)]$ for $0 < R(y) < \min(\tau_f, \tau_g)$ then in the sense of equality in $D(I)$, $f = g$.

Proof Let $\phi(x)$ be an arbitrary member of $D(I)$, then using the inversion theorem, we have

$$< f-g, \phi > = \lim_{r \to \infty} < \frac{1}{2\pi i} \int_{\sigma-ir}^{\sigma+ir} \frac{x^{-k} \rho^{1-k} F_1(k) \, dk}{A_{p+1,q}^{m+1,n}[w/\rho^\lambda]}, \phi(x) >$$

where

$$F_1(k) = \int_0^\infty y^{-k} \{LA[f(x)] - LA[g(x)]\} dy = 0, \quad \text{since}$$

$LA[f(x)] = LA[g(x)]$, $0 < R(y) < \min(\tau_f, \tau_g)$.

Thus $< f-g, \phi > = 0$, $\phi(x) \in D(I)$, $\therefore f = g$. □

The following result will be useful

$$x^\mu A[w(yx)] = \frac{1}{y^\mu w^{\mu/\lambda}} A_{p,q}^{m,n}\left[w(yx)^\lambda \Bigg| \begin{array}{c}((a_p - \frac{\mu}{\lambda} \alpha_p, \alpha_p)) \\ ((b_q - \frac{\mu}{\lambda} \beta_q, \beta_q))\end{array}\right] \quad (8.1)$$

Proof is trivial.

9. OPERATION DIFFERENTIATION

Theorem

$$i_{c,d,\ell}^{u+1,v}[-D_x\phi(x))] = i_{c,d,\ell+1}^{u,v}\phi(x).$$

Proof

$$i_{c,d,\ell}^{u+1,v}[-D_x\phi(x)] = \sup_{0<x<\infty} |T^*(x) x D_x^\ell - D_x\phi(x)|$$

$$= \sup_{0<x<\infty} |T^*(x) x D_x^{\ell+1}[\phi(x)]| = i_{c,d,\ell+1}^{u,v}[\phi(x)].$$

Thus the operation differentiation shows $\phi(x) \to -D_x\phi(x)$ is a continuous linear mapping from $LA(u,v;c,d)$ onto $LA(u+1,v;c,d)$. By Zemanian [[30], Theorem 1.10, p. 7] the adjoint mapping $f(x) \to D_x f(x)$ is also linear and continuous from $LA(u+1,v;c,d)$ onto $LA(u,v;c,d)$. Thus the operation transform formula is obtained as

$$\langle D_x f(x), \phi(x) \rangle = \langle f(x), -D_x \phi(x) \rangle$$

where $f(x)$ LA'$(u+1,v;c,d)$.

Theorem If $f(x) \in LA'(u,v;c,d)$ and $LA[f(x)]$ be the Laplace A-transform of $f(x)$ then

$$\frac{\delta^r}{\delta y^r}\{LA[f(x)]\} = \langle f(x), \frac{\delta^r}{\delta y^r}\{\exp(-\rho y x) A[w(yx)^\lambda]\} \rangle$$

for any non-negative integer r.

Proof To prove the right hand side of the theorem, we consider

$$i_{c,d,\ell}^{u,v}[\frac{\delta^r}{\delta y^r}\exp(-\rho y x)A[w(yx)^\lambda]\}]$$

$$= \sup_{0<x<\infty}|T^*(x)D_x^\ell \frac{\delta^r}{\delta y^r}\{\exp(-\rho y x)A[w(yx)^\lambda]\}]|$$

$$< \theta \sup_{0<x<\infty}|T^*(x)x^{-\ell}(-\rho y x)^{r-j+k}\exp(-\rho y x)A_{p+2,q+2}^{m+2,n} \cdot$$

$$\cdot \left[w(yx)^\lambda \middle| \begin{array}{l}(1,\lambda)(\ell,\lambda),((a_p,\alpha_p))\\((b_q,\beta_q)),(1-j,\lambda)(1-i,\lambda)\end{array}\right]|$$

where

$$\theta = \sum_{j=0}^{r}\sum_{i=0}^{\ell}\sum_{k=0}^{\ell-i} r!\ell!y^{-r}/j!i!k!(\ell-i-k)!(r-\ell-j+i+v)!.$$

Now for any j,i,k such that $0 \leq j \leq r$, $0 \leq k \leq \ell$ and $0 \leq k \leq \ell-i$ respectively. The right hand side of the above result is bounded for

(i) $0 < x < 1$ as $x \to 0$ and $A(x) = O(|x|^\delta)$ for small x where $\xi = 0$, $\eta > 0$ and $\delta = \min\{R(b_j-1/\beta_j)\}$, $j = 1,\ldots,n$ and $|\arg(\zeta x)| < \pi\eta/2$.

(ii) $1 < x < \infty$ as $x \to \infty$ and $A(x) = O(|x|^\gamma)$ for large x where $\xi = 0, \eta \geq 0$ and $\gamma = \max R(a_j/\alpha_j)$, $j = 1,\ldots,m$.

Therefore, we conclude that

$$\frac{\delta^r}{\delta y^r}\{\exp(-\rho y x) A[w(yx)^\lambda]\} \in LA(u,v;c,d).$$

Now using inductive argument the right hand side of the theorem follows. Since, it is true for $r = 0$, and let it be true for r replaced by $(r-1)$. Let y be fixed and $\Delta y \neq 0$, consider

$$\frac{1}{\Delta y}[F^{r-1}(y+\Delta y) - F^{r-1}(y)] - <f(x), k^r(y,x)> = <f(x)\theta_{\Delta y}(x)>$$

where $k(x,y) = \exp(-\rho yx)A[w(yx)^\lambda]$ and

$$\theta_{\Delta y}(x) = \frac{1}{\Delta y}[k^{r-1}(y+\Delta y, x) - k^{r-1}(y,x)] - k^r(y,x).$$

Now we prove that

$$\theta_{\Delta y}(x) \to 0 \text{ in } LA(u,v;c,d).$$

Therefore, for any non-negative integer ℓ

$$\theta_{\Delta y}^\ell(x) = \frac{1}{\Delta y} k^{\ell+r-1}(y+\Delta y, x) - k^{\ell+r-1}(y,x) - k^{\ell+r}(y,x)$$

$$= \frac{1}{\Delta y} \int_y^{y+\Delta y} dt \int_y^t k^{\ell+r+1}(\tau,x) d\tau.$$

If z denotes the interval

$$y - |\Delta y| < \tau < y + |\Delta y|, \text{ then } |\theta_{\Delta y}(x)| < \frac{|\Delta y|}{2} \sup_{\tau \in z} |k^{j+r+1}(\tau,x)|$$

considering now $i_{c,d,\ell}^{u,v}[\theta_{\Delta y}(x)]$ and using the above result, we see that it $\to 0$ as $\Delta y \to 0$ in $LA(u,v;c,d)$.

The theorem is thus proved. □

Multiplication by x^μ; where $0 < \mu < u$ and $0 < x < \infty$.

The mapping $\phi(x) \to x^\mu \phi(x)$ is an isomorphism from $LA(u,v;c,d)$ onto $LA(u-\mu,v;c,d)$. Linearity and one-one ness is obvious. Consider

$$i_{c,d,\ell}^{u-\mu,v}[x^\mu \phi(x)] = \sup_{0<x<\infty} |T^*(x) x^{-\ell-\mu+1} D_x^\ell \{x^\mu \phi(x)\}|$$

$$= \sup_{0<x<\infty} |\sum_{j=0}^\ell \ell_{c_j} \mu(\mu-1) \cdots (\mu-\ell+j+1) T_{c,d}(x) e^{-dx+vx^c+u-\mu+\mu+1} D_x^j \phi(x)|$$

$$\leq M \sum_{j=0}^\ell \sup_{0<x<\infty} |T_{c,d}(x) e^{-dx+vx^c+u+j} D_x^j \phi(x)|$$

where

$$|\ell_{c_j} \mu(\mu-1) \cdots (\mu-\ell+j+1)| < M.$$

Thus

$$i_{c,d,\ell}^{u-\mu,v}[x^\mu \phi(x)] < M \sum_{j=0}^\ell i_{c,d,j}^{u,v}[\phi(x)] < \infty.$$

It is clear that $x^\mu \phi(x) \in LA(u-\mu,v;c,d)$ and the mapping $\phi(x) \to x^\mu \phi(x)$ is linear and continuous. The inverse mapping is given by $\psi(x) \to x^{-\mu}\psi(x)$ from $LA(u-\mu,v;c,d)$ onto $LA(u,v;c,d)$. Similarly it can also be proved that

$$i^{u,v}_{c,d,\ell}[x^{-\mu}\psi(x)] \le M_1 \sum_{j=1}^{\ell} i^{u-\mu}_{c,d,j}[\psi(x)]$$

where

$$|\ell_{c_j}(-1)^{\ell-j}\mu(\mu+1) \ldots (\mu+\ell-1)| < M_1.$$

Therefore, the inverse mapping is also linear and continuous. Thus the mapping $\phi(x) \to x^{\mu}\phi(x)$ is an isomorphism. Consequently, the adjoint mapping $f(x) \to x^{\mu}f(x)$ is also an isomorphism from $LA(u-\mu,v;c,d)$ onto $LA(u,v;c,d)$ and the operation transform formula

$$< x^{\mu}f(x), \phi(x) > = < f(x), x^{\mu}\phi(x) >.$$

Multiplication by exponential function

Here we prove that the mapping $\phi(x) \to \exp(-\beta yx)\phi(x)$ is an automorphism from $LA(u,v;c,d)$ onto itself, where β and y are fixed such that $R(\beta) = r$ and $R(y) > \frac{d-v}{-\rho}$. Clearly the mapping is linear and one-one. It is easy to prove that

$$i[\exp(-\beta yx)\phi(x)] \le M \sum_{j=0}^{\ell} \sup_{0<x<\infty} |T^*(x) x^{-\ell+j} D^j_x \phi(x)|$$

$$= M \, i^{u,v}_{c,d,j}\phi(x) \text{ where } |\ell_{c_j}(-\beta yx)^{\ell-j} \exp(-\beta yx) < M.$$

Thus the mapping $\phi(x) \to \exp(-\beta yx) \to \exp(-\beta yx)\phi(x)$ is continuous and linear from $LA(u,v;c,d)$ onto itself. Consequently $\psi(x) \to \exp(\beta yx)\psi(x)$ is a unique inverse mapping of $\phi(x) \to \exp(-\beta yx)\phi(x)$. Similarly it can also shown that $\psi(x) \to \exp(\beta yx)\psi(x)$ is linear and continuous from $LA(u,v;c,d)$ onto itself. Therefore the mapping $\phi(x) \to \exp(-\beta yx)\phi(x)$ is also an automorphism. Thus we get the operation transform formula as

$$< \exp(-\beta yx)f(x), \phi(x) > = < f(x), \exp(-\beta yx)\phi(x) >.$$

Because the adjoint mapping is also an automorphism.

ACKNOWLEDGEMENT

Thanks are due to Professor R. P. Agarwal, Vice-Chancellor, Rajasthan

University, Jaipur for kindly giving the travel grant (to A.N.G.) which has made this visit possible.

REFERENCES

1. C. S. Meijer, Nederl. Akad. Wetensch., Proc. 44, 1062-1070, (1941).
2. C. Fox, Trans. Amer. Math. Soc. 98, 395-429, (1961).
3. R. P. Agarwal, Proc. Nat. Inst. Sci. India, Part A, 31,536-546, (1965).
4. B. L. Sharma, Ann. Soc. Sci. Bruxelles, Ser. 1, 79, 26-40, (1965).
5. D. P. Mourya, Indian J. Pure Appl. Math., 464-469, (1970).
6. S. S. Khadiya and A. N. Goyal, Vijnana Parishad Anusandhan Patrika, 13, 191-201, (1970).
7. J. Gopal Krishana and Mohammed Ghouse, Indian Math. Soc. Annual Conf., Abstract No. 102., (1976).
8. R. G. Buschman, Jnanabha, 7, 107-118, (1977); Indian J. Math, 20, 105-116, (1978); Pure Appl. Math. Sci., 9, 13-18, (1979).
9. O. P. Tandon, Inst. J. Math. Educ. Sci. Tech., No. 1, 14, 655, (1983).
10. K. C. Sharma, Ph. D. Thesis Raj. Univ., 21, (1959).
11. R. K. Saxena, Proc. Nat. Inst. Sci. India, Part. A, 26, 400-413, (1960).
12. K. K. Chaturvedi and A. N. Goyal, Indian J. Pure Appl. Math., 3, 357-360, (1972).
13. P. K. Mittal and K. C. Gupta, Proc. Indian Acad. Sci., Sect. A, 75, 117-123, (1972).
14. G. K. Goyal, Raj. Univ. Studies. Math., 1, 37-46, (1971).
15. P. C. Munot and S. L. Kalla, Univ. Nac. Tucuman Rev., Ser. A, 21, 67-84, (1971).
16. R. U. Verma, An. Sti. Univ. 'Al. I Cuza'Iasi, Sect. Ia Mat. (N. S.), 17, 103-109, (1971).
17. R. S. Pathak, Bull. Calcutta Math. Soc., 62, 97-106, (1970).
18. C. M. Joshi, Indian J. Pure Appl. Math., 8, 103-116, (1977).
19. Y. N. Prasad, Vijnana Parishad Anusandhan Patrika, 23, 83-92, (1980).
20. R. K. Saxena, Kyungpook Math. J., 14, 255-259, (1974).
21. R. K. Saxena, Kyungpook Math. J., 17, 221-226, (1977).
22. H. M. Srivastava and R. Panda, J. Reine Angew Math, 283/284, 265-274, (1976).
23. S. P. Goyal, Kyungpook Math. J., 15, 117-131, (1975).
24. K. K. Chaturvedi, Ph. D. Thesis Raj. Univ., (1969).

25. A. M. Mathai and R. K. Saxena, "The H-function with Applications in Statistics and other Disciplines", Wiley Eastern, (1978).
26. H. M. Srivastava, K. C. Gupta and S. P. Goyal, "The H-functions of One and Two Variables with Applications", South Asian Publishers, New Delhi, (1982).
27. G. P. Gautam and A. N. Goyal, Indian J. Pure Appl. Math., 12, 1094-1105, (1981).
28. G. P. Gautam and A. N. Goyal, Vijanana Parishad Anusandhan Patrika, 24, 217-232, (1981).
29. E. D. Rainville, "Special Functions", Chelsa Publ. Co. Bronx, (1971).
30. A. H. Zemanian, "Generalized Integral Transformations", Interscience Publishers, New York, (1968).

AN ALGEBRAIC APPROACH TO DISTRIBUTION THEORIES

J. de Graaf and A. F. M. ter Elst

Eindhoven University of Technology

Eindhoven, The Netherlands

In this note Mikusiński's idea of convolution quotients is generalized in two directions simultaneously:
- The ring R is not merely acting on itself but acts also on a separate vector space V.
- The ring R need not be commutative.

This approach makes it possible to bring many theories of generalized functions under the same viewpoint. So e.g. Schwartz' tempered distribution space, many (all?) distribution spaces of Gelfand-Šilov and Mikusiński's space of convolution quitients (cf. the Examples) are all very special examples of the construction that we present here. The authors expect that many more locally convex topological vector spaces can be brought under the same viewpoint. The connection with (non-commutative) harmonic analysis, especially the case that R is a (subset of) the convolution algebra of a Lie group is now being studied by the second author. The algebra of this paper is inspired by Ore's construction of non-commutative fields. See [5], p. 119.

Let R be a (possibly non-commutative) ring which acts on a vector space V. In other words V is an R-module. We consider a subset $Q \subset R$, $Q \neq \emptyset$, with the following

<u>Properties 1</u> (i) Q is a semi-group. Q does not necessarily contain a unity.

(ii) Q has the left common multiple property, i.e.

$$\forall_{K \in Q} \forall_{L \in Q} \exists_{K_1 \in Q} \exists_{L_1 \in Q} [L_1 K = K_1 L].$$

(iii) Q has the right cancellation property, i.e.

$\forall_{K,L,M \in Q} [KL = ML \Rightarrow K = M]$.

(iv) Q acts injectively on the vector space V.

Definition 2 On $X \times Q$ we introduce the relation $\sim$ by

$$(f,K) \sim (g,L) \Leftrightarrow \forall_{L_1 \in Q} \forall_{K_1 \in Q} [L_1 K = K_1 L \Rightarrow L_1 f = K_1 g]$$

$(f, g \in V, K, L \in Q)$.

Theorem 3 Let $f, g \in V$ and $K, L \in Q$. Then

$$(f,K) \sim (g,L) \Leftrightarrow [\exists_{L_1 \in Q} \exists_{k_1 \in Q}(L_1 K = K_1 L) \wedge L_1 f = K_1 g].$$

Proof

$\Rightarrow$) Trivial.

$\Leftarrow$) Let $L_2, K_2 \in Q$ and suppose that $L_2 K = K_2 L$, then we must show that $L_2 f = K_2 g$. Because of 1.(ii) there are $M_1, M_2 \in Q$ such that

$$M_1 L_1 K = M_2 L_2 K \text{ and hence } M_1 L_1 = M_2 L_2.$$

Then

$$M_1 K_1 L = M_2 K_2 L \text{ and hence } M_1 K_1 = M_2 K_2.$$

Since $M_1 L_1 f = M_1 K_1 g$ we find

$$M_2 L_2 f = M_2 K_2 g.$$

Finally, because of 1.(iv)

$$M_2 (L_2 f - K_2 g) = 0 \text{ hence } L_2 f - K_2 g = 0. \square$$

Lemma 4 Let $f \in V$ and $K, L \in Q$. Then
(i) $(f, K) \sim (Lf, LK)$.
(ii) $(Kf, K) \sim (Lf, L)$.

Proof

(i) Let $A, B \in Q$ be such that $AK = BLK$. Then $A = BL$. So indeed $Af = BLf$.

(ii) From $AK = BL$ it follows that $AKf = BLf$, $(A, B \in Q)$. $\square$

Theorem 5 The relation $\sim$ is an equivalence relation.

Proof

(a) Let $f \in V$ and $A,B,K \in Q$. If $AK = BK$, then $A = B$ and hence $Af = Bf$. So $(f,K) \sim (f,K)$.

(b) It is trivial that $\sim$ is a symmetric relation.

(c) Now reflexity follows with Theorem 3.

Let $f,g,h \in V$ and $K,L,M \in Q$ and suppose that $(f,K) \sim (g,L)$ and $(g,L) \sim (h,M)$. From this we must show that $(f,K) \sim (h,M)$. By Theorem 3 there exist $K_1, K_2, L_1, L_2 \in Q$ such that $L_1 K = K_1 L$, $L_2 M = M_2 L$, $L_1 f = K_1 g$ and $M_2 g = L_2 h$. Let $K_3, M_3 \in Q$ and suppose that $M_3 K = K_3 M$. Because of 1.(i) and 1.(ii) there are $A,B,C \in Q$ such that

$$AL_1 K = AK_1 L = BL_2 M = BM_2 L = CM_3 K = CK_3 M.$$

Application of 1.(iii) yields

$$BL_2 = CK_3, \quad AK_1 = BM_2, \quad AL_1 = CM_3.$$

Hence

$$CM_3 f = AL_1 f = AK_1 g = BM_2 g = BL_2 h = CK_3 h$$

and finally, because of 1.(iv)

$$C(M_3 f - K_3 h) = 0 \text{ hence } M_3 f = K_3 h. \quad \square$$

Definition 6 The equivalence class of (f,K) with $f \in V$ and $K \in Q$ is denoted by $[(f,K)]$.

Theorem 7 The addition formula

(i) $[(f,K)] + [(g,L)] = [(L_1 f + K_1 g, K_1 L)]$ with $K_1, L_1 \in Q$

such that $K_1 L = L_1 K$ and the scalar multiplication formula

(ii) $\alpha[(f,K)] = [(\alpha f, K)]$

make sense ($f,g \in V$, $K,L \in Q$, α a scalar).

(iii) $[(f,K)] + [(g,K)] = [(f + g, K)]$ for all $f,g \in V$, $K \in Q$.

Hence, the addition in associative.

Proof (i) First we show that the definition does not depend on the choice of K_1, L_1. Indeed, let $K_2, L_2 \in Q$ and suppose that $K_2 L = L_2 K$. There are $A, B \in Q$ with $AK_1 L = AL_1 K = BK_2 L = BL_2 K$. Hence, $AK_1 = BK_2$ and $AL_1 = BL_2$.

So $A(L_1f + K_1g) = B(L_2f + K_2g)$. It follows that

$$(L_1f + K_1g, K_1L) \sim (L_2f + K_2g, K_2L).$$

Secondly, we show that the definition amounts to the same if different representatives are taken. Let $\hat{f}, \hat{g} \in V$, $\hat{K}, \hat{L} \in Q$ and suppose that $(\hat{f}, \hat{K}) \sim (f, K)$ and $(\hat{g}, \hat{L}) \sim (g, L)$. Let $\hat{K}_1, \hat{L}_1 \in Q$ be such that $\hat{K}_1\hat{L} = \hat{L}_1\hat{K}$. We must show that $(\hat{L}_1\hat{f} + \hat{K}_1\hat{g}, \hat{K}_1\hat{L}) \sim (L_1f + K_1g, K_1L)$. Let $R, S \in Q$ such that $RK_1L = S\hat{K}_1\hat{L}$. From $(RK_1)L = (S\hat{K}_1)\hat{L}$ follows $RK_1g = S\hat{K}_1\hat{g}$. From $(RL_1)K = (S\hat{L}_1)\hat{K}$ follows $RL_1f = S\hat{L}_1\hat{f}$. So $R(L_1f + K_1g) = S(\hat{L}_1\hat{f} + \hat{K}_1\hat{g})$.

(ii) We omit the proof of the scalar multiplication property.

(iii) Follows from Lemma 4.(i). □

Definition 8 $T(V;Q)$ denotes the vector space of all equivalence classes $[(f,K)]$. So $T(V;Q) = \{[(f,K)] : f \in V, K \in Q\}$.

Theorem 9 (i) The action of $M \in Q$ on $T(V;Q)$ is given by

$$M[(f,K)] = [(M_1f, K_1)] \text{ with } K_1, M_1 \in Q \text{ such that } K_1M = M_1K$$

$(f \in V, K \in Q)$.

This makes sense and defines a linear mapping.

Let $f \in V$ and $K, M \in Q$. Then

(ii) $M[(Kf,K)] = [(M_1Kf, K_1)] = [(K_1Mf, K_1)]$

with $K_1, M_1 \in Q$ such that $K_1M = M_1K$.

(iii) $M[(f,M)] = [(Mf,M)]$.

(iv) $M[(f,KM)] = [(f,K)]$.

Proof (i) First observe that $(M_1f, K_1) \sim (M_2f, K_2)$ if $K_2M = M_2K$ ($K_2, M_2 \in Q$). Indeed, let $A, B \in Q$ and suppose that $AK_1 = BK_2$. Then $AM_1K = AK_1M = BK_2M = BM_2K$. So from 1.(iii), $AM_1 = BM_2$, hence $A(M_1f) = B(M_2f)$.

Secondly, we must show that the definition does not depend on the representant. Let $\hat{f} \in V$ and $\hat{K} \in Q$ and suppose that $(\hat{f}, \hat{K}) \in (f, K)$. Let $\hat{K}_1, \hat{M}_1 \in Q$ such that $\hat{K}_1\hat{M} = \hat{M}_1\hat{K}$. We must show that $(M_1f, K_1) \sim (\hat{M}_1\hat{f}, \hat{K}_1)$. Let $\hat{A}, A \in Q$ be such that $\hat{A}\hat{K}_1 = AK_1$. Then $(\hat{A}\hat{M}_1)\hat{K} = \hat{A}\hat{K}_1M = (AM_1)K$, hence $\hat{A}\hat{M}_1\hat{f} = AM_1f$.

(ii) Replace f by Kf.

(iii) Take in (i): $M = M_1 = K_1 = K$.

(iv) Let $E, F \in Q$ such that $FM = EKM$. Then $F = EK$ and by Lemma 4.(i):

$M[(f,KM)] = [(Ef,F)] = [(Ef,EK)] = [(f,K)]$. □

In the next Theorem we introduce a second action of Q on T(V;Q).

Theorem 10 (i) The action L^{-1} of $L \in Q$ on T(V;Q) is given by

$$L^{-1}[(f,K)] = [(f,KL)], \quad (f \in V, K \in Q).$$

This makes sense and defines a linear mapping.

(ii) $L^{-1} \circ L = L \circ L^{-1} = I$.

(iii) $L \circ M[(f,K)] = (LM)[(f,K)], \quad (f \in V, K,L,M \in Q)$.

Proof (i) We show consistency. Let $\hat{f} \in V$ and $\hat{K} \in Q$ and suppose that $(\hat{f},\hat{K}) \sim (f,K)$. Let $E,\hat{E} \in Q$ and suppose $EKL = \hat{E}\hat{K}L$. Then $EK = \hat{E}\hat{K}$ and hence $Ef = \hat{E}\hat{f}$. So $(\hat{f},\hat{K}L) \sim (f,KL)$.

(ii) Let $f \in V$ and $K \in Q$. Let $K_1, L_1 \in Q$ such that $K_1L = L_1K$. Then

$$L^{-1} \circ L[(f,K)] = L^{-1}[(L_1f,K_1)] = [(L_1f,K_1L)] = [(L_1f,L_1K)] = [(f,K)].$$

By Theorem 9.(iv) we obtain $L \circ L^{-1}[(f,K)] = L[(f,KL)] = [(f,K)]$.

(iii) Since $M^{-1} \circ L^{-1}[(f,K)] = [(f,KLM)] = (LM)^{-1}[(f,K)]$ for all $f \in V$ and $K \in Q$, it follows that $M^{-1} \circ L^{-1} = (LM)^{-1}$ and $I = L \circ M \circ (LM)^{-1}$. By (ii), $L \circ M[f,K] = (LM)[(f,K)]$ for all $f \in V$ and $K \in Q$.

Definition 11 V is considered as a subset of T(V;Q) by the mapping $f \to [(Kf,K)]$. Cf. Lemma 4.(ii). By Property 1.(iv) this map is injective. The action of $M \in Q$ on $f \in V$ agrees with the action of M on the corresponding element of T(V:Q). (See Theorem 9.(ii).)

Theorem 12 (i) Let $f \in V$ and $K \in Q$. Then $K[(f,K)] \in V$.
(ii) Let $K,M \in Q$ and $f \in KM(V)$. Then $M^{-1}[(f,K)] \in V$.

Proof (i) Trivial, Theorem 9.(iii).
(ii) Let $g \in V$ be such that $f = KMg$. Then

$$M^{-1}[(f,K)] = [(f,KM)] = [(KMg,KM)] = g \in V. \square$$

Finally, we introduce a test space consisteng of very smooth elements.

Definition 13 $\tau(V;Q) = \bigcap_{R \in Q} R(V)$.

If V is a Hilbert space and the necessary precautions are taken, there is a natural duality between $\tau(V;Q)$ and $T(V;Q)$ given by

$$\langle h, [(f,K)] \rangle = ((K^{-1})*h, f)_V.$$

By Theorem 3 this pairing does not depend on the choice of the representant (f,K).

APPLICATIONS

I. Mikusiński's Field.

Take $Q = R = C([0,\infty))$ with the convolution product. Take $V = C([0,\infty))$. The action of R on $C([0,\infty))$ is again the convolution product. Cf. [4], Ch. VI.6.

II. The $T_{X,A}$ and $\sigma(X,A)$ spaces of [1] and [2]. Take respectively

$$Q = \left\{ \psi^{-1} : \psi \text{ Borel measurable function on } \mathbb{R}, \psi \geq 1 \text{ and } \forall_{t>0} \left[\sup_{x \geq 0} \psi(x) e^{-tx} < \infty \right] \right\} \text{ and } Q = \{e^{-tx} | t > 0\}$$

with pointwise multiplication. Take $V = X$, a separable Hilbert space. For any unbounded nonnegative self-adjoint operator A in X we define the action of $\phi \in Q$ by

$$\phi f = \phi(A) f.$$

Note that $\sigma\left(L_2(\mathbb{R}), x^2 - \frac{d^2}{dx^2}\right)$ is the Schwartz space of tempered distributions. See Example 4 in the Epilogue of [1].

III. A general construction based on commutative harmonic analysis.

Let G be a locally compact Abelian topological group with Haar measure μ. Let U be a continuous unitary representation of G in a Hilbert space V. Let $Q \subset L^1(G)$. The Hilbert space V becomes a $L^1(G)$ module with $Kf := U(K)f := \int_G K(x) U_x f \, d\mu(x)$ ($K \in L^1(G), f \in V$). Under suitable conditions on the set Q there can be defined the topological vector spaces $S_{U,Q} := \bigcup_{K \in Q} K(V)$ and $T_{U,Q}$. The latter consists of all functions F from Q into V such that $F(K * L) = [U(L)](F(K))$ for all $K \in Q$ and $L \in L^1(G)$ with $K * L \in Q$. It follows that $S_{U,Q} \subset V \subset T_{U,Q}$ and there is a duality between $S_{U,Q}$ and $T_{U,Q}$ such that the topologies on $S_{U,Q}$ are compatible with the duality. With a little more structure on the set Q, it can be proved that $S_{U,Q}$ is complete iff $T_{U,Q}$ is bornological, iff $T_{U,Q}$ is reflexive iff $S_{U,Q}$ and $T_{U,Q}$ are topologically isomorphic with the strong dual of each other. It turns out that $S_{U,Q}$ and $T_{U,Q}$ are both inductive and projective limits of Hilbert spaces. Topological tensor products between two T-spaces have been determined completely. For the details see [3].

IV. A construction based on a non-commutative semi-group of matrices.

Consider

$$Q = \left\{ \begin{pmatrix} a & b & e \\ c & d & f \\ 0 & 0 & 1 \end{pmatrix} \,\middle|\, a,b,c,d,e,f \in \mathbb{C},\ ad - bc = 1,\ \mathrm{Re}\, z > 0 \Rightarrow \mathrm{Re}\,\frac{az+b}{cz+d} > 0 \right\}$$

Take $V = L_2(\mathbb{R})$. The action is, roughly, given by an integral operator on $L_2(\mathbb{R})$

$$\pm \int_{-\infty}^{\infty} \exp\left\{ -\frac{\pi}{c}(az^2 - 2zt + 2(ce - af) + 2ft + af^2 - cef \right\} g(t)\,dt.$$

The obtained distribution space is the dual of the testspace

$$\left\{ f \,\middle|\, f \text{ entire},\ \forall_{A, 0<A<1},\ \sup_{x+iy \in \mathbb{C}} |f(x+iy)| \exp\left(\frac{1}{2} A x^2 - \frac{1}{2}\frac{1}{A} y^2 \right) < \infty \right\}.$$

It is a little bit larger than $(S_{1/2}^{1/2})'$, a Gelfand-Šilov distribution space, cf. [6].

V. Subject of present study.

For R we take the Heisenberg convolution algebra acting on $V = L_2(\mathbb{R})$. For suitable $Q \subset R$ the abstract space $T(V;Q)$ coincides with classical distribution spaces of Schwartz and Gelfand-Šilov. For the commutative case, with examples, see [3].

REFERENCES

1. S. J. L. Eijndhoven, van, J. de Graaf, "Trajectory spaces, generalized functions and unbounded operators", Lect. Notes in Math. 1162, Springer-Verlag, Berlin etc., (1985).

2. S. J. L. Eijndhoven, van, J. de Graaf, "A mathematical introduction to Dirac's formalism", North-Holland Mathematical Library, Vol. 36, Amsterdam etc., (1986).

3. A. F. M. Elst, ter, Distribution theories based on representations of locally compact Abelian topological groups. To appear as EUT-Report, Eindhoven University of Technology, Eindhoven, (1988).

4. K. Yosida, "Functional Analysis", 3rd Edition, Grundlehren Band 123, Springer-Verlag, Berlin etc., (1971).

5. N. Jacobson, "Basic Algebra 1", 2nd Edition, W.H. Freeman, New York, (1985).

6. I. M. Gelfand and G. E. Šilov, Generalized functions, Vol. 2, Ac. Press, New York, (1968).

PRODUCTS OF WIENER FUNCTIONALS ON AN ABSTRACT WIENER SPACE

Shiro Ishikawa

Department of Mathematics
Keio University, Japan

1. INTRODUCTION

Mikusiński in [1] has proved that the product of the distributions $\delta(x)$ and $\text{pf}.\frac{1}{x}$ on the one-dimensional Euclidean space $\mathbb{R}$ exists in the sense of generalized operations and equals $-\frac{1}{2}\delta'(x)$. This result can be easily extended to the case of an n-dimensional Euclidean space $\mathbb{R}^n$, i.e. for any $\ell = (\ell_1, \ell_2, \ldots, \ell_n) \in \mathbb{R}^n$, $(\ell \neq 0)$,

$$\delta((\ell,x)) \cdot \text{pf}.\frac{1}{(\ell,x)} = -\frac{1}{2}\delta'((\ell,x)) \qquad x = (x_1, \ldots, x_n) \in \mathbb{R}^n,$$

where $(\ell, x) = \sum_{k=1}^{n} \ell_k x_k$.

In this paper we shall try to extend the above results to the case of an infinite dimensional space i.e. an abstract Wiener space.

2. WIENER FUNCTIONALS

Let (W, H, μ) be an abstract Wiener space; W is a real separable Banach space, H is a real separable Hilbert space that densely and continuously imbedding in W and μ is a Gaussian measure on W with mean 0 such that

$$\int_W e^{\sqrt{-1}(\ell,w)} \mu(dw) = \exp\{-\frac{1}{2}|\ell|_H^2\} \qquad \ell \in W^*(\subset H^* = H \subset W),$$

where $(\cdot,\cdot)$ is the natural bilinear form on $W^* \times W$ and $|\cdot|_H$ is the Hilbert space H norm. The space $L_p(\mu)$ $(1 \leq p < \infty)$ of p-integral functions on W is usually defined. Let $\{\ell_i\}_{i=1}^{\infty}$ be a complete orthonormal system in H such that $\{\ell_i\}_{i=1}^{\infty} \subset W^*$. A function $F : W \to$ is a polynomial, if there exists a nonnegative integer n and $p(x_1,\ldots,x_n)$ a real polynomial in n variables such that $F(w) = p((\ell_1,w),\ldots,(\ell_n,w))$ $(w \in W)$. We denote by P the set of such polynomials and by P_n the set of the polynomials of degree $\leq n$. For $1 \leq p < \infty$, $P \subset L_p(\mu)$ and its inclusion is dense. We define an operator $T_t : L_1(\mu) \to L_1(\mu)$, $t \geq 0$, by

$$T_t F(w) = \int_W F(e^{-t}w + \sqrt{1-e^{-2t}}\, u)\mu(du) \qquad (F(w) \in L_1(\mu)). \qquad (1)$$

It is well known that the family $\{T_t\}_{t \geq 0}$ forms a contraction semigroup on $L_p(\mu)$ $(1 \leq p < \infty)$. We call it the Ornstein-Uhlenbeck semigroup.

Note that, if $F(w) = p((\ell_1,w),\ldots,(\ell_n,w))$ is a polynomial, (1) is equivalent to the following condition,

$$T_t F(w) = \int \cdots \int_{\mathbb{R}^n} p(e^{-t}(\ell_1,w) + \sqrt{1-e^{-2t}}\, \eta_1,\ldots,e^{-t}(\ell_n,w)$$
$$+ \sqrt{1-e^{-2t}}\, \eta_n) \frac{1}{(\sqrt{2\pi})^n} \cdot e^{-(\eta_1^2+\ldots+\eta_n^2)/2} d\eta_1 \cdots d\eta_n. \qquad (2)$$

And note that for $L_2(\mu)$, T_t can be represented by

$$T_t F = \sum_{n=0}^{\infty} e^{-nt} J_n F \qquad (F \in L_2(\mu)),$$

where J_n $(n = 0,1,\ldots)$ is a projection on the subspace C_n of $L_2(\mu)$, such that $\bigoplus_{k=0}^{n} C_k$ is an $L_2(\mu)$-closure of P_n and $L_2(\mu) = \bigoplus_{n=0}^{\infty} C_n$ is called the Wiener-Ito decomposition. Now, we shall define the infinitesimal generator L of the Ornstein-Uhlenbeck semi-group T_t, which is called the Ornstein-Uhlenbeck operator, as follows: For $F(w) \in P$, define

$$LF = \frac{d}{dt} T_t F \Big|_{t=0} = \sum_{n=0}^{\infty} (-n) J_n F.$$

Definition 1 (Sobolev spaces on an abstract Wiener space)

(i) Let $F \in P$, $1 < p < \infty$, $-\infty < s < \infty$. Then

$$\|F\|_{p,s} := \|(1-L)^{s/2} F\|_{L_p(\mu)},$$

where

$$(1-L)^{s/2} F = \sum_{n=0}^{\infty} (1+n)^{s/2} J_n F \in P.$$

(ii) Let $1 < p < \infty$, $-\infty < s < \infty$. Define

$$\mathcal{D}_{p,s} := \text{"the completion of } P \text{ by the } \|\cdot\|_{p,s}\text{"},$$

(iii) $\mathcal{D}_\infty := \bigcap_{p,s} \mathcal{D}_{p,s}$ and $\mathcal{D}_{-\infty} := \bigcup_{p,s} \mathcal{D}_{p,s}$.

The element of $\mathcal{D}_{-\infty}$ is called a Wiener functional on an abstract Wiener space. If $p \leq p'$ and $s \leq s'$, then $\mathcal{D}_{p',s'} \subset \mathcal{D}_{p,s}$ and its imbedding is densely continuous. Note that $\mathcal{D}^*_{p,s} = \mathcal{D}_{q,-s}$ ($1/p + 1/q = 1$, $-\infty < s < \infty$) and $\mathcal{D}^*_\infty = \mathcal{D}_{-\infty}$.

3. MAIN RESULTS

It is known that, for any $u \in \mathcal{D}_{p,s}$, $T_t u$ can be defined in $\mathcal{D}_\infty$ and $T_t u$ converges to u as $t \to 0_+$ in the topology of $\mathcal{D}_{p,s}$. From this result and the fact that the mapping $\mathcal{D}_\infty \times \mathcal{D}_\infty \ni (u,v) \to u \cdot v \in \mathcal{D}_\infty$ is continuous, we have the following definition:

<u>Definition 2</u> Let u,v be elements in $\mathcal{D}_{-\infty}$. The produvt $u \cdot v$ of u and v is defined by

$$(u \cdot v)(\phi) = \lim_{\varepsilon \to 0_+} \int_W T_\varepsilon u(w) \cdot T_\varepsilon u(w) \cdot \phi(w) \mu(dw) \qquad (\phi \in \mathcal{D}_\infty),$$

if the limit exists for any $\phi \in \mathcal{D}_\infty$.

From the theorem of Banach-Steinhaus, it is easily seen that $u \cdot v \in \mathcal{D}_{-\infty}$.

Let $S(\mathbb{R})$ be the Schwartz space of rapidly decreasing C^∞-functions on $\mathbb{R}$ and $S'(\mathbb{R})$ be its dual space. Define, for any integer k,

$$\|\phi\|_{T_{2k}} = \sup_{x \in \mathbb{R}} \left|\left(1 + x^2 - \frac{d^2}{dx^2}\right)^k \phi(x)\right| \qquad (\phi \in S(\mathbb{R})).$$

Let $T_{2k} := "\|\cdot\|_{T_{2k}} -$ a completion of $S(\mathbb{R})$". If $k' < k$, then $T_{2k} \subset T_{2k'}$. And $S(\mathbb{R}) = \bigcap_k T_{2k}$ and $S'(\mathbb{R}) = \bigcup_k T_{2k}$. Let $\delta(x)$ be a Dirac function at $0 \in \mathbb{R}$ and let pf.$(1/x)$ be defined by

$$\int_\mathbb{R} \text{pf.} \frac{1}{x} \phi(x) dx = \text{P.V.} \int_\mathbb{R} \frac{\phi(x)}{x} dx \qquad (\phi \in S(\mathbb{R})).$$

Since $\delta(x)$ and pf.$\frac{1}{x}$ belong to T_{-2}, the composed function $\delta((\ell,w))$

and pf. $\frac{1}{(\ell,w)}$ can be defined $\mathcal{D}_{p,-2}$ for any $\ell \in W^*$ ($\ell \neq 0$). Similarly $\delta'((\ell,w)) \in \mathcal{D}_{p,-4}$ (see S. Watanabe [2], pp. 57).

Now we have the following theorem.

Theorem Let $\ell \in W^*$ such that $\ell \neq 0$. Then
$$\delta((\ell,w)) \cdot pf. \frac{1}{(\ell,w)} = -\frac{1}{2} \delta'((\ell,w)) \quad (w \in W).$$

Proof Let $\varepsilon > 0$. Put $\sigma = (e^\varepsilon) \sqrt{1-e^{-2\varepsilon}}$ and $\rho(\sigma) = \sigma/\sqrt{1+\sigma^2}$ ($\sigma > 0$).

Note that $\varepsilon \to 0_+$ iff $\sigma \to 0_+$. Without loss of generality, we assume that $\|\ell\|_H = 1$. Then we see that

$$T_\varepsilon[\delta((\ell,w))] = \int_R \delta(e^{-\varepsilon}(\ell,w) + \sqrt{1-e^{-2\varepsilon}}\eta) \cdot \frac{1}{\sqrt{2\pi}} e^{-\eta^2/2} d\eta$$
$$= \frac{1}{\sqrt{2\pi} \cdot \sqrt{1-e^{-2\varepsilon}}} \cdot e^{-\frac{1}{2} \cdot \frac{e^{-2\varepsilon}(\ell,w)^2}{1-e^{-2\varepsilon}}} = \frac{1}{\sqrt{2\pi}\rho(\sigma)} e^{-\frac{(\ell,w)^2}{2\sigma^2}} \quad (3)$$

and

$$T_\varepsilon[pf. \frac{1}{(\ell,w)}] = P.V. \int_R \frac{1}{(e^\varepsilon(\ell,w) + \sqrt{1-e^{-2\varepsilon}}\eta)} \frac{1}{\sqrt{2\pi}} \cdot e^{-\eta^2/2} d\eta$$
$$= P.V. \int \frac{1}{y} \frac{1}{\sqrt{2\pi}\rho(\sigma)} \cdot e^{-\frac{(y-(\ell,w))^2}{2\sigma^2}} dy. \quad (4)$$

Let $\{\ell_i\}_{i=1}^\infty$ be a complete orthonormal system in H, such that $\{\ell_i\}_{i=1}^\infty \subset W^*$ and $\ell_1 = \ell$. Since

$$\frac{1}{\sqrt{2\pi}\rho(\sigma)} \cdot e^{-\eta^2/2\sigma^2} \in S(R^n) \quad (\sigma > 0) \text{ and } \int_R \frac{1}{\sqrt{2\pi}\rho(\sigma)} \cdot e^{-\eta^2/2\sigma^2} d\eta \to 1$$

($\sigma \to 0$), we see, by the theorem Mikusiński, that, for any $p((\ell_1,w),\ldots,(\ell_n,w)) \in P$,

$$\lim_{\varepsilon \to 0_+} \int_W T_\varepsilon[\delta((\ell,w))] \cdot T_\varepsilon[pf. \frac{1}{(\ell,w)}] p((\ell_1,w),\ldots,(\ell_n,w)) \mu(dw)$$

$$= \lim_{\varepsilon \to 0_+} \int_W \frac{1}{\sqrt{2\pi}\rho(\sigma)} \cdot e^{-(\ell_1,w)^2/2\sigma^2} \cdot P.V. \int_R \frac{1}{y} \frac{1}{\sqrt{2\pi}\rho(\sigma)} \cdot e^{-(y-(\ell_1,w))^2/2\sigma^2} dy$$

$$p((\ell_1,w),\ldots,(\ell_n,w))\mu(dw) =$$

$$= \lim_{\sigma \to 0_+} \int \cdots \int_{\mathbb{R}^n} \frac{1}{\sqrt{2\pi}\rho(\sigma)} \cdot e^{-\eta_1^2/2\sigma^2} \text{ P.V.} \int_{\mathbb{R}} \frac{1}{y} \frac{1}{\sqrt{2\pi}\rho(\sigma)} \cdot e^{-(y-\eta_1)^2/2\sigma^2} dy$$

$$\cdot p(\eta_1, \ldots, \eta_n) \frac{1}{(\sqrt{2\pi})^n} \cdot e^{-(\eta_1^2 + \ldots + \eta_n^2)/2} d\eta_1 \ldots d\eta_n$$

$$= \int \cdots \int_{\mathbb{R}^n} \delta'(\eta_1) \cdot p(\eta_1, \ldots, \eta_n) \frac{1}{(\sqrt{2\pi})^n} \cdot e^{-(\eta_1^2 + \ldots + \eta_n^2)/2} d\eta_1 \ldots d\eta_n$$

$$= \int \cdots \int_{\mathbb{R}^{n-1}} \frac{1}{2} \partial_1 p(0, \eta_2, \ldots, \eta_n) \frac{1}{(\sqrt{2\pi})^{n-1}} \cdot e^{-(\eta_2^2 + \ldots + \eta_n^2)/2} d\eta_2 \ldots d\eta_n. \tag{5}$$

Also we see that, for any $p((\ell_1, w), \ldots, (\ell_n, w)) \in P$,

$$\int_W -\frac{1}{2} \delta'((\ell, w)) p(\ell_1, w), \ldots, (\ell_n, w)) \mu(dw)$$

$$= \int \cdots \int_{\mathbb{R}^n} -\frac{1}{2} \delta'(\eta_1) \cdot p(\eta_1, \ldots, \eta_n) \cdot \frac{1}{(\sqrt{2\pi})^n} \cdot e^{-(\eta_1^2 + \ldots + \eta_n^2)/2} d\eta_1 \ldots d\eta_n$$

$$= \int \cdots \int_{\mathbb{R}^{n-1}} \frac{1}{2} \partial_1 p(0, \eta_2, \ldots, \eta_n) \frac{1}{(\sqrt{2\pi})^{n-1}} \cdot e^{-(\eta_2^2 + \ldots + \eta_n^2)/2} d\eta_2 \ldots d\eta_n. \tag{6}$$

Also, by the theorem of Mikusiński and the theorem of Banach-Steinhaus, there exists $M > 0$ such that, for any $\varepsilon > 0$ and $\phi \in S(\mathbb{R}^n)$,

$$\left| \int_{\mathbb{R}} \frac{1}{\sqrt{2\pi}\rho(\sigma)} \cdot e^{-\eta^2/2\sigma^2} \cdot (\text{P.V.} \int_{\mathbb{R}} \frac{1}{y} \frac{1}{\sqrt{2\pi}\rho(\sigma)} \cdot e^{-(y-\eta)^2/2\sigma^2} dy) \cdot \phi(\eta) d\eta \right|$$

$$\leq M(\sup_\eta |\phi(\eta)| + \sup_\eta |\phi'(\eta)|). \tag{7}$$

From (3), (4) and (7), we see that, for any $t > 0$,

$$\left| T_t(T_\varepsilon[\delta((\ell, w))] \cdot T_\varepsilon[\text{pf.} \frac{1}{(\ell, w)}]) \right|$$

$$= \left| \int \left[\int_{\mathbb{R}} \frac{1}{\sqrt{2\pi}\rho(\sigma)} \cdot e^{-((e^{-t}(\ell,w) + \sqrt{1-e^{-2t}}\eta)^2)/2\sigma^2} \right. \right.$$

$$\left. \cdot \left[\text{P.V.} \int \frac{1}{y} \frac{1}{\sqrt{2\pi}\rho(\sigma)} \cdot e^{-(y-(e^{-t}(\ell,w) + \sqrt{1-e^{-2t}}\eta))^2/2\sigma^2} dy \right] \cdot \frac{1}{\sqrt{2\pi}} e^{-\eta^2/2} d\eta \right|$$

$$= \left| \int \left[\frac{1}{\sqrt{2\pi}\rho(\sigma)} \cdot e^{-\eta^2/2\sigma^2} \cdot (\text{P.V.} \int \frac{1}{\sqrt{2\sigma}\rho(\sigma)} \cdot e^{-(y-\eta)^2/2\sigma^2} dy) \right] \cdot \right.$$

$$\cdot \frac{1}{\sqrt{2\pi}\sqrt{1-e^{-2t}}} e^{-(\eta-e^{-t}(\ell,w))^2/2(1-e^{-2t})} d\eta|$$

$$\leq M \left(\sup_{\eta} \left| \frac{1}{\sqrt{2\pi}\sqrt{1-e^{-2t}}} \cdot e^{-(\eta-e^{-t}(\ell,w))^2/2(1-e^{-2t})} \right| + \right.$$

$$\left. + \sup_{\eta} \left| \frac{1}{\sqrt{2\pi}(\sqrt{1-e^{-2t}})^3} \cdot e^{-(\eta-e^{-t}(\ell,w))^2/2(1-e^{-2t})} \right| \right)$$

$$= M \left(\frac{1}{\sqrt{2\pi}\sqrt{1-e^{-2t}}} + \frac{1}{\sqrt{2\pi}(1-e^{-2t})} e^{-1/2} \right). \qquad (8)$$

Since, for $s > 1$,

$$(1-L)^{-s} = \sum_{n=0}^{\infty} \frac{1}{(1+n)^s} J_n = \sum_{n=0}^{\infty} \frac{1}{\Gamma(s)} \int_0^{\infty} e^{-(1+n)t} t^{s-1} dt \, J_n$$

$$= \frac{1}{\Gamma(s)} \int_0^{\infty} e^{-t} t^{s-1} \sum_{n=0}^{\infty} e^{-nt} J_n \, dt = \frac{1}{\Gamma(s)} \int_0^{\infty} e^{-t} t^{s-1} T_t dt,$$

we see from (8) that,

$$\left\| T_\varepsilon[\delta((\ell,w))] \cdot T_\varepsilon[\text{pf.} \frac{1}{(\ell,w)}] \right\|_{p,-s}$$

$$\leq [\int_W | \frac{1}{\Gamma(s)} \int_0^{\infty} e^{-t} t^{s-1} M \left(\frac{1}{\sqrt{2\pi}\sqrt{1-e^{-2t}}} + \frac{1}{\sqrt{2\pi}(1-e^{-2t})} \right) e^{-1/2} dt |^p \mu(dw)]^{1/p}$$

$$\leq M'. \qquad (9)$$

Let $\psi(w)$ be any element in $\mathcal{D}_\infty$. Let $\gamma > 0$. By (9), there exists $\psi_0(w) \in P$ such that

$$\left| \int_W T_\varepsilon[\delta((\ell,w))] \cdot T_\varepsilon[\text{pf.} \frac{1}{(\ell,w)}] (\psi(w) - \psi(w_0)) \mu(dw) \right| \leq \frac{\gamma}{2} \text{ for any } \varepsilon > 0$$

and

$$\left| \int_W -\frac{1}{2} \delta'((\ell,w))(\psi(w) - \psi(w_0)) \mu(dw) \right| \leq \frac{\gamma}{2}.$$

Then, we see, from (5) and (6), that

$$\left| \int_W T_\varepsilon[\delta((\ell,w))] \cdot T_\varepsilon[\text{pf.} \frac{1}{(\ell,w)}] \psi(w) \mu(dw) - \int_W -\frac{1}{2} \delta'((\ell,w)) \psi(w) \mu(dw) \right| \leq$$

$$\leq \gamma + \left| \int T_\varepsilon[\delta((\ell,w))] \cdot T_\varepsilon[pf.\frac{1}{(\ell,w)}] \psi_0(w) \mu(dw) \right.$$

$$\left. - \int_W -\frac{1}{2} \delta'((\ell,w)) \psi_0(w) \mu(dw) \right| \to \gamma \quad (\text{as } \varepsilon \to 0_+).$$

Since γ is any positive number, this implies, from Definition 2, that

$$\delta((\ell,w)) \cdot pf.\frac{1}{(\ell,w)} = -\frac{1}{2} \delta'((\ell,w)) \quad (w \in W).$$

This completes the proof. □

REFERENCES

1. J. Mikusiński, On the square of Dirac delta-distribution, <u>Bull. Acad. Polon. Sci., Ser. Sci. Math. Astronom. Phys.</u>, 14, 511–513, (1966).
2. S. Watanabe, "Stochastic differential equations and Malliavin calculus", Tata Institute of Fundamental Research, Bombay, (1984).

CONVOLUTION IN $K'\{M_p\}$-SPACES

A. Kamiński and J. Uryga

Institute of Mathematics
Polish Academy of Sciences
Katowice

1. I. M. Gelfand and G. E. Shilov introduced in [3] (see p. 78) spaces of generalized functions, dual to the spaces $K\{M_p\}$ defined by means of an arbitrary non-decreasing sequence $\{M_p\}$ of functions $M_p : \mathbb{R}^d \to [1,\infty]$, which are supposed to be continuous on the set $S = S_p = \{x \in \mathbb{R}^d : M_p(x) < \infty\}$ ($p \in \mathbb{N}$). We shall assume that S is open and the following conditions are valid:

(P) For each $p \in \mathbb{N}$ there exists a $q > p$ such that for every $\varepsilon > 0$ there is a $T > 0$ with the property $m_{pq}(x) = M_p(x)M_q^{-1}(x) < \varepsilon$ whenever $x \in S$ and $|x| > T$;

(F) For each p $\mathbb{N}$ there are a $q > p$ and $C_p > 0$ such that $M_p(x+y)$ $\leq C_p M_q(x) M_q(y)$ for $x, y \in \mathbb{R}^d$ (cf. [3], p. 87; [13]; [8]).

The space $K\{M_p\}$ is defined to consist of all smooth functions ϕ on $\mathbb{R}^d$ such that $\phi(x) = 0$ for $x \notin S$ and

$$\|\phi\|_p := \sup_{|a|\leq p} \sup_{x \in S} M_p(x) |D^a \phi(x)| < \infty \quad (p \in \mathbb{N}), \tag{1}$$

where $|a| = \alpha_1 + \ldots + \alpha_d$ for the multi-index $a = (\alpha_1, \ldots, \alpha_d)$. The set $K\{M_p\}$ endowed with the sequence of norms (1) is a complete locally convex space. The spaces $K'\{M_p\}$, dual to $K\{M_p\}$, embrace many known spaces of distributions, e.g. the Schwartz spaces $\mathcal{D}'_K$, S' and the spaces $(S_{\alpha,A})'$, $(W_{M,a})'$ (see [3], pp. 176, 247). It is easy to see that $\mathcal{D}$ is a dense subspace iff $S = \mathbb{R}^d$ and in this case $K'\{M_p\} \subset \mathcal{D}'$. We assume that $S = \mathbb{R}^d$ in the sequel.

2. The convolution in $K'\{M_p\}$-spaces can be defined in a similar way as V. S. Vladimirov defines in [14] (pp. 137-138) the convolution of dis-

tributions, i.e. by using unit-sequences (approximate units).

By a unit-sequence we mean every sequence $\{\eta_n\}$ of functions of the class $\mathcal{D}$ on $\mathbb{R}^d$ such that

(α) For any compact $K \subset \mathbb{R}^d$ there is an index n_0 such that $\eta_n(x) = 1$ for $n > n_0$ and $x \in K$;

(β) For any multi-index a,

$$\sup_{x} \sup_{k \in \mathbb{N}} |D^a \eta_k(x)| \leq c_a < \infty.$$

Given $f, g \in \mathcal{D}'$ [$f, g \in K'\{M_p\}$] on $\mathbb{R}^d$, the convolution $f * g$ exists in $\mathcal{D}'$ [in $K'\{M_p\}$] on $\mathbb{R}^d$ if the limit in the equation

$$\lim_{n \to \infty} < f(x) \otimes g(y), \eta_n(x,y) \phi(x+y) > = < f * g, \phi >$$

exists for any unit-sequence $\{\eta_n\}$ on $\mathbb{R}^{2d}$ and ϕ in $\mathcal{D}$ [$K\{M_p\}$] on $\mathbb{R}^d$. Then the equation defines $f * g$ in $\mathcal{D}'$ [in $K'\{M_p\}$] on $\mathbb{R}^d$.

There exist in literature very well known equivalent conditions for the existence of the convolution $f * g$ in $\mathcal{D}'$ on $\mathbb{R}^d$ expressed in terms of the supports $A, B \subset \mathbb{R}^d$ of f, g, respectively (see e.g. [4], p. 383):

(i) For each compact set $K \subset \mathbb{R}^d$, the set $(K - A) \cap B$ is compact in $\mathbb{R}^d$;

(ii) For each compact set $K \subset \mathbb{R}^d$, the set $A \cap (K - B)$ is compact in $\mathbb{R}^d$;

(iii) For each compact set $K \subset \mathbb{R}^d$, the set $(A \times B) \cap K^\Delta$ is compact in $\mathbb{R}^{2d}$,

where $K - A = \{x - y \in \mathbb{R}^d : x \in K, y \in A\}$ and $K^\Delta = \{(x,y) \in \mathbb{R}^{2d} : x + y \in K\}$.

If the sets A and B are closed in $\mathbb{R}^d$ (which is true in case of supports), it is easy to see that the word "compact" in conditions (i) - (iii) can be equivalently replaced by the word "bounded". J. Mikusiński proved in [10] (see also [1], p. 125) that conditions (i) - (ii) in a similarly modified form are equivalent to the following very convenient condition:

(iv) If $x_n \in A$, $y_n \in B$ ($n \in \mathbb{N}$) and $|x_n| + |y_n| \to \infty$, then $|x_n + y_n| \to \infty$.

Following [1] (p. 125), we shall call sets $A, B \subset \mathbb{R}^d$ compatible if they satisfy (iv).

The following result is classical:

(*) If $f, g \in \mathcal{D}'(\mathbb{R}^d)$ and supp f, supp g are compatible, then $f * g$ exists in $\mathcal{D}'(\mathbb{R}^d)$ and supp $f * g \subset A + B$,

where $A + B$ is the algebraic sum of A and B (see [4], p. 368; [1], p. 124). These results cannot be improved as it was shown in [7], namely:

(**) If $A, B \subset \mathbb{R}^d$ are two sets with the property that $f * g$ exists in $\mathcal{D}'(\mathbb{R}^d)$ for arbitrary $f, g \in \mathcal{D}'(\mathbb{R}^d)$ such that supp $f \subset A$, supp $g \subset B$, then A and B are compatible.

In [5], [6] (see also [2]), it was shown that for $f, g \in S'(\mathbb{R}^d)$ the compatibility of the supports of f and g does not guarantee the existence $f * g$ in $S'(\mathbb{R}^d)$ and Theorem 1 in [6] shows that the same is true for the convolution in $K'\{M_p\}$.

In case of the convolution in S', the condition of so-called polynomial compatibility of supports of $f, g \in S'$, introduced in [5], [6], is sufficient for the existence of $f * g$ in S'. Also an analogue of (**) holds (see [7]).

The notion of polynomial compatibility was modified for $K'\{M_p\}$-spaces first in [11] and then, in a simpler form, in [12] and [13]. Namely, sets $A, B \subset \mathbb{R}^d$ are called in [12], [13] M_p-compatible if for every p there exist an index $q > p$ and a constant $c_p > 0$ such that $M_p(x) M_p(y) \leq c_p M_q(x + y)$ for arbitrary $x \in A$, $y \in B$. For this notion, analogues of results (*) and (**) also hold (see [12], [13]).

In this paper, we are going to generalize the notion of the M_p-compatibility for the case of n sets in $\mathbb{R}^d$ and prove some results about the existence of the convolution $f_1 * \ldots * f_k$ in $K'\{M_p\}$, where $f_i \in K'\{M_p\}$ for $i = 1, 2, \ldots, k$. This leads us to conditions for the commutativity of the convolution in $K'\{M_p\}$. In particular, we obtain the results proved in [5] for the convolution in S'. We also recall analogous results for the convolution in $\mathcal{D}'$, given in [5].

Some results presented here were announced in [8].

3. Let $f_1, \ldots, f_k \in \mathcal{D}'[f_1, \ldots, f_k \in K'\{M_p\}]$. We say that the convolution $f_1 * \ldots * f_k$ exists in $\mathcal{D}'$ [in $K'\{M_p\}$] on $\mathbb{R}^d$ if the limit in the equation

$$\lim_{n \to \infty} < f_1(x_1) \otimes \ldots \otimes f_k(x_k), \eta_n(x_1, \ldots, x_k) \phi(x_1 + \ldots + x_k) > =$$

$$= < f_1 * \ldots * f_k, \phi >$$

exists for any unit-sequence $\{\eta_n\}$ on $\mathbb{R}^{kd}$ and $\phi \in \mathcal{D}[\phi \in K\{M_p\}]$ on $\mathbb{R}^d$. Then, the equation defines $f_1 * \ldots * f_k$ in $\mathcal{D}'[K'\{M_p\}]$ on $\mathbb{R}^d$.

We say that the sets $A_1, \ldots, A_k \subset \mathbb{R}^d$ are compatible if

$$x_{in} \in A_i \ (1 \leq i \leq k) \text{ and } \sum_{i=1}^{k} |x_{in}| \to \infty \text{ implies } |\sum_{i=1}^{k} x_{in}| \to \infty;$$

and M_p-compatible if for each $p \in \mathbb{N}$, there are an index $q > p$ and a constant $c_p > 0$ such that

$$\sum_{i=1}^{k} M_p(x_i) \le c_p M_q\left(\sum_{i=1}^{k} x_i\right) \text{ for } x_i \in A_i \ (i = 1,\ldots,k). \qquad (2)$$

It can be shown that the M_p-compatibility for $M_p(x) = (1 + |x|)^p$ coincides with the polynomial compatibility of sets $A_1,\ldots,A_k$ introduced in [5].

The compatibility and M_p-compatibility of n sets can be reduced to the respective property of smaller number of sets. Namely the following statement formulated parallelly for compatible and M_p-compatible sets is true. For compatible and polynomially compatible sets the statement can be found in [5].

Theorem 1 If sets $A_1,\ldots,A_k \subset \mathbb{R}^d$ are compatible [M_p-compatible], then for every m, $1 \le m \le k$, and every partition $\{K_j\}_{1 \le j \le m}$ of $K = \{1,\ldots,k\}$ the two properties hold:

(a) the sets A_i ($i \in K_j$) are compatible [M_p-compatible] for $j = 1,\ldots,m$;

(b) the sets $B_j = \sum_{i \in K_j} A_i = \{\sum_{i \in K_j} x_i : x_i \in A_i\}$ ($j = 1,\ldots,m$) are compatible [M_p-compatible].

Conversely, if for some m, $1 \le m \le k$, there is a partition $\{K_j\}_{1 \le j \le m}$ of K such that conditions (a) - (b) are satisfied, then the sets $A_1,\ldots,A_k$ are compatible [M_p-compatible].

Proof Suppose first that the sets $A_1,\ldots,A_k \subset \mathbb{R}^d$ are compatible and consider a partition $\{K_j\}_{1 \le j \le m}$ of K for a fixed m, $2 \le m \le k$.

Given j ($1 \le j \le m$), assume that $\sum_{i \in K_j} |x_{in}| \to \infty$ as $n \to \infty$ for $x_{in} \in A_i$ ($i \in K_j$). Put $x_{in} = x_i^o$ ($n \in \mathbb{N}$), where x_i^o is a fixed element of A_i, for $i \notin K_j$. Since

$$\sum_{i=1}^{k} |x_{in}| = \sum_{i \in K_j} |x_{in}| + \sum_{i \notin K_j} |x_i^o| \to \infty,$$

it follows that

$$\sum_{i \in K_j} |x_{in}| \ge \left|\sum_{i=1}^{k} x_{in}\right| - \left|\sum_{i \notin K_j} x_i^o\right| \to \infty,$$

by the compatibility of the sets $A_1,\ldots,A_K$. Consequently, the sets A_i ($i \in K_j$) are compatible. The compatibility of the sets $B_1,\ldots,B_m$ follows directly from the compatibility of $A_1,\ldots,A_k$ and the triangle inequality. We have thus proved the first part of Theorem 1 for compatible sets.

Assume now that $A_1,\ldots,A_k \subset \mathbb{R}^d$ are M_p-compatible and consider

$\{K_j\}_{1 \le j \le m}$ as before, fixing m ($2 \le m \le k$) and j ($1 \le j \le m$). Let x_i vary in A_i for $i \in K_j$ and let $x_i = x_i^o$ be fixed elements of A_i for $1 \notin K_j$. Then

$$\prod_{i \in K_j} M_p(x_i) \le \prod_{i \in K_j} M_p(x_i) \le c_p M_q\left(\sum_{i \in K_j} x_i + \sum_{i \notin K_j} x_i^o\right) \le c_p' M_r\left(\sum_{i \in K_j} x_i\right)$$

for some $q > p$, $r > q$, $c_p, c_p' > 0$, and arbitrary $x_i \in A_i$ ($i \in K_j$), by virtue of (2) and (F). The sets A_i ($i \in K_j$) are therefore M_p-compatible. Now, if $y_j = \sum_{i \in K_j} x_i$ vary in B_j for $j = 1,\ldots,m$, then

$$\prod_{j=1}^m M_p(y_j) \le c_p \prod_{i=1}^k M_q(x_i) \le c_p' M_r\left(\sum_{i=1}^k x_i\right) = c_p' M_r\left(\sum_{j=1}^m y_j\right)$$

for some $q > p$, $r > q$, $c_p, c_p' > 0$ and arbitrary $y_j \in B_j$ ($j = 1,\ldots,m$), in view of (F) and (2). Consequently, the sets B_j ($j = 1,\ldots,m$) are M_p-compatible and the first part of Theorem 1 for M_p-compatible sets is proved.

The second part of Theorem 1 will be proved simultaneously for compatible and M_p-compatible sets. Considering an arbitrary partition $\{K_j\}_{j \le 1 \le m}$ of K, we shall apply induction with respect to m ($1 \le m \le k$).

Suppose that the assertion is true for some m ($1 \le m \le k$) and let $\{L_j\}_{1 \le j \le m+1}$ be a partition of K such that

(a_1) the sets A_i ($i \in L_j$) are compatible [M_p-compatible] for $j = 1,\ldots,m+1$;

(b_1) the sets $B_j = \sum_{i \in L_j} A_i$ ($j = 1,\ldots,m+1$) are compatible [M_p-compatible].

The second condition, due to the first part of Theorem 1 implies the two conditions:

(b_2) the sets $B_1,\ldots,B_{m-1}, B_m + B_{m+1}$ are compatible [M_p-compatible].

(b_3) the sets B_m, B_{m+1} are compatible [M_p-compatible].

Assume that

$$y_n + z_n \not\to \infty \text{ as } n \to \infty, \qquad (3)$$

where

$$y_n = \sum_{i \in L_m} x_{in}, \quad z_n = \sum_{i \in L_{m+1}} x_{in}.$$

By (b_3) and (3), there exists an increasing sequence $\{k_n\}$ of positive integers such that the sequences $\{y_{k_n}\}$ and $\{z_{k_n}\}$ are bounded. But this and (b_3) for $j = m, m+1$ imply that $\sum_{i \in K_m} |x_{in}| \to \infty$ as $n \to \infty$, where $K_m = L_m \cup L_{m+1}$. Since (3) means that $\sum_{i \in K_m} |x_{in}| \to \infty$ as $n \to \infty$, we have proved that A_i ($i \in K_m$) are compatible.

On the other hand, for every $p \in \mathbb{N}$ there exist indices $q > p$ and

$r > q$ and constants $c_p, c_p' > 0$ such that for arbitrary $x_i \in A_i$ ($i \in K_m$) we have

$$\prod_{i \in K_m} M_p(x_i) \le c_p M_q\left(\sum_{i \in L_m} x_i\right) M_q\left(\sum_{i \in L_{m+1}} x_i\right) \le c_p' M_r\left(\sum_{i \in K_m} x_i\right),$$

according to (a_1) for $j = m$, $m+1$ and (b_3). The above means that the sets A_i ($i \in K_m$) are M_p-compatible.

By (a_1), also the sets A_i ($i \in K_j = L_j$) are compatible [M_p-compatible] for $j = 1, \ldots, m-1$. This and (b_2) yield the compatibility [M_p-compatibility] of the sets $A_1, \ldots, A_k$, in view of the induction hypothesis.

The proof is thus finished. □

4. Let us formulate some corollaries of Theorem 1.

Corollary 1 Let $A_1, \ldots, A_k \subset \mathbb{R}^d$. The following are equivalent:
 (i) $A_1, \ldots, A_k$ are compatible [M_p-compatible];
 (ii) $A_1 + \ldots + A_j$ and A_{j+1} are compatible [M_p-compatible] for all $j = 1, \ldots, k$.

Corollary 2 Let $A, B, C \subset \mathbb{R}^d$. The following are equivalent:
 (i) A, B, C are compatible [M_p-compatible];
 (ii) $A+B, C$ are compatible [M_p-compatible] and A, B are compatible [M_p-compatible].

The above for compatible sets is proved e.g. in [1], pp. 126-127, and for polynomially compatible sets in [5].

Assume that $K_0 = \{i_1, \ldots, i_\ell\}$, where $i_1 < \ldots < i_\ell$. Then the convolution $f_{i_1} * \ldots * f_{i_\ell}$ in $\mathcal{D}'$ [in $K'\{M_p\}$] will be also denoted by $\underset{i \in K_0}{*} f_i$.

Theorem 1 and the results (*), (**) about the convolution in $\mathcal{D}'$ and, analogously, in $K'\{M_p\}$ lead to the following

Theorem 2 Let $f_1, \ldots, f_k \in \mathcal{D}'$ [$K'\{M_p\}$] on $\mathbb{R}^d$. Suppose that supp $f_i \subset A_i$ and A_i ($1 \le i \le k$) are compatible [M_p-compatible]. Then for arbitrary partition $\{K_j\}_{1 \le j \le m}$ of $K = \{1, \ldots, k\}$ all the convolutions in the equality

$$f_1 * \ldots * f_k = \underset{1 \le j \le m}{*}\left(\underset{i \in K_j}{*} f_i\right) \tag{4}$$

exist in $\mathcal{D}'$ [in $K'\{M_p\}$] and (4) holds. Moreover

$$\text{supp } f_1 * \ldots * f_k \subset A_1 + \ldots + A_k.$$

In particular, we have

$$f_1 * \ldots * f_k = (\ldots((f_1 * f_2) * f_3) \ldots) * f_k$$

and, if $k = 3$,

$$f_1 * f_2 * f_3 = (f_1 * f_2) * f_3 = f_1 * (f_2 * f_3) = (f_1 * f_3) * f_2.$$

For $f_1, f_2, f_3 \in \mathcal{D}'$ with compatible supports (see [1], p. 127 and [4], p. 390); for $f_1, f_2, f_3 \in S'$ with polynomially compatible supports (see [5]).

5. In this section we shall show that a theorem analogous to the statement (**) in section 1 holds for the convolutions $f_1 * \ldots * f_k$, i.e. that the converse (in some sense) to Theorem 2 is true. In case of compatibility we shall prove even a little more. Namely,

<u>Theorem 3</u> (cf. Theorem 5 in [12])   Let $A_1, \ldots, A_k \subset \mathbb{R}^d$. Suppose that the convolution $f_1 * \ldots * f_k$ exists in $\mathcal{D}'$ on $\mathbb{R}^d$ for arbitrary $f_i \in K'\{M_p\}$ such that supp $f_i \subset A_i$ ($i = 1, \ldots, k$). Then the sets $A_1, \ldots, A_k$ are compatible.

<u>Proof</u>   Suppose that $A_1, \ldots, A_k$ are not compatible. By Corollary 1, the sets $A_1 + A_2 + \ldots + A_j$ and A_{j+1} are not compatible for some j, $1 \leq j \leq k-1$. This means that there exist sequences $\{x_{in}\}$ of elements of A_i ($i = 1, \ldots, j+1$) such that

$$|\sum_{i=1}^{j} x_{in}| + |x_{j+1,n}| \to \infty; \quad \sum_{i=1}^{j+1} x_{in} \to x_0 \in \mathbb{R}^d \text{ as } n \to \infty. \qquad (5)$$

The above conditions imply that

$$|\sum_{i=1}^{j} x_{in}| \to \infty \text{ and } |x_{j+1,n}| \to \infty \text{ as } n \to \infty.$$

We infer that at least one of the sequences $\{|x_{in}|\}$ ($i = 1, \ldots, j$) is unbounded. Clearly, we may assume that still $|x_{j+1,n}| \to \infty$ and

$$|x_{in}| \to \infty \ (1 \leq i \leq i_0); \quad x_{in} \to x_i \ (i_0 < i \leq j) \text{ as } n \to \infty$$

for some $x_i \in \mathbb{R}^d$ ($i_0 < i \le j$). In addition, it can be assumed that $|x_{i,n+1}| - |x_{in}| > 1$ and

$$\sum_{i=1}^{n} m_{1p}(x_{in}) < \infty \text{ for } i \in I = \{1,\ldots,i_0,j+1\} \text{ and } n \in \mathbb{N},$$

where p is an index for which $m_{1p}(x) \to 0$ as $|x| \to \infty$, by (P).

Define

$$f_i(x) = \sum_{n=1}^{\infty} \delta(x - x_{in}) \text{ for } i \in I; \tag{7}$$

$$f_i(x) = \delta(x - x_{i1}) \text{ for } i \in J = \{i_0+1,\ldots,j,j+2,\ldots,k\}, \tag{8}$$

where x_{i1} for $i = j+2,\ldots,k$ are arbitrarily fixed elements of A_i, respectively. It is evident that f_i defined for $i \in J$ by (8) are members of $K'\{M_p\}$. In case of (7), we have, for $i \in I$,

$$|<f_i,\phi>| \le \sum_{n=1}^{\infty} |\phi(x_{in})| \le \|\phi\|_p \cdot \sum_{n=1}^{\infty} m_{1p}(x_{in}) < \infty,$$

so $f_i \in K'\{M_p\}$ for all $i = 1,\ldots,k$. Also supp $f_i \subset A_i$ for $i = 1,\ldots,k$.

Notice that the second parts of (5) and (6) yield

$$\sum_{i \in I} x_{in} + \sum_{i \in J} x_{i1} \to y_0 \text{ as } n \to \infty, \tag{9}$$

where $y_0 = x_0 + \sum_{i \in J} x_{i1} - \sum_{i_0 < i \le j} x_i \in \mathbb{R}^d$.

Consider an arbitrary non-negative function $\phi \in \mathcal{D}(\mathbb{R}^d)$ such that $\phi(y) > 0$ in some neighbourhood of y_0. We have

$$<f_1 * \ldots * f_k, \phi> \ge \sum_{n=1}^{\infty} \phi(\sum_{i \in I} x_{in} + \sum_{i \in J} x_{i1}) = \infty,$$

in view of (9). The contradiction proves our assertion. □

Theorem 4 Let $A_1,\ldots,A_k \subset \mathbb{R}^d$. Suppose that the convolution $f_1 * \ldots * f_k$ exists in $K'\{M_p\}$ on $\mathbb{R}^d$ for arbitrary $f_i \in K'\{M_p\}$ such that supp $f_i \subset A_i$ ($i = 1,\ldots,k$). Then the sets $A_1,\ldots,A_k$ are M_p-compatible.

Proof By Theorem 3, the sets $A_1,\ldots,A_k$ are compatible. Suppose that they are not M_p-compatible. This means that there exist a $p \in \mathbb{N}$ and sequences $\{x_{in}\}$ of elements of A_i ($i = 1,\ldots,k$) such that

$$\sum_{i=1}^{k} M_p(x_{in}) > 2^{2kn} M_{p+n}\left(\sum_{i=1}^{k} x_{in}\right) \text{ for all } n \in \mathbb{N}. \tag{10}$$

Hence, at least one of the sequences $\{M_p(x_{in})\}$ $(i = 1,\ldots,k)$ and so, by the continuity of M_p, at least one of the sequences $\{|x_{in}|\}$ $(i = 1,\ldots,k)$ is unbounded. Now, the compatibility of $A_1,\ldots,A_k$ implies that the sequence $\{|y_n|\}$ is bounded, where $y_n = \sum_{i=1}^{k} x_{in}$. We can assume that $|y_{n+1}| - |y_n| > 2$ for $n \in \mathbb{N}$. Define

$$f_i(x) = \sum_{n=1}^{\infty} 2^{-n} M_p(x_{in}) \delta(x - x_{in}) \quad (i = 1,\ldots,k).$$

Since, for each $\phi \in K\{M_p\}$,

$$|<f_i,\phi>| \leq \sum_{n=1}^{\infty} 2^{-n} M_p(x_{in}) |\phi(x_{in})| \leq \|\phi\|_p < \infty,$$

we have $f_i \in K'\{M_p\}$ for $i = 1,\ldots,k$.

Put $\phi = \sum_{n=1}^{\infty} \phi_n$, where $\phi_n(x) = M_{p+n}^{-1}(y_n)\omega(x - y_n)$ and ω is a non-negative function of the class $\mathcal{D}$ such that $\omega(0) = 1$ and $\omega(x) = 0$ for $|x| > 1$. Notice that for each $x \in \mathbb{R}^d$ there is at most one index $n_0 = n_0(x)$ such that $|x - y_{n_0}| \leq 1$ and $\phi_{n_0}(x) \neq 0$. Therefore, is such an index n_0 exists, we have

$$M_p(x)|D^a\phi(x)| = M_p(x) M_{p+n_0}^{-1}(y_{n_0}) |D^a\omega(x - y_{n_0})|$$

$$\leq c_p \sup_{|y|\leq 1} |D^a\phi(y)| \sup_{|y|\leq 1} M_1(y) \sup_{n\leq q-p} \{M_q(y_n) M_{p+n}^{-1}(y_n)\} < \infty.$$

Otherwise, $M_p(x)|D^a\phi(x)| = 0$. Thus $\phi \in K\{M_p\}$.

Now, since $\phi \geq 0$ and $\phi(y_n) = M_{p+n}^{-1}(y_n)$, we have

$$<f_1 * \ldots * f_k, \phi> \geq \sum_{n=1}^{\infty} 2^{-kn} \prod_{i=1}^{k} M_p(x_{in}) \phi(y_n)$$

$$\geq \sum_{n=1}^{\infty} M_{p+n}(y_n) \cdot \phi(y_n) = \infty,$$

in view of (10). This completes the proof of our assertion. □

REFERENCES

1. P. Antosik, J. Mikusiński, R. Sikorski, "Theory of distributions. The sequential approach", PNW-Elsevier, Warszawa - Amsterdam (1973).

2. P. Dierolf, J. Voigt, Convolution and S'-convolution of distributions, Collect. Math. 29(3), 185-196 (1978).

3. I. M. Gelfand, G. E. Shilov, "Generalized functions", vol. 2, Academic Press, New York - London (1968).
4. J. Horvath, "Topological Vector Spaces and Distributions", Addison-Wesley, Reading - London (1966).
5. A. Kamiński, Całkowanie i operacje nieregularne (Operacje regularne i nieregularne na dystrybucjach), Doctor Thesis, Warsaw (1975), (Preprint 11, Ser. B, Institute of Mathematics, Polish Academy of Sciences, Warsaw (1981)).
6. A. Kamiński, On convolutions, products and Fourier transforms of distributions, Bull. Acad. Polon. Sci. Sér. Sci. Math. Astronom. Phys. 25, 369-374 (1977).
7. A. Kamiński, On the Rényi theory of conditional probabilities, Studia Math. 74, 151-191 (1984).
8. A. Kamiński and J. Uryga, On the existence and associativity of the convolution of generalized functions in the space $K\{M_p\}'$ of Gel'fand-Shilov, Abstracts Amer. Math. Soc. 42, 93 (1986).
9. L. Kitchens and C. Swartz, Convergence in the dual of certain $K\{M_p\}$-spaces, Colloq. Math. 30, 149-155 (1974).
10. J. Mikusinski, Sequential theory of convolutions of distributions, Studia Math. 29, 151-160 (1968).
11. S. Pilipović, On the convolution in the space of $K\{M_p\}'$-type, Math. Nachr. 120, 103-112 (1985).
12. J. Uryga, O pewnym kryterium istnienia splotu funkcji uogólnionych w przestrzeniach typu $K\{M_p\}'$, Preprint 15, Ser. B, Institute of Mathematics, Polish Academy of Sciences, Warsaw (1986).
13. J. Uryga, On compatibility of supports of generalized functions of Gelfand-Shilov type (submitted to Bull. Ac. Pol.: Math.).
14. V. S. Vladimirov, "Equations of Mathematical Physics", Nauka, Moscow (1967) (in Russian).

THE PROBLEM OF THE JUMP AND THE SOKHOTSKI FORMULAS IN THE SPACE OF

GENERALIZED FUNCTIONS ON A SEGMENT OF THE REAL AXIS

L. V. Kartashova and V. S. Rogozhin

Rostov State University

344090 Rostov on Don, USSR

Let $S_{m,n}$ (m,n are fixed, $m \geq 0$, $n \geq 0$) denote the linear countable normed space of smooth functions that can be represented with their derivatives in the form

$$\phi^{(k)}(t) = \frac{\phi_k^0(t) \ln^{\ell_k}(t-a) \ln^{q_k}(b-t)}{(t-a)^{m+\alpha_k+k}(b-t)^{r+\beta_k+k}}, \quad k = 0,1,2,\ldots, \qquad (1)$$

where $0 \leq \alpha_k < 1$, $0 \leq \beta_k < 1$, $\ell_k, q_k \geq 0$, $\phi_k^0(t)$ ($k = 0,1,2,\ldots$) are smooth functions on (a,b) and H-continuous on $[a,b]$; a function ψ is an H-function or Hölder's function, from H_λ, $\lambda > 0$, if there is a constant A so that $|\psi(t_1) - \psi(t_2)| < A|t_1 - t_2|^\lambda$ for all $t_1, t_2 \in [a,b]$.

The topology in $S_{m,n}$ is introduced by using the system of norms

$$\|\phi(t)\|_r = \max \{\|\phi(t)\|_{L_p(\rho_1)}, \|\phi'(t)\|_{L_p(\rho_2)}, \ldots \|\phi^{(r-1)}(t)\|_{L_p(\rho_2)}\},$$

$$\rho_k(t) = (t-a)^{p(m+k)}(b-t)^{p(n+k)}, \quad (k = 1,2,\ldots,r), \quad p > 1,$$

$$\|\phi(t)\|_{L_p(\rho)} = \left[\int_a^b \rho(t)|\phi(t)|^p dt\right]^{1/p}.$$

Let $\tilde{S}_{m,n}$ denote the lineal of the functions $\phi(t)$ for which in (1) $\ell_k = q_k = \alpha_k = \beta_k = 0$, ($k = 0,1,2,\ldots$). Clearly, $\tilde{S}_{m,n} \subset S_{m,n}$ and therefore $S'_{m,n} \subset \tilde{S}'_{m,n}$. The Cauchy representation of a generalized function $f \in S'_{m,n}$ is given by

$$\hat{f}(z) = \frac{1}{2\pi i} (f_t, \frac{1}{t-z}), \quad z \in \mathbf{Z} \setminus [a,b].$$

Clearly, $\frac{1}{t-z} \in S_{m,n}$, for $z \in \mathbf{Z} \setminus [a,b]$ and $f(z)$ is analytic in $\mathbf{Z} \setminus [a,b]$.

G. Bremermann [1] and D. Mitrović [3] proved the Sokhotski formulas in classical spaces of generalized functions $\mathcal{D}'$ and σ'_α. We shall make use of the Bremermann's scheme bearing in mind that the functions of the $S_{m,n}$-space have been determined on a segment and cannot be integrated. Following Adamar (see [2, pp. 424-425]), the integrals will be conceived in terms of finite part.

Let the following notation be introduced

$$f*(x+ih) = \hat{f}(x+ih) - \hat{f}(x-ih) = \frac{h}{\pi}(f_t, \frac{1}{(t-x)^2 + h^2}), \quad f \in S'_{m,n},$$

$$\phi*(x+ih) = \frac{h}{\pi} \text{ F.P.} \int_a^b \frac{\phi(x)dx}{(t-x)^2 + h^2}, \quad \phi \in \tilde{S}_{m,n}.$$

From the definition of the integral in the sense of Adamar, there follows:

Lemma 1 If $f(t) \in S'_{m,n}$, $\delta_1, \delta_2 > 0$, then

$$\text{F.P.}^{1)} \int_a^b f*(x+ih)\phi(x)dx = \text{F.P.} \lim_{\delta_1,\delta_2 \to 0}{}^{2)} (f_t, \frac{h}{\pi}\int_{a+\delta_1}^{b-\delta_2} \frac{\phi(x)dx}{(x-t)^2 + h^2})$$

at any fixed $h > 0$, $\phi(t) \in \tilde{S}_{m,n}$.

Lemma 2 Under the conditions of Lemma 1

$$\text{F.P.} \lim_{\delta_1,\delta_2 \to 0} (f_t, \frac{h}{\pi}\int_{a+\delta_1}^{b-\delta_2} \frac{\phi(x)dx}{(x-t)^2 + h^2}) = (f_t, \phi*(t+ih)). \tag{2}$$

Proof Consider the expressions

$$\phi^*_{\delta_1,\delta_2}(t+ih) = \frac{h}{\pi}\int_{a+\delta_1}^{b-\delta_2} \frac{\phi(x)dx}{(x-t)^2 + h^2}, \quad \phi*(t+ih) = \frac{h}{\pi} \text{ F.P.} \int_a^b \frac{\phi(x)dx}{(x-t)^2 + h^2}.$$

$\phi^*_{\delta_1,\delta_2}(t+ih)$ may be represented in the form

$$\phi^*_{\delta_1,\delta_2}(t+ih) = \frac{1}{2\pi i}\int_{a+\delta_1}^{b-\delta_2} \frac{\phi^0_0(x)(x-a)^{-m}(b-x)^{-n}}{x-t-ih} dx$$

$$- \frac{1}{2\pi i}\int_{a+\delta_1}^{b-\delta_2} \frac{\phi^0_0(x)(x-a)^{-m}(b-x)^{-n}}{x-t+ih} dx.$$

---

[1] F.P. Indicates that the integral is considered in terms of the finite part.

[2] In this case, F.P. means that the finite part of the limit is to be taken similarly as when defining the F.P. of the integral.

Now use will be made of the identities

$$\frac{(x-a)^{-m}}{x-t \mp ih} = -\sum_{k=0}^{m-1} \frac{(t\pm ih-a)^{k-m}}{(x-a)^{k+1}} + \frac{(t\pm ih-a)^{-m}}{x-t \mp ih},$$

$$\frac{(b-x)^{-n}}{x-t \mp ih} = \sum_{k=0}^{n-1} \frac{(b-t\pm ih)^{k-n}}{(b-x)^{k+1}} + \frac{(b-t\pm ih)^{-n}}{x-t \mp ih}.$$

We have

$$\frac{(x-a)^{-m}(b-x)^{-n}}{x-t \mp ih} = -\sum_{k=0}^{m-1} \frac{(t\pm ih-a)^{k-m}(b-x)^{-n}}{(x-a)^{k+1}} + \frac{(t\pm ih-a)^{-m}}{(b-x)^n(x-t \mp ih)}$$

and

$$\frac{(b-x)^{-n}(t\pm ih-a)^{-m}}{x-t \mp ih} = \sum_{k=0}^{n-1} \frac{(b-t-ih)^{k-n}(t\pm ih-a)^{-m}}{(b-x)^{k+1}} + \frac{(b-t\mp ih)^{-n}(t\pm ih-a)^{-m}}{x-t \mp ih}.$$

With the notation $z_1 = t+ih$, $z_2 = t-ih$, $(h > 0)$

$$c_k = \frac{1}{2\pi i}\int_{a+\delta_1}^{b-\delta_2} \frac{\phi_0^0(x)\,dx}{(x-a)^{k+1}(b-x)^n}, \quad \tilde{c}_k = \frac{1}{2\pi i}\,F.P.\int_a^b \frac{\phi_0^0(x)\,dx}{(x-a)^{k+1}(b-x)^n},$$

$$d_k = \frac{1}{2\pi i}\int_{a+\delta_1}^{b-\delta_2} \frac{\phi_0^0(x)\,dx}{(b-x)^{k+1}}, \quad \tilde{d}_k = \frac{1}{2\pi i}\,F.P.\int_a^b \frac{\phi_0^0(x)\,dx}{(b-x)^{k+1}}.$$

$\phi^*_{\delta_1,\delta_2}(t+ih)$ and $\phi^*(t+ih)$ can be written, respectively, in the form

$$\phi^*_{\delta_1,\delta_2}(t+ih) = -\sum_{k=0}^{m-1} \frac{c_k}{(z_1-a)^{m-k}} + \sum_{k=0}^{n-1} \frac{(z_1-a)^{-m}d_k}{(b-z_1)^{n-k}}$$

$$+ \frac{(z_1-a)^{-m}}{(b-z_1)^n}\int_{a+\delta_1}^{b-\delta_2} \frac{\phi_0^0(x)\,dx}{x-z_1} + \sum_{k=0}^{m-1} \frac{c_k}{(z_2-a)^{m-k}} - \sum_{k=0}^{n-1} \frac{(z_2-a)^{-m}d_k}{(b-z_2)^{n-k}}$$

$$- \frac{(z_2-a)^{-m}}{(b-z_2)^n}\int_{a+\delta_1}^{b-\delta_2} \frac{\phi_0^0(x)\,dx}{x-z_2}; \qquad (3)$$

$$\phi^*(t+ih) = -\sum_{k=0}^{m-1} \frac{\tilde{c}_k}{(z_1-a)^{m-k}} + \sum_{k=0}^{n-1} \frac{(z_1-a)^{-m}\tilde{d}_k}{(b-z_1)^{n-k}} + \frac{(z_1-a)^{-m}}{(b-z_1)^n}\int_a^b \frac{\phi_0^0(x)\,dx}{x-z_1}$$

$$+ \sum_{k=0}^{m-1} \frac{\tilde{c}_k}{(z_2-a)^{m-k}} - \sum_{k=0}^{n-1} \frac{(z_2-a)^{-m}\tilde{d}_k}{(b-z_2)^{n-k}} - \frac{(z_2-a)^{-m}}{(b-z_2)^n}\int_a^b \frac{\phi_0^0(x)\,dx}{x-z_2}. \qquad (4)$$

Since

$$\text{F.P.} \lim_{\delta_1,\delta_2\to 0} (f_t, \frac{h}{\pi} \int_{a+\delta_1}^{b-\delta_2} \frac{\phi_0^0(x)\,dx}{(x-t)^2 + h^2})$$

$$= \text{F.P.} \lim_{\delta_1,\delta_2\to 0} (f_t, \phi^*_{\delta_1,\delta_2}(t+ih) - \phi^*(t+ih)) + (f_t, \phi^*(t+ih)),$$

in order to prove assertion (2), it will be sufficient to show that

$$\text{F.P.} \lim_{\delta_1,\delta_2\to 0} (f_t, \phi^*_{\delta_1,\delta_2}(t+ih) - \phi^*(t+ih)) = 0. \tag{5}$$

To prove assertion (5), there are used the representations (3) and (4) as well as the definition of the integral in the sense of Adamar. □

Lemma 3 The following relationship holds

$$\frac{h}{\pi} \int_a^b \frac{dx}{(x-t)^2 + h^2} \xrightarrow[h\to 0]{S_{m,n}} 1 \quad ^{3)}.$$

Theorem 1 If $\phi(x) \in \tilde{S}_{m,n}$, then

$$\frac{h}{\pi} \text{F.P.} \int_a^b \frac{\phi(x) - \phi(t)}{(x-t)^2 + h^2}\,dx \xrightarrow[h\to 0]{S_{m,n}} 0.$$

Theorem 2 If $f(t) \in S'_{m,n}$, $\phi(t) \in \tilde{S}_{m,n}$, then

$$\lim_{h\to +0} (f_t, \phi^*(t+ih)) = (f,\phi). \tag{6}$$

Proof Since the expression $\lim_{h\to +0}(f_t, \phi^*(t+ih)) = (f,\phi)$ is equivalent to $\lim_{h\to +0}(f_t, \phi^*(t+ih) - \phi(t)) = 0$, in order to prove assertion (6) it would be sufficient to show that

$$(\phi^*(t+ih) - \phi(t)) \xrightarrow[h\to 0]{S_{m,n}} 0.$$

Since, in accordanse with Lemma 3,

$$\frac{h}{\pi} \int_a^b \frac{dx}{(x-t)^2 + h^2} \xrightarrow[h\to 0]{S_{m,n}} 1$$

---

[3] The notation $\psi_N \xrightarrow[N\to 0]{S_{m,n}} \phi$ signifies that ψ_N converges to ϕ in the $S_{m,n}$ topology when $N \to 0$.

we have

$$\lim_{h \to +0} [\phi^*(t+ih) - \phi(t)] = \lim_{h \to +0} \frac{h}{\pi} \text{ F.P.} \int_a^b \frac{\phi(x)-\phi(t)}{(x-t)^2 + h^2} dx.$$

But, by Theorem 1,

$$\frac{h}{\pi} \text{ F.P.} \int_a^b \frac{\phi(x)-\phi(t)}{(x-t)^2 + h^2} dx \xrightarrow[h \to +0]{\tilde{S}_{m,n}} 0, \text{ hence } (\phi^*(t+ih) - \phi(t)) \xrightarrow[h \to 0]{\tilde{S}_{m,n}} 0.$$

Thus,

$$\lim_{h \to +0} (f_t, \phi^*(t+ih)) = (f, \phi),$$

which completes the proof. □

Lemma 4 If $f(t) \in S'_{m,n}$, $\phi(t) \in \tilde{S}_{m,n}$, then

$$\text{F.P.} \int_a^b \tilde{f}(x+ih)\phi(x) dx = (f_t, -\tilde{\phi}(t+ih)),$$

$$\tilde{\phi}(t+ih) = \frac{1}{\pi i} \text{ F.P.} \int_a^b \frac{(x-t)\phi(x) dx}{(x-t-ih)(x-t+ih)}.$$

Lemma 5 Under the conditions of Lemma 4

$$\lim_{h \to +0} (f, -\tilde{\phi}(t+ih)) = (f, -s\phi) = (sf, \phi),$$

where

$$s\phi = \frac{1}{\pi i} \text{ F.P.} \int_a^b \frac{\phi(x)}{x-t} dx.$$

Proof It is sufficient to show that $\tilde{\phi}(t+ih)$ converges to

$$2J(t) \equiv \frac{1}{\pi i} \text{ F.P.} \int_a^b \frac{\phi(x)}{x-t} dx \text{ in } \tilde{S}_{m,n}, \text{ if } h \to +0.$$

For the proof, use is made of the structure of the space of basic functions. □

Let

$$f^{\pm}(x) = \lim_{h \to +0} \hat{f}(x \pm ih) = \lim_{h \to +0} \frac{1}{2\pi i} (f_t, \frac{1}{t-(x \pm ih)}).$$

Theorem 3 If $f(t) \in S'_{m,n}$, $\phi(t) \in \tilde{S}_{m,n}$, there exists $\lim_{h \to +0} (f(x \pm ih), \phi(x)) = (f^{\pm}(x), \phi(x))$ and the following relationships are satisfied

$$f^+(x) + f^-(x) = sf, \quad f^+(x) - f^-(x) = f(x).$$

Proof   We have

$$\lim_{h\to +0} (\hat{f}(x+ih) - \hat{f}(x-ih), \phi(x)) = \lim_{h\to +0} (f^*(x+ih), \phi(x)) =$$

$$= \lim_{h\to +0} \text{F.P.} \int_a^b b^*(x+ih)\phi(x)dx.$$

According to Lemmas 1 and 2

$$\text{F.P.} \int_a^b f^*(x+ih)\phi(x)dx = (f_t, \phi^*(t+ih));$$

by Theorem 2

$$\lim_{h\to +0} (f, \phi^*(t+ih)) = (f, \phi). \tag{7}$$

We know that

$$\lim_{h\to +0} (\hat{f}(x+ih) + \hat{f}(x-ih), \phi(x)) = \lim_{h\to +0} (\tilde{f}(x+ih), \phi(x))$$

$$= \lim_{h\to +0} \text{F.P.} \int_a^b \tilde{f}(x+ih)\phi(x)dx.$$

From Lemma 4

$$\text{F.P.} \int_a^b \hat{f}(x+ih)\phi(x)dx = (f, -\tilde{\phi}(t+ih)),$$

while from Lemma 5

$$\lim_{h\to +0} (f, -\tilde{\phi}(t+ih)) = (f, -s\phi).$$

Thus,

$$\lim_{h\to +0} (\hat{f}(x+ih) + \hat{f}(x-ih), \phi(x)) = (f, -s\phi) = (sf, \phi). \tag{8}$$

Taking into account (7) and (8), we obtain

$$\lim_{h\to +0} (2\hat{f}(x+ih), \phi(x)) = \lim_{h\to +0} ([\hat{f}(x+ih) - \hat{f}(x-ih)]$$

$$+ [\hat{f}(x+ih) + \hat{f}(x-ih)], \phi(x)) = (f, \phi) + (sf, \phi).$$

Hence, $\lim_{h\to +0} (\hat{f}(x+ih),\phi(x))$ exists and

$$\lim_{h\to +0} (\hat{f}(x+ih),\phi(x)) = (\tfrac{1}{2} f + \tfrac{1}{2} sf, \phi).$$

Similarly,

$$\lim_{h\to +0} (\hat{f}(x-ih),\phi(x)) = \left(-\tfrac{1}{2} f + \tfrac{1}{2} sf, \phi\right).$$

Hence,

$$f^{\pm}(x) = \pm\tfrac{1}{2} f + \tfrac{1}{2} sf, \quad f^{+}(x) - f^{-}(x) = f, \quad f^{+}(x) + f^{-}(x) = sf(x)$$

on $S_{m,n}$. □

Now, it is easy to obtain a solution of the following problem of the jump.

Function $\Phi(z)$ can be found locally holomorphic everywhere except on segment $[a,b]$ whose limiting values belong to the space $S'_{m,n}$ satisfying the boundary condition: $(\Phi^{+}(t) - \Phi^{-}(t),\phi(t)) = (g,\phi(t))$ where $\phi(t) \in \tilde{S}_{m,n}$, $g(t) \in S'_{m,n}$. The solution of the formulated problem has the form

$$\Phi(z) = \frac{1}{2\pi i} (g(x), \frac{1}{x-z}), \quad z = t \pm ih.$$

REFERENCES

1. G. Bremermann, "Distributions, Complex Variables, and Fourier Transforms", Mir, Moscow, (1968), (in Russian).
2. R. Edwards, "Functional Analysis, Theory and Applications", Mir, Moscow, (1969), (in Russian).
3. D. Mitrović, A singular convolution equation in the space of distributions, Publ. Inst. Math. Beograd, 21(35), 151-163, (1977).

A GENERALIZED FRACTIONAL CALCULUS AND INTEGRAL TRANSFORMS

Virginia Kiryakova

Institute of Mathematics
Bulgarian Academy of Sciences
Sofia 1090, Bulgaria

In this paper a generalized fractional calculus and its applications to different topics in analysis, especially to some integral transforms, are discussed. The kernel-function of the generalized operators of integration of fractional multiorder considered here is a suitably chosen case of Meijer's G-function:

$$G_{pq}^{mn}\left[\sigma \Big| \begin{matrix} a_1,\ldots,a_p \\ b_1,\ldots,b_q \end{matrix}\right] = \frac{1}{2\pi i}\int_L \frac{\prod_{k=1}^{m}\Gamma(b_k - s)\prod_{j=1}^{n}\Gamma(1 - a_j + s)}{\prod_{k=m+1}^{q}\Gamma(1 - b_k + s)\prod_{j=n+1}^{p}\Gamma(a_j - s)} \sigma^s ds \quad (1)$$

([1],[2]).

Definition 1 ([3], [4]) Let $m \geq 1$ be an integer, $\beta > 0$; $\gamma_1,\ldots,\gamma_m$ and $\delta_1 \geq 0,\ldots,\delta_m \geq 0$ be arbitrary real numbers. Consider $\gamma = (\gamma_1,\ldots,\gamma_m)$ as a "multiweight" and $\delta = (\delta_1,\ldots,\delta_m)$ as a "multiorder" of integration. Every operator of the form

$$Rf(x) = x^{\beta\delta_0} R_{\beta,m}^{(\gamma_k),(\delta_k)} f(x), \quad (2)$$

where $\delta_0 \geq 0$ is arbitrary and

$$R_{\beta,m}^{(\gamma_k),(\delta_k)} f(x) = \int_0^1 G_{mm}^{m0}\left[\sigma \Big| \begin{matrix} (\gamma_k + \delta_k)_1^m \\ (\gamma_k)_1^m \end{matrix}\right] f(x\sigma^{1/\beta}) d\sigma$$

$$= x^{-\beta}\int_0^x G_{mm}^{m0}\left[(\tau/x)^{\beta} \Big| \begin{matrix} (\gamma_k + \delta_k)_1^m \\ (\gamma_k)_1^m \end{matrix}\right] f(\tau) d(\tau^{\beta}) \quad (3)$$

is said to be a <u>generalized</u> (m-dimensional) <u>operator of fractional integra-</u>

tion of the Riemann-Liouville type, or briefly: a generalized R.-L. fractional integral.

It is convenient to consider operators (2) in the linear sets of functions $C_\alpha^{(\ell)}$, $\ell \geq 0$ defined as follows:

$$C_\alpha^{(\ell)} = \{f(x) = x^p \tilde{f}(x); \; p > \alpha, \; \tilde{f} \in C^{(\ell)}[0,\infty)\}; \quad C_\alpha^{(0)} := C_\alpha.$$

So, (2) are well-defined in C_α with $\alpha \geq \max_k [-\beta(\gamma_k + 1)]$ and

$$R_{\beta,m}^{(\gamma_k),(\delta_k)} : C_\alpha \to C_\alpha^{(\eta_1 + \ldots + \eta_m)},$$

where

$$\eta_k = \begin{cases} [\delta_k] + 1 & \text{for noninteger } \delta_k, \\ \delta_k & \text{for integer } \delta_k, \; k = 1, \ldots, m. \end{cases}$$

In [3], [4] the corresponding <u>fractional derivatives</u>

$$Df(x) = D_{\beta,m}^{(\gamma_k),(\delta_k)} x^{-\beta\delta_0} f(x)$$

are defined too:

$$D_{\beta,m}^{(\gamma_k),(\delta_k)} f(x) = \left[\prod_{k=1}^{m} \prod_{j=1}^{\eta_k} \left(\frac{1}{\beta} x \frac{d}{dx} + \gamma_k + j \right) \right] R_{\beta,m}^{(\gamma_k+\delta_k),(\eta_k-\delta_k)} f(x), \quad (4)$$

such that

$$D_{\beta,m}^{(\gamma_k),(\delta_k)} : C_\alpha^{(\eta_1+\ldots+\eta_m)} \to C_\alpha$$

and (for the proof see [5]):

$$D_{\beta,m}^{(\gamma_k),(\delta_k)} R_{\beta,m}^{(\gamma_k),(\delta_k)} f(x) = f(x) \quad \text{for every } f \in C_\alpha.$$

The generalized fractional integrals (3) are specializations of the generalized operators of the fractional integration of Kalla [6]:

$$Rf(x) = \int_0^1 \Phi(\sigma)\sigma^\gamma f(x\sigma) d\sigma = x^{-\gamma-1} \int_0^x \Phi(\tau/x)\tau^\gamma f(\tau) d\tau$$

with arbitrary kernel-function $\Phi(\sigma) \in C(0,1)$. It seems that the appropriate choice of this function to be a G_{mm}^{m0}-function is important for obtaining fractional integrals with many useful applications. Most of the known operators of fractional integration and differentiation investigated and used by different authors till now are quite <u>special cases</u> of our operators (2), (4). Indeed:

(i) Let $\underline{m = 1}$. Since

$$G_{11}^{10}\left[\sigma \left| \begin{array}{c} \gamma+\delta \\ \gamma \end{array}\right.\right] = \frac{(1-\sigma)^{\delta-1}}{\Gamma(\delta)} \sigma^\gamma, \quad 0 < \sigma < 1,$$

particular cases of (2), (3) are the well-known:

<u>Riemann-Liouville</u> (R.-L.) fractional integral of order $\delta \geq 0$:

$$R^\delta f(x) = \int_0^x \frac{(x-\tau)^{\delta-1}}{\Gamma(\delta)} f(\tau) d\tau = x^\delta R_{1,1}^{0,\delta} f(x); \quad R^0 f(x) = f(x),$$

<u>Erdélyi-Kober</u> (E.-K.) fractional integral:

$$I_\beta^{\gamma,\delta} f(x) = \frac{x^{-\beta(\gamma+\delta)}}{\Gamma(\delta)} \int_0^x (x^\beta - \tau^\beta)^{\delta-1} \tau^{\beta\gamma} f(\tau) d(\tau^\beta)$$

$$= \int_0^1 \frac{(1-\sigma)^{\delta-1}}{\Gamma(\delta)} \sigma^\gamma f(x\sigma^{1/\beta}) d\sigma = R_{\beta,1}^{\gamma,\delta} f(x) \quad \text{(see (13) below)} \tag{5}$$

as well as many other "one-dimensional" fractional integrals as <u>Džrbasjan Gelfond-Leontijev</u> (D.-G.-L.) operators of integration ([7], [8], [9]):

$$\ell_{\rho,\mu} f(x) = \frac{x}{\Gamma(1/\rho)} \int_0^1 (1-\sigma)^{\frac{1}{\rho}-1} \sigma^{\mu-1} f(z\sigma^{1/\rho}) d\sigma = xR_{\rho,1}^{\mu-1,1/\rho} f(x), \tag{6}$$

defined, firstly, for power series as

$$\ell_{\rho,\mu}\left\{\sum_{k=0}^\infty a_k x^k\right\} = \sum_{k=0}^\infty a_k \frac{\Gamma(\mu + k/\rho)}{\Gamma(\mu + (k+1)/\rho)} x^{k+1} \quad (\rho > 0, \mu \geq 1).$$

Examples of functional derivatives $D_{\beta,1}^{\gamma,\delta}$ ($m = 1$) are: the classical fractional derivative

$$D^\delta = \left(\frac{d}{dx}\right)^\delta = D_{1,1}^{0,\delta} x^{-\delta},$$

the differintegral operator

$$x^{-\gamma}\left(\frac{d}{dx}\right)^\delta x^{\gamma+\delta} = D_{\beta,1}^{\gamma,\delta},$$

the D.-G.-L. differentiation operator

$$D_{\rho,\mu} f(x) = D_{\rho,1}^{\mu-1,1/\rho} x^{-1} f(x) \quad \left[D_{\rho,\mu}\left\{\sum_{k=0}^\infty a_k x^k\right\} = \sum_{k=1}^\infty a_k \frac{\Gamma(\mu + \frac{k}{\rho})}{\Gamma(\mu + \frac{k-1}{\rho})} x^{k-1}\right]$$

(6')

for which (6) is a linear right inverse one: $D_{\rho,\mu}\ell_{\rho,\mu} f = f$, $\forall f \in C_{-\mu\rho}$.

(ii) Let $\underline{m = 2}$. Then ([2], p. 64)

$$G_{22}^{20}\left[\sigma\left|\begin{matrix}\gamma_1+\delta_1,\gamma_2+\delta_2\\ \gamma_1,\gamma_2\end{matrix}\right.\right]=\frac{\sigma^{\gamma_2}(1-\sigma)^{\delta_1+\delta_2-1}}{\Gamma(\delta_1+\delta_2)}\ {}_2F_1(\gamma_2+\delta_2-\gamma_1,\delta_1;\delta_1+\delta_2;1-\sigma),$$

$$0<\sigma<1, \tag{7}$$

and, therefore, the "two-dimensional" fractional integrals (2) are in essence the "hypergeometric fractional integrals" involving the Gauss ${}_2F_1$-function (see E. R. Love, M. Saigo [10], S. Kalla [6] etc.).

(iii) For <u>arbitrary m ≥ 1</u> one of the best examples of m-dimensional fractional integrals and derivatives are the <u>hyper-Bessel operators of Dimovski</u> [11], [12], [13] and the corresponding transmutation operators (<u>Poisson-Sonine type transformations</u> [12], [13], [14]). Indeed, the Bessel-type differential operator of order m:

$$B=x^{-\beta}Q_m\left(x\frac{d}{dx}\right)\quad(\beta>0,\ Q_m(\mu)\text{ is m-th degree polynomial}) \tag{8}$$

$$=x^{-\beta}\prod_{k=1}^{m}\left(x\frac{d}{dx}+\beta\gamma_k\right)=x^{\alpha_0}\frac{d}{dx}x^{\alpha_1}\cdots\frac{d}{dx}x^{\alpha_{m-1}}\frac{d}{dx}x^{\alpha_m},$$

where $\alpha_0=-\beta-\beta\gamma_1+1$, $\alpha_k=\beta\gamma_k-\beta\gamma_{k+1}+1$, $k=1,\ldots,m-1$, $\alpha_m=\beta\gamma_m$, is a generalized "fractional" derivative of multiorder $(1,\ldots,1)$:

$$B=\beta^m\ D_{\beta,m}^{(\gamma_k),(1)}\ x^{-\beta},$$

while the linear right inverse operator L of B (BLf = f, $\forall f\in C_\alpha$), the so-called hyper-Bessel integral operator

$$Lf(x)=\frac{x^\beta}{\beta^m}\underbrace{\int_0^1\cdots\int_0^1}_{(m)}\prod_{k=1}^{m}\sigma_k^{\gamma_k}\ f[x(\sigma_1\cdots\sigma_m)^{1/\beta}]d\sigma_1\cdots d\sigma_m, \tag{9}$$

its fractional powers L^λ, $\lambda>0$, the generalized Sonine transformation (12, 14)

$$\phi f(x)=x^{m(\gamma_m+1)-1}\underbrace{\int_0^1\cdots\int_0^1}_{(m-1)}\left[\prod_{k=1}^{m-1}\frac{(1-\sigma_k)^{\gamma_m-\gamma_k+\frac{k}{m}-1}}{\Gamma(\gamma_m-\gamma_k+\frac{k}{m})}\sigma_k^{\gamma_k}\right]\cdot \tag{10}$$

$$f[x^{m/\beta}(\sigma_1\cdots\sigma_{m-1})^{1/\beta}]d\sigma_1\cdots d\sigma_{n-1},$$

and the other transmutations are generalized fractional integrals:

$$L^\lambda f(x) = \left[\frac{x^\beta}{\beta^m}\right]^\lambda R_{\beta,m}^{(\gamma_k),(\lambda)} f(x)$$

$$\phi f(x^{\beta/m}) = x^{\beta(\gamma_m + \frac{m-1}{m})} R_{\beta,m-1}^{(\gamma_k),(\gamma_m - \gamma_k + \frac{k}{m})} f(x).$$
(11)

It is worth mentioning that the idea to represent multiple integral operators (9), (10) as simple integrals with Meijer's G-function as kernel, was the inspiration to introduce in [14] the generalized operators of fractional integration (3). More generaly,

Theorem 1 Every generalized operator of fractional integration

$$R_{\beta,m}^{(\gamma_k),(\delta_k)} f(x) = \int_0^1 G_{mm}^{m0}\left[\sigma \left| \begin{matrix}(\gamma_k+\delta_k)_1^m \\ (\gamma_k)_1^m\end{matrix}\right.\right] f(x\sigma^{1/\beta}) d\sigma$$

can be represented as m-dimensional composition of E.-K. fractional integrals (6):

$$R*f(x) = \left[\prod_{k=1}^m R_{\beta,1}^{\gamma_k,\delta_k}\right] f(x) = \left[\prod_{k=1}^m I_\beta^{\gamma_k,\delta_k}\right] f(x)$$

$$= \underbrace{\int_0^1 \ldots \int_0^1}_{(m)} \left[\prod_{k=1}^m \frac{(1-\sigma_k)^{\delta_k-1}}{\Gamma(\delta_k)} \sigma_k^{\gamma_k}\right] f\, x(\sigma_1 \ldots \sigma_m)^{1/\beta} d\sigma_1 \ldots d\sigma_m, \quad (12)$$

and conversely.

Proof To prove that representations (12) and (3) are equivalent, we shall use mathematical induction. Assume $\underline{\beta = 1}$ (if $\beta \neq 1$, Th. 2, g) below is to be used). For $m = 1$:

$$I_1^{\gamma_1,\delta_1} f(x) = \int_0^1 \frac{(1-\sigma_1)^{\delta_1-1}}{\Gamma(\delta_1)} \sigma_1^{\gamma_1} f(x\sigma_1) d\sigma_1 = \int_0^1 G_{11}^{10}\left[\sigma_1 \left|\begin{matrix}\gamma_1+\delta_1 \\ \gamma_1\end{matrix}\right.\right] f(x\sigma_1) d\sigma_1$$

$$= R_{1,1}^{\gamma_1,\delta_1} f(x). \quad (13)$$

Consider the case $\underline{m = 2}$. Writing

$$R*f(x) = \prod_{k=1}^2 I_1^{\gamma_k,\delta_k} f(x)$$

in the form

$$x^{-(\gamma_2+\delta_2)} \int_0^x \frac{(x-\tau_2)^{\delta_2-1}}{\Gamma(\delta_2)} \tau_2^{\delta_2} \left\{\tau_2^{-(\gamma_1+\delta_1)} \int_0^{\tau_2} \frac{(\tau_2-\tau_1)^{\delta_1-1}}{\Gamma(\delta_1)} \tau_1^{\gamma_1} f(\tau_1) d\tau_1\right\} d\tau_2$$

$$= x^{-(\gamma_2+\delta_2)} \int_0^x f(\tau_1)\tau_1^{\gamma_1} d\tau_1 \int_{\tau_1}^x \frac{(x-\tau_2)^{\delta_2-1}}{\Gamma(\delta_2)} \tau_2^{\gamma_2-\gamma_1-\delta_1} \frac{(\tau_2-\tau_1)^{\delta_1-1}}{\Gamma(\delta_1)} d\tau_2,$$

we can evaluate the inner integral occuring after changing the order of integrations. Its value is

$$\frac{(x-\tau_1)^{\delta_1+\delta_2-1}}{\Gamma(\delta_1+\delta_2)} \tau_1^{\gamma_2-\gamma_1-\delta_1} \left(\frac{\tau_1}{x}\right)^{\delta_1} {}_2F_1(\gamma_2+\delta_2-\gamma_1,\delta_1;\delta_1+\delta_2;1-(\tau_1/x)).$$

To obtain (3) there remains to change the variable and to use (7):

$$R*f(x) = \int_0^1 G_{22}^{20}\left[\sigma \,\Big|\, \begin{matrix}\gamma_1+\delta_1,\gamma_2+\delta_2\\ \gamma_1,\gamma_2\end{matrix}\right] f(x\sigma)d\sigma = R_{1,2}^{(\gamma_1,\gamma_2),(\delta_1,\delta_2)} f(x).$$

Suppose now that representations (12) and (3) are equivalent <u>for the (m-1)-dimensional</u> fractional integrals:

$$\left[\prod_{k=1}^{m-1} I_1^{\gamma_k,\delta_k}\right] f(x) = R_{1,m-1}^{(\gamma_k)(\delta_k)} f(x).$$

Then, <u>for arbitrary m > 1</u> we have

$$R*f(x) = I_1^{\gamma_m,\delta_m}\left[\prod_{1}^{m-1} I_1^{\gamma_k,\delta_k} f(x)\right] = I_1^{\gamma_m,\delta_m}\left[R_{1,m-1}^{(\gamma_k)_1^{m-1},(\delta_k)_1^{m-1}} f(x)\right]$$

$$= x^{-(\gamma_m+\delta_m)} \int_0^x \frac{(x-\tau_m)^{\delta_m-1}}{\Gamma(\delta_m)} \tau_m^{\gamma_m} \Bigg\{\tau_m^{-1} \int_0^{\tau_m} G_{m-1,m-1}^{m-1,0}\left[\frac{\tau_{m-1}}{\tau_m}\,\Big|\,\begin{matrix}(\gamma_k+\delta_k)_1^{m-1}\\(\gamma_k)_1^{m-1}\end{matrix}\right]$$

$$\cdot f(\tau_{m-1})d\tau_{m-1}\Bigg\} d\tau_m.$$

Changing the order of integrations and using properties (8), (9), p. 205, [1], we receive

$$R*f(x) = x^{-(\gamma_m+\delta_m)} \int_0^x f(\tau_{m-1})\tau_{m-1}^{\gamma_m-1} d\tau_{m-1} \int_{\tau_{m-1}}^x \frac{(x-\tau_m)^{\delta_m-1}}{\Gamma(\delta_m)}$$

$$\cdot G_{m-1,m-1}^{0,m-1}\left[\frac{\tau_m}{\tau_{m-1}}\,\Big|\,\begin{matrix}(\gamma_m-\delta_k)\\(\gamma_m-\gamma_k-\delta_k)\end{matrix}\right] d\tau_m.$$

Further, the inner integral is evaluated as $\int_0^x$ according to (5), p. 209, [1], since

$$G_{m-1,m-1}^{0,m-1}\left[\frac{\tau_m}{\tau_{m-1}}\right] \equiv 0 \text{ for } \tau_m < \tau_{m-1} \quad (\text{see (5), p. 204, [1]}).$$

Its value is:

$$\tau_{m-1}^{\delta_m} G_{m,m}^{m,0}\left[\frac{\tau_{m-1}}{x}\bigg|\begin{array}{c}(1-\gamma_m-\delta_m+\gamma_k+\delta_k)_1^{m-1},1\\(1-\gamma_m-\delta_m+\gamma_k)_1^{m-1},1-\delta_m\end{array}\right].$$

In this manner

$$R*f(x) = x^{-1}\int_0^x G_{mm}^{m0}\left[\frac{\tau_{m-1}}{x}\bigg|\begin{array}{c}(+\delta_k)_1^{m-1},(\gamma_m+\gamma_m)\\(\gamma_k)_1^{m-1},\gamma_m\end{array}\right] f(\tau_{m-1})d\tau_{m-1}$$

$$= \int_0^1 G_{mm}^{m0}\left[\sigma\bigg|\begin{array}{c}(\gamma_k+\delta_k)_1^m\\(\gamma_k)_1^m\end{array}\right] f(x\sigma)\,d\sigma$$

$$= R_{1,m}^{(\gamma_k)_1^m,(\delta_k)_1^m} f(x),$$

i.e. operators (12) and (3) coincide in C_α, $\alpha \geq \max_k[-\beta(\gamma_k+1)]$, both of them being well defined there (see e.g. [14]). The proof is completed.

Due to the simple and useful properties of Meijer's G-function as a kernel-function of (3), it is easy to establish the following analogons of the well-known results for the classical R.-L. and E.-K. fractional integrals and derivatives. For example,

<u>Theorem 2</u> ([3], [4], [5]) Let $\alpha \geq \max_k[-\beta(\gamma_k+1)]$, $f \in C_\alpha$.

a) $R_{\beta,m}^{(\gamma_1,\ldots,\gamma_m),(0,\ldots,0)} f(x) = f(x);$

b) $R_{\beta,m}^{(\gamma_k),(\delta_k)}\{x^p\} = x^p \prod_{k=1}^m \frac{\Gamma(\gamma_k+\frac{p}{\beta}+1)}{\Gamma(\gamma_k+\delta_k+\frac{p}{\beta}+1)}$, $p > \alpha;$

c) $R_{\beta,m}^{(\gamma_k),(\delta_k)} x^{\beta\lambda} f(x) = x^{\beta\lambda} R_{\beta,m}^{(\gamma_k+\lambda),(\delta_k)} f(x);$

d) $R_{\beta,m_1}^{(\tau_j),(\sigma_j)} R_{\beta,m_2}^{(\gamma_k),(\delta_k)} f(x) = R_{\beta,m_2}^{(\gamma_k),(\delta_k)} R_{\beta,m_1}^{(\tau_j),(\sigma_j)} f(x);$

(commutability of operators of form (3) with the same $\beta > 0$)

e) $R_{\beta,m}^{(\gamma_k+\delta_k),(\sigma_k)} R_{\beta,m}^{(\gamma_k),(\delta_k)} f(x) = R_{\beta,m}^{(\gamma_k),(\sigma_k+\delta_k)} f(x);$

(law of indices)

f) $R_{\beta,m_1}^{(\gamma_k'),(\delta_k')} R_{\beta,m_2}^{(\gamma_k''),(\delta_k'')} f(x) = R_{\beta,m_1+m_2}^{((\gamma_k'),(\gamma_k'')),((\delta_k'),(\delta_k''))} f(x);$

g) $R_{\beta,m}^{(\gamma_k),(\delta_k)} f(x) = W^{-1}\left[R_{\beta,\omega,m}^{(\gamma_k),(\delta_k)} Wf\right]$, where $Wf(x) = f(x^\omega)$, $\omega > 0;$

h) $\quad m\left\{R_{\beta,m}^{(\gamma_k),(\delta_k)} f; s\right\} = \prod_{k=1}^{m} \frac{\Gamma(\gamma_k - \frac{s}{\beta} + 1)}{\Gamma(\gamma_k + \delta_k - \frac{s}{\beta} + 1)} m\{f;s\}$

($m\{f;s\}$ denotes the Mellin-transform)

i) $\quad \left\{R_{\beta,m}^{(\gamma_k),(\delta_k)}\right\}^{-1} g(x) \quad$ (inversion formula, $n_k = \left\{\begin{matrix}[\delta_k]+1\\ \delta_k\end{matrix}\right.$)

$= D_{\beta,m}^{(\gamma_k),(\delta_k)} g(x) =$

$= \left[\prod_{k=1}^{m} \prod_{j=1}^{n_k} \left(\frac{1}{\beta} x \frac{d}{dx} + \gamma_k + j\right)\right] \int_0^1 G_{mm}^{m0}\left[\sigma \,\Big|\, \begin{matrix}(\gamma_k+n_k)\\ (\gamma_k+\delta_k)\end{matrix}\right] g(x\sigma^{1/\beta})d\sigma.$

In [4], [14] and in a number of other recently published papers we propose **many different applications** of the generalized fractional integrals and derivatives (3), (4), for example: to obtain new and simpler representations of the hyper-Bessel operators (9), of their fractional powers, convolutions and transmutation operators; to propose convolutions of the D.-G.-L. operator $\ell_{\rho,\mu}$ and of the more general E.-K. operator $L = x^{\beta\delta}I_\beta^{\gamma,\delta}$; to solve Bessel type ordinary differential equations of arbitrary order $m > 1$ and dual integral equations of a quite general nature; to obtain new Poisson type integral and Rodrigues type differential representations of the generalized hypergeometric functions ${}_pF_q(x)$ in the cases $p < q$, $p = q$ and $p = q+1$ (including the hyper-Bessel functions ${}_0F_m$) etc.

Here, we wish to discuss the two-fold **connection** of the generalized fractional integrals and derivatives **with a class of integral transformations** generalizing the Laplace transformation. Firstly, every fractional integral (2) generates such a transformation. On the other hand, these integral transformations turn some fractional integrals and derivatives into algebraic operations.

Denote by $L\{f;z\}$ the well-known Laplace integral transform

$$L\{f(x);z\} = \int_0^\infty e^{-zx} f(x)dx, \quad \text{Re}(z) > \mu, \qquad (14)$$

considered in C^{exp}, the linear space of functions $f(x)$ continuous in $[0,\infty)$ which are $O(e^{\mu x})$ for $x \to \infty$, $\mu \in \mathbb{R}$. More generally, denote by $C_{\alpha,\omega}^{exp}$ the set of functions

$$C_{\alpha,\omega}^{exp} = \{f \in C_\alpha; f(x) = O(\exp\mu x^\omega) \text{ for } x \to \infty, \text{ with some real } \mu\}.$$

Let us define the **transmutation operator** ϕ as the following (m-1)-

dimensional fractional integration operator:

$$\phi f(x^{\beta/m}) = x^{\beta(\gamma_m+1) - \frac{\beta}{m}} R_{\beta,m-1}^{(\gamma_k),(\lambda_k)} f(x), \quad f' \in C_\alpha, \tag{15}$$

where $\beta > 0$, $\gamma_1, \ldots, \gamma_m$ and $\lambda_1 \geq 0, \ldots, \lambda_{m-1} \geq 0$ are arbitrary.

Definition 2 The composition of the transmutation operator and of the Laplace transform:

$$K\{f(x); z\} = L\{\phi f(x); z\}, \quad f \in C_{\alpha,\beta/m}^{\exp}, \quad \text{Re}(z) > \mu, \tag{16}$$

we shall call a __generalized integral transformation__ of the Laplace-type, generated by the transmutation operator ϕ.

Theorem 3 The integral transformation (16) has the following explicit representation

$$K\{f(x); z\} = Mz^{-m\gamma_m} \int_0^\infty G_{m-1,2m-1}^{2m-1,0}\left[\left(\frac{z}{m}\right)^m \tau^\beta \Big| \begin{array}{c} (\gamma_k + \lambda_k)_1^{m-1} \\ (\gamma_k)_1^{m-1}, (\gamma_m + \frac{k}{m})_0^{m-1} \end{array}\right] f(\tau) d\tau \tag{17}$$

$(M = \sqrt{m}(2\pi)^{1-m})$ as a special case of the so-called __G-transforms__.

Proof We can write $\phi f(x)$ in the form

$$\phi f(x) = x^{m\gamma_m - 1} \int_0^{x^{m/\beta}} G_{m-1,m-1}^{m-1,0}\left[\frac{\tau^\beta}{x^m} \Big| \begin{array}{c} (\gamma_k + \lambda_k)_1^{m-1} \\ (\gamma_k)_1^{m-1} \end{array}\right] f(\tau) d(\tau^\beta).$$

Then,

$$K\{f(x); z\} = L\{\phi f(x); z\} = \int_0^\infty e^{-zx} \phi f(x) dx$$

$$= \int_0^\infty e^{-zx} x^{m\gamma_m - 1} dx \int_0^{x^{m/\beta}} G_{m-1,m-1}^{m-1,0}\left[\frac{\tau^\beta}{x^m} \Big| \begin{array}{c} (\gamma_k + \lambda_k) \\ (\gamma_k) \end{array}\right] f(\tau) d(\tau^\beta)$$

$$= \int_0^\infty f(\tau) d(\tau^\beta) \int_{\tau^{\beta/m}}^\infty x^{m\gamma_m - 1} e^{-zx} G_{m-1,m-1}^{0,m-1}\left[\frac{1}{\tau^\beta} x^m \Big| \begin{array}{c} (1 - \gamma_k) \\ (1 - \gamma_k - \lambda_k) \end{array}\right] dx.$$

The inner integral occuring after changing the order of integrations can be evaluated as an integral from 0 at ∞ of a product of two G-functions (f-la (3.2.2), p. 80, ⌊2⌋) if we take into account that

$$e^{-zx} = G_{01}^{10}[zx|0] \quad \text{and} \quad G_{m-1,m-1}^{0,m-1}\left[\frac{x^m}{\tau^\beta}\right] \equiv 0 \quad \text{for} \quad x < \tau^{\beta/m}.$$

So,

$$K\{f;z\} = Mz^{-m\gamma_m} \int_0^\infty G_{m-1,2m-1}^{2m-1,0}\left[(z/m)^m \tau^\beta \Big| \begin{matrix}(\gamma_k + \lambda_k)_1^{m-1}\\ (\gamma_k)_1^{m-1}, (\gamma_m + \frac{k}{m})_0^{m-1}\end{matrix}\right] f(\tau) d(\tau^\beta)$$

and represent a function analytical in the half-plane $\text{Re}(z) > \mu$. □

Integral transformation (17) <u>generalizes</u> both the Laplace and Meijer transforms, the Borel-Džrbasjan transform and a number of Bessel-type integral transforms related to operators (8). Indeed, it is sufficient to choose $\beta = m$, $\gamma_m = \frac{1}{m} - 1$ and $\lambda_1 = \ldots = \lambda_{m-1} = 0$ to obtain a transmutation operator ϕ which is the identity operator. Then $K\{f;z\} = L\{f;z\}$ will be the <u>Laplace transform</u> (this can be verified directly from (17), too). Further,

Corollary 1   Let $\underline{m = 1}$. Then the transmutation operator ϕ reduces to the simple transformation $\phi f(x) = x^\gamma f(x^{1/\beta})$ (all the integrations disappear) and the integral transformation

$$K\{f;z\} = L\{\phi f;z\} = \int_0^\infty e^{-zx} x^\gamma f(x^{1/\beta}) dx = \beta \int_0^\infty e^{-zt^\beta} t^{\beta(\gamma+1)-1} f(t) dt$$

is a modification of the so-called <u>Borel-Džrbasjan transform</u> ([4], [9])

$$B_{\rho,\mu}\{f;z\} = \rho \int_0^\infty e^{-zt^\rho} t^{\mu\rho-1} f(t) dt \quad (\beta := \rho > 0, \gamma := \mu-1). \tag{18}$$

As it was shown in [9], (18) transforms the D.-G.-L. operators of differentiation and integration (6'), (6) into algebraic operations.

Corollary 2   Let $\underline{m > 1}$ be an arbitrary integer and let us choose $\lambda_k = \gamma_m - \gamma_k + \frac{k}{m}$, $k = 1,\ldots,m-1$. Then the transmutation (15) reduces to the <u>generalized Sonine transformation</u> (10) proposed by Dimovski as a similarity from the hyper-Bessel integral operator (9) to the m-typle integration operator ℓ^m, viz.

$$\phi L = (m/\beta)^m \ell^m \phi \text{ in } C_\alpha, \alpha = \max_x [-\beta(\gamma_k + 1)] \quad (\text{see } [12], [13]).$$

In this case, the general transform (17) takes the form

$$K\{f;z\} = Mz^{-m\gamma_m} \int_0^\infty G_{m-1,2m-1}^{2m-1,0}\left[(z/m)^m \tau^\beta \Big| \begin{matrix}(\gamma_m + \frac{k}{m})_1^{m-1}\\ (\gamma_k)_1^{m-1}, (\gamma_m + \frac{k}{m})_1^{m-1}\end{matrix}\right] f(\tau) d(\tau)$$

$$= Mz^{-m\gamma_m} \int_0^\infty G_{0m}^{m0} \left[(z/m)^m \tau^\beta \bigg| (\gamma_k)_1^m \right] f(\tau) d(\tau^\beta).$$

Up to a constant multiplier and to the substitution $(z/m)^{m/\beta} := Z$ it coincides with the modified Obrechkoff integral transformation

$$\mathcal{O}\{f;Z\} = \beta Z^{-(\gamma_m+1)-1} \int_0^\infty G_{0m}^{m0} \left[(Z\tau)^\beta \bigg| (\gamma_k - \tfrac{1}{\beta} + 1)_1^m \right] f(\tau) d\tau \qquad (19)$$

(see [4], [14], [15]), proposed by Dimovski [16], [12], [13] in the form

$$\mathcal{O}\{f;Z\} = \beta \int_0^\infty x^{\beta(\gamma_m+1)-1} f(x) dx \underbrace{\int_0^\infty \cdots \int_0^\infty}_{(m-1)} e^{-u_1 - \cdots - u_{m-1} - \frac{(Zx)^\beta}{u_1 \cdots u_{m-1}}}$$

$$\cdot \prod_{k=1}^{m-1} u_k^{\gamma_k-\gamma_m-1} du_1 \cdots du_{m-1}, \qquad (19')$$

as a transform basis of an operational calculus for hyper-Bessel operators (8), (9).

A number of operational properties, convolution, real and complex inversion formulas for (19 - 19') were established in [16], [12], [17], [15], [14]. The relationship between the Obrechkoff and Laplace integral transforms by means of the Sonine transformation ϕ:

$$\mathcal{O}\{f(x); (z/m)^{m/\beta}\} = \sqrt{m(2\pi)^{1-m}} \; L\{\phi f(x); z\} \qquad (20)$$

is a special case of (16) assumed here as a definition. In particular, it turns in some relations between particular Bessel type transforms and Laplace transform (see [15], [14]). For example, if $m = \beta = 2$, $\gamma_1 = -\tfrac{\nu}{2}$, $\gamma_2 = \tfrac{\nu}{2}$, then (16), resp. (20) gives the known relation

$$K_\nu\{f;z\} = \frac{\sqrt{\pi} 2^{-\nu} z^{\nu+\frac{1}{2}}}{\Gamma(\nu+\tfrac{1}{2})} L\left\{ \int_0^x (x^2-t^2)^{\nu-\frac{1}{2}} t^{-\nu+\frac{1}{2}} f(t) dt; z \right\}$$

for the classical Meijer transform to which (17), (19) reduce:

$$K\{f;z\} = K_\nu\{f;z\} = 2 \int_0^\infty (zx)^{\nu/2} K_\nu(2\sqrt{zx}) f(x) dx.$$

Acknowledgement

The author is thankful to Prof. Dimovski under whose guidance thesis [4] and most of the results in the present paper were accomplished.

REFERENCES

1. H. Bateman, A. Erdely, "Higher transcedental functions", Moscow, (1978), (in Russian).
2. A. M. Mathai, R. K. Saxena, "Generalized hypergeometric functions with applications ...", Lecture Notes in Math. 348, Berlin, (1973).
3. V. Kiryakova, On operators of fractional integration involving Mejer's G-function, C. R. Acad. Bulg. Sci., 39, 10, 25-28, (1986).
4. V. Kiryakova, Generalized operators of fractional integration and differentiation and applications, Author's summary of Ph. D. Thesis, Sofia, (1986).
5. V. Kiryakova, On a class of generalized operators of fractional integration, Proc. Jubilee Sess. dev. to acad. Chakalov '86 (to appear).
6. S. L. Kalla, "Operators of fractional integration", in Lecture Notes in Math., 798, Springer-Verlag, (1980).
7. I. Dimovski, Convolutional representation of the commutant of Gelfond-Leontiev integr. operator, C. R. Acad. Bulg. Sci., 34, 12, 1643, (1981).
8. I. Dimovski, V. Kiryakova, "Convolution and commutant of Gelfond-Leontiev integr. operator", in Function Theory '81, Sofia, (1982).
9. I. Dimovski, V. Kiryakova, "Convolution and differential property of Borel-Džrbasjan transform", in Complex Anal. and Appl. '81, Sofia, (1984).
10. M. Saigo, "A generalization of fractional calculus", in Fractional calculus, London, Pitman, (1985).
11. I. Dimovski, Operational calculus for a class of differential operators, C. R. Acad. Bulg. Sci., 19, 12, 1111-1114, (1966).
12. I. Dimovski, Foundations of operational calculi for the Bessel-type differential operators, Serdica, 1, 51-63, (1975).
13. I. Dimovski, A convolutional method in operational calculus, Author's summary of Ph. D. Thesis, Sofia, (1977).
14. I. Dimovski, V. Kiryakova, "Transmutations, convolutions and fractional powers of Bessel-type operators via Meijer's G-function", in Complex Anal. and Appl. '83, Sofia, (1985).
15. I. Dimovski, V. Kiryakova, Complex inversion formulas for the Obreckhoff transform, Pliska, 4, 110-116.

16. I. Dimovski, On a Bessel-type integral transformation, due to N. Obrechkoff, C. R. Acad. Bulg. Sci., 27, 1, 23-26, (1974).
17. I. Dimovski, V. Kiryakova, "On an integral transformation, due to N. Obrechkoff", in Lecture Notes in Math., 798, Springer-Verlag, (1980).

ON THE GENERALIZED MEIJER TRANSFORMATION

E. L. Koh*, E. Y. Deeba** and M. A. Ali***

* Department of Mathematics and Statistics
University of Regina, Regina, Canada S4S 0A2

** Department of Applied Mathematical Sciences
University of Houston-Downtown, Houston, Texas 77002, USA

*** #1598, Way 510, Muharraq 205, Bahrain

1. INTRODUCTION

Following the method of Mikusiński [1], Ditkin [2] and later with Prudnikov [3] developed an operational calculus for the operator DtD. In the 60's, Meller [4] generalized Ditkin's calculus to the operator $B_\alpha = t^{-\alpha}Dt^{1+\alpha}D$ with $\alpha \in (-1,1)$. Generalizations to Bessel operators of a higher order were made by Botashev [5], Dimovski [6], Krätzel [7] and others. Koh [8] extended Meller's results to $\alpha > 1$ by using fractional calculus. Later [9], a direct extension was achieved in which the convolution of $\phi(t)$ and $\psi(t)$ is given by

$$\phi * \psi = \frac{1}{\Gamma(\alpha+1)} D^\alpha DtD \int_0^t \int_0^1 \eta^\alpha (1-x)^\alpha \phi(x\eta)[\psi \ (1-x)(t-\eta)]dxd\eta, \qquad (1)$$

where $\alpha \geq 0$, $D^\alpha = D^n I^{n-\alpha}$ and I^ν is the Riemann-Liouville integral,

$$I^\nu f(t) = \frac{1}{\Gamma(\nu)} \int_0^t (t-\xi)^{\nu-1} f(\xi)d\xi. \qquad (2)$$

Meller's operational calculus may also be developed by means of the Meijer transformation, M_μ, given by

$$M_\mu\{f\}(p) = \frac{2p}{\Gamma(\mu+1)} \int_0^\infty (pt)^{\mu/2} K_\mu(2\sqrt{pt}) f(t)dt, \ \text{Re}(\mu) > -1. \qquad (3)$$

This is a slight departure from Krätzel's integral transform [10].

The Meijer transform (3) is well-defined for certain locally integrable functions on $(0,\infty)$. Recently [11], we were able to extend it to certain generalized functions. We will outline this extension, derive some transforms and apply the generalized Meijer transform to some boundary value problems. Proofs of the properties in Section 2 and 3 are given in [11].

2. THE SPACE $M_{\mu,\gamma}$ AND ITS DUAL

Let γ be any real number and $\mu \in (-\infty, 1)$. Let $B_{-\mu}^k$ be the kth iterate of the operator $B_{-\mu}$ defined earlier. Let $M_{\mu,\gamma}$ be the set of infinitely differentiable functions on $I = (0,\infty)$ defined as

$$M_{\mu,\gamma} = \{\phi \in C^\infty(I) \,|\, \lambda_{\gamma,k}^\mu(\phi) < \infty\},$$

where $\lambda_{\gamma,k}^\mu(\phi) = \sup_{0<t<\infty} |e^{\gamma\sqrt{t}} t^{1-\mu} B_{-\mu}^k(\phi(t))|$, $k = 0,1,2,\ldots$.

In view of the weight $e^{\gamma\sqrt{t}} t^{1-\mu}$, the elements of $M_{\mu,\gamma}$ are those smooth functions on $(0,\infty)$ that grow no faster than $e^{\gamma\sqrt{t}}$ for large t and behave like power functions near the origin. The kernel of the Maijer transform is of this type. Indeed for a fixed complex number p such that $p \neq 0$, $-\pi < \arg p < \pi$, $\text{Re} 2\sqrt{p} > \gamma$, $(pt)^{\mu/2} K_\mu(2\sqrt{pt}) \in M_{\mu,\gamma}$. Furthermore, the derivative of the kernel with respect to either p or t belongs to $M_{\mu,\gamma}$.

The family $\{\lambda_{\gamma,k}^\mu\}_{k=0}^\infty$ is a countable multinorm, since each member is a seminorm and $\lambda_{\gamma,0}^\mu$ is a norm on $M_{\mu,\gamma}$. We assign to $M_{\mu,\gamma}$ the topology generated by this separating family of seminorms. A sequence $\{\phi_\nu\}$ is Cauchy in $M_{\mu,\gamma}$ if $\phi_\nu \in M_{\mu,\gamma}$ for all ν and for every $k = 0,1,2,\ldots,\lambda_{\gamma,k}^\mu(\phi_\nu - \phi_\xi) \to 0$ as $\nu,\xi \to \infty$ independently.

The space $M_{\mu,\gamma}$ is complete. Its dual space is denoted by $M'_{\mu,\gamma}$ and is supplied with the usual weak topology. The dual space is also complete since $M_{\mu,\gamma}$ is. Other properties of these spaces follow:

(i) The operator $B_{-\mu}^k$ is a continuous linear mapping on $M_{\mu,\gamma}$. The adjoint $B_{-\mu}^{*k}$ of $B_{-\mu}^k$ is defined by

$$\langle B_{-\mu}^{*k} f, \phi \rangle = \langle f, B_{-\mu}^k \phi \rangle, \quad \phi \in M_{\mu,\gamma}, \; f \in M'_{\mu,\gamma}.$$

It follows that $B_{-\mu}^{*k}$ is a continuous linear operator on $M'_{\mu,\gamma}$. It can be shown that $B_{-\mu}^{*k} = B_\mu^k$.

(ii) If $f(t)$ is locally integrable on $(0,\infty)$ and $f(t)e^{-\gamma\sqrt{t}} t^{-1+\mu}$ is absolutely integrable on $(0,\infty)$, then $f(t)$ generates a regular distribution f of $M'_{\mu,\gamma}$ by means of

$$< f, \phi > = \int_0^\infty f(t)\phi(t)dt, \quad \phi \in M_{\mu,\gamma}.$$

(iii) The space $M_{\mu,\gamma}$ contains the Schwartz space $\mathcal{D}(I)$, and is in turn contained in the space $\mathcal{E}(I)$. Thus the restriction of any $f \in M'_{\mu,\gamma}$ to $\mathcal{D}(I)$ belongs to $\mathcal{D}'(I)$ and in turn the restriction of $f \in \mathcal{E}'(I)$ to $M_{\mu,\gamma}$ is in $M'_{\mu,\gamma}$.

(iv) If $\gamma < \alpha$, then $M_{\mu,\alpha} \subset M_{\mu,\gamma}$ and the restriction of $f \in M'_{\mu,\gamma}$ to $M_{\mu,\gamma}$ is in $M'_{\mu,\gamma}$. Therefore, given a generalized function f, there is a σ_f such that the restriction of f belongs to $M'_{\mu,\gamma}$ if $\gamma > \sigma_f$ and otherwise if $\gamma < \sigma_f$. This number σ_f is called the abscissa of the definition of f.

3. THE GENERALIZED MEIJER TRANSFORMATION

For $f \in M'_{\mu,\gamma}$, $\mu \in (-\infty, 1)$, $p \in \Omega_f = \{p \in \mathbb{C} \mid \text{Re}2\sqrt{p} > \gamma > \sigma_f, p \neq 0, |\arg p| < \pi\}$ we define the Meijer transform of f by

$$(\bar{M}_\mu f)(p) = \frac{2p}{\Gamma(1-\mu)} < f(t), (pt)^{\mu/2} K_\mu(2\sqrt{pt}) >.$$

The generalized Meijer transform satisfies the following properties:

(i) If $f(t)$ is a regular gf as in 2(ii), then

$$(\bar{M}_\mu f)(p) = \frac{2p}{\Gamma(1-\mu)} \int_0^\infty f(t)(pt)^{\mu/2} K_\mu(2\sqrt{pt}) dt.$$

(ii) Let $f \in M'_{\mu,\gamma}$ and $(\bar{M}_\mu f)(p)$ be the Meijer transform of f for $p \in \Omega_f$. Then

$$\bar{M}_\mu(B_\mu^k f)(p) = p^k \bar{M}_\mu(f)(p).$$

This is the basis for an operational calculus for B_μ.

(iii) The transform $(\bar{M}_\mu f)(p)$ is analytic in Ω_f and

$$\frac{d}{dp}(\bar{M}_\mu f)(p) = \frac{2}{\Gamma(1-\mu)} < f(t), \frac{d}{dp} p(pt)^{\mu/2} K_\mu(2\sqrt{pt}) >.$$

(iv) <u>Inversion Theorem</u>: Let $f \in M'_{\mu,\gamma}$ and let $(\bar{M}_\mu f)(p)$ be its Meijer transform for $p \in \Omega_f$. Then, in the sense of convergence in $\mathcal{D}'(I)$,

$$f(t) = \lim_{\theta_1 \to \pi} \frac{\Gamma(1-\mu)}{2\pi i} \int_{-\theta_1}^{\theta_1} (\bar{M}_\mu f)(p) p^{-1}(pt)^{-\mu/2} I_\mu(2\sqrt{pt}) dp(\theta),$$

where $p(\theta) = \frac{r^2}{2} e^{i\theta} \sec^2\frac{\theta}{2}$ and γ is a fixed real number in Ω_f.

4. APPLICATIONS

We will first give some simple examples of $\bar{M}_\mu f$. Next we will apply $\bar{M}_\mu$ to some problems.

(i) <u>Some Generalized Meijer Transforms</u>.

(a) For $\tau \in (0,\infty)$, $p \neq 0$, $\bar{M}_\mu \delta(t-\tau) = \dfrac{2p}{\Gamma(1-\mu)} (p\tau)^{\mu/2} K_\mu(2\sqrt{p\tau})$.

(b) Let $1_+(t-\tau)$ be the Heaviside step-function

$$\bar{M}_\mu 1_+(t-\tau) = -\dfrac{2}{\Gamma(1-\mu)} (p\tau)^{(\mu+1)/2} K_{\mu+1}(2\sqrt{p\tau}).$$

(c) $\bar{M}_\mu(\delta^{(k)}(t-\tau)) = \dfrac{2p^{k+1}}{\Gamma(1-\mu)} (p\tau)^{(\mu-k)/2} K_{\mu-k}(2\sqrt{p\tau})$.

(d) $\bar{M}_\mu(B_\mu^k \delta(t-\tau)) = \dfrac{2p^{k+1}}{\Gamma(1-\mu)} (p\tau)^{\mu/2} K_\mu(2\sqrt{p\tau})$.

(ii) <u>An Electric Circuit</u>. Consider a series circuit which consists of a voltage source $v(t)$, a capacitor $C(t) = 1/ct$, a resistor $R(t) = r/t$ and an inductor L. Here, the constants c, r and L are real numbers and the variable t is restricted to $0 < t < \infty$. Letting $q(t)$ be the circulating charge, we can apply Kirchhoff's voltage law to yield

$$v(t) = L D_t^2 q(t) + \dfrac{r}{t} D_t q(t) + \dfrac{1}{t} \dfrac{q(t)}{c}. \qquad (4)$$

If we set $v(t) = \dfrac{L}{t} g(t)$, then (4) becomes

$$g(t) = \left[B_{(r/L)-1} + \dfrac{1}{Lc} \right] q(t). \qquad (5)$$

Assuming that $g(t)$ is a transformable generalized function and taking the generalized Meijer transform of (5), we obtain

$$Q(p) = \dfrac{Lc}{pLc+1} G(p), \quad \text{Re} p \geq \max(\sigma_g, \text{Re}\sqrt{-Lc}), \qquad (6)$$

where $Q(p)$ and $G(p)$ are the Meijer transforms of $q(t)$ and $g(t)$, respectively. One obtains $q(t)$ from (6) by the inversion theorem as a distribution in $\mathcal{D}'(I)$.

(iii) <u>A Boundary Value Problem</u>. Consider the problem of finding $h(x,t)$ on the domain $\{(t,x): 0 < t < \infty, 0 < x < \ell\}$ which satisfies the one-dimensional wave equation

$$\frac{\partial}{\partial t} t^{1+\mu} \frac{\partial}{\partial t} t^{-\mu} h(x,t) - \frac{\partial^2}{\partial x^2} h(x,t) = 0 \tag{7}$$

subject to the conditions

(a) as $x \to 0_+$, $\frac{\partial h(x,t)}{\partial x}$ converges in $M'_{\mu,\gamma}$ to $f(t)$;

(b) as $x \to \ell_-$, $\frac{\partial h(x,t)}{\partial x}$ converges in $M'_{\mu,\gamma}$ to zero.

Equation (7) can be written as

$$B_\mu h(x,t) - \frac{\partial^2}{\partial x^2} h(x,t) = 0. \tag{8}$$

If $H(x,p) = \bar{M}_\mu h(x,t)$, then equation (8) transforms to

$$pH(x,p) - \frac{\partial^2}{\partial x^2} H(x,p) = 0. \tag{9}$$

The solution of the linear differential equation (9) is

$$H(x,p) = A(p)e^{\sqrt{p}x} + C(p)e^{-\sqrt{p}x}. \tag{10}$$

Now, condition (a) says that for each $\phi \in M_{\mu,\gamma}$, $\langle \frac{\partial h}{\partial x}, \phi(t) \rangle$ converges to $\langle f(t), \phi(t) \rangle$ as $x \to 0_+$. Since $(pt)^{\mu/2} K_\mu(2\sqrt{pt})$ belongs to $M_{\mu,\gamma}$ for $\text{Re} 2\sqrt{p} > \gamma$, it follows that for each p,

(a') $\frac{\partial}{\partial x} H(x,p)$ converges to $F(p)$ as $x \to 0_+$ where $F(p) = \bar{M}_\mu f(t)$.

Similarly,

(b') $\frac{\partial}{\partial x} H(x,p)$ converges to zero as $x \to \ell_-$.

If we differentiate (10) and apply conditions (a') and (b'), we obtain

$$F(p) = \sqrt{p}\, A(p) - \sqrt{p}\, C(p)$$

$$0 = \sqrt{p}\, A(p) \cdot e^{\ell\sqrt{p}} - \sqrt{p}\, C(p) \cdot e^{-\ell\sqrt{p}}.$$

Thus we can solve for $A(p)$ and $C(p)$ to yield

$$H(x,p) = F(p) \frac{1}{\sqrt{p}} \left[\frac{e^{(x-2L)\sqrt{p}} + e^{-x\sqrt{p}}}{e^{-2L\sqrt{p}} - 1} \right]. \tag{11}$$

Equation (11) can be written as

$$H(x,p) = -F(p) \frac{1}{\sqrt{p}} \frac{\cosh((L-x)\sqrt{p})}{\sinh L\sqrt{p}} \tag{12}$$

It can be shown that a necessary and sufficient condition for a function $F(p)$ to be the μth order Meijer transform of some generalized function in $M'_{\mu,\gamma}$ is that there exists a region Ω_f of the form given in section 3 on which $F(p)$ is analytic and $|F(p)|$ is bounded by a polynomial in $|p|$. (The proof of this structure theorem is lengthy and will be given in a subsequent paper.) The right side of (12) is clearly analytic and, so, bounded since $F(p)$ is. Thus, $h(x,t)$ exists for which $H(x,p) = \bar{M}_\mu h(x,t)$, and is obtained by means of the Inversion Theorem in §3 (iv).

5. REMARKS

(i) In [12], Zemaninan extended the K-transformation given by

$$F(s) = \int_0^\infty f(t)\sqrt{st}\, K_\mu(st)dt$$

to certain distributions. As well, Rao and Debnath [13] extended Krätzel's integral transformation. In both cases, one does not obtain explicitly an operational calculus for the operator B_μ.

(ii) An alternative approach to the extension of the Meijer transformation can be made by means of a Parseval type formula. In this case, a Fréchet space S is constructed such that for $f \in S'$, $\bar{M}_\mu$ is defined by

$$< \bar{M}_\mu f, \phi > = < f, \bar{M}_\mu \phi >, \quad \phi \in S.$$

For this to be meaningful, the classical Meijer transformation M_μ must be an isomorphism on S.

(iii) In previous works including [12] and [13], a convolution theorem has not yet been proved. To this end, a convolution process for generalized functions has to be defined. One way is to use tensor products that will reduce to eq. (1) above for regular distributions f and g. Consider $\mu = 0$. For regular members f and g of $M'_{0,\gamma}$, we have

$$< f * g, \phi > = \int_0^\infty \phi DtD \int_0^t \int_0^1 f(x\xi)g[(1-x)(t-\xi)]dxd\xi$$

$$= \int_0^\infty f(y) \int_0^\infty g(\tau) \int_0^\infty \frac{1}{x} B_0\phi(\frac{1}{x}(y+x)(\tau+x))dxd\tau dy$$

$$= < f(y), < g(\tau), \int_0^\infty \frac{1}{x} B_0\phi\left[\frac{1}{x}(y+x)(\tau+x)\right]dx >>.$$

The last can be used as the definition for $*$, provided it is meaningful in $M_{0,\gamma}$. From this one has to show that

$$\bar{M}_0(f * g) = \bar{M}_0(f)\bar{M}_0(g).$$

For $\mu \neq 0$, one has to investigate the action of the fractional differentiation operator D^μ on $M_{\mu,\gamma}$. The convolution process is necessarily more involved than for $\mu = 0$.

REFERENCES

1. J. Mikusiński, "Operational Calculus", Pergamon Press, Oxford, (1959).
2. V. A. Ditkin, Operational Calculus Theory, Dokl. Acad. Nauk. SSSR, 116, 15-17, (1957).
3. V. A. Ditkin and A. P. Prudnikov, "Integral Transforms and Operational Calculus", Pergamon Press, New York, (1965).
4. N. A. Meller, On an Operational Calculus for the Operator $B_\alpha = t^{-\alpha}\frac{d}{dt}t^{\alpha+1}\frac{d}{dt}$, Vichis. Mat, 6, 161-168, (1960).
5. A. I. Botashev, Operational Calculus Theory, Issled. Integrodiff. Uravn. Kirgisu, 2, 297-304, (1962).
6. I. Dimovski, On an Operational Calculus for a Differential Operator, C. R. Acad. Bulg. Sc., 21, 513-516, (1968).
7. E. Krätzel, Differentiationssatze der L transformation und Differentialgleichungen nach dem Operator $\frac{d}{dt}t^{(1/n)-\nu}(t^{1-(1/n)}\frac{d}{dt})^{n-1}t^{\nu-1-(2/n)}$, Math. Nachr. 35, 105-114, (1967).
8. E. L. Koh, A Mikusiński Calculus for the Bessel Operator B_μ, Springer-Verlag Lecture Notes #564, 291-300, (1976).
9. E. L. Koh, A direct Extension of Meller's Calculus, Inter. J. Math. & Math. Sc., 5, 785-791, (1982).
10. E. Krätzel, Eine Verallgemeinerung der Laplace und Meijer Transformation, Wiss. Z. Univ. Jena, Math. Naturw. Reihe, 5, 369-381, (1965).
11. E. L. Koh, E. Y. Deeba and M. A. Ali, The Meijer Transformation of Generalized Functions, Inter. J. Math. & Math. Sc., 10, 267-286, (1987).
12. A. H. Zemanian, "Generalized Integral Transformations", Interscience, New York, (1968).
13. G. L. N. Rao and L. Debnath, A Generalized Meijer Transformation, Int. J. Math. & Math. Sc, 8, 359-365, (1985).

THE CONSTRUCTION OF REGULAR SPACES AND HYPERSPACES WITH RESPECT TO A PARTICULAR OPERATOR

G. Liu

Eindhoven University of Technology

5600 MB Eindhoven, The Netherlands

Let H be a Hilbert space with an inner product $(\cdot,\cdot)$ and corresponding norm $\|\cdot\|$. There is given an <u>unbounded operator</u> $B : D(B) \subset H \to H$ such that $0 \in p(B)$ and $-B$ is a generator of an analytic semigroup. That is,

$$p(B) \supset \Sigma^+ = \{\lambda | \lambda \in \mathbb{C}, 0 < \omega < |\arg\lambda| \leq \pi\} \cup \{0\} \tag{1.1}$$

and

$$\|R(\lambda;B)\| \leq \frac{M}{|\lambda|}, \quad \forall \lambda \in \Sigma^+ \tag{1.2}$$

where $\omega < \pi/2$ and M is a positive constant, see, e.g., [4]. Note that, in this case, B* also satisfies conditions (1.1) and (1.2), hence $-B*$ also generates an analytic semigroup $e^{-tB*} = (e^{-tB})*$. For such an operator B, powers of arbitrary order can be defined an they enjoy nice properties. We shall collect some of them as follows (Conf., e.g., [4])

(a) $B^{-\alpha}$ is injective for each $\alpha > 0$ and $B^{\alpha} = (B^{-\alpha})^{-1}$

(b) $\{B^{-\alpha} | \alpha \geq 0 \}$ is a C_0-semigroup of bounded operators on H.

(c) $\alpha \geq \beta \geq 0$ implies $D(B^{\alpha}) \subset D(B^{\beta})$, where $B^0 = I$.

(d) $D(B^{\infty}) = \bigcap_{\alpha \geq 0} D(B^{\alpha})$ is dense in H.

(e) If α, β are real then

$$B^{\alpha+\beta}u = B^{\alpha} B^{\beta} u$$

for every $u \in D(B^{\nu})$ vhere $\nu = \max\{\alpha,\beta,\alpha+\beta\}$.

(f) For $1 \geq \alpha > \beta > 0$ there is a constant $C_{\alpha,\beta}$ such that

$$\|B^\beta u\| \le C_{\alpha,\beta} \|B^\alpha u\|^{\alpha/\beta} \|u\|^{1-\beta/\alpha}, \quad \forall u \in D(B^\alpha).$$

For $t > 0$, $D(B^t)$ equipped with the inner product

$$(u,v)_t = (B^t u, B^t v), \quad \forall u,v \in D(B^t) \tag{1.3}$$

and the corresponding norm

$$\|u\|_t = (u,u)_t^{1/2}, \quad \forall u \in D(B^t) \tag{1.4}$$

is denoted by W_B^t, which is a Hilbert space since the mapping B^t is an isometric operator from W_B^t into H. We have another isometric mapping B^{-t} : $(H, \|\cdot\|_{-t}) \to (D(B^t), \|\cdot\|)$ where the norm $\|\cdot\|_{-t}$ and the corresponding inner product $(\cdot,\cdot)_{-t}$ on H is defined as follows.

$$(u,v)_{-t} = (B^{-t} u, B^{-t} v), \quad \forall u,v \in H \tag{1.5}$$

$$\|u\|_{-t} = (u,u)_{-t}^{1/2}, \quad \forall u \in H. \tag{1.6}$$

Let a completion of $(H, \|\cdot\|_{-t})$ be denoted by W_B^{-t}, with the corresponding inner product $(\cdot,\cdot)_{-t}$. Then, it is easy to see from the denseness of $D(B^t)$ in H that operator B^{-t} : $(H, \|\cdot\|_{-t}) \to (D(B^t), \|\cdot\|)$ is extended by B^{-t}. It is obvious that W_B^{-t} could be taken to be all the families of mappings F : $(0,\infty) \to H$ satisfying

$$\|B^{-t}(F(t') - F(t''))\| \to 0 \text{ as } t', t'' \to 0. \tag{1.7}$$

Two mappings F and G, satisfying (1.7), are in the same family (referred to as being equivalent) if

$$\|B^{-t}(G(\tau) - F(\tau))\| \to 0 \text{ as } \tau \to 0. \tag{1.8}$$

Note that in each family we can always choose a representive F such that

$$B^{-t} F(\tau) = F(t+\tau), \quad \forall \tau > 0, \tag{1.9}$$

and we shall always do so. Thus, for $F,G \in W_B^{-t}$

$$(F,G)_{-t} = (B^{-t} F, B^{-t} G) = \lim_{\tau \to 0} (B^{-t} F(\tau), B^{-t} G(\tau))$$

$$= \lim_{\tau \to 0} (B^{-t} F(t+\tau), G(t+\tau)) \tag{1.10}$$

and

$$\|F\|_{-t} = \|B^{-t} F\| = \lim_{\tau \to 0} \|B^{-t} F(\tau)\| = \lim_{\tau \to 0} \|F(t+\tau)\| \tag{1.11}$$

It is easy to see that

$$W_B^{-t} \subset W_B^{-\tau} \text{ if } \tau > t > 0 \qquad (1.12)$$

and

$$|u|_{-\tau} \le |B^{-(\tau-t)}| \, |u|_{-t}, \quad \forall u \in W_B^{-t}. \qquad (1.13)$$

Since both the operator B and B* satisfy conditions (1.1) and (1.2), we could have constructed spaces W_{B*}^t and W_{B*}^{-t} together with W_B^t and W_B^{-t}. Now, we shall establish a pairing $\langle \cdot, \cdot \rangle_t$ between the spaces W_B^t and W_{B*}^{-t}, as follows

$$\langle u, F \rangle_t = (C^t u, (B*)^{-t} F) = \lim_{\tau \to 0} (B^t u, (B*)^{-t} F(\tau))$$

$$= \lim_{\tau \to 0} (B^t u, F(t+\tau)), \quad \forall u \in W_B^t, F \in W_{B*}^{-t}. \qquad (1.14)$$

In particular, when $F \in H$

$$\langle u, F \rangle_t = (B^t u, (B*)^{-t} \cdot) = (B^{-t} \cdot B^t u, F) = (u, F). \qquad (1.15)$$

It is evident that each $F \in W_{B*}^{-t}$ gives a continuous linear functional, i.e. an element in $(W_B^t)'$, via the duality (1.14), the norm of which is exactly $|(B*)^{-t} \cdot F| = |F|_{-t}^*$ (Here and afterwards $|\cdot|_{-t}^*$ and $(\cdot,\cdot)_{-t}^*$ stand for the norm and inner product in the space W_{B*}^{-t}) conversely, each continuous linear functional on W_B^t could be written in the form (Riesz's representation theorem)

$$(u,v)_t = (B^t u, B^t v) \text{ for some } v \in W_B^t.$$

Therefore,

$$(u,v)_t = \langle u, F \rangle_t$$

where $F = (B*)^t B^t v \in W_{B*}^{-t}$. Thus, W_{B*}^{-t} can be identified to be $(W_B^t)'$ via an antilinear mapping. We use the symbol $(W_B^t)' = W_{B*}^{-t}$ to express this fact. On the other hand, every $u \in W_B^t$ gives rise to a continuous linear functional on W_B^t via the same duality (1.14), the norm of which is just $|B^t u| = |u|_t$. According to Riesz's representation theorem, each element in $(W_{B*}^{-t})'$ must have the form

$$(F,G)_{-t}^* = ((B*)^{-t} F, (B*)^{-t} G) \text{ for some } G \in W_{B*}^{-t}.$$

Hence

$$(F,G)_{-t}^* = \overline{\langle u, F \rangle_t}$$

229

where $u = B^{-t}(B*)^{-t} F \in W_B^t$. Therefore, W_B^t can be identified to be $(W_{B*}^{-t})'$ via an antilinear mapping, denoted by $(W_{B*}^{-t})' = W_B^t$. In summary, we have

Theorem 1.1 For each $t > 0$

$$(W_B^t)' = W_{B*}^{-t}, \quad (W_{B*}^{-t})' = W_B^t. \tag{1.16}$$

From the Hilbert spaces W_B^t, we construct two types of locally topological spaces

$$W_B^\infty = \bigcap_{t>0} W_B^t = \bigcap_{\nu=1}^\infty W_B^\nu \text{ with projective topology}$$

$$W_B^{0+} = \bigcup_{t>0} W_B^t = \bigcup_{\nu=1}^\infty W_B^{1/\nu} \text{ with inductive limit topology.}$$

Similarly, from the Hilbert spaces W_{B*}^{-t}, we define

$$W_{B*}^{-\infty} = \bigcup_{t>0} W_{B*}^{-t} = \bigcup_{\nu=1}^\infty W_{B*}^{-\nu} \text{ with inductive limit topology}$$

$$W_{B*}^{0-} = \bigcap_{t>0} W_{B*}^{-t} = \bigcap_{\nu=1}^\infty W_{B*}^{-1/\nu} \text{ with projective topology.}$$

Explicitly, $W_{B*}^{-\infty}$ consists of all the families of mappings $F : (0,\infty) \to H$ satisfying (1.7) with $B*$ instead of B for some $t > 0$, while W_{B*}^{0-} the families of mappings with representives $F : (0,\infty) \to H$ satisfying (1.9) for all $t, \nu > 0$ with $B*$ instead of B, for in this case (1.7) is valid automatically, as is seen from the following

$$(B*)^{-t} (F(t') - F(t'')) = [(B*)^{-t'} - (B*)^{-t''}]F(t) \to 0$$

as $t', t'' \to 0$. We call the spaces W_B^t with $t > 0$ (including ∞ and 0_+) regular spaces (with respect to operator B) and W_{B*}^t with $t < 0$ (including $-\infty$ and 0_-) hyperspaces (with respect to operator B). From the denseness of $D(B^\infty)$ in H and H in W_{B*}^{-t} ($t > 0$), it is easy to see that in the following diagram

$$W_B^\infty \subset \ldots W_B^t \subset \ldots W_B^{0-} \xrightarrow{B^t} H \subset W_{B*}^{0-} \subset \ldots \subset W_{B*}^{-t} \xrightarrow{(B*)^t} \ldots \subset W_{B*}^{-\infty} \tag{1.17}$$

each smaller space is dense in another larger space and the embedding in continuous.

Suppose $0 < t < \tau$. From Theorem 1.1 it follows that every element in $(W_B^t)'$ corresponsing to an $F \in W_{B*}^{-t}$ such that

$$\langle u, F \rangle_t = (B^t u, (B*)^{-t} F), \quad \forall u \in W_B^t.$$

Since for $u \in W_B^\tau \subset W_B^t$, we have

$$(B^t u, (B*)^{-t}) = (B^{-(\tau-t)} \cdot B^\tau u, (B*)^{-t} F) = (B^\tau u, (B*)^{-\tau} F)$$

it follows that as an element in $(W_B^\tau)'$ the linear functional $<u, F>_t$ corresponds to the same $F \in W_{B*}^{-\tau}$. The situation is exactly the same for the spaces $(W_{B*}^{-t})'$, $(t > 0)$. Thus we obtain

Theorem 1.2

$$(W_B^\infty)' = (\bigcap_{t>0} W_B^t)' = \bigcup_{t>0} (W_B^t)' = \bigcup_{t>0} W_{B*}^{-t} = W_{B*}^{-\infty} \qquad (1.18)$$

$$(W_{B*}^{-\infty})' = (\bigcup_{t>0} W_{B*}^{-t})' = \bigcap_{t>0} (W_{B*}^{-t})' = \bigcap_{t>0} W_B^t = W_B^\infty \qquad (1.19)$$

$$(W_B^{0+})' = (\bigcup_{t>0} W_B^t)' = \bigcap_{t>0} (W_B^t)' = \bigcap_{t>0} W_{B*}^{-t} = W_{B*}^{0-} \qquad (1.20)$$

$$(W_{B*}^{0-})' = (\bigcap_{t>0} W_{B*}^{-t})' = \bigcup_{t>0} (W_{B*}^{-t})' = \bigcup_{t>0} W_B^t = W_B^{0+}. \qquad (1.21)$$

We emphasize that the dualities $<\cdot,\cdot>_\infty$ and $<\cdot,\cdot>_0$ between W_B^∞ and $W_{B*}^{-\infty}$ and between W_B^{0+} and W_{B*}^{0-}, respectively, are defined as follows

$$<u, F>_\infty = (B^t u, B^{-t} F) \text{ for } t \text{ so large that } F \in W_{B*}^{-t} \qquad (1.22)$$

$$<u, F>_0 = (B^t u, B^{-t} F) \text{ for } t \text{ so small that } u \in W_B^t \qquad (1.23)$$

which are not dependent on t.

We should point out that when $B = e^A$ is taken for some nonnegative self-adjoint operator A, the spaces $S_{H,A}$, $T_{H,A}$ of de Graaf and spaces $\tau(H,A)$ and $\sigma(H,A)$ of van Eindhoven are obtained. Actually,

$$W_{e^A}^{0+} = S_{H,A}, \quad W_{e^A}^{0-} = T_{H,A},$$

$$W_{e^A} = \tau(H,A) \text{ and } W_{e^A}^- = \sigma(H,A).$$

Note also that although we always have $W_B^\infty \subset W_B^{0-}$, it could happen, of course, that $W_{B_1}^{0+} \subset W_{B_2}^\infty$ for different operators B_1 and B_2. For instance, let $B_1 = e^c$ and $B_2 = c$ for some self-adjoint operators $c \geq I$. Concretely, when $H = L_2(R)$ and $C = x^2 - (\partial^2/\partial x^2) + 1$ we have

$$S_{1/2}^{1/2} = W_{e^C}^{0+} \subset W_C^\infty = S,$$

where S is the space of a rapidly decreasing function of Schwartz.

231

Finally, for convenience of later citation we shall formulate a result for the extension of continuous operators on W_B^t to continuous operators on W_{B*}^{-t}.

Theorem 1.3 Suppose that $T : D(T) \subset H \to H$ is a densely defined operator. If $D(T*) \supset W_B^t$, $T*W_B^t \subset W_B^t$ and $T* \mid_{W_B^t}$ is continuous on W_B^t, then T is uniquely extended to a continuous operator $\tilde{T}$ on W_{B*}^{-t} and $\|\tilde{T}\| = \|T*\mid_{W_B^t}\|$.

Proof Define the operator $\tilde{T}$ on W_{B*}^{-t} via the following duality relation as the dual of $T*\mid_{W_B^t}$

$$< u, \tilde{T} \cdot F >_t \; = \; < T* \cdot u, F >_t, \quad \forall u \in W_B^t, \; F \in W_{B*}^{-t}. \qquad (1.24)$$

By the standard arguments or using the well known fact that "the norm of the dual equals that of the original", it easily follows that $\|\tilde{T}\| = \|T*\mid_{W_B^t}\|$. Now, we shall prove that $\tilde{T}$ is indeed an extension of T. So let $F \in D(T)$. Then, from (1.15) and (1.24), we have

$$(B^t u, (B*)^{-t} \cdot \tilde{T} \; F) \; = \; (u, T \; F) \; = \; (B^t u, (B*)^{-t} \cdot T \; F) \quad \text{for all } u \in W_B^t.$$

Thus

$$(w, (B*)^{-t} \cdot \tilde{T} \; F) \; = \; (w, (B*)^{-t} \cdot T \; F), \quad \forall w \in H.$$

Therefore, $(B*)^{-t} \cdot \tilde{T} \; F = (B*)^{-t} \cdot T \; F$ from which follows that $\tilde{T} \cdot F = T \cdot F$. Hence $\tilde{T}$ is an extension of T and the uniqueness follows from the denseness of $D(T)$ in W_{B*}^{-t}. □

ACKNOWLEDGEMENT

The author would like to express his gratitude to Prof. Dr. Jan de Graaf for his quidance.

REFERENCES

1. J. de Graaf, A theory of generalized functions based on holomorphic semigroups, T. H. Report, 79, Wsk. 02, Eindhoven, (1979).
2. J. de Graaf, A theory of generalized functions based on holomorphic semigroups, Proc. K. N. A. W. A., (87) 2, 155-171, (1987).

3. S. J. L. van Eindhoven, A theory of generalized functions based on one parameter groups of bounded self-adjoint operators, <u>T.H. Report</u>, 81, Wsk. 03, Eindhoven, (1981).
4. A. Pazy, "Semigroups of linear operators and applications to partial differential equations", Springer Verlag, (1980).

OPERATIONAL CALCULUS WITH DERIVATIVE $\hat{S} = S^2$

Eligiusz Mieloszyk

Gdańsk Technical University

Majakowskiego 11/12, 80-952 Gdańsk, Poland

Boundary value problems for abstract differential equations were considered among others in [7], [12]. In [10] the author has constructed an operational calculus by operation s_q S and by linear operation $B : L^2 \to$ Ker S under some special assumptions (see [10], p. 252). This paper is a generalization of some conclusions presented in [10].

Let there be a given operational calculus $CO(L^0, L^1, S, T_q, s_q, Q)$ in which L^0, L^1 are linear spaces, S, T_q, s_q are linear operations such that

$S : L^1 \to L^0$ (onto),

$T_q : L^0 \to L^1$,

$s_q : L^1 \to$ Ker S

for $q \in Q$ (Q a set of indices).
Moreover, operations S, T_q, s_q satisfy the conditions

$ST_q f = f$ for $f \in L^0$, $q \in Q$,

$T_q Sx = x - s_q x$ for $x \in L^1$, $q \in Q$.

Operation S is called derivative, operation T_q is called integral and operation s_q is called a limit condition. (Axioms of the operational calculus and its properties can be found in [2, 3, 4, 7, 13]).

Let be $L^1 \subset L^0$.

Definition 1 (See, for example [3, 4]) L^m is a set defined in the following way

$$L^m \stackrel{df}{=} \{x \in L^{m-1} : Sx \in L^{m-1}\}, \quad m = 2, 3, \ldots .$$

Let $B : L^2 \to \operatorname{Ker} S$ be a linear operation.

Theorem 1 If $Bc = c$ for $c \in \operatorname{Ker} S$ then three operations $\hat{S}$, $\hat{T}_q$, $\hat{s}_q$ defined by the following formulas

$$\hat{S}x \stackrel{df}{=} S^2 x, \quad x \in L^2, \tag{1}$$

$$\hat{T}_q f \stackrel{df}{=} T_q^2 f - B T_q^2 f, \quad f \in L^0, \quad q \in Q \tag{2}$$

$$\hat{s}_q x \stackrel{df}{=} T_q s_q Sx - B T_q s_q Sx + Bx, \quad x \in L^2, \quad q \in Q \tag{3}$$

satisfy the axioms of the operational calculus. Operation $\hat{S}$ is a derivative, operation $\hat{T}_q$ is an integral, operation $\hat{s}_q$ is a limit condition.

Proof Of course, operations $\hat{S}$, $\hat{T}_q$, $\hat{s}_q$ are linear operations. From the definition of operations $\hat{S}$, $\hat{T}_q$ it follows that

$$\hat{S}\hat{T}_q f = f \quad \text{for} \quad f \in L^0, \quad q \in Q.$$

One has to prove that $\hat{T}_q \hat{S} x = x - \hat{s}_q x$ for $x \in L^2$, $q \in Q$. From the definition of operation $\hat{S}$, $\hat{T}_q$ we have

$$\hat{T}_q \hat{S} x = T_q^2 S^2 x - B T_q^2 S^2 x = x - s_q x - T_q s_q Sx - Bx + B s_q x + B T_q s_q Sx$$

$$= x - T_q s_q Sx + B T_q s_q Sx - Bx,$$

i.e. in fact we have

$$\hat{T}_q \hat{S} x = x - \hat{s}_q x \quad \text{for} \quad x \in L^2, \quad q \in Q. \quad \square$$

Corollary 1 Operation $\hat{T}_q$ is an injection. Operation $\hat{s}_q$ is a projection L^2 onto $\operatorname{Ker} S^2$, i.e. $\hat{s}_q$ is a surjection L^2 onto $\operatorname{Ker} S^2$ and $\hat{s}_q^2 = \hat{s}_q$.

Corollary 2 If $Bc = c$ for $c \in \operatorname{Ker} S$ then abstract differential equation

$$S^2 x = f$$

with the conditions

$$s_q Sx = x_q, \quad Bx = x_B,$$

where $x \in L^2$, $f \in L^0$, $x_q, x_B \in \operatorname{Ker} S$

has only one solution given by the formula

$$x = T_q x_q - BT_q x_q + x_B + T_q^2 f - BT_q^2 f \quad \text{(see also [12])}. \tag{4}$$

<u>Remark</u> If $B = s_q$ then we obtain initial-value problem.

The obtained in Theorem 1 operational calculus makes it possible, for example, to solve abstract differential equations of the type

$$\sum_{i=0}^{n} A_i S^{2i} x = f \tag{5}$$

with conditions

$$s_q S^{2i+1} x = x_{i,q}, \quad BS^{2i} x = x_{i,B} \tag{6}$$

for $i = 0,1,\ldots,n-1$ on the basis of methods presented in [1, 2, 3, 4, 7, 12, 13, 14]. Coefficients A_i, $i = 0,1,\ldots,n$ which appear in equation (5) can be scalars (numbers) or commutative or non-commutative operations with derivative S, integral T_q and operation B. For example abstract differential equation

$$S^{2n} x = f$$

with conditions (6), where $x \in L^{2n}$, $f \in L^0$ and $x_{i,q}, x_{i,B} \in \text{Ker } S$ for $i = 0,1,\ldots,n-1$ has only one solution given by the formula

$$x = \sum_{i=0}^{n-1} (T_q^2 - BT_q^2)^i (T_q x_{i,q} - BT_q x_{i,q} + x_{i,B}) + (T_q^2 - BT_q^2)^n f.$$

Let L^0 be a Mikusiński's space with partial order $\leq$ and modulus $|\cdot|$ (see [4]).

<u>Definition 2</u> ([3], [4]) $x_n \to x$ iff there exists a non-negative element $f \in L^0$ such that for each $\varepsilon > 0$ ($\varepsilon \in \mathbb{R}$) the inequality $|x_n - x| \leq \varepsilon f$ is satisfied for $n > N(f,\varepsilon)$.

<u>Definition 3</u> ([3], [4]) The linear operation $U : L^0 \to L^0$ will be called non-negative iff $Ux \geq 0$ for each $x \geq 0$.

<u>Definition 4</u> ([3]) The linear operation $U : L^0 \to L^0$ is regular operation iff there exist a non-negative linear operation M such that

$$|Ux| \leq M|x|.$$

<u>Lemma 1</u> ([4]) Operation U is a regular operation iff $U = U_1 - U_2$ where U_1 and U_2 are non-negative operation.

Lemma 2 ([3], [4]) Regular operation U is continuous, i.e. $x_n \to x$ implies $Ux_n \to Ux$.

Theorem 2 If T_q and B are non-negative operations then operation $\hat{T}_q$ is a regular operation.

Operation $\hat{T}_q$ is continuous.

Proof Theorem 2 follows directly from Lemma 1 and from Lemma 2.

Definition 5 ([3], [4]) We shall call the problem of solution of an abstract differential equation with the limit condition

$$Sx = f, \quad s_q x = x_q, \quad x \in L^1, \quad f \in L^0, \quad x_q \in \text{Ker } S \qquad (7)$$

the well-defined solution problem if the following conditions are satisfied:

1^0 For each $f \in L^0$, $x_q \in \text{Ker } S$ there exists an element $x \in L^1$ such that (7) is satisfied.

2^0 The solution x of problem (7) is uniquely determined.

3^0 The solution is continuous to the right side of the equation and to the limit conditions, i.e.

$$Sx_n = f_n \in L^0, \quad s_q x_n = x_{q,n} \in \text{Ker } S, \quad f_n \to f, \quad x_{q,n} \to x_q$$

imply $x_n \to x$ where $Sx = f$, $s_q x = x_q$.

Directly from Theorem 2 and Definition 5 follows the theorem

Theorem 3 If T_q and B are non-negative operations then problem

$$\hat{S}x = f, \quad \hat{s}_q x = \hat{x}_q, \quad x \in L^2, \quad f \in L^0, \quad \hat{x}_q \in \text{Ker } \hat{S}^2$$

is well-defined solution problem.

Examples

A. In the case of an operational calculus with derivative

$$S\{x(t)\} \stackrel{df}{=} \{x'(t) + p(t)x(t)\},$$

integral

$$T_{t_0}\{f(t)\} \stackrel{df}{=} \left\{ e^{-\int_{t_0}^{t} p(\tau)d\tau} \int_{t_0}^{t} f(\tau) e^{\int_{t_0}^{\tau} p(\xi)d\xi} d\tau \right\}$$

and limit condition

$$s_{t_0}\{x(t)\} \stackrel{df}{=} \left\{ x(t_0) e^{-\int_{t_0}^{t} p(\tau)d\tau} \right\},$$

where

$$x \in C^1(\langle t_1, t_2 \rangle, \mathbb{R}) \stackrel{df}{=} L^1, f, p \in C^0(\langle t_1, t_2 \rangle, \mathbb{R}) \stackrel{df}{=} L^0, t_0 \in \langle t_1, t_2 \rangle$$

(see [8]), the derivative $\hat{S}$, the integral $\hat{T}_{t_0}$ and the limit condition $\hat{s}_{t_0}$ are defined by the formulas

$$\hat{S}\{y(t)\} \stackrel{df}{=} \{y''(t) + 2p(t)y'(t) + (p'(t) + p^2(t))y(t)\},$$

$$y \in C^2(\langle t_1, t_2 \rangle, \mathbb{R}),$$

$$\hat{T}_{t_0}\{f(t)\} \stackrel{df}{=} \left\{ \left[\int_{t_0}^{t} (t-\tau) e^{\int_{t_0}^{\tau} p(\xi)d\xi} f(\tau)d\tau \right] e^{-\int_{t_0}^{t} p(\tau)d\tau} \right\}$$

$$- B\left\{ \left[\int_{t_0}^{t} (t-\tau) e^{\int_{t_0}^{\tau} p(\xi)d\xi} f(\tau)d\tau \right] e^{-\int_{t_0}^{t} p(\tau)d\tau} \right\}, f \in C^0(\langle t_1, t_2 \rangle, \mathbb{R}),$$

$$\hat{s}_{t_0}\{y(t)\} \stackrel{df}{=} \left\{ (t-t_0)(y'(t_0) + p(t_0)y(t_0)) e^{-\int_{t_0}^{t} p(\tau)d\tau} \right\}$$

$$- B\left\{ (t-t_0)(y'(t_0) + p(t_0)y(t_0)) e^{-\int_{t_0}^{t} p(\tau)d\tau} \right\} + By, \, y \in C^2(\langle t_1, t_2 \rangle, \mathbb{R})$$

if $Bc = c$ for $c \in \text{Ker}\left(\frac{d}{dt} + p\right)$.

The differential equation

$$x'' + 2px' + (p' + p^2)x = \{f(t)\} \tag{8}$$

with conditions

$$x'(t_0) + p(t_0)x(t_0) = \alpha, \tag{9}$$

$$\int_{t_1}^{t_2} e^{\int_{t_1}^{\tau} p(\xi)d\xi} x(\tau)d\tau = \beta, \tag{10}$$

where $x \in C^2(\langle t_1, t_2 \rangle, \mathbb{R})$, $p, f \in C^0(\langle t_1, t_2 \rangle, \mathbb{R})$, $\alpha, \beta \in \mathbb{R}$, $t_0 \in \langle t_1, t_2 \rangle$ has only one solution given by the formula

$$\{x(t)\} = \left\{ \alpha(t-t_0) e^{-\int_{t_0}^{t} p(\tau)d\tau} \right\} - \alpha B\left\{ (t-t_0) e^{-\int_{t_0}^{t} p(\tau)d\tau} \right\} +$$

$$+ \left\{ \frac{\beta}{t_2-t_1} e^{-\int\limits_{t_1}^{t} p(\tau)d\tau} \right\} + \left\{ \left[\int\limits_{t_0}^{t} (t-\tau) e^{\int\limits_{t_0}^{\tau} p(\xi)d\xi} f(\tau)d\tau \right] e^{-\int\limits_{t_0}^{t} p(\tau)d\tau} \right\}$$

$$- B \left\{ \left[\int\limits_{t_0}^{t} (t-\tau) e^{\int\limits_{t_0}^{\tau} p(\xi)d\xi} f(\tau)d\tau \right] e^{-\int\limits_{t_0}^{t} p(\tau)d\tau} \right\}.$$

In this case operation B is defined by the formula

$$Bx \stackrel{df}{=} \left\{ \frac{1}{t_2-t_1} e^{-\int\limits_{t_1}^{t} p(\tau)d\tau} \int\limits_{t_1}^{t_2} e^{\int\limits_{t_1}^{\tau} p(\xi)d\xi} x(\tau)d\tau \right\},$$

where $x \in C^2(<t_1,t_2>,\mathbb{R})$.

Of course, operation B satisfies the condition $Bc = c$ for $c \in \mathrm{Ker}\left(\frac{d}{dt} + p\right)$. $C^0(<t_1,t_2>,\mathbb{R})$ is Mikusiński's space given the definite order $\{f(t)\} = f \geq 0$ iff $f(t) \geq 0$ for $t \in <t_1,t_2>$ and the modulus $|f| = \{|f(t)|\}$.

Problem (8), (9), (10) is well-defined solution problem if $t_0 = t_1$.

B. The difference equation

$$\{x_{k+2} + (-p_{k+1} - p_k)x_{k+1} + p_k^2 x_k\} = \{f_k\} \tag{11}$$

with conditions

$$x_{k_0+1} - p_{k_0} x_{k_0} = \alpha, \tag{12}$$

$$\frac{x_{k_2}}{\prod\limits_{i=0}^{k_2-1} p_i} + \frac{x_{k_1}}{\prod\limits_{i=0}^{k_1-1} p_i} = \beta, \tag{13}$$

where x_k, f_k, p_k are real sequences, $\alpha, \beta \in \mathbb{R}$ and $p_k \neq 0$ for $k = 0,1,2,\ldots$ has only one solution.

In this case it is necessary to take operations determined by the following formulas as derivative S, integral T_q and the limit condition:

$$S\{x_k\} \stackrel{df}{=} \{x_{k+1} - p_k x_k\} \stackrel{df}{=} \Delta_{p_k}\{x_k\}, \tag{14}$$

$$T_q\{f_k\} \stackrel{df}{=} \left\{ \prod\limits_{i=0}^{k-1} p_i \right\} T_{k_0} \left\{ \frac{f_k}{\prod\limits_{i=0}^{k} p_i} \right\}, \tag{15}$$

$$s_q\{x_k\} \stackrel{df}{=} \left\{ \frac{\prod_{i=0}^{k-1} p_i}{\prod_{i=0}^{k_0-1} p_i} x_{k_0} \right\}. \qquad (16)$$

where

$$T_{k_0}\{y_k\} \stackrel{df}{=} \begin{cases} 0 & \text{for } k = k_0, \\ \sum_{i=k_0}^{k-1} y_i & \text{for } k_0 < k, \\ -\sum_{i=k}^{k_0-1} y_i & \text{for } k_0 > k \end{cases} \qquad \text{(see [9])}.$$

(The operational calculus with the defined in this way derivative, integral and the limit condition is in paper [11].) Operation B is defined by the formula

$$B\{x_k\} \stackrel{df}{=} \left\{ \left(\frac{1}{2} \prod_{i=0}^{k-1} p_i \right) \left(\frac{x_{k_2}}{\prod_{i=0}^{k_2-1} p_i} + \frac{x_{k_1}}{\prod_{i=0}^{k_1-1} p_i} \right) \right\}. \qquad (17)$$

Of course the operation satisfies condition $B\{c_k\} = \{c_k\}$ for $\{c_k\} \in \text{Ker}\Delta_{Pk}$. If so then from Corollary 2 it follows that the solution of equation (11) with conditions (12), (13) is determined by formula (4).

In this formula in place of T_q, s_q and B one should adopt the operations defined by formulas (15) - (17) respectively and substitute

$$x_q = \alpha \left\{ \frac{\prod_{i=0}^{k-1} p_i}{\prod_{i=0}^{k_0-1} p_i} \right\} \quad \text{and} \quad x_B = \frac{1}{2}\beta \left\{ \prod_{i=0}^{k-1} p_i \right\}.$$

Space of real sequences is Mikusiński's space given the definite order

$$\{x_k\} = x \geq 0 \text{ iff } x_k \geq 0 \text{ for } k = 0,1,\ldots \text{ and the modulus } |x| = \{|x_k|\}.$$

Problem (11), (12), (13) is well-defined solution problem if $k_0 = 0$.

C. Similarly it is possible to show that the partial differential equation

$$\left(\sum_{i=1}^{n} b_i \frac{\partial}{\partial x_i} + p(x_1, x_2, \ldots, x_n) \right)^2 \{u(x_1, x_2, \ldots, x_n)\} = \{f(x_1, x_2, \ldots, x_n)\} \tag{18}$$

with conditions

$$\left\{ \sum_{i=1}^{n} b_i \frac{\partial u(x_1, x_2, \ldots, x_{n-1}, x_n^0)}{\partial x_i} + p(x_1, x_2, \ldots, x_{n-1}, x_n^0) u(x_1, x_2, \ldots, x_n^0) \right\} =$$

$$= \{\phi(x_1, x_2, \ldots, x_{n-1})\}, \tag{19}$$

$$\sum_{j=1}^{k} \left\{ \frac{u(x_1 - \frac{b_1}{b_n}(x_n - x_n^j), \ldots, x_{n-1} - \frac{b_{n-1}}{b_n}(x_n - x_n^j), x_n^j)}{e^{-\frac{1}{b_n} \int_{\alpha}^{x_n^j} p(x_1 - \frac{b_1}{b_n}(x_n - \tau), \ldots, x_{n-1} - \frac{b_{n-1}}{b_n}(x_n - \tau), \tau) d\tau}} \right\} = \psi \tag{20}$$

where $u \in C^3(\mathbb{R}^{n-1} \times <\alpha, \beta>, \mathbb{R})$, $f \in C^1(\mathbb{R}^{n-1} \times <\alpha, \beta>, \mathbb{R})$, $\psi \in \text{Ker}\left(\sum_{i=1}^{n} b_i \frac{\partial}{\partial x_i}\right)$, $\phi \in C^2(\mathbb{R}^{n-1}, \mathbb{R})$, $b_i \in \mathbb{R}$ for $i = 1, 2, \ldots, n$, $b_n \neq 0$, $x_n^j \in <\alpha, \beta>$ for $j = 0, 1, 2, \ldots, k$ has only one solution defined by the formula of type (4). $C^0(\mathbb{R}^{n-1} \times <\alpha, \beta>, \mathbb{R})$ is Mikusiński's space given the definite order

$$\{f(x_1, x_2, \ldots, x_n)\} = f \geq 0 \text{ iff } f(x_1, x_2, \ldots, x_n) \geq 0$$

for $(x_1, x_2, \ldots, x_n) \in \mathbb{R}^{n-1} \times <\alpha, \beta>$ and the modulus $|f| = \{|f(x_1, x_2, \ldots, x_n)|\}$. Problem (18), (19), (20) is well-defined solution problem if $x_n^0 = \alpha$.

I should like to express my sincere thanks to Prof. Dr Hab. R. Bittner for his directions and remarks on this paper.

REFERENCES

1. L. Berg, "Operatorenrechnung I", Algebraische Methoden, Verlag der Wissenschaften, Berlin, (1972).
2. R. Bittner, Operational calculus in linear spaces, <u>Studia Math.</u>, 20, 1-18, (1961).
3. R. Bittner, Algebraic and analytic properties of solution of abstract differential equations, <u>Rozprawy Matematyczne</u>, XLI, 1-63, (1964).
4. R. Bittner, "Rachunek operatorów w przestrzeniach liniowych, Polish Scientific Publishers, Warszawa, (1974).
5. R. Bittner, E. Mieloszyk, Properties of eigenvalues and eigenelements of some difference equations in a given operational

calculus, Zeszyty Naukowe UG w Gdańsku, Matematyka, 5, 5-18, (1981).

6. R. Bittner, E. Mieloszyk, Application of the operational calculus to solving non-homogeneous linear partial differential equations of the first order with real coefficients, Zeszyty Naukowe PG w Gdańsku, Matematyka, XII, 33-45, (1982).

7. I. Dimovski, "Convolutional calculus", Publishing House of the Bulgarian Academy of Sciences, Sofia, (1982).

8. E. Mieloszyk, Operational calculus in algebras, Publ. Math. Debrecen, 34,(1-2), 137-143, (1987).

9. E. Mieloszyk, Application of the operational calculus in solving partial difference equation. Acta Mathematica Hungarica, 48, (1-2), 118-130, (1986).

10. E. Mieloszyk, Operational Calculus and Boundary Value Problem for an Abstract Differential Equation, Zeitschrift für Analysis und ihre Anwendungen, Bd. 6, (3), 251-255, (1987).

11. E. Mieloszyk, Example of operational calculus, Zeszyty Naukowe PG w Gdańsku, Matematyka, XIII, 151-157, (1985).

12. D. Przeworska-Rolewicz, Concerning boundary value problems for equations with right invertible operators, Demonstratio Mathematica, Vol. VII, (3), 365-380, (1974).

13. D. Przeworska-Rolewicz, "Shifts and periodicity for right invertible operators", Research Notes in Mathematics 43, Pitman Advanced Publishing Program, Boston, Mass., (1980).

14. M. Tasche, Funktionalanalytische Methoden in der Operatorenrechnung, Nova Acta Leopoldina, 231, (1978).

SOLVABILITY OF NONLINEAR OPERATOR EQUATIONS WITH APPLICATIONS TO HYPERBOLIC EQUATIONS

P. S. Milojević

Department of Mathematics
New Jersey Institute of Technology
Newark, NJ 07102, USA

1. INTRODUCTION

Consider nonlinear equations of the form

$$Au - F(x,u) = f(x), \quad x \in Q \tag{1}$$

in $H = L_2(Q, \mathbb{R}^m)$, where Q is a bounded domain in $\mathbb{R}^n$, $f \in H$ is given, $F : Q \times \mathbb{R}^m \to \mathbb{R}^m$ is a Caratheodory function and $A : D(A) \subset H \to H$ is a selfadjoint map with possibly ∞-dimensional null space.

In this paper, we will study the solvability of Eq. (1) when there is no resonance at infinity, i.e., when F stays away from the spectrum $\sigma(A)$ of A at infinity. Unlike the approaches used by other authors, we will use the pseudo A-proper mapping approach based on finite dimensional approximations of (1) and the Brouwer degree theory.

Regarding A we require

$0 \in \sigma(A)$ and $\sigma(A) \cap (0, \infty) \neq \emptyset$ and consists of isolated eigenvalues having finite multiplicity; (2)

The set of eigenvectors of A forms a complete orthonormal system in H. (3)

Suppose that $F = F_1 + F_2$ and

There are $M_1 > 0$, $\ell \in (0,1)$ and $h_1 \in H$ such that
$|F_1(x,y)| \le M_1 |y|^\ell + h_1(x)$ for a.e. $x \in Q$, $y \in \mathbb{R}^m$; (4)

There are $h_2 \in H$ and, for some consecutive eigenvalues λ_i and λ_{i+1} of A, $\lambda \in (\lambda_i, \lambda_{i+1})$ and $0 < \gamma < \min\{\lambda - \lambda_i, \lambda_{i+1} - \lambda\}$ such that $|F_2(x,y) - \lambda y| \le \gamma |y| + h_2(x)$ for a.e. $x \in Q$, $y \in \mathbb{R}^m$. (5)

Our basic result is

Theorem 1 Let $A : D(A) \subset H \to H$ be a selfadjoint map with closed range and conditions (2) - (5) hold. Suppose that F is a monotone function in y, i.e. $(F(x,y_1) - F(x,y_2)) \cdot (y_1 - y_2) \geq 0$ for all $x \in Q$ and all $y_1, y_2 \in \mathbb{R}^m$. Then Eq. (1) is solvable for each $f \in H$.

Theorem 1 is applicable to periodic boundary value problems for nonlinear hyperbolic equations and Hamiltonian systems. For example, let $Q = (0,\pi) \times (0,2\pi)$ and consider the nonlinear wave equation

$$\begin{aligned} u_{tt} - u_{xx} - F(x,t,u) &= f(x,t), (x,t) \in Q \\ u(0,t) = u(\pi,t) &= 0, \ t \in \mathbb{R} \\ u(x,t+2\pi) &= u(x,t), \ t \in \mathbb{R}, \ x \in (0,\pi) \end{aligned} \qquad (6)$$

and the telegraph equation

$$\begin{aligned} u_{tt} + u_{xxxx} - F(x,t,u) &= f(x,t), \ (x,t) \in Q \\ u(0,t) = u_{xx}(0,t) = u(\pi,t) &= u_{xx}(\pi,t) = 0, \ t \in \mathbb{R} \\ u(x,0) - u(x,2\pi) = u_t(x,0) - u_t(x,2\pi) &= 0, \ x \in (0,\pi). \end{aligned} \qquad (7)$$

Let A denote the abstract realisation in $L_2(Q,\mathbb{R}^m)$ of $D_t^2 - D_x^2$ (respectively, $D_t^2 + D_x^4$) with the above boundary conditions. Then, it is well known that A is selfadjoint, has closed range and a discrete spectrum $\{j^2 - k^2 \mid j = 1,2,\ldots, k \in Z\}$ (respectively, $\{j^4 - k^2 \mid j = 1,2,\ldots, k \in Z\}$). Hence, 0 is the only eigenvalue with infinite multiplicity. Thus, if (4) - (5) hold and F is monotone, Problems (6) and (7) have at least one weak solution for each $f \in L_2(Q,\mathbb{R}^m)$ by Theorem 1. Of course, one can also consider more general linear parts in (6) and (7), like $u_{tt} - Lu$, where L is a strongly elliptic linear operator, subject to appropriate boundary and periodicity conditions. The proofs of these results are based on some abstract results treated next. We add that Problems (1), (6) and (7) in resonance have been studied by many authors, in particular when n = m = 1, (cf. e.g. [1 - 8, 10] and the literature in there).

2. ABSTRACT RESULTS

Let H be a real Hilbert space with a projectionally complete scheme $\Gamma = \{H_n, P_n\}$, where $H_1 \subset H_2 \subset \ldots$ are finite dimensional subspaces of H whose union is dense in H and $P_n : H \to H_n$ are orthogonal projections. Recall that ([9]) a map $T : D \subset H \to H$ is (pseudo)A-proper w.r.t. Γ if

$T_n = P_n T : D \cap H_n \to H_n$ is continuous for each n and if $\{x_{n_k} \in D \cap H_{n_k}\}$ is bounded in H and $P_{n_k} T x_{n_k} \to f$ in H as $k \to \infty$, then some subsequence $x_{n_{k(i)}} \to x$ (respectively, there is $x \in D$) such that $Tx = f$.

Let $A : D(A) \subset H \to H$ be a linear densely defined closed map with a possibly infinite dimensional kernel and $N : H \to H$ be a nonlinear map. Equations of the form

$$Ax - Nx = f \quad (x \in D(A), f \in H) \qquad (8)$$

have been studied by many authors usually assuming additionally that the partial inverse $A^{-1} : R(A) \to R(A)$ is compact (cf. [1-3, 5, 10]). Here $R(A)$ denotes the range of A. For the study of Eq. (8) in a more general framework we refer to [4, 7, 8].

When studying, say, systems of nonlinear hyperbolic equations, A^{-1} is not compact and it is this situation that we will treat here.

For our applications in mind, we restrict ourselves to the study of Eq. (8) when there is no resonance at infinity. We have

Theorem 2 Let $A : D(A) \subset H \to H$ be selfadjoint, $N : H \to H$ be a nonlinear map and $\lambda \notin \sigma(A)$ such that

$$\left. \begin{array}{l} \text{There are positive constants } a,b,c,\ell \in (0,1) \text{ and } r \text{ such that} \\ \|Nx - \lambda x\| \leq a\|x\| + b\|x\|^\ell + c \text{ for } \|x\| \geq r, \end{array} \right\} \qquad (9)$$

$$0 < a < \min\{|\mu| \mid \mu \in \sigma(A - \lambda I)\}. \qquad (10)$$

Then, if $A - N : D(A) \subset H \to H$ is pseudo A-proper w.r.t. Γ with $P_n Ax = Ax$ for $x \in H_n$, Eq. (8) is solvable for each $f \in H$.

Proof We note first that since $A - \lambda I$ is selfadjoint, it follows that $\min\{|\mu| \mid \mu \in \sigma(A - \lambda I)\} = \|(A - \lambda I)^{-1}\|^{-1}$ and

$$\|(A - \lambda I)x\| \geq \|(A - \lambda I)\|^{-1} \|x\|, \quad x \in H. \qquad (11)$$

Now, let $f \in H$ be fixed and show that Eq. (8) is solvable. Consider the homotopy $H(t,x) = (A - \lambda I)x - t(N - \lambda I)x$ on $[0,1] \times D(A)$. We claim that there are $\gamma > 0$, $R \geq r$ and $n_0 \geq 1$ such that

$$\|P_n H(t,x) - t P_n f\| \geq \gamma \text{ for } t \in [0,1], x \in \partial B(0,R) \cap H_n, n \geq n_0. \qquad (12)$$

If not, then there would exist $t_k \in [0,1]$, $t_k \to t_0$, and $x_{n_k} \in H_{n_k}$ such that $\|x_{n_k}\| \to \infty$ and $z_k = P_{n_k} H(t_k, x_{n_k}) - t_k P_{n_k} f \to 0$ as $k \to \infty$. Set $y_k = (A - \lambda I) x_{n_k}$. Then $\|y_k\| \to \infty$ as $k \to \infty$ by (11) and

$$1 \leq (\|(N - \lambda I)x_{n_k}\| + \|z_k\| + \|f\|)/\|y_k\| \leq a\|(A - \lambda I)^{-1}\|$$
$$+ (b\|x_{n_k}\|^\ell + c + \|z_k\| + \|f\|)/\|y_k\|.$$

Hence, passing to the limit, we obtain $1 \le a\|(A - \lambda I)^{-1}\|$, in contradiction to (10), and so (12) holds.

Now, by (12) the Brouwer degree

$$\deg(P_n(A - N), B(0,R) \cap H_n, P_n f) = \deg(P_n(A - \lambda I), B(0,R) \cap H_n, 0) \ne 0$$

for $n \ge n_0$.

Hence, there are $x_n \in B(0,R) \cap H_n$ such that $P_n(A - N)x_n = P_n f$ for each $n \ge n_0$. Moreover, by the pseudo A-properness of $A - N$, there is an $x \in D(A)$ such that $Ax - Nx = f$. □

For the purposes of our applications of Theorem 2 to (1), we will now exibit a new class of maps $A - N$ which is pseudo A-proper.

Theorem 3 Let $A : D(A) \subset H \to H$ be a selfadjoint map with closed range $R(A)$ and such that

$$0 \in \sigma(A) \text{ and } \sigma(A) \cap (0,+\infty) \ne \emptyset \text{ and consists of isolated eigenvalues having finite multiplicities.} \quad (13)$$

If $N : H \to H$ is bounded and monotone, i.e. $(Nx - Ny, x - y) \ge 0$ for all $x,y \in H$, then $A - N : D(A) \subset H \to H$ is pseudo A-proper w.r.t. $\Gamma = \{H_n, P_n\}$ with $P_n Ax = Ax$ for $x \in H_n$.

Proof Let λ_1 be the smallest positive eigenvalue of A and $K : R(A) \to R(A)$ be the right inverse of $-A$, i.e., $K = (-A \mid D(A) \cap R(A))^{-1}$. Then, by the closed graph theorem, K is a bounded linear map on $R(A)$ and $\mu \in \sigma(K) \setminus \{0\}$ if and only if $-\mu^{-1} \in \sigma(A)$. Let $\{P_\lambda \mid \lambda \in R\}$ be the spectral resolution of $-A$, and set

$$P^- = \int_{-\infty}^{-\lambda_1/2} dP_\lambda, P^+ = \int_{-\lambda_1/2}^{\infty} dP_\lambda, H^\pm = P^\pm(R(A)).$$

Then $P^\pm$ are orthogonal projections, $R(A) = H^+ \oplus H^-$ (orthogonal direct sum), $KH^\pm \subset H^\pm$, KP^+ is semi-positive definite on $R(A)$ and, by (13), KP^- is compact on $R(A)$. Since K is selfadjoint, by (13) we get

$$(Kx,x) \ge - \lambda_1^{-1}\|x\|^2 \text{ for all } x \in R(A).$$

Now, we will show that $A - N$ is pseudo A-proper. Let $\{x_{n_k} \in D(A) \cap H_{n_k}\}$ be bounded and $y_k = P_{n_k}(A - N)x_{n_k} \to f$ in H as $k \to \infty$. Then, since N is bounded such is $\{Ax_{n_k}\}$ and we may assume that $x_{n_k} \rightharpoonup x$ and $Ax_{n_k} \rightharpoonup y$ weakly. By the weak closedness of the graph of A, it follows that $x \in D(A)$ and $Ax = y$. By the monotonicity of N, we have for each $z \in D(A)$

$$0 \le (Nx_{n_k} - Nz, x_{n_k} - z) = (Nx_{n_k}, x_{n_k}) - (Nx_{n_k}, z) - (Nz, x_{n_k} - z) =$$

$$= (AP^-x_{n_k}, P^-x_{n_k}) + (AP^+x_{n_k}, P^+x_{n_k}) - (y_k, x_{n_k}) - (Nx_{n_k}, z) - (Nz, x_{n_k} - z).$$

Then
$$(y_k, x_{n_k}) \to (f, x), \quad (Nz, x_{n_k} - z) \to (Nz, x - z)$$
and
$$(Nx_{n_k}, z) = (P_{n_k} Nx_{n_k}, z) + ((I - P_{n_k}) Nx_{n_k}, z) =$$
$$(Ax_{n_k}, z) - (y_{n_k}, z) + (Nx_{n_k}, (I - P_{n_k})z) \to (Ax, z) - (f, z).$$

Moreover, since KP^- is compact,
$$P^-x_{n_k} = P^- K(-A) x_{n_k} = -KP^- Ax_{n_k} \to -KP^- Ax = P^-x,$$
and therefore, $(AP^-x_{n_k}, P^-x_{n_k}) \to (AP^-x, P^-x)$.
Hence,
$$\liminf (-AP^+x_{n_k}, P^+x_{n_k}) \le (AP^-x, P^-x) - (f, x-z) - (Ax, z) - (Nz, x-z). \tag{14}$$

But, since
$$0 \le (-AP^+(x - x_{n_k}), x - x_{n_k}) = -(AP^+x, x) + (AP^+x, x_{n_k}) + (AP^+x_{n_k}, x)$$
$$- (AP^+x_{n_k}, x_{n_k}) = -(AP^+x, x) + 2(AP^+x, x_{n_k}) - (AP^+x_{n_k}, x_{n_k}),$$
we get
$$0 \le (AP^+x, x) + \liminf (-AP^+x_{n_k}, x_{n_k}).$$

Hence,
$$\liminf (-AP^+x_{n_k}, x_{n_k}) = \liminf (-AP^+x_{n_k}, x_{n_k}) \ge -(AP^+x, x) =$$
$$= -(AP^+x, P^+x).$$

Thus, (14) implies that
$$(Ax - Nz - f, x - z) \ge 0 \text{ for all } z \in D(A). \tag{15}$$

Now, we will use Minty's trick. Let $z = x - tu$, $u \in D(A)$, $t > 0$ in (15), and dividing by t, we get
$$(Ax - N(x - tu) - f, u) \ge 0, \quad t > 0, \quad u \in D(A).$$
Taking $t \to 0^+$, we get
$$(Ax - Nx - f, u) \ge 0 \text{ for all } u \in D(A).$$

Since $D(A)$ is dense in H, we obtain $Ax - Nx = f$, and therefore $A - N$ is pseudo A-proper. □

<u>Corollary 1</u> Let $A : D(A) \subset H \to H$ be a selfadjoint map with closed range, $N : H \to H$ be a bounded monotone map and conditions (9), (10) and (13) hold. Then Eq. (8) is solvable for each $f \in H$.

3. PROOF OF THEOREM 1

We observe first that condition (10) holds with $a = \gamma$. Next, define $N = N_1 + N_2$, where $N_i u = F_i(x,u)$ for $u \in H = L_2(Q,R^m)$, $i = 1,2$. Then, it is easy to see that condition (9) holds in view of (4) and (5). Hence, Theorem 1 is a direct consequence of Corollary 1. □

REFERENCES

1. A. Bahri and S. Sanchez, Periodic solutions of a nonlinear telegraph equation in one dimension, Bull. Un. Mat. Ital., t. 5, 18-B, 709-720 (1981).
2. H. Brezis, Periodic solutions of nonlinear vibrating strings and duality principles, Bull. Amer. Math. Soc., 8, 409-426 (1983).
3. H. Brezis and L. Nirenberg, Characterizations of the ranges of some nonlinear operators and applications to boundary value problems, Ann. Scuola Norm. Sup. Pisa, 5, 225-326 (1978).
4. L. Cesari and R. Kannan, Solutions of nonlinear hyperbolic equations at resonance, Nonlinear Anal. TMA 6, 751-805 (1982).
5. J. Mawhin, Compacticite, monotonie et convexite dans l'etude de problems aux limites semi-lineaires, Lecture notes, Univ. de Sherbrooke, vol. 19, (1981).
6. J. Mawhin and M. Willem, Convex perturbations of quadratic forms, Annales de l'Institut Henri Poincare' - Analyse non lineaire, 3(6), 431-453 (1986).
7. P. S. Milojević, On the index and the covering dimension of the solution set of semilinear equations, Proc. Symp. in Pure Math., Amer. Math. Soc. vol. 45, part 2, 183-205 (1986).
8. P. S. Milojević, Solvability of nonlinear hyperbolic equations at resonance, in preparation.
9. W. V. Petryshyn, On the approximation-solvability of equations involving A-proper and pseudo A-proper mappings, Bull. Amer. Math. Soc. 81, 223-312 (1975).
10. P. Rabinowitz, Periodic solutions of nonlinear hyperbolic partial differential equations, Comm. Pure Appl. Math., 20, 145-205 (1967).

SOME IMPORTANT RESULTS OF DISTRIBUTION THEORY

O. P. Misra

Department of Mathematics
Indian Institute of Technology
Hauz Khas, New Delhi - 110 016, India

The theme of the conference entitled 'Generalized Functions Convergence, Structures and their Applications' is based on the vital mathematical theory of Functional Analysis envolved in present century, hailed by Browder (1972) as the century of functional analysis. The opinion of some mathematicians about functional analysis as a purely mathematical abstraction is true to some extent. In response to this question, Dieudonne (1972) has referred to the applications of the theory of distributions created by the French mathematician Laurent Schwartz since his theory is based on functional analysis, but further work is necessary, keeping in mind the historical perspective pointed out by Dieudonne.

1. THE SPACE

It is a well known fact that L. Schwartz has formulated the theory of distributions on the spaces of test functions and consequently distributions in the duals of test function spaces. We call these spaces Schwartz spaces. Thus, one might ask whether the results obtained for distributions in Schwartz spaces hold good in other constructed spaces of generalized functions (distributions). This was the question raised by the late Proffesor Garnir whille commenting on some of my work in 1971. A brief resume of this work is given below:

The space I   During (1965-70), Professor A. H. Zemanian defined the distributional setting of his posed problems by constructing the space of test functions and their duals. Kepping in mind this procedure, Misra (1972) has given the distributional setting of the Stieltjes transform as given below:

We make considerable use of the following differentiation operator:

$$S^k = t^{p+k} D_t^k$$

where D_t^k is the differentiation operator with respect to t of order k (k ∈ ℕ, the set of non-negative integers).

Let K_α be a continuous function on $0 < t < \infty$ defined by

$$K_\alpha(t) = \begin{cases} t^\alpha & 1 \le t < \infty, \ \alpha \le 1, \\ 1 & 0 < t < 1. \end{cases}$$

We define the functional $\gamma_k(\phi)$ for $k \in N$ on certain smooth functions $\phi(t)$ by

$$\gamma_k(\phi) \triangleq \sup_{0<t<\infty} |K_\alpha(t) S^k \phi(t)| < \infty. \tag{1.1}$$

The space I_α is defined as the linear space of all the complex valued smooth functions $\phi(t)$ on $0 < t < \infty$, for which $\gamma_k(\phi)$ exists (i.e. finite) for every $k \in N$. For more properties of I_α (see Misra, [8]).

The dual of I_α is denoted by I_α'. It is the space of continuous linear functionals on I_α. We shall denote the complex number that $f(t) \in I_\alpha'$ assigns to $\phi \in I_\alpha$ by

$$< f(t), \phi(t) >.$$

The space $\mathcal{D}(I)$ (the space of smooth functions of I (I = $(0,\infty)$) whose supports are a compact subset of I is the subspace of I and the topology of $\mathcal{D}(I)$ is stronger than the topology induced on $\mathcal{D}(I)$ by I_α. Consequently, $I_\alpha' \subset \mathcal{D}'(I)$ i.e. I_α' is the smaller space to that of $\mathcal{D}'(I)$ of Schwartz. According to Proffesor Garnir [3] such distributions are ordinary distributions.

Łojasiewicz definition of value and limit at a point of a distribution [7] According to Łojasiewicz we say that $T_x \in D'$ has a limit at the right side of the origin equal to c, if as $\lambda \to 0_+$,

$$< T_x, \tfrac{1}{\lambda}\phi(x/\lambda) > \to < c, \phi(x) >, \; \forall \phi \in D.$$

That is, we write $\lim_{x \to 0_+} T_x = c$.

As we see, Łojasiewicz has formulated his results by taking distributions on Schwartz spaces. This result has been taken by Misra [8] to establish his initial value theorem for the Stieltjes transform which is stated as follows (see [8], p. 595, Theorem 3.1.2).

Theorem 1 If $f(t) \in I'_\alpha$ and if $(f(t)/t^n) \to \alpha$ as $t \to 0_+$ in the sense of Łojasiewicz, then for a given $\varepsilon > 0$,

$$|< f(t) - \alpha t^n, \frac{1}{(a+t)^{p+1}} >| < \varepsilon \text{ as } a \to 0_+.$$

However, in 1971, Garnir 3 had pointed out that one cannot take the Łojasiewicz definition in I'_α directly, as taken by me in the above theorem. Work in this direction has been further developed by Carmichael and Milton (1979), Pandey and others without considering the fact of this space.

Lavoine and Misra [4 - 6] have given the representation of distributions on Schwartz spaces and used the Łojasiewicz definition to establish their Abelian theorems.

It took me 12 years to give the correct formulation of Theorem 3.1.2 of Misra [8]. This is discussed at some lenght in Misra [9 - 10].

2. THE SPACE $\Pi'(r)$

As we have seen in Section 1, distribution theory is meant for applications. But, Temple [12] has pointed out the disadvantages of test functions in distributions. He stated that in the first place test functions do not appear in the final results of many applications of distributions in mathematical physics, and a more direct treatment is possible without test functions. Secondly, if test functions are employed, then, different species have to be provided according to the nature of the space in which the problem is posed. A number of mathematicians such as Temple, Lighthill, Mikusiński etc. have developed the theory without the use of test functions.

It is impossible to delete test functions from distributions. Keeping in mind the above two disadvantages of test functions in distribution theory, Lavoine and Misra ([4 - 6] and finally [10]) have formulated the structure of distributions in the following manner for their posed problems.

Let $\Pi'(r)$ (r is any complex number such that Re r > -1) denote the space of distributions of the form:

$$F_t = D^k f(t) \tag{2.1}$$

where D^k denotes the differential operator of order $k \in N$, $f(t)$ is a locally summable function, zero for $t < 0$ and such that $t^{-r-k+\alpha}f(t)$ is bounded as $t \to \infty$ for a certain $\alpha > 0$. If $r = 0$, we denote Π' instead of $\Pi'(0)$. Let $T_t \in \Pi'(r)$. Then, its Stieltjes transform of index r defined according to (2.1)

$$S_s^r T_t = \langle T_t, (s+t)^{-r-1} \rangle = \frac{\Gamma(r+k+1)}{\Gamma(r+1)} \int_0^\infty f(t)(s+t)^{-r-k-1} dt \ldots, \quad (2.2)$$

where s is a complex number such that $|\arg s| < \pi$.

If $T_t \in \Pi'$. Then, its Stieltjes transform of index zero is given by

$$S_s T_t = \langle T_t, (s+t)^{-1} \rangle = \Gamma(k+1) \int_0^\infty f(t)(s+t)^{-k-1} dt.$$

In (2.1) and (2.2) we see that the right hand sides of these equations exists. Accordingly, we do not bother about test functions in such a structure of distributions. Thus, the objections to test functions raised by Temple, stated above, can possibly be removed by adopting the above structure of distributions. Such a structure of distributions can be seen in the work of Misra and Lavoine [10].

Some aspects of distribution theory have engaged the attention of mathematicians working in the field, but in their work, the matrix has been used without discussing the distributional settling of matrix analysis. The distributional setting of the matrix has been by Zemanian [13].

Conclusions First, as pointed out in Section 1, one should find out the conditions under which the results obtained for distributions in Schwartz spaces hold good in other constructed spaces of distributions (generalized functions). Further, one should take care to mention the limit of a 'distribution', in the sense of Łojasiewicz etc., instead of writing simply $t \to 0$ or $t \to \infty$ for the distribution $f(t)$, whenever such phenomena occur. This deficiency is found in existing literature. Second, for the use of distributions in mathematical physics, the representation of distribution should be shown, without excessive reliance on test functions, as stated in Section 2. Finally, one should not take direct results of classical theory without performing a distributional setting. For such reasons, Laurent Schwartz has expressed the view that distribution has a 'very high meaning'.

REFERENCES

1. F. E. Browder, Hist. Math. 2, 575-590, (1975).
2. J. Dieudonne, Hist. Math. 2, 537-548, (1975).
3. H. G. Garnir, Personal discussion, (1971).
4. J. Lavoine and O.P. Misra, C. R. Ac. Sc. Paris, (1974).
5. J. Lavoine and O. P. Misra, Math. Proc. Cambridge Phil. Soc., 86, 287-293, (1979).
6. J. Lavoine and O.P. Misra, C. R. Ac. Sc. Paris, 290, 139-142, (1980).
7. S. Łojasiewicz, Studia Maths., 16, 1-36, (1957).
8. O. P. Misra, J. Math. Anal. Appls., 39, 590-599, (1972).
9. O. P. Misra, Simon Stevin, 60, 269-275, (1986).
10. O. P. Misra and J. Lavoine, "Transform Analysis of Generalized Functions", North-Holland, (1986).
11. L. Schwartz, "Theorie des distributions", Hermann, Paris, (1966).
12. G. Temple, Proc. Royal Soc. London, 276, 149-177, (1963).
13. A. H. Zemaninan, SIAM, 13, 463-468, (1965).
14. A. H. Zemanian, "Generalized Integral Transformations", John Wiley, New York, (1967).

HYPERBOLIC SYSTEMS WITH DISCONTINUOUS COEFFICIENTS: EXAMPLES

Michael Oberguggenberger

Institut für Mathematik und Geometrie
Universität Innsbruck
Technikerstr. 13, A-6020 Innsbruck, Austria

INTRODUCTION

Consider the initial value problem for a linear hyperbolic (n×n)-system in two variables

$$(\partial_t + \Lambda(x,t)\partial_x)V = F(x,t)V + G(x,t), \quad (x,t) \in \mathbb{R}^2$$

$$V(x,0) = A(x), \quad x \in \mathbb{R}$$

(1)

where Λ and F are (n×n)-matrices, Λ real valued and diagonal, and V, G, A are n-vectors. In the case where Λ is a discontinuous function, multiplicative products of distributions appear in system (1), and so there is no general way of giving a meaning to system (1) in the sense of distributions. Moreover, the example of Hurd and Sattinger [3] shows that such a system may fail to have distributional solutions even if it is in the form of a conservation law. Observe that if one formally transforms a general strictly hyperbolic (n×n)-system with discontinuous coefficients into diagonal form, then the matrix F will contain measures as coefficients, or even products of measures and discontinuous functions. The purpose of this note is to discuss the solvability of system (1) in the Colombeau algebra $G(\mathbb{R}^2)$. $G(\mathbb{R}^2)$ is a commutative differential algebra containing the distributions (and having the algebra of infinitely differentiable functions as a subalgebra), see [1]. Thus the notion of a solution to system (1) makes sense in the framework of $G(\mathbb{R}^2)$, even if the coefficients are themselves members of this algebra. Moreover, discontinuous functions which are bounded away from zero are invertible in $G(\mathbb{R}^2)$; this fact enables one to transform general (n×n)-system into systems which are in diagonal form and, more importantly, are equivalent to the original ones in the algebra $G(\mathbb{R}^2)$.

In Section 1 we present a general existence and uniqueness result in $G(\mathbb{R}^2)$ for system (1) and demonstrate the coherence of the abstract solution with possibly existing classical solutions in a transmission problem from acoustics. The details of these results can be found in [7]. Section 2 is devoted to the Hurd-Sattinger example. This concerns the single equation

$$\partial_t v(x,t) - \partial_x (H(x) v(x,t)) = 0$$
$$v(x,0) = 1 \qquad (2)$$

where H is the Heaviside function. Let us recall why there does not exist a member $v \in L^1_{loc}(\mathbb{R}^2)$ which satisfies the equation in the sense of distributions. Indeed, for $x < 0$ we have $\partial_t v = 0$; for $x > 0$ we have $(\partial_t - \partial_x)v = 0$. From the constancy theorem of distribution theory, v must be equal to one on $\{t > 0\}$. But then $\partial_t v = \partial_x (Hv) = \delta(x)$, where δ denotes the Dirac measure, and

$$v(x,t) = 1 + t\delta(x) \quad \text{for} \quad x \in \mathbb{R}, t > 0 \qquad (3)$$

contradicting $v \in L^1_{loc}(\mathbb{R}^2)$.

We shall show that problem (2) has a uniqe solution in $G(\mathbb{R}^2)$ for arbitrary initial data in $G(\mathbb{R})$ and that the solution admits an associated distribution [1, Def. 2.5.5] in case the initial data belong to $L^1_{loc}(\mathbb{R})$. In the special case $v(x,0) = 1$, it is seen that the associated distribution is given by (3) on $\{t > 0\}$. In an appendix we give some examples with even more singular coefficients, which also serve to illuminate the hypothesis for the existence-uniqueness result of Section 1.

2. A GENERAL EXISTENCE-UNIQUENESS RESULT

The notion of the sets $A_q(\mathbb{R}^m)$, the algebra $\mathcal{E}_M[\mathbb{R}^m]$, the ideal $N(\mathbb{R}^m)$ and the Colombeau algebra $G(\mathbb{R}^m) = \mathcal{E}_M[\mathbb{R}^m]/N(\mathbb{R}^m)$ we employ here are described in detail in [6]; see also [2]. To each $T \in \mathcal{D}'(\mathbb{R}^m)$ we assign the class of the map

$$\phi^{\otimes m} \to T * \phi^{\otimes m} \qquad (4)$$

from $A_q(\mathbb{R}^m)$ to $C^\infty(\mathbb{R}^m)$; this defines the inclusion of $\mathcal{D}'(\mathbb{R}^m)$ into $G(\mathbb{R}^m)$. In order to ensure hyperbolicity of the diagonalized system (1), we shall assume in this article that all functions and generalized functions are real valued. Given matrices Λ, F and vectors G, A in $G(\mathbb{R}^2)$, respectively $G(\mathbb{R})$, an element $V \in G(\mathbb{R}^2)$ is called a solution to system (1) if it satisfies it in the sense of differentiation, multiplication, and restriction to the line $\{t = 0\}$ in the algebra $G(\mathbb{R}^2)$. We have the following general result:

Proposition 1 Let $\Lambda, F, G \in G(\mathbb{R}^2)$. Assume that Λ is globally bounded and that $\partial_x \Lambda$ as well as F are locally of logarithmic growth. Then given initial data $A \in G(\mathbb{R})$, problem (1) has a unique solution $V \in G(\mathbb{R}^2)$.

Here an element $U \in G(\mathbb{R}^2)$ is called globally bounded, if it has a representative $u \in E_M[\mathbb{R}^2]$ such that

$$\sup_{(x,t) \in \mathbb{R}^2} |u(\phi_\varepsilon \otimes \phi_\varepsilon, x, t)| \leq C$$

for all $\phi \otimes \phi \in A_N(\mathbb{R}^2)$ with N large enough, some $C > 0$ which may depend on $\phi \otimes \phi$, and for all $\varepsilon > 0$ small enough. As usual, $\phi_\varepsilon(x) = \frac{1}{\varepsilon} \phi(\frac{x}{\varepsilon})$. The generalized function U is called locally of logarithmic growth, if it has a representative u such that for every compact subset $K \subset \mathbb{R}^2$

$$\sup_{(x,t) \in K} |u(\phi_\varepsilon \otimes \phi_\varepsilon, x, t)| \leq N \log \frac{C}{\varepsilon}$$

for $\phi \otimes \phi \in A_N(\mathbb{R}^2)$ with N large enough (depending on K), some $C > 0$ (depending on K and ϕ), and all $\varepsilon > 0$ small enough.

The proof of Proposition 1 follows the lines of the proof of [6, Prop. 2]. The global boundedness condition gives control over the characteristic curves, the logarithmic growth condition enters in a Gronwall-type estimate. The details are given in [7]; a similar result has been obtained independently by Lafon [5]. Here we recall only that the solution V has a representative v such that $v(\phi \otimes \phi, x, t)$ is the classical solution to the system

$$(\partial_t + \lambda(\phi \otimes \phi, x, t)\partial_x)v(\phi \otimes \phi, x, t) = f(\phi \otimes \phi, x, t)v(\phi \otimes \phi, x, t) + g(\phi \otimes \phi, x, t)$$
$$v(\phi \otimes \phi, x, 0) = a(\phi, x) \tag{5}$$

where λ, f, g, and a are representatives of Λ, F, G, and A. What concerns the hypotheses on Λ and F in Proposition 1 we remark that existence or uniqueness may fail to hold if the conditions are dropped, in general. An example for nonuniqueness is given in the Appendix.

When solving hyperbolic systems with discontinuous coefficients we wish to allow $\Lambda \in L^\infty(\mathbb{R}^2)$, $F \in W^{-1,\infty}_{loc}(\mathbb{R}^2)$. So the question is whether in this case Λ and F, viewed as elements of $G(\mathbb{R}^2)$, will satisfy the growth requirements of Proposition 1. It is immediately seen that this may not be the case if the canonical imbedding (4) of $\mathcal{D}'(\mathbb{R}^2)$ into $G(\mathbb{R}^2)$ is employed. For instance, the member of $G(\mathbb{R}^2)$ corresponding to the Heaviside function

$H(x) \otimes 1(t)$ satisfies the global boundedness assumption, but its derivative with respect to x, the class of $\phi_\varepsilon \otimes \phi_\varepsilon \to \phi_\varepsilon(x) = \frac{1}{\varepsilon}\phi(\frac{x}{\varepsilon})$, does not satisfy the condition of logarithmic growth. The way around this obstacle is to use the concept of association. Recall that an element $U \in G(\mathbb{R}^2)$ is said to admit $T \in \mathcal{D}'(\mathbb{R}^2)$ as associated distribution, if for some representative u of U and every $\psi \in \mathcal{D}(\mathbb{R}^2)$

$$\lim_{\varepsilon \to 0} \iint u(\phi_\varepsilon \otimes \phi_\varepsilon, x, t)\psi(x,t)dxdt = <T, \psi>$$

for all $\phi \otimes \phi \in A_N(\mathbb{R}^2)$ with N large enough (depending on ψ). This means that the nets of smooth approximants defining the element $U \in G(\mathbb{R}^2)$ converge to the distribution T. It is possible to show [7] that given $\Lambda \in L^\infty(\mathbb{R}^2)$, $F \in W_{loc}^{-1,\infty}(\mathbb{R}^2)$, there are elements in $G(\mathbb{R}^2)$ which are associated to Λ and F, respectively, and satisfy the required growth conditions. The method, how to construct such elements, is demonstrated in the example of Section 2. In this sense hyperbolic systems with discontinuous coefficients are solvable in $G(\mathbb{R}^2)$.

The following question arises: What happens in the case of a system which has a classical or distributional solution? Will the abstract solution in $G(\mathbb{R}^2)$ be associated to this solution? A typical problem of that kind is the following transmission problem from linear acoustics,

$$\partial_t \rho(x,t) + \rho_0(x)\partial_x u(x,t) = 0$$
$$\rho_0(x)\partial_t u(x,t) + \partial_x p(x,t) = 0$$
$$p(x,t) = c_0^2(x)\rho(x,t)$$

which describes the propagation of an acoustic wave in a medium in Lagrangian coordinates, see Poiree [8, 9]; (ρ, u, p) are first order approximations to the density, velocity, and pressure. ρ_0 is the density, c_0 the sound speed of the medium at rest. It is assumed that ρ_0, c_0 are strictly positive functions which suffer a jump discontinuity at $x = 0$. Assuming that the acoustic field (ρ, u, p) is known at some point $x_0 < 0$ for all times t, the system may be solved classically on both sides of the line $\{x = 0\}$ by imposing a transmission condition along this line. The physically meaningful condition is continuity of u and p. This way one obtains a unique field (ρ, u, p) with u, p continuous which solves the equations on $\{x < 0\}$ and $\{x > 0\}$. For this example it has been shown [7] that the unique solution in $G(\mathbb{R}^2)$ admits precisely the above classical solution as associated distribution. A similar result has been obtained by Lafon [5] in the case of an equation describing electron transport in a layered medium [4].

It should be noted that in the above example, the classical solution (u,p) is a distributional solution to the system

$$\rho_0(x)^{-1} c_0(x)^{-2} \partial_t p(x,t) + \partial_x u(x,t) = 0$$

$$\rho_0(x) \partial_t u(x,t) + \partial_x p(x,t) = 0,$$

and this is the main ingredient in the proof. We shall now meet a distributionally unsolvable equation which has a unique solution in $G(\mathbb{R}^2)$ admitting an associated distribution.

2. THE HURD-SATTINGER EXAMPLE REVISITED

In the algebra $G(\mathbb{R}^2)$, the Hurd-Sattinger equation may be written as

$$(\partial_t + \Lambda \partial_x) V = - (\partial_x \Lambda) V \qquad (6)$$

$$V(x,0) = A(x)$$

where Λ is to be an element of $G(\mathbb{R}^2)$ associated to $-H(x) \otimes 1(t)$. To ensure that Λ satisfies the hypotheses of Proposition 1 we proceed as follows. We fix a function $\chi \in \mathcal{D}(\mathbb{R})$ with $\int \chi(x) dx = 1$ and define an element $\lambda \in E_M[\mathbb{R}^2]$ by

$$\lambda(\phi \otimes \phi, x, t) = - \int_{-\infty}^{x} \mu \chi(\mu \xi) d\xi$$

where

$$\mu = \mu(\phi \otimes \phi) = \log \left[\int_{-\infty}^{\infty} \phi^2(\xi) d\xi \right].$$

Observing that $\mu(\phi_\varepsilon \otimes \phi_\varepsilon) = \log \frac{1}{\varepsilon} + \mu(\phi \otimes \phi)$, we infer immediately that Λ, the class of λ in $G(\mathbb{R}^2)$, is globally bounded and $\partial_x \Lambda$ is locally of logarithmic growth; moreover, Λ is associated to $-H(x) \otimes 1(t)$. From Proposition 1 we may therefore assert:

<u>Proposition 2</u> With Λ as above, problem (6) has a unique solution $V \in G(\mathbb{R}^2)$, given arbitrary initial data $A \in G(\mathbb{R})$.

In fact, a representative v of the solution V is given explicitly as follows. For a fixed $\phi \otimes \phi \in A_0(\mathbb{R}^2)$, denote $\lambda(\phi_\varepsilon \otimes \phi_\varepsilon, x, t)$ by $\lambda^\varepsilon(x,t)$. Let a be a representative of A and set $a^\varepsilon(x) = a(\phi_\varepsilon, x)$. According to formula (5), the solution V has a representative v such that $v^\varepsilon(x,t) = v(\phi_\varepsilon \otimes \phi_\varepsilon, x, t)$ is the classical smooth solution to the system

$$(\partial_t + \lambda^\varepsilon \partial_x) v^\varepsilon = - (\partial_x \lambda^\varepsilon) v^\varepsilon \qquad (7)$$

$$v^\varepsilon(x,0) = a^\varepsilon(x).$$

The characteristic curve $\gamma^\varepsilon(x,t,\tau)$ passing through (x,t) at time $t = \tau$ is the solution of the equation

$$\frac{\partial}{\partial \tau} \gamma^\varepsilon(x,t,\tau) = \lambda^\varepsilon(\gamma^\varepsilon(x,t,\tau),\tau)$$

$$\gamma^\varepsilon(x,t,t) = x.$$

Equation (7) is an ordinary differential equation along the characteristic curves, whose solution is easily calculated as

$$v^\varepsilon(x,t) = \exp\left[-\int_0^t \partial_x \lambda^\varepsilon(\gamma^\varepsilon(x,t,\tau),\tau) d\tau\right] a^\varepsilon(\gamma^\varepsilon(x,t,0)). \qquad (8)$$

We shall now show that the solution V admits an associated distribution in case the initial belong to $L^1_{loc}(\mathbb{R})$.

<u>Proposition 3</u>   Let $a \in L^1_{loc}(\mathbb{R})$ and denote by A its class in $G(\mathbb{R})$ obtained from the imbedding (4). Then the solution $V \in G(\mathbb{R}^2)$ to problem (6) constructed in Proposition 2 admits an associated distribution, which is given by

$$H(-x)a(x) + H(x)H(x+t)a(x+t) + \delta(x)H(t)\int_0^t a(\xi)d\xi$$

where H denotes the Heaviside function and δ the Dirac measure.

The values of the associated distribution are depicted in Figure 1.

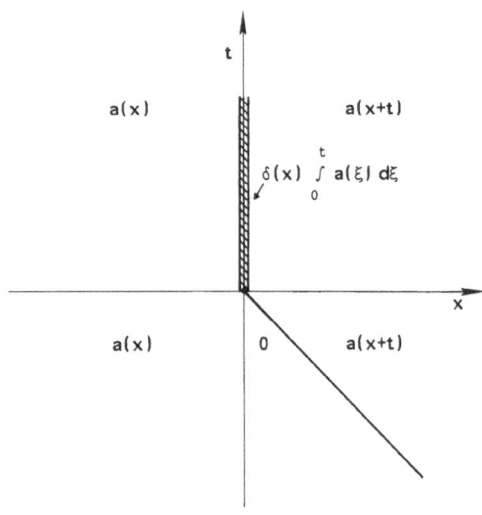

Figure 1

Proof Let $\psi \in \mathcal{D}(\mathbb{R}^2)$. Introducing the change of coordinates $y = \gamma^\varepsilon(x,t,0)$, $s = t$, respectively $x = \gamma^\varepsilon(y,0,s)$, $t = s$, one calculates from the differential equation defining γ^ε that

$$\frac{\partial}{\partial y} \gamma^\varepsilon(y,0,s) = \exp\left(\int_0^s \partial_x \lambda^\varepsilon(\gamma^\varepsilon(y,0,\sigma),\sigma)d\sigma\right).$$

Observing the relation $\gamma^\varepsilon(x,t,\tau) = \gamma^\varepsilon(y,0,\tau)$ one obtains from formula (8) that

$$\iint v^\varepsilon(x,t)\psi(x,t)dxdt = \iint a^\varepsilon(y)\psi(\gamma^\varepsilon(y,0,s),s)dyds.$$

Recall that

$$\lambda^\varepsilon(x,t) = -\int_{-\infty}^x \mu(\varepsilon)\chi(\mu(\varepsilon)\xi)d\xi$$

with $\mu(\varepsilon) = \log\frac{1}{\varepsilon} + \mu(\phi \times \phi)$. Let $\eta(\varepsilon) = \max\{|\xi| : \mu(\varepsilon)\chi(\mu(\varepsilon)\xi) \neq 0\}$. Then $\eta(\varepsilon) \to 0$ as $\varepsilon \to 0$, and $\lambda^\varepsilon(x,t) = 0$ for $x \leq -\eta(\varepsilon)$, $\lambda^\varepsilon(x,t) = -1$ for $x \geq \eta(\varepsilon)$. Consider a characteristic curve $\gamma^\varepsilon(y,0,s)$ starting at $(y,0)$. If $y \leq -\eta(\varepsilon)$, then it will be vertical, i.e. $\gamma^\varepsilon(y,0,s) = y$. If y is positive, then the characteristic curve moves to the left with speed one until it is trapped in the strip $\{-\eta(\varepsilon) \leq x \leq \eta(\varepsilon)\}$. More precisely, $\gamma^\varepsilon(y,0,s) = y - s$ if $y \geq \eta(\varepsilon)$ and $s \leq y - \eta(\varepsilon)$, while $|\gamma^\varepsilon(y,0,s)| \leq \eta(\varepsilon)$ whenever $-\eta(\varepsilon) \leq y \leq s+\eta(\varepsilon)$. We conclude that, if $y \neq 0$ and $s \neq y$, then $\gamma^\varepsilon(y,0,s)$ tends to $\gamma(y,0,s)$, where

$$\gamma(y,0,s) = \begin{cases} y & \text{for } y < 0 \\ y-s & \text{for } y > 0 \text{ and } s < y \\ 0 & \text{for } 0 < y < s. \end{cases}$$

From this convergence almost everywhere, the uniform boundedness of $\psi(\gamma^\varepsilon(y,0,s),s)$, and the L^1_{loc}-convergence of a^ε to a we infer that

$$\lim_{\varepsilon \to 0} \iint v^\varepsilon(x,t)\psi(x,t)dxdt = \iint a(y)\psi(\gamma(y,0,s),s)dyds =$$

$$\int_{-\infty}^{\infty}\int_{-\infty}^{0} a(y)\psi(y,s)dyds + \int_{-\infty}^{0}\int_{0}^{\infty} a(y)\psi(y-s,s)dyds +$$

$$\int_{0}^{\infty}\int_{s}^{\infty} a(y)\psi(y-\sigma,\sigma)dyd\sigma + \int_{0}^{\infty}\int_{0}^{s} a(y)dy\psi(0,\sigma)d\sigma,$$

from where the assertion follows. □

In the Hurd-Sattinger example, a is taken to be identically one. We see that the corresponding solution $V \in G(\mathbb{R}^2)$ admits an associated distribution which equals $1 + t\delta(x)$ for $t > 0$; this is precisely the expression (3). For $t < 0$ it equals $H(-x) + H(x)H(x+t)$, and this is the unique locally integrable solution to problem (2) for negative time. Indeed, if $v \in L^1_{loc}(\mathbb{R}^2_-)$ solves (2), then $v(x,t) = 1$ for $x < 0$ and $x > -t$, and $v(x,t)$ is given by some function $f(x+t)$ for $0 < x < -t$. But near $x = 0$, $\partial_t v(x,t) - \partial_x(H(x)v(x,t)) = -f(t)\delta(x)$, so f must vanish. Hence $v(x,t) = H(-x) + H(x)H(x+t)$, and this solves (2) classically for $t < 0$.

APPENDIX

Here we consider some examples with even more singular coefficients than covered by Proposition 1. The first one is an equation with unique solutions in $G(\mathbb{R}^2)$ whose coefficient does not satisfy the global boundedness assumption. We denote by $\delta(t-1)$ the member of $G(\mathbb{R}^2)$ which is obtained from the canonical imbedding (4) of the Dirac measure along the line $\{t = 1\}$. That is, $\delta(t-1)$ is the class of the element of $E_M[\mathbb{R}^2]$ which has the value $\phi(t-1)$ on each $\phi \otimes \phi \in A_0(\mathbb{R}^2)$.

Example 1 Given $A \in G(\mathbb{R})$, the problem

$$(\partial_t + \delta(t-1)\partial_x)V = 0$$

$$V(x,0) = A(x)$$

has a unique solution $V \in G(\mathbb{R}^2)$. If A corresponds to a function $a \in L^1_{loc}(\mathbb{R})$, then V admits an associated distribution given by $a(x - H(t-1))$.

Indeed, if $V \in G(\mathbb{R}^2)$ is a solution, then for every representative v of V and a of A there are $d_1 \in N(\mathbb{R})$, $d_2 \in N(\mathbb{R}^2)$ such that

$$(\partial_t + \phi(t-1)\partial_x)v(\phi \otimes \phi, x, t) = d_2(\phi \otimes \phi, x, t)$$

$$v(\phi \otimes \phi, x, t) = a(\phi, x) + d_1(\phi, x).$$

It follows that

$$v(\phi \otimes \phi, x, t) = a(\phi, x - \int_0^t \phi(s-1)ds) + d_1(\phi, x - \int_0^t \phi(s-1)ds) + \int_0^t d_2(\phi \otimes \phi, x - \int_\tau^t \phi(s-1)ds, \tau)d\tau.$$

Setting $d_1 = 0$, $d_2 = 0$, it is clear that the v thus constructed will belong to $E_M[\mathbb{R}^2]$. This proves existence of a solution. On the other hand, if $a = 0$ and $d_1 \in N(\mathbb{R})$, $d_2 \in N(\mathbb{R}^2)$, then v will belong to $N(\mathbb{R}^2)$, which proves uniqueness.

If $a \in L^1_{loc}(\mathbb{R})$ we interpret it as the class of $a(\phi,x) = (a * \phi)(x)$. Let $\psi \in \mathcal{D}(\mathbb{R}^2)$. Then

$$\iint v(\phi_\varepsilon \otimes \phi_\varepsilon, x, t)\psi(x,t)dxdt = \iint a(\phi_\varepsilon, x)\psi(x + \int_0^t \phi_\varepsilon(s-1)ds, t)dxdt$$

converges to

$$\iint a(x)\psi(x + H(t-1), t)dxdt.$$

The next example shows that, on the other hand, uniqueness may fail to hold if the global boundedness assumption is dropped, the reason for this being that then the characteristic curves may become horizontal as $\varepsilon \to 0$.

Example 2 The solution $V \in \mathcal{G}(\mathbb{R}^2)$ to the problem

$$(\partial_t + \delta^2(t-1)\partial_x)V = 0$$
$$V(x,0) = 0$$
(9)

is not unique.

Indeed, we shall construct a solution V different from the zero solution. Let $\chi \in \mathcal{D}(\mathbb{R})$ with $\chi(0) \neq 0$ and consider the map $a(\phi,x) = \chi(x + \int \phi^2(\xi)d\xi)$. Since the support of $a(\phi_\varepsilon, x) = \chi(x + \frac{1}{\varepsilon}\int \phi^2(\xi)d\xi)$ moves to the left indefinitely as $\varepsilon \to 0$, $a(\phi,x)$ is a representative of zero in $\mathcal{G}(\mathbb{R})$. The classical solution $v(\phi \otimes \phi, x, t)$ to

$$(\partial_t + \phi^2(t-1)\partial_x)v(\phi \otimes \phi, x, t) = 0$$
$$v(\phi \otimes \phi, x, t) = a(\phi, x)$$

is given by

$$v(\phi \otimes \phi, x, t) = a(\phi, x - \int_0^t \phi^2(s-1)ds),$$

so

$$v(\phi_\varepsilon \otimes \phi_\varepsilon, x, t) = \chi(x + \frac{1}{\varepsilon}\int_{-\infty}^\infty \phi^2(s)ds - \frac{1}{\varepsilon}\int_{-\frac{1}{\varepsilon}}^{\frac{t}{\varepsilon}-\frac{1}{\varepsilon}} \phi^2(s)ds).$$

It is clear that v belongs to $\mathcal{E}_M[\mathbb{R}^2]$, thus its class V is a solution to (9). On the other hand, $v(\phi_\varepsilon \otimes \phi_\varepsilon, 0, 2) \to \chi(0) \neq 0$ for every $\phi \otimes \phi \in A_0(\mathbb{R}^2)$, and so $V \neq 0$ in $\mathcal{G}(\mathbb{R}^2)$.

We remark that a similar argument shows nonuniqueness for the equation $(\partial_t + \delta'(t-1)\partial_x)V = 0$. An example of nonexistence, in case the condition of logarithmic growth in Proposition 1 is violated, is given in [7].

REFERENCES

1. J. F. Colombeau, "Elementary Introduction to New Generalized Functions", Amsterdam - New York - Oxford, North Holland (1985).
2. J. F. Colombeau, Nouvelles solutions d'équations aux dérivées partielles, C. R. Acad. Sc. Paris, 301, Série I, 281-283 (1985).
3. A. E. Hurd und D. H. Sattinger, Questions of existence and uniqueness for hyperbolic equations with discontinuous coefficients, Trans. Am. Math. Soc., 132, 159-174 (1968).
4. F. Lafon, Transport d'electrons dans un materiaux multicouche, Rapport de DEA, Bordeaux (1986).
5. F. Lafon, personal communication.
6. M. Oberguggenberger, Generalized solutions to semilinear hyperbolic systems, Monatshefte Math., 103, 133-144 (1987).
7. M. Oberguggenberger, Hyperbolic systems with discontinuous coefficients: generalized solutions and a transmission problem in acoustics, Preprint (1987).
8. B. Poirée, Les équations de l'acoustique linéaire dans un fluide parfait au repos à caracteristiques indéfiniment différentiables par morceaux, Rev. Cethedec, 52, 69-79 (1977).
9. B. Poirée, Les équations de l'acoustique linéaire et non linéaire dans les fluides en mouvement, Thèse, Université de Paris VI, (1982).

ESTIMATIONS FOR THE SOLUTIONS OF OPERATOR LINEAR DIFFERENTIAL EQUATIONS

Endre Pap and Đurđica Takači

Institute of Mathematics
University of Novi Sad
dr I. Đuričića 4, 21000 Novi Sad, Yugoslavia

ABSTRACT

In this paper we observe the approximate solution of the linear operator differential equation and estimate the error of approximation. For this purpose we use the results from [6]. They enable us to introduce some measures of approximation on the space L of locally integrable functions on $[0,\infty)$ and on the field of Mikusiński operators.

1. INTRODUCTION

The linear partial differential equation with constant coefficients

$$\sum_{k=0}^{m} \sum_{j=0}^{n} \alpha_{k,j} \frac{\partial^{k+j} x(\lambda,t)}{\partial \lambda^k \partial t^j} = f_1(\lambda,t)$$

corresponds to the equation in the field of Mikusiński's operators

$$\sum_{k=0}^{m} \sum_{j=0}^{n} \alpha_{k,j} s^j x^{(k)}(\lambda) = f(\lambda)$$

where s is the differential operator and $f(\lambda)$ is the corresponding operator to the function $f_1(\lambda,t)$ (see [5] and [6]).

In this paper we shall observe, using the results of Boehme [1], Burzyk [2] and [6] on the characterization of the I' type convergence in F_0 (see section 2 for notations), the approximate solution of the mentioned operator equation and construct some measures of approximation.

The approximate solutions $x_n(\lambda)$ are obtained as the partial sums of the series expansion of the exact solution $x(\lambda)$ with respect to degrees of the integral operator ℓ. The estimations depend only on the number n and not on the length T of the interval $[0,T]$. The dependence of T makes problems in paper [7]. In our case, also, the measure of approximation is in direct connection with the type I' convergence in the field F_0.

If the solution of the operator differential equation represents a locally integrable function, then the obtained estimation is in direct connection with the convergence in the ring of locally integrable functions on the interval $[0,\infty)$ ([6]).

2. PRELIMINARY NOTATIONS AND NOTIONS

In this section we shall explain some important definitions, notations and results which are necessary for this paper. More details about this can be seen in [6].

The field of Mikusiński operators F can be considered as the quotient field of the ring L of locally integrable functions on $[0,\infty)$ (or as the quotient field of the ring C of continuous complex valued functions on $[0,\infty)$), both with the usual addition and multiplication given by the convolution

$$(f \cdot g)(t) = \int_0^t f(t - \tau) g(\tau) \, d\tau, \quad f, g \in L \quad (\text{or } f, g \in C).$$

The space L is endowed with the topology defined by the family of seminorms:

$$\{\|f\|_T = \int_0^t |f(t)| \, dt; \; T > 0\}, \quad f \in L. \tag{1}$$

The convergence in all the seminorms $\|\cdot\|_T$, $T > 0$, will be called the convergence in L.

The subspace of L, consisting of all the functions f such that $\|f\|_T > 0$ for any $T > 0$ will be denoted by L_0.

We say that the sequence $\{x_n\}$, $x_n \in F$, converges to $x \in F$ type I, if there exist representations $x_n = f_n/g$, $x = f/g$, $(f_n, f, g \in L, g \neq 0)$ such that $f_n \to f$ as $n \to \infty$ in L. The sequence $\{x_n\}$, $x_n \in F$ converges to x type I' of each subsequence of $\{x_n\}$ possesses a subsequence which converges to x type I.

As in papers [1], [2] let us denote by F_0 the algebra of all the operators of the form f/g, $f \in L$ and $g \in L_0$. The functional $B_{T,\varepsilon}(x)$ for $x \in F_0$, is introduced as

$$B_{T,\varepsilon}(x) = \inf\{ \|f\|_T : x = f/g, \|g\|_T < 1, \|\ell - \ell g\|_T < \varepsilon\}.$$

In [2] J. Burzyk has proved that a sequence $\{x_n\}$ converges type I' to x, $x_n, x \in F_0$, iff $B_{T,\varepsilon}(x_n - x) \to 0$ as $n \to \infty$ for every $T, \varepsilon > 0$,

We consider the linear homogeneous differential equation in the field F

$$\sum_{k=0}^{m} \sum_{j=0}^{n} \alpha_{k,j} s^j x^{(k)}(\lambda) = 0, \quad 0 \leq \lambda \leq \lambda_0, \tag{2}$$

with the conditions

$$x^{(k)}(0) = \phi_k, \quad k = 0, \ldots, m-1. \tag{3}$$

The solution of equation (2) is of the form (see pages 269-272 and 446 in [5]):

$$x(\lambda) = \sum_{j=1}^{m} b_j \exp(\lambda \omega_j), \tag{4}$$

where operators b_j are determined by (3) and ω_j, $j = 1, \ldots, m$, are the simple solutions of the characteristic equation of (2)

$$\sum_{k=0}^{m} \sum_{j=0}^{n} \alpha_{k,j} s^j \omega^k = 0.$$

In [5] p. 448, it was shown that ω_j can be written as:

$$\omega_j = \sum_{i=0}^{\infty} c_{i,j} \ell^{i\alpha_j - \beta_j},$$

α_j, β_j are rational numbers, $\alpha_j > 0$ and $\beta_j \leq 1$ for $j = 1, \ldots, m$, $c_{i,j}$ are complex numbers. Since ω_j are the solutions of an equation of order m, very often it is not possible to find them in the exact form, so we are supposed to find an approximate solution of the characteristic equation.

In paper [7] was observed (without loss of generality) only one of the linearly independent solutions of equation (2)

$$x(\lambda) = b \exp(\lambda \omega), \text{ where } \omega = \sum_{i=0}^{\infty} c_i \ell^{i\alpha - \beta}$$

(for $\alpha = 1/q$ and $\beta = p/q$ in [7]).

The approximate solutions of the characteristic equation were treated in the form

$$\omega_n = \sum_{i=0}^{n} C_i \ell^{i\alpha - \beta}. \tag{6}$$

So, the approximate solution of equation (2) has the form:

$$x_n(\lambda) = b \exp(\lambda \omega_n). \tag{7}$$

In this paper we analyze the case $\alpha > 0$ and $\beta \leq 1$ and we always suppose that the operators of the type $\exp(\lambda C_i \ell^{-(\beta - \alpha i)})$ for $0 < \beta - \alpha i < 1$, satisfy the condition $|\arg(-\lambda C_i)| < (\pi/2)(1-\alpha)$ (see [6]).

3. DEFINITIONS OF THE ESTIMATIONS IN F_0 AND L

The following theorem is proved in [3]:

Theorem 1 Let X be a group with convergence. If there is a function $A : X \to R^+$ such that:

(i) $A(x_n) \to 0$ and $A(y_n) \to 0$ implies $A(x_n - y_n) \to 0$

(ii) $A(x) = 0$ iff $x = 0$

then there exists a quasi-norm $\|\cdot\|$ in X (i.e. the conditions: (1) $\|0\| = 0$, (2) $\|-x\| = \|x\|$, (3) $\|x+y\| \leq \|x\| + \|y\|$ are satisfied) such that $A(x_n) \to 0$ iff $\|x_n\| \to 0$.

The functional A introduced as:

$$A(x) = \sum_{i=0}^{\infty} \frac{B_{i,1/i}(x)}{e^{ie^{i^2}}(1 + B_{i,1/i}(x))}, \quad (x \in F_0)$$

satisfies the conditions of the previous theorem. Then there exist a quasi-norm $\|\cdot\|$ in F_0 such that $A(x_n - x) \to 0$ iff $\|x_n - x\| \to 0$. However $A(x_n - x) \to 0$ iff $B_{T,\varepsilon}(x_n - x) \to 0$ for every $T, \varepsilon > 0$, which means that type I' convergence is equivalent to the convergence defined by A, given by (8).

Now, we can give the following

Definition 1 Operator $\tilde{x} \in F_0$ is the approximation of the operator $x \in F_0$ according to the functional A (given by (8)) with the measure of approximation $\delta > 0$, if $A(x - \tilde{x}) < \delta$.

Let us now observe the functional

$$F(f) = \sum_{i=1}^{\infty} \frac{1}{e^{ie^{i^2}}} \frac{\|f\|_i}{1 + \|f\|_i} \qquad (9)$$

for $f \in L$, where $\|f\|_i$ is the seminorm defined by (1). The sequence $\{x_n\}$ from L converges to a function $x \in L$ iff $F(x_n - x) \to 0$ as $n \to \infty$.

Definition 2 The function $\tilde{f}$ from L is the approximation of the operator f from L according to the functional F given by (9) with the measure of approximation $\delta_L > 0$, if $F(f - \tilde{f}) < \delta_L$.

Remark 1 If we introduce for every $\varepsilon > 0$ the functional

$$A_\varepsilon(x) = \sum_{i=1}^{\infty} \frac{1}{e^{ie^{i^2}}} \frac{B_{i,\varepsilon}(x)}{1 + B_{i,\varepsilon}(x)},$$

then

$$A(x_n) \to 0 \text{ iff } A_\varepsilon(x_n) \to 0 \text{ for every } \varepsilon > 0.$$

Then by the properties of $B_{T,\varepsilon}$ we obtain

$$\lim_{\varepsilon \to 0} A_\varepsilon(f) = F(f), \quad f \in L.$$

It is known that there exist sequences $\{f_n\}$ from L such that $B_{T,\varepsilon}(f_n) \to 0$ as $n \to \infty$, but $\|f_n\| \not\to 0$ as $n \to \infty$. For this reason we had to introduce the function F from (9).

Remark 2 Let us observe the following family of seminorms on C

$$\{|g|_T = \sup_{t \leq T} |g(t)|, \; T > 0\}, \quad g \in C.$$

There exist some sequences $\{g_n\}$ from C such that $\|g_n\|_T \to 0$ for every T but $|g_n|_T \not\to 0$.

We can introduce the functional

$$G(h) = \sum_{i=1}^{\infty} \frac{1}{e^{ie^{i^2}}} \frac{|h|_i}{1 + |h|_i}$$

which by Theorem 1 involves a quasi-norm $\|\cdot\|$ in C such that

$$G(x_n) \to 0 \text{ iff } \|x_n\| \to 0 \text{ as } n \to \infty, \; x_n \in C.$$

4. ESTIMATION IN L

Let us suppose that operators $b \exp(\lambda \omega_n)$ for $n = 1, 2, \ldots$, $b \exp(\lambda \omega)$ represent functions from L, $b \in L$.

We shall need the following theorem from [6].

Theorem 2 If $x_n(\lambda)$ (given by (7)) and $x(\lambda)$ (given by (5)) are functions from L then, it holds that

$$\|x(\lambda) - x_n(\lambda)\|_T \leq k(\lambda, T, \alpha, \beta) \frac{1}{\Gamma\left(\frac{(n+1)\alpha - \beta - 1}{2} + 1\right)} \tag{10}$$

where

$$k(\lambda, T, \alpha, \beta) = \begin{cases} \tilde{R}_0(\lambda, T) \ldots \tilde{R}_{r+1}(\lambda, T) \cdot E(\lambda, T) & \text{for } 0 \leq \beta < 1 \\ & \text{and } \alpha \text{ arbitrary} \\ \tilde{R}_1(\lambda, T + \lambda C_0) \ldots \tilde{R}_r(\lambda, T) R_{r+1}(\lambda, T) \cdot E(\lambda, T) & \text{for } \beta = 1 \\ & \text{and } \alpha \leq 1 \\ \tilde{R}_1(\lambda, T + \lambda C_0) \cdot E(\lambda, T) & \text{for } \beta = 1 \text{ and } \alpha > 1 \\ \tilde{R}_0(\lambda, T) \cdot E(\lambda, T) & \text{for } \beta < 0 \text{ and } \alpha \text{ arbitrary} \end{cases}$$

with notations from [6]

$$\frac{T^{2n+1}}{(2n)!} C(2n+1) + \frac{T^{2n+2}}{(2n+1)!} C(2n+2) \equiv \tilde{R}_i(\lambda, T), \quad i = 1, \ldots, r \text{ or } i = 0$$

for every $n \in \mathbb{N}$, $0 \leq t \leq T$, where

$$C(k) = \frac{2}{\pi(\beta - i\alpha)} \Gamma\left(\frac{k}{\beta - i\alpha}\right),$$

r is chosen such that $r\alpha - \beta \leq 0$ and $(r+1)\alpha - \beta > 0$,

$$B(T) \cdot e^{\lambda D(T)} \equiv \tilde{R}_{r+1}(\lambda, t),$$

where

(i) $B(T) = \int_0^T |b(t)| dt$, and

(ii) $D(t) = \sum_{i=r+1}^{\infty} |c_i| \frac{T^{i\alpha - \beta}}{\Gamma(i\alpha - \beta + 1)},$

$$E(\lambda, T) = \frac{T^{(n+1)\alpha - \beta - 1} \nu(T)}{\Gamma\left(\frac{(n+1)\alpha - \beta - 1}{2} + 1\right)} \exp\left[|\lambda|\nu(T) \frac{T^{(n+1)\alpha - \beta - 1}}{\Gamma((n+1)\alpha - \beta)}\right],$$

$$\nu(T) \geq \sum_{i=0}^{\infty} |c_{n+1+i}| \frac{T^{i\alpha+1}}{\Gamma(i\alpha+2)}.$$

Now, we can prove the following

Theorem 3 If $x(\lambda)$ and $x_n(\lambda)$ are given by (5) and (7) respectively and belong to L, then the measure of approximation according to $F(f)$, $f \in L$, is

$$F(x(\lambda) - x_n(\lambda)) \leq \frac{1}{\Gamma\left(\frac{(n+1)\alpha-\beta-1}{2}+1\right)} Q(\lambda,\alpha,\beta) \qquad (11)$$

where $Q(\lambda,\alpha,\beta)$ is a real number such that

$$Q(\lambda,\alpha,\beta) \geq \sum_{i=1}^{\infty} \frac{k_1 e^{k_2 e^{k_3 i}}}{e^{i e^{i^2}}} \quad \text{and} \quad k_1, k_2, k_3 \text{ are constants.}$$

Proof Since the function $y = x/(1+x)$ is monotonically increasing for $x > 0$, we can write by (10)

$$\frac{\|x(\lambda) - x_n(\lambda)\|_T}{1+\|x(\lambda)-x_n(\lambda)\|_T} \leq \frac{k(\lambda,T,\alpha,\beta) \frac{1}{\Gamma\left(\frac{(n+1)\alpha-\beta-1}{2}+1\right)}}{1 + k(\lambda,T,\alpha,\beta) \frac{1}{\Gamma\left(\frac{(n+1)\alpha-\beta-1}{2}+1\right)}} \leq \frac{1}{\Gamma\left(\frac{(n+1)\alpha-\beta-1}{2}+1\right)} k(\lambda,T,\alpha,\beta). \qquad (12)$$

It is obvious that there exist such constants k_1, k_2 and k_3, depending of α, β and λ, which satisfy the inequality

$$k(\lambda,T,\alpha,\beta) \leq k_1 e^{k_2 e^{k_3 T}}. \qquad (13)$$

The series

$$\sum_{i=1}^{\infty} \frac{k_1 e^{k_2 e^{k_3 i}}}{e^{i e^{i^2}}}$$

is convergent, and therefore the series

$$\sum_{i=1}^{\infty} \frac{k(\lambda,i,\alpha,\beta)}{e^{i e^{i^2}}}$$

is also convergent.

From relations (9), (12) and (13) we get:

$$F(x(\lambda) - x_n(\lambda)) \leq \frac{1}{\Gamma\left(\frac{(n+1)\alpha-\beta-1}{2} + 1\right)} \sum_{i=1}^{\infty} \frac{k(\lambda,i,\alpha,\beta)}{e^{ie^{i^2}}} \leq$$

$$\frac{1}{\Gamma\left(\frac{(n+1)\alpha-\beta-1}{2} + 1\right)} \sum_{i=1}^{\infty} \frac{k_1 e^{k_2 e^{k_3 i}}}{e^{ie^{i^2}}} \quad (14)$$

Example 1 Starting from the same equation as in Example 1 in [6]

$$(s - 1)x'(\lambda) - x(\lambda) = 0$$

with the condition

$$x(0) = \ell,$$

we obtain by Theorem 3 the measure of approximation according to $F(f)$ for $\lambda = 1$.

$$F(x(\lambda) - x_n(\lambda)) \leq \frac{1}{\Gamma\left(\frac{n+1}{2} + 1\right)} Q(1,1,-1) \equiv \Delta_L,$$

where

$$\sum_{i=1}^{\infty} \frac{2e^{2i}}{e^{ie^{i^2}}} \leq \left(\frac{2e^2}{e^e} + \frac{-e}{1-e}\right) \equiv Q(1,1,-1).$$

5. ESTIMATIONS IN F_0

Let us suppose that the operators $x(\lambda)$ and $x_n(\lambda)$, given by (5) and (7) respectively, are from F_0. Let

$$z(\lambda) = \frac{gx(\lambda)}{g} \quad \text{and} \quad z_n(\lambda) = \frac{gx_n(\lambda)}{g},$$

where $g \in L_0$. Then the operators $z(\lambda)$ and $z_n(\lambda)$ belong to F_0.
It is easy to prove the following:

Lemma 1 For each $T > 0$ and $k > 0$ there exists an operator $g_2 = (\ell/(I+k\ell))k$ representing the function $k \cdot e^{-kt}$ from L_0 and satisfying the inequalities

$$\|g_2\|_T < 1 \quad \text{and} \quad \|\ell - \ell g_2\|_T < 1/k.$$

Similarly as in section 4 but now using Lemma 1 and modified Theorem 2 from [6] (taking g_2 instead of g_1) we have

Theorem 4 If $x(\lambda)$ and $x_n(\lambda)$ are given by (5) and (7) respectively, then the measure of approximation according to the functional A is estimated by

$$A(x(\lambda) - x_n(\lambda)) = \sum_{i=0}^{\infty} \frac{B_{i,1/i}(x(\lambda) - x_n(\lambda))}{1 + B_{i,1/i}(x(\lambda) - x_n(\lambda))} e^{-ie^{i^2}} \leq$$

$$\frac{1}{\Gamma\left(\frac{(n+1)\alpha - \beta - 1}{2} + 1\right)} Q_{g_2}(\lambda, \alpha, \beta)$$

where $Q_{g_2}(\lambda, \alpha, \beta)$ is a real number such that

$$Q_{g_2}(\lambda, \alpha, \beta) \geq \sum_{i=1}^{\infty} \frac{k_{1,g_2} e^{k_{2,g_2} e^{k_3 i}}}{e^{ie^{i^2}}} ;$$

the constants k_{1,g_2} and k_{2,g_2} depend of λ, α, β and g_2 and the constant k_3 depends on α, β.

Example 2 The differential equation $(s+1)x'(\lambda) - x(\lambda) = 0$ with condition $x(0) = I$ has the exact solution

$$x(\lambda) = \exp(\lambda \omega), \quad \text{where} \quad \omega = \sum_{i=0}^{\infty} (-1)^i \ell^{i+1},$$

while the approximate one is

$$x_n(\lambda) = \exp(\lambda \omega_n), \quad \text{where} \quad \omega_n = \sum_{i=0}^{n} (-1)^i \ell^{i+1}.$$

The measure of approximation is

$$A(x_n(\lambda) - x(\lambda)) \leq \frac{1}{\Gamma\left(\frac{n+1}{2} + 1\right)} Q_{g_2}(\lambda, 1, -1) \equiv \Delta_A$$

for $g_2 = \{k \cdot e^{-kt}\}$ and

$$\sum_{i=1}^{\infty} \frac{3e^{2i}e^{2i}}{e^{ie^{i^2}}} \leq \frac{2e^2}{e^e} + \frac{e}{e-1} \equiv Q_{g_2}(1, 1, -1).$$

Example 3 The differential equation

$$u''(x) - s^2 u(x)(I + G) = -(I + G)$$

where

$$G = \sum_{i=1}^{\infty} a_i \ell^{\alpha i}, \quad \alpha > 0,$$

represent in the field of Mikusiński operators, the generalized equation of the oscillation of the viscoelastic bar. In paper [6] it was proved that the approximate solution

$$u_n(x) = (1/2)\ell \exp(-x \sum_{i=0}^{n} c_i \ell^{i\alpha-1}), \quad \text{for } c_i < \frac{\sqrt{2}}{r}, \; r > 0,$$

converges in L to the exact solution. So, the error of approximation can be treated as:

$$L(u(x) - u_n(x)) \leq \frac{1}{\Gamma((n+1)/4 + 1)} Q(1,1/2,1) \equiv \Delta_L$$

where

$$Q(1,1/2,1) = 7!4(e^{e^2}/e^e + e/e-1), \quad c_0 = 1 \text{ and } \alpha = 1/2.$$

Appendix The corresponding tables are
a) for Example 1 (n = 2k+1)
b) for Example 2 we obtain the same table as Table 1 taking only Δ_A instead of Δ_L
c) for Example 3 (n = 4k+3) (Table 2).

Table 1

k	Δ_L
8	0.476166376569 E+00
9	0.476166376569 E-01
10	0.432878524153 E-02
11	0.360732103461 E-03
12	0.277485233432 E-04
13	0.198204452451 E-05
14	0.132136301634 E-06
15	0.825851885213 E-08
16	0.485795226596 E-09
17	0.269886236998 E-10
18	0.142045387894 E-11
19	0.710226939468 E-13
20	0.338203304508 E-14
21	0.153728774777 E-15
22	0.668385977289 E-17
23	0.276494157204 E-18
24	0.111397662882 E-19
25	0.428452549544 E-21

Table 2

k \ Δ_L	Δ_L
10	0.899312973022 E+01
11	0.817557275295 E+00
12	0.681297704577 E-01
13	0.524075143039 E-02
14	0.374339375412 E-03
15	0.249559579970 E-04
16	0.155974737481 E-05
17	0.917498468311 E-07
18	0.509721376218 E-08
19	0.268274402693 E-09
20	0.134137197877 E-10
21	0.638748546411 E-12
22	0.290340261598 E-13
23	0.126234897728 E-14
24	0.525978740534 E-16
25	0.210391486287 E-17
26	0.809198029154 E-19
27	0.299702964785 E-20
28	0.107036774941 E-21
29	0.369092319221 E-23
30	0.123030771430 E-24
31	0.396873456226 E-26

REFERENCES

1. T. Boehme, The Mikusiński Operator as a Topological Space, Amer. J. Math., 98, 55-66 (1976).

2. J. Burzyk, On Convergence in the Mikusiński Operational Calculus, Stud. Math., 75, 313-333 (1983).

3. J. Burzyk and P. Mikusiński, On normability of semigroups, Bull. Acad. Polon. Sci., Ser. Math. Astronom. Phys., 1-2, 33-35 (1980).

4. A. A. Lokšin, V. E. Rok, Automodal solution of the wave equation with delayed time, Uspehi Mat. Nauk., T. 33. N°. 6 (204), 221 - 222, (1980), (in Russian).

5. J. Mikusiński, "Operational calculus", Pergamon Press, Warszawa (1959).

6. E. Pap and Đ. Takači, Convergences of the solutions of operator linear differential equations (to appear).

7. B. Stanković, Approximate solution of the operator linear differential equation I, Publ. de l'Inst. Math., T. 21(35), 185-196 (1977).

INVARIANCE OF THE CAUCHY PROBLEM FOR DISTRIBUTION DIFFERENTIAL EQUATIONS

Jan Persson

Matematiska Institutionen
Lunds Universitet
Box 118, S-22100 Lund, Sweden

1. INTRODUCTION

Let n > 2 be an integer. In Persson [2] and [3], the Cauchy problem for the equation

$$u^{(n)} + a_{n-1} u^{(n-1)} + \ldots + a_0 u = f \tag{1.1}$$

is treated. The new thing is that some coefficients and f are allowed to be distributions and not necessarily measures. Then some of the derivatives $u^{(j)}$, $0 \leq j < n$, may not be pointwise defined. Still, a Cauchy problem can be defined for (1.1) with n initial data as in the ordinary Cauchy problem for measure differential equations. The function u is defined as a solution of an integral equation. Here, iterated primitive distributions of some of the coefficients are involved. One also chooses an iterated primitive distribution of f. As long as the primitive distribution is not pointwise defined, one makes a choice differing from another choice by a constant. In case of iterated primitive distributions, the difference is a polynomial. As soon as the primitive distribution is pointwise defined, one chooses the primitive distribution to be zero at the initial point of the Cauchy problem, just as one does in the measure differential equation case. We prove that the affine space of solutions of (1.1) is invariant under the choice of the iterated primitive distributions of the coefficients and of f.

In order to give a meaning to (1.1), we shall repeat the definition of the spaces P^j and B^j in Persson [3].

<u>Definition 1.1</u> If g is a complex Borel measure on $\mathbb{R}$, then g is said to be in P^0. If g is a complex valued Borel measurable function on $\mathbb{R}$,

then g is said to be in B^0. We agree that two functions in B^0 are different at least at one point. Let D be distribution differentiation and let $D^{-1}g$ denote a primitive distribution of $g \in \mathcal{D}'(\mathbb{R})$. If for some integer j $D^j g \in P^0$, then g is said to be in P^j. Outside distribution theory we agree that all functions of P^1 are right continuous and that all functions of P^j, $j > 1$, are continuous. In the same way, if for some integer j $D^j g \in B^0$, then g is said to be in B^j. Outside distribution theory we agree that all functions of B^j, $j > 0$, are continuous. Further, if $f \in B^j$, then we fix a certain $g \in B^0$ such that $D^j f = g$. When we write $f \in B^j$, then we mean the pair (f,g). By convention we mean (f,f) when $f \in B^0$. Let $f_1 \in B^j$ with $D^j f_1 = g_1$. Let $j > 0$. Then $f = f_1$ in B^j if $(f,g) = (f_1, g_1)$ pointwise. If $j < 0$, then $f = f_1$ in B^j if $f = f_1$ in $\mathcal{D}'(\mathbb{R})$ and $g - g_1$ is a polynomial of order at most $-1-j$.

We shall define the multiplication of elements in B^j and P^j.

Definition 1.2 Let j be an integer, let $a \in P^{-j}$ and let $f \in B^j$. We choose $b \in P^0$ and $g \in B^0$ such that $D^j b = a$ and $D^j f = g$. If $j \geq 0$, then the distribution fa at $\phi \in \mathcal{D}(\mathbb{R})$ is defined as

$$\langle fa, \phi \rangle = (-1)^j \int D^j(f\phi) db. \tag{1.2}$$

If $j < 0$, then

$$\langle fa, \phi \rangle = (-1)^j \int gd(D^{-1}(\phi a)). \tag{1.3}$$

Definition 1.2 gives

Proposition 1.3 Let j be an integer. If $j \geq 0$, then P^{-j} is a B^j module. If $j > 0$, then B^{-j} is a P^j module.

Remark We regard the modules of Proposition 1.3 as two-sided to make the book-keping easier. We also notice that one has $P^{j+1} \in B^j$, if for $f \in P^{j+1}$ one agrees that in $(f,g) \in B^j$ $D^j f = g \in P^1$, i.e. g is unique and right continuous. In the same way for $(f,g) \in B^j$, $j \geq 0$, g is unique in P^0. It is unique modulo a polynomial of order at most $-1-j$, if $j < 0$. Thus we get $B^j \in P^j$ for all j. Proposition 1.3 shows that P^j is a P^{-j+1} module, $j \leq 0$, and that P^{-j+1} is a P^j module, $j > 0$. In the following the ambiguity with the space B^j is removed, if one replaces it by the P^{j+1} space.

Proposition 1.4 Let $j > 0$ be an integer, let $a \in P^{-j}$, $f \in B^j$ and let $b \in P^0$ be such that $D^j b = a$. Then,

$$D^{-1}(fa) = D^{-1}(fD^j b) = fD^{j-1}b - D^{-1}((Df)D^{j-1}b) + \text{constant}. \tag{1.4}$$

Proposition 1.5 Let $j > 0$ be an integer, let $a \in P^j$, $f \in B^{-j}$ and let $g \in B^0$ be such that $D^j g = f$. Then,

$$D^{-1}(af) = D^{-1}(aD^j g) = aD^{j-1}g - D^{-1}((Da)D^{j-1}g) + \text{constant}. \tag{1.5}$$

These propositions are proved in [2] and the proof is not repeated here. Induction, then, gives as in [2].

Proposition 1.6 Let j, b and f be as in Proposition 1.4. Then,

$$D^{-1}(fD^j b) = \sum_{k=0}^{j} (-1)^k \binom{j}{k} D^{-k}((D^k f)b) + p, \tag{1.6}$$

where p is a polynomial of at most degree $j-1$.

Proposition 1.7 Let j, a, and g be as in Proposition 1.5. Then,

$$D^{-j}(aD^j g) = \sum_{k=0}^{j} (-1)^k \binom{j}{k} D^{-k}(gD^k a) + p, \tag{1.7}$$

where p is a polynomial of at most degree $j-1$.

The last remark in [3] shows that one could as well define a two-sided Cauchy problem for (1.1). For simplicity we assume that the initial point is $x = 0$. Let $a < b$, and let $\int_{a+}^{b} = \int_{(a,b]}$. From now on, we let

$$D^{-1} = \int_{0+}^{x}, \quad x \geq 0, \text{ and } D^{-1} = -\int_{x+}^{0}, \quad x < 0,$$

when D^{-1} is applied to P^0. Correspondingly, if $f = (f, f_1) \in B^{-1}$, then $D^{-1}f = f_1 - f_1(0)$. We reformulate the theorems of [3] to the two-sided case. At the same time we exclude the measure differential equations and add the invariance part not exhibited in [3]. The measure differential equations are treated in Persson [1].

Theorem 1.8 Let m, ℓ, and n be integers such that $1 \leq m < \ell$, and $n = 2\ell-1$. Let $a_j \in P^{m-\ell}$, $0 \leq j \leq 2m-2$, let $a_j \in P^{j+2-\ell-m}$, $2m-2 < j \leq 2\ell-2$, and let $f \in P^{m-\ell}(B^{m-\ell-1})$. Choose $b_j \in P^0$ such that $D^{\ell-m}b_j = a_j$, $0 \leq j \leq 2m-2$, and such that $D^{\ell+m-2-1}b_j = a_j$, $2m-1 \leq j \leq m+\ell-2$. Choose $D^{1-2\ell}f$ such that $D^k(D^{1-2\ell}f)(0) = 0$, $0 \leq k \leq \ell+m-2$. Then, for each choice of $c = (c_0, \ldots, c_{n-1}) \in \mathbb{C}$, there is a unique $u \in P^{\ell+m-1}(B^{\ell+m-2})$ fulfilling

281

$$u + \sum_{j=0}^{2m-2} \sum_{k=0}^{\ell-m} (-1)^k \binom{\ell-m}{k} D^{1-\ell-m-k}((D^{j+k}u)b_j)$$

$$+ \sum_{j=2m-1}^{m+\ell-2} \sum_{k=0}^{m+\ell-j-2} (-1)^k \binom{\ell+m-j-2}{k} D^{m-\ell-j-k-1}((D^{j+k}u)b_j)$$

$$+ \sum_{j=m+\ell-1}^{2\ell-2} \sum_{k=0}^{j+2-\ell-m} (-1)^k \binom{j+2-\ell-m}{k} D^{j+3-3\ell-m-k}((D^k a_j) D^{\ell+m-2}u)$$

$$= D^{1-2\ell}f + \sum_{k=0}^{2\ell-2} c_k x^k/k!. \tag{1.8}$$

Then, for another choice of b_j, $0 \le j \le m+\ell-2$, and $D^{1-2\ell}f$ still satisfying the same hypothesis, there is one choice of a new $c \in \mathbb{C}^n$ such that the original u solving the original (1.8) is also the unique solution of the modified (1.8). In the new $c, c_0, \ldots, c_{\ell+m-2}$ are unaltered since they correspond to $D^k u(0)$, $0 \le k \le \ell+m-2$. The difference between the new and old constants $c_{\ell+m-1}, \ldots, c_{2\ell-2}$ only depends on $D^k u(0)$, $0 \le k \le \ell+m-2$, and the differences between the choices of b_j, $0 \le j \le m+\ell-2$, and of $D^{1-2\ell}f$.

<u>Theorem 1.9</u> Let m and ℓ be integers such that $1 \le m < \ell$, and let $n = 2\ell$. Let $a_j \in P^{m-\ell}$, $0 \le j \le 2m-1$, let $a_j \in P^{j+1-\ell-m}$, $2m \le j \le 2\ell$, and let $f \in P^{m-\ell}(B^{m-\ell-1})$. Choose $b_j \in P^0$ such that $D^{\ell-m}b_j = a_j$, $0 \le j < 2m$, and such that $D^{\ell+m-j-1}b_j = a_j$, $2m \le j < m+1$. Choose $D^{-2\ell}f$ such that $D^k(D^{-2\ell}f)(0) = 0$, $0 \le k \le m+\ell-1$. Then, for each choice of $c = (c_0, \ldots, c_{n-1}) \in \mathbb{C}^n$ there is a unique $u \in P^{\ell+m}(B^{\ell+m-1})$ fulfilling

$$u + \sum_{j=0}^{2m-\ell} \sum_{k=0}^{\ell-m} (-1)^k \binom{\ell-m}{k} D^{-\ell-m-k}((D^{j+k}u)b_j)$$

$$+ \sum_{j=2m}^{m+\ell-1} \sum_{k=0}^{\ell+m-j-1} (-1)^k \binom{\ell+m-j-1}{k} D^{m-\ell-j-k-1}((D^{j+k}u)b_j)$$

$$+ \sum_{j=m+\ell}^{2\ell-1} \sum_{k=0}^{j+1-\ell-m} (-1)^k \binom{j+1-\ell-m}{k} D^{j+1-3\ell-m-k}((D^k a_j) D^{\ell+m-1}u)$$

$$= D^{-2\ell}f + \sum_{k=0}^{2\ell-1} c_k x^k/k!. \tag{1.9}$$

Then, for another choice of b_j, $0 \le j < m+1$, and of $D^{-2\ell}f$ still satisfying the same hypothesis there is one choice of a new $c \in \mathbb{C}^n$ such that the original u solving the original (1.8) is also the unique solution of the modified (1.8). In the new c $c_0,\ldots,c_{\ell+m-1}$ are unaltered since they correspond to $D^k u(0)$, $0 \le k \le \ell+m-1$. The difference between the new and old constants $c_{\ell+m},\ldots,c_{2\ell-1}$ only depends on $D^k u(0)$, $0 \le k \le \ell+m-1$, and the differences between the choices of b_j, $0 \le j \le m+\ell-1$, and of $D^{-2\ell}f$.

It follows from [2] and [3] that the only thing that there remains to prove is the last part of the two theorems. This proof is given in Section 2 for Theorem 1.8. We do not write down the corresponding proof for Theorem 1.9 since that would mean repeating the same arguments.

2. PROOF OF THE INVARIANCE

We shall start with a lemma.

Lemma 2.1 Let $r \ge 0$ be an integer. Let $f \in P^r$ and let $v \in B^r$. Then,

$$\sum_{k=0}^{r} (-1)^k \binom{r}{k} D^{-1-k}(fD^k v) = D^{-r-1}(vD^r f) + p. \qquad (2.1)$$

Here p is a polynomial of order at most r such that $p(0) = 0$. Further, the coefficinets of p are polynomials in $D^k v(0)$ and $D^k f(0)$, $0 \le k < r$.

The lemma is proved by induction over r. In order to prove the last part of Theorem 1.9, we replace b_j by $b_j' = b_j + q_j$, where q_j is a polynomial of order at most $\ell-m-1$, $0 \le j \le 2m-2$, and of order at most $\ell+m-j-3$, $2m-1 \le j \le m+\ell-3$. To the original choice of $D^{1-2\ell}f$ we add a polynomial $\sum_{k=\ell+m-1}^{2\ell-2} d_k x^k$. Then, to original left member of (1.8) we add the following sum on which we use Lemma 2.1

$$\sum_{j=0}^{2m-\ell} \sum_{k=0}^{\ell-m} (-1)^k \binom{\ell-m}{k} D^{1-\ell-m-k}((D^{j+k}u)q_j)$$

$$+ \sum_{j=2m-1}^{m+\ell-3} \sum_{k=0}^{\ell+m-j-2} (-1)^k \binom{\ell+m-j-2}{k} D^{m-\ell-k-j-1}((D^{j+k}u)q_j)$$

$$+ \sum_{j=0}^{2m-2} D^{2-\ell-m}(D^{-\ell+m-1}((D^j u)D^{\ell-m}q_j) + p_j)$$

$$+ \sum_{j=2m-1}^{m+\ell-3} D^{m-\ell-j}(D^{-\ell-m+j+1}((D^j u)D^{\ell+m-j-2}q_j) + p_j) =$$

$$= \sum_{j=0}^{2m-2} D^{2-\ell-m} p_j + \sum_{j=2m-1}^{m+\ell-3} D^{m-\ell-j} p_j = p.$$

We notice that order $p_j \leq \ell-m$, $0 \leq j \leq 2m-2$, and order $p_j \leq \ell+m-j-2$, $2m-1 \leq j \leq m+\ell-3$, with $p_j(0) = 0$, $0 \leq j \leq m+\ell-3$. It follows that order $p \leq 2\ell-2$ and that $D^k p(0) = 0$, $0 \leq k \leq m+\ell-2$. Then, we just choose new c_k, $\ell+m-1 \leq k \leq 2\ell-2$, to compensate for the added polynomials. The last part of Theorem 1.8 is proved.

REFERENCES

1. J. Persson, Fundamental theorems for linear measure differential equations, to appear in Math. Scand.
2. J. Persson, Linear distribution differential equations, Comment. Math. St. Paul., 33, 119-126, (1984).
3. J. Persson, The Cauchy problem for linear distribution differential equations, to appear in Funkcial. Ekvac.

ON THE SPACE $\mathcal{D}'^{(M_p)}_{L^q}$, $q \in [1,\infty]$

S. Pilipović

Institute of Mathematics
University of Novi Sad
dr I. Đuričića 4, 21000 Novi Sad, Yugoslavia

1. INTRODUCTION

We shall present in this paper mainly the results for $q = 2$. Namely, we shall present the results from [3 - 7] concerning the space of Beurling ultradistributions $\mathcal{D}'^{(M_p)}_{L^2}$. In our investigations we follow the Komatsu approach to spaces of ultradistrinutions [2], so for the notions and the basic results of ultradistribution theory we refer the reader to this paper.

In the last two sections we shall present new results. In the first of them we prove that $\mathcal{D}' \cap \mathcal{D}'^{(M_p)}_{L^q} \neq \mathcal{D}'_{L^q}$, $2 \leq q \leq \infty$, and in the second we introduce the convolution in spaces $\mathcal{D}'^{(M_p)}_{L^q}$, $q > 1$ and give some basic properties of the operation.

2. BASIC SPACES

We assume that M_p, $p \in \mathbb{N}_0$, ($\mathbb{N}_0 = \mathbb{N} \cup \{0\}$) is a sequence of positive numbers, such that

(M.1) $M_p^2 \leq M_{p-1} M_{p+1}$, $p \in \mathbb{N}$;

(M.2) There are constants A nad H, such that

$$M_p \leq AH^p \min_{0 \leq q \leq p} \{M_q M_{p-q}\}, \quad p \in \mathbb{N};$$

(M.3) There is a constant A, such that

$$\sum_{q=p+1}^{\infty} M_{q-1}/M_q \leq A p M_p/M_{p+1}, \quad p \in \mathbb{N}.$$

The functions M and M* are defined by

$$M(\rho) = \sup_{p \in \mathbb{N}_0} \{\log(\rho^p M_0/M_p)\}, \quad M^*(\rho) = \sup_{p \in \mathbb{N}_0} \{\log(\rho^p p! \, M_0/M_p)\}, \quad \rho > 0.$$

An operator of the form

$$P(D) = \sum_{\alpha \in \mathbb{N}_0^n} a_\alpha D^\alpha, \quad D^\alpha = \frac{\partial_j^{|\alpha|}}{i^{|\alpha|} \partial x_1^{\alpha_1} \cdots \partial x_n^{\alpha_n}} \quad (|\alpha| = \alpha_1 + \ldots + \alpha_n)$$

is called an ultradifferential operator of class (M_p), iff for some constants C and L

$$|a_{|\alpha|}| < C L^{|\alpha|}/M_{|\alpha|}, \quad \alpha \in \mathbb{N}_0^n.$$

For the definition and the properties of $\mathcal{D}^{(M_p)}$ and its strong dual $\mathcal{D}'^{(M_p)}$, we refer the reader to [2].

Let $r \in [1,\infty)$. We define the space $\mathcal{D}_{L^r}^{(M_p)}(\mathbb{R}^n) = \mathcal{D}_{L^r}^{(M_p)}$ by

$$\mathcal{D}_{L^r}^{(M_p)} = \text{proj} \lim_{h \in \mathbb{N}_0} \mathcal{D}_{L^r, h}^{(M_p)},$$

where

$$\mathcal{D}_{L^r, h}^{(M_p)} = \{\phi \in C^\infty; \|\phi\|_{r,h} = \sum_{\alpha \in \mathbb{N}_0^q} h^{|\alpha|} \|\phi^{(\alpha)}\|_r / M_{|\alpha|} < \infty \}.$$

($\| \ \|_r$ denotes the norm in L^r.)

The space $\mathcal{D}_{L^r}^{(M_p)}$ is an (FG)-space ([1]) and

$$\mathcal{D}^{(M_p)} \hookrightarrow \mathcal{D}_{L^r}^{(M_p)} \hookrightarrow \mathcal{D}_{L^r},$$

where "A $\hookrightarrow$ B" means that space A is a dense subspace of B and that the inclusion mapping is continuous. ($\mathcal{D}_{L^r}$ is the well-known Schwartz space.) Moreover, we have ([6])

$$\mathcal{D}_{L^1}^{(M_p)} \hookrightarrow \mathcal{D}_{L^r}^{(M_p)} \quad (r \in [1,\infty)).$$

The strong dual of $\mathcal{D}_{L^r}^{(M_p)}$ is denoted by $\mathcal{D}'_{L^s}^{(M_p)}$ where $\frac{1}{r} + \frac{1}{s} = 1$.

Theorem 1 [3]   If $f \in \mathcal{D}'^{(M_p)}_{L^s}$, $s \in (1,\infty]$, then there are functions f_α, $\alpha \in \mathbb{N}_0^n$ from L^s and $k > 0$ so that

$$f = \sum_{\alpha \in \mathbb{N}_0^n} f_\alpha^{(\alpha)} \quad \text{weakly in } \mathcal{D}'^{(M_p)}_{L^s} \tag{1}$$

and

$$\sum_\alpha \frac{M_{|\alpha|}}{k^{|\alpha|}} \|f_\alpha\|_s < \infty. \tag{2}$$

Conversely, if f_α, $\alpha \in \mathbb{N}_0^n$ are functions from L^s so that (2) holds, then the series in (1) converges in $\mathcal{D}'^{(M_p)}_{L^s}$ to some element from $\mathcal{D}'^{(M_p)}_{L^s}$.

Let us denote by $\mathcal{D}^{(M_p)}_{\infty,0}$ a space of smooth functions ϕ for which all the norms

$$\|\phi\|_{\infty,k} = \sup_{\alpha \in \mathbb{N}_0^n}\left\{\frac{k^{|\alpha|}}{M_{|\alpha|}}\|\phi^{(\alpha)}\|_\infty\right\}, \quad k \in \mathbb{N}$$

are finite and for any $\alpha \in \mathbb{N}_0^n$

$$|\phi^{(\alpha)}(x)| \to 0 \text{ as } |x| \to \infty.$$

This is an (FG)-space with the dual $\mathcal{D}'^{(M_p)}_{L^1}$.

Theorem 1' Theorem 1 holds with $s = 1$.

3. SPACE $\mathcal{D}'^{(M_p)}_{L^2}$

By using the L^2-theory, we studied in [3 – 5] and [7] the subspace $\mathcal{D}'^{(M_p)}_{L^2}$ of the space of Beurling ultradistributions. First, let us recall that the Fourier and inverse Fourier transforms from L^2 are defined by

$$(Ff)(\xi) = \ell.i.m. \int_{-A}^{+A} \cdots \int_{-A}^{+A} f(x) \exp(i<x,\xi>) \, dx_1 \cdots dx_n, \quad \xi \in \mathbb{R}^n$$

$$(<x,\xi> = <x_1\xi_1 + \ldots + x_n\xi_n>),$$

$$(F^{-1}f)(x) = (2\pi)^{-n}(Ff)(-x), \quad x \in \mathbb{R}^n.$$

The next theorem is the so-called second structural theorem for an $f \in \mathcal{D}'^{(M_p)}_{L^2}$.

Theorem 2 [5] There is an ultradifferentiable operator of class (M_p) and an $F \in L^2$ so that

$$f = P(D)F. \tag{3}$$

Obviously, by the left-hand side of (3), an element from $\mathcal{D}'^{(M_p)}_{L^2}$ is defined.

The Cauchy kernel of the tubular region $T^n = \mathbb{R}^n + i\mathbb{R}^n_+$, where $\mathbb{R}^n_\pm = \{x \in \mathbb{R}^n, x_i \gtrless 0, i = 1,\ldots,n\}$, is defined by

$$K_n(z) = i^n(z_1 \cdot z_2 \cdots z_n)^{-1}, \quad z = x + iy \in T^n.$$

The function $t \mapsto K_n(z-t)$, $z \in T^q$ is fixed, belongs to $\mathcal{D}'^{(M_p)}_{L^2}$

Theorem 3 [5] Let $C(f,z) = F(z) = (2\pi)^{-n} \langle f(t), K_n(z-t) \rangle$, $z \in T^n$.
Then:

(i) $C(f,z)$ is analytic in T^n.

(ii) There are $k > 0$ and $C > 0$, so that

$$\int_{\mathbb{R}^n} |F(x+iy)|^2 dx \leq C \exp\left(\sum_{i=1}^{n} M^*(k/|y_i|)\right), \quad y \in \mathbb{R}^n_+. \quad (*)$$

The Poisson kernel is defined by

$$P_n(x,y) = \frac{y_1 \cdots y_n}{\pi^n |z_1|^2 \cdots |z_n|^2}, \quad z = x + iy \in T^n = \mathbb{R}^n + i\mathbb{R}^n_+.$$

Theoreme 4 [5]

(i) For every $z \in T^n$, $t \to P_n(x-t,y)$ belongs to $\mathcal{D}'^{(M_p)}_{L^2}$.

(ii) If $f \in \mathcal{D}'^{(M_p)}_{L^2}$, then

$$\langle f(t), P_q(x-t,y) \rangle \to f(x) \text{ as } y \to 0, \, y \in \mathbb{R}^n_+, \text{ weakly in } \mathcal{D}'^{(M_p)}_{L^2}.$$

Denote by Ω the set of all the variations of order n of elements $\{-1,1\}$. If $s = \{s_1, s_2, \ldots, s_n\} \in \Omega$, we put

$$\mathbb{R}^n_s = \{x, \, x_i \neq 0, \, \sum_{i=1}^n x_i s_i = \sum_{i=1}^n |x_i|\}.$$

For example $\mathbb{R}^n_+$, in the previous notation, is now $\mathbb{R}^n_{(1,\ldots,1)}$ and $\mathbb{R}^n_- = \mathbb{R}^n_{(-1,\ldots,-1)}$.

The Cauchy kernel for the n-octant $\mathbb{R}^n_s$ is defined by

$$K_{ns}(z) = F(H_s(\xi) \exp(-\langle y, \xi \rangle))(x), \quad z \in \mathbb{R}^n + i\mathbb{R}^n_s = T^n_s,$$

where H_s is the characteristic function for $\mathbb{R}^n_s$.

We put

$$C_s(f,z) = F_s(z) = (2\pi)^{-n} \langle f(t), K_{ns}(z-t) \rangle, \quad z \in T^n_s, \, s \in \Omega.$$

Theorem 3' [5] Theorem 3 holds with $C_s(f,z)$, K_{ns}, $\mathbb{R}^n_s$ and T^n_s instead of $C(f,z)$, K_n, $\mathbb{R}^n$ and T^n, $s \in \Omega$.

We shall denote by $(*)'$ condition $(*)$ in the case of Theorem 3'.
The boundary value representation of the elements from $\mathcal{D}'^{(M_p)}_{L^2}$ is given by:

Theorem 5 [5] Let $f \in \mathcal{D}'^{(M_p)}_{L^2}$. Then there are holomorphic functions $F_s(x+iy)$, $x+iy \in T^n_s$, $s \in \Omega$ which satisfy $(*)'$, such that

$$F(x) = \sum_{s \in \Omega} \lim_{\substack{y \to 0 \\ y \in \mathbb{R}^n_s}} F_s(x + iy), \quad x \in \mathbb{R}^n, \text{ weakly in } \mathcal{D}'^{(M_p)}_{L^2}.$$

The converse assertion is given in the following theorem.

Theorem 6 [5] Let $F(x + iy)$ be a holomorphic function in $T^n = \mathbb{R}^n + i\mathbb{R}^n_+$, such that for some C and k (*) holds. Then there is an $\tilde{F} \in \mathcal{D}'^{(M_p)}_{L^2}$ such that

$$F(x + iy) \to \tilde{F}(x), \quad y \to 0, \quad y \in \mathbb{R}^q_+, \text{ weakly in } \mathcal{D}'^{(M_p)}_{L^2}.$$

For the assertions which are to follow we need the definition and the properties of a convolution and the Fourier transformation on $\mathcal{D}'^{(M_p)}_{L^2}$. Let $f \in \mathcal{D}'^{(M_p)}_{L^2}$ and $\phi \in \mathcal{D}^{(M_p)}_{L^2}$. We put

$$(f \circledast \phi)(x) = \langle f(t), \phi(x-t) \rangle, \quad x \in \mathbb{R}^n.$$

If for a $g \in \mathcal{D}'^{(M_p)}_{L^2}$, $g \circledast \phi \in \mathcal{D}^{(M_p)}_{L^2}$ for any $\phi \in \mathcal{D}^{(M_p)}_{L^2}$, then g is called the convolutor on $\mathcal{D}'^{(M_p)}_{L^2}$. The space of convolutors is denoted by $\mathcal{O}'^{(M_p)}_{C,L^2}$ and the convolution of an $f \in \mathcal{D}'^{(M_p)}_{L^2}$ and $g \in \mathcal{O}'^{(M_p)}_{C,L^2}$ is defined by

$$\langle f * g, \phi \rangle = \langle f, \check{g} \circledast \phi \rangle, \quad (\check{g}(x) = g(-x)), \quad \phi \in \mathcal{D}^{(M_p)}_{L^2}.$$

If g and ϕ are as above, then $g * \phi = g \circledast \phi$ ([4]), so we shall always use the symbol $*$.

The Fourier and inverse Fourier transformations map $\mathcal{D}^{(M_p)}_{L^2}$ into the space

$$D^{(M_p)}_{L^2} = \{\phi \in L^2; \|\phi\|_{k,L^2} = \sup_\alpha \{\frac{k^{|\alpha|}}{M_{|\alpha|}} \|x^\alpha \phi(x)\|_2 < \infty, \ k \in \mathbb{N}\}.$$

The Fourier and inverse Fourier transformations on $\mathcal{D}'^{(M_p)}_{L^2}$ are defined as the corresponding adjoint mappings. So, we have:

$$\langle Ff, \phi \rangle = \langle f, F^{-1}\phi \rangle, \quad \phi \in D^{(M_p)}_{L^2},$$

$$\langle F^{-1}f, \phi \rangle = \langle f, F^{-1}\phi \rangle, \quad \phi \in \mathcal{D}^{(M_p)}_{L^2}.$$

All the "expected" properties for the convolution and the Fourier transformation in $\mathcal{D}'^{(M_p)}_{L^2}$ hold ([4]). Let us only quote the following four theorems.

Theorem 7 [4] A $g \in \mathcal{D}'^{(M_p)}_{L^2}$ belongs to $\mathcal{O}'^{(M_p)}_{C,L^2}$, iff for some $k > 0$ $\|(Fg) \exp(-M(k|\cdot|))\|_\infty < \infty$.

Observe, now, the convolution equation

$$S * U = V, \quad (**)$$

where $S \in O'^{(M_p)}_{C,L^2}$, $V \in D'^{(M_p)}_{L^2}$ and U is an unknown distribution.

Theorem 8 [4] The necessary condition for the solvability of (**) for any $V \in D'^{(M_p)}_{L^2}$ is the following:

There exist $C > 0$, $D > 0$ and $k > 0$, such that

$$|(FS)(\xi)| \geq C \exp(-M(k|\xi|)), |\xi| \geq D. \quad (***)$$

Theorem 7 implies that $(FS)(\xi) = s(\xi) \exp M(k|\xi|)$ for some $k > 0$ and some $s \in L^\infty$; Theorem 2 implies that $FV = P(\xi)v(\xi)$ for some ultradifferential operator of class (M_p) and some $v \in L^2$. So, we get:

Theorem 9 [4] A sufficient condition that $S * U = V$ is solvable for a given V is that the equation $su = v$ has a solution u in L^2.

It is said that equation (**) is hypoelliptic iff it is solvable and if $V \in D'^{(M_p)}_{L^2}$, then $U \in D'^{(M_p)}_{L^2}$, as well. We have:

Theorem 10 [4] Equation (**) is hypoelliptic iff (***) holds.

For an $f \in D'^{(M_p)}_{L^2}$ we define ([3]):

$$f_1 = f * \text{Im } K_n.$$

(Im means an imaginary part.)

Theorem 11 [3]

(i) $f_1 \in D'^{(M_p)}_{L^2}$.

(ii) The following two conditions are equivalent:

  a) $f = -(2/(2\pi)^n)^2 f_1 * \text{Im} K_n$.

  b) $f = (2/(2\pi)^n) f * \text{Re} K_n$.

If one of the conditions (a) and (b) holds, we say that f and f_1 form a pair of Hilbert transformations and that f_1 is the Hilbert transformation of f: $Hf = f_1$.

Theorem 12 [7] The following conditions are equivalent:

(i) f is a boundary value of some holomorphic function $F(z)$, $z \in T^n$ which satisfies (*).

(ii) $f \in \mathcal{D}'^{(M_p)}_{L^2}$ and

$$\text{Re} f = (-2/(2\pi)^n) \, \text{Im} f * \text{Im} K_n,$$

$$\text{Im} f = (2/(2\pi)^n) \, \text{Re} f * \text{Im} K_n.$$

(iii) $f \in \mathcal{D}'^{(M_p)}_{L^2}$ and supp $(F^{-1}f) \subset \overline{\mathbb{R}^n_+}$.

So, we get that $-2(2\pi)^{-n} \text{Im} f$ and $\text{Re} f$ form a pair of Hilbert transformations.

4. RELATION BETWEEN $\mathcal{D}'^{(M_p)}_{L^q}$ AND $\mathcal{D}'_{L^q}$, $q \in [1,\infty]$

We shall show in this section that

$$\mathcal{D}' \cap \mathcal{D}'^{(M_p)}_{L^q} \not\subset \mathcal{D}'_{L^q}, \quad q \in [2,\infty].$$

Let

$$f = \sum_{\alpha=0}^{\infty} \frac{\delta^{(\alpha)}(x-\alpha)}{M_\alpha}.$$

Clearly, it belongs to $\mathcal{D}'$ and does not belong to $\mathcal{D}'_{L^q}$ because it is of an infinite order. We are going to prove that $f \in \mathcal{D}'^{(M_p)}_{L^q}$. We shall show that for any $\phi \in \mathcal{D}^{(M_p)}_{L^r}$, $\frac{1}{q} + \frac{1}{r} = 1$, $|<f,\phi>| < \infty$. The closed graph theorem implies that $f \in \mathcal{D}'^{(M_p)}_{L^q}$. We have

$$<f,\phi> = \sum_\alpha <\frac{\delta^{(\alpha)}(x-\alpha)}{M_\alpha}, \phi(x)> = \sum_{\alpha=0}^{\infty} \frac{1}{M_\alpha} (-1)^\alpha \phi^{(\alpha)}(\alpha).$$

Since $\phi^{(\alpha)}(x) \in L^r$, $\alpha \in \mathbb{N}_0$, we have $\xi^\alpha \hat{\phi}(\xi) \in L^q$, $\alpha \in \mathbb{N}_0$ and, thus, by using the Cauchy inequality, we get $\xi^\alpha \hat{\phi}(\xi) \in L^1$. From this fact and [8, Theorem 8.12.(iii)], it follows that

$$\phi^{(\alpha)}(\eta) = (2\pi)^{-1} \int_{\mathbb{R}} e^{-i\xi\eta} \xi^\alpha \hat{\phi}(\xi) \, d\xi \quad \text{for any } \eta \in \mathbb{R}.$$

Thus, with $\eta = \alpha$

$$|<f,\phi>| \leq (2\pi)^{-1} \sum_\alpha \frac{1}{M_\alpha} \int_{\mathbb{R}} \xi^\alpha |\hat{\phi}(\xi)| \, d\xi$$

$$\leq (2\pi)^{-1} \sum_\alpha \frac{1}{M_\alpha} \int_{\mathbb{R}} \frac{1}{1+\xi^2} (|\xi^\alpha \hat{\phi}(\xi)| + |\xi^{\alpha+2}\hat{\phi}(\xi)|) d\xi$$

$$\leq (2\pi)^{-1} \left\| \frac{1}{1+\xi^2} \right\|_r \sum_\alpha \frac{1}{M_\alpha} (\|\xi^\alpha \hat{\phi}(\xi)\|_q + \|\xi^{\alpha+2}\hat{\phi}(\xi)\|_q).$$

Since $M_{\alpha+2} \leq AH^{\alpha+1} M_{\alpha+1} \leq \frac{A^2}{H^3}(H^2)^{\alpha+1} M_\alpha$ (this follows from (M.2)), and by using the Young inequality,

$$|<f,\phi>| \leq C \left(\sum_{\alpha=0}^\infty \frac{1}{M_\alpha} \|\xi^\alpha \hat{\phi}(\xi)\|_q + \frac{A^2}{H^2} \sum_{\alpha=0}^\infty \frac{(H^2)^{\alpha+2}}{M_{\alpha+2}} \|\xi^{\alpha+2}\hat{\phi}(\xi)\|_q \right)$$

$$\leq C_1 \left(\sum_{\alpha=0}^\infty \frac{1}{M_\alpha} \|\phi^{(\alpha)}\|_r + \sum_{\alpha=0}^\infty \frac{(H^2)^\alpha}{M_\alpha} \|\phi^{(\alpha)}\|_r \right).$$

This completes the proof.

5. CONVOLUTION IN $\mathcal{D}'^{(M_p)}_{L^q}$, $q \in (1,\infty]$

We shall give, in this section, the definition of a convolution in $\mathcal{D}'^{(M_p)}_{L^q}$, $q \in (1,\infty]$ and also some of its properties. Let us remark that investigations of a convolution for $q \neq 2$ are much more complicated. For $1 \leq q \leq 2$, we shall investigate this operation in a forthcoming paper.

We shall introduce the convolution in a slightly diffrent way than for $q = 2$ in [3].

Let $q \in (1,\infty]$ be fixed and let $\frac{1}{r} + \frac{1}{q} = 1$.

Definition An $f \in \mathcal{D}'^{(M_p)}$ is a convolution operator, convolutor, for $\mathcal{D}'^{(M_p)}_{L^q}$, if

$$\mathcal{D}^{(M_p)} \ni \phi \mapsto f * \phi = <f(t), \phi(\cdot - t)> \tag{1}$$

is a linear continuous mapping from $\mathcal{D}^{(M_p)}$ into $\mathcal{D}^{(M_p)}_{L^r}$ which can be extended onto $\mathcal{D}^{(M_p)}_{L^r}$ to be linear and continuous.

We shall denote by $\mathcal{O}'^{(M_p)}_{C,L^q}$ the space of all the convolutors for $\mathcal{D}'^{(M_p)}_{L^q}$.

Proposition 1

$$\mathcal{O}'^{(M_p)}_{C,L^q} \subset \mathcal{D}'^{(M_p)}_{L^q}.$$

Proof Let $f \in O'^{(M_p)}_{C,L^q}$. The mapping

$$\mathcal{D}^{(M_p)}_{L^r} \ni \phi \mapsto f * \check{\phi} \mapsto (f * \phi)\check{\,}(0) = \langle f(t), \phi(t) \rangle$$

can be extended as a continuous linear mapping from $\mathcal{D}^{(M_p)}_{L^r}$ into $\mathbb{C}$. Denote this extension by E_f. $E_f \in \mathcal{D}'^{(M_p)}_{L^q}$ and since on a dense subspace $\mathcal{D}^{(M_p)}$ of $\mathcal{D}^{(M_p)}_{L^r}$ we have $E_f(\phi) = \langle f, \phi \rangle$, it follows that $f = E_f$.

Obviously, if $f \in O'^{(M_p)}_{C,L^q}$, then for any $\phi \in \mathcal{D}^{(M_p)}_{L^r}$

$$\phi \mapsto \langle f(t), \phi(\cdot - t) \rangle = (f * \phi)$$

is the continuous linear extension quoted in the definition.

Let $f \in O'^{(M_p)}_{C,L^q}$. Define

$$f_\circledast : g \mapsto f \circledast g \quad \text{from } \mathcal{D}'^{(M_p)}_{L^q} \text{ into } \mathcal{D}'^{(M_p)}_{L^q} \text{ by} \tag{2}$$

$$\langle f \circledast g, \phi \rangle = \langle g, \check{f} * \phi \rangle, \quad \phi \in \mathcal{D}^{(M_p)}_{L^r}.$$

This mapping is obviously weakly continuous.

Proposition 2 Let $f, g, h \in O'^{(M_p)}_{C,L^q}$. Then

a) $f \circledast g \in O'^{(M_p)}_{C,L^q}$;

b) $(f \circledast g) \circledast h = f \circledast (g \circledast h)$.

Proof We shall prove only a). Clearly, $f \circledast g \in \mathcal{D}'^{(M_p)}_{L^q}$. For fixed $x \in \mathbb{R}$ and $\phi \in \mathcal{D}^{(M_p)}$ we have

$$((f \circledast g) * \phi)(x) = \langle (f \circledast g)(t), \phi(x-t) \rangle = \langle g(t), (\check{f}(u) * \phi(x-u))(t) \rangle$$

$$= \langle g(t), \langle f(-u), \phi(x - t+u) \rangle \rangle = \langle g(t), (f * \phi)(x-t) \rangle =$$

$$= g * (f * \phi)(x).$$

Now, by the definition of a convolutor, the assertion follows.

Proposition 3 For any $q \in (1, \infty]$ $\mathcal{D}'^{(M_p)}_{L^1} \subset O'^{(M_p)}_{C,L^q}$.

Proof Let $f \in \mathcal{D}'^{(M_p)}_{L^1}$ be of the form (1) such that (2) holds for some $k > 0$ (and $s = 1$). Then, for $p > 0$, $\beta \in \mathbb{N}_0$ and $\phi \in \mathcal{D}^{(M_p)}_{L^r}$

$$\frac{p^{|\beta|}}{M_{|\beta|}} \|(f * \phi)(x)\|_r \leq \sum_{\alpha \in \mathbb{N}_0} \frac{p^{|\beta|}}{M_{|\beta|}} \| \int_{\mathbb{R}^n} |f_\alpha(t)| |\phi^{(\alpha+\beta)}(x-t)| dt \|_r \leq$$

$$\leq \sum_{\alpha} \frac{p^{\beta}}{M_{|\beta|}} \frac{k^{\alpha}}{M_{|\alpha|}} \frac{M_{|\alpha|}}{k^{\alpha}} \|\phi^{(\alpha+\beta)}\|_{r} \|f_{\alpha}\|_{1}$$

$$\leq A \sup_{\alpha,\beta} \left\{ \frac{(H(p+k))^{|\alpha+\beta|}}{M_{|\alpha+\beta|}} \right\} \|\phi^{(\alpha+\beta)}\|_{r} \sum_{\alpha} \frac{M_{\alpha}}{k^{\alpha}} \|f_{\alpha}\|_{1} \leq A_{1} \|\phi\|_{L^{r},H(p+k)}.$$

In general, the assertions with general convolutors from $\mathcal{O}'^{(M_p)}_{C,L^q}$, $q \in (1,\infty]$, are difficult. But, if we observe only those which belong to $\mathcal{D}'^{(M_p)}_{L^1}$, we have some simple assertions.

Proposition 4 Let $\phi \in \mathcal{D}^{(M_p)}_{L^q}$, $q \in (1,\infty]$, and $f \in \mathcal{D}'^{(M_p)}_{L^1}$. Then,

$$f \circledast \phi = f * \phi.$$

On the left-hand side we have a convolution between a convolutor and an element from $\mathcal{D}'^{(M_p)}_{L^q}$, and on the right-hand side we have an element from $\mathcal{D}'^{(M_p)}_{L^q}$ determined by a test function from $\mathcal{D}^{(M_p)}_{L^q}$.

Proof Let $\psi \in \mathcal{D}^{(M_p)}_{L^r}$. By using the representation theorem for f, we have

$$< f \circledast \phi, \psi > = < \phi, \check{f} * \psi > = < \phi(x), < \tilde{f}(t), \psi(x-t) >$$

$$= < \phi(x), < f(t), \psi(x+t) >> = \sum_{\alpha} < \phi(x), < f_{\alpha}(t), (-1)^{\alpha} \psi^{(\alpha)}(x+t) >>$$

$$= \sum_{\alpha} (-1)^{\alpha} \int_{\mathbb{R}} \phi(x) \left[\int_{\mathbb{R}} f_{\alpha}(t) \frac{d^{\alpha}}{dt^{\alpha}} \psi(x+t) dt \right] dx = \ldots .$$

Since $\frac{d^{\alpha}}{dt^{\alpha}} \psi(x+t)$, $\alpha \in \mathbb{N}_0$, as a function on t, are from L^r, we have

$$\ldots = \sum_{\alpha} (-1)^{\alpha} \int_{\mathbb{R}} \phi(x) \frac{d^{\alpha}}{dx^{\alpha}} \left[\int_{\mathbb{R}} f_{\alpha}(t) \psi(x+t) dt \right] dx = \ldots$$

$$= \sum_{\alpha} (-1)^{\alpha} \int_{\mathbb{R}} (-1)^{\alpha} \phi^{(\alpha)}(x) \left[\int_{\mathbb{R}} f_{\alpha}(t) \psi(x+t) dt \right] dx = \ldots ,$$

where we used the fact that $\phi^{(\alpha)}(x) \to 0$, $|x| \to \infty$, $\alpha \in \mathbb{N}_0$.

Now, by Fubini's theorem

$$\ldots = \sum_{\alpha} \int_{\mathbb{R}} \int_{\mathbb{R}} \left[\int_{\mathbb{R}} f_{\alpha}(u-x) \phi^{(\alpha)}(x) dx \right] \psi(u) du$$

$$= \sum_{\alpha} << f_{\alpha}(u-x), \phi^{(\alpha)}(x) >, \psi(u) > = \sum_{\alpha} << f_{\alpha}^{(\alpha)}(u-x), \phi(x) >, \psi(u) >$$

$$= << \sum_{\alpha} f_{\alpha}^{(\alpha)}(t), \phi(u-t) >, \psi(u) > = < f * \phi, \psi > .$$

In a similar way, one can prove

Proposition 5 If $f, g \in \mathcal{D}'^{(M_p)}_{L^1}$, then $f \circledast g = g \circledast f$.

REFERENCES

1. K. Floret, J. Wloka, "Einführung in die Theorie der lokalkonvexen Räume", Lect. Not. Math. 71, Springer, Berlin - Heidelberg - New York, (1968).
2. H. Komatsu, Ultradistributions I, Structure theorems and a characterization, J. Fac. Sci. Univ. Tokyo, Sect. IA Math., 20, 25-105 (1973).
3. S. Pilipović, Hilbert transformation of Beurling ultradistributions, Rend. Math. Univ. Padova, 77, 1-13 (1987).
4. S. Pilipović, On the convolution in the space $\mathcal{D}'^{(M)}_{L^2}$, Rend. Math. Univ. Padova, 78, (1987) (to appear).
5. S. Pilipović, Boundary value representation for a class of Beurling ultradistributions, Port. Math. (to appear)
6. S. Pilipović, Ultradistributional boundary values for a class of holomorphic functions, Comm. Math. Univ. St. Pauli, (to appear).
7. S. Pilipović, The generalized Cauchy-Bochner representation for elements of $\mathcal{D}'^{(M)}_2$ and pairs of Hilbert transformations (to appear).
8. D. C. Champeney, "A handbook of Fourier theorems", Cambridge Univ. Press, Cambridge - New York - New Rochelle - Melbourne - Sydney, (1987).

PEETRE'S THEOREM AND GENERALIZED FUNCTIONS

J. W. de Roever

University of Twente
Dept. of Mathematics
P.O. Box 217, 7500 AE Enschede, The Netherlands

ABSTRACT

Sheaf morphisms are considered in sheaves of generalized functions. It is proved that for (ultra)distributions they must be continuous outside discrete points. Contrary to Peetre's original theorem, which applies to sheaves of test functions, an example makes clear that these points can really be points of discontinuity. Finally, it is shown that in the sheaf of hyperfunctions there are more general discontinuous sheaf morphisms.

Peetre's theorem says that any sheaf morphism in the sheaf of C^∞-functions is a differential operator. We shall investigate sheaf morphisms in sheaves of generalized functions, in particular distributions, ultradistributions of the Beurling and of the Roumieu type, and hyperfunctions. All these sheaves are soft so that their sections with a compact support form flabby cosheaves which are the duals, with respect to a certain topology, of the sheaves of their associated test functions. The main point is to investigate the continuity of a cosheaf morphism P (= local operator) in one of these cosheaves. At places where P is continuous its transposed $^t P$ is a continuous sheaf morphism in the sheaf of test functions and it follows that $^t P$, and hence P itself, are appropriate differential operators there. In this paper we shall only briefly mention these results, as well as the generalization of Peetre's theorem to the soft sheaves of test functions. Our main attention will be on the continuity of a local operator in a space of generalized functions and we shall indicate what possibilities there are for a discontinuous sheaf morphism.

1. TEST FUNCTIONS AND GENERALIZED FUNCTIONS

As our results are local we can just as well consider an open set $\Omega \subset \mathbb{R}^n$ instead of manifolds. For the notions and properties of sheaves and cosheaves we refer to Bredon [1]. Let F' be a soft sheaf of generalized functions and $\tilde{F}'_c$ the flabby cosheaf of their sections with a compact support. By F itself we denote the sheaf of test functions of which F'_c is the dual with respect to a certain topology. The topology in $F(\Omega)$, and hence in $F'_c(\Omega)$, is determined by certain norms, which we shall describe in more detail in this section.

Let K be a compact set in Ω and let there be a sequence of norms, denoted by $\|\cdot\|_{K,m}$, depending only on the restriction of a test function and its derivatives to K. We have

$$\|\phi\|_{K_1,m} \leq \|\phi\|_{K_2,m}, \quad K_1 \subset K_2, \quad \phi \in F(\Omega). \tag{1}$$

There are two types of topologies in the spaces of test functions depending on whether for fixed K the sequence of norms is increasing or decreasing for growing m. Let us refer to these types as *type* I or *type* II, respectively.

In the following examples of *type* I spaces of test functions FS-spaces are obtained by taking the projective limit for $m \to \infty$ and fixed K (where K is regular). One example is $F = E$, the sheaf of C^∞-functions with the usual norms and the other examples are $F = E^{(M_p)}$, sheaves of ultradifferentiable functions of the Beurling type with the norms given in Komatsu [2, cf. (2.11) where h is replaced by 1/m]. They are the test functions for distributions or ultradistributions of the Beurling type, respectively. Then the topology in $F'_c(\Omega)$ is determined by the dual norms defined for sections f with a support in the interior K^o of K:

$$\|f\|_{K,m} = \sup \{|<f,\phi>| \,\big|\, \phi \in F(\Omega), \|\phi\|_{K,m} \leq 1\} \tag{2}$$

where these norms, in *type* I spaces of generalized functions, are decreasing for a fixed K and growing m.

Examples of *type* II spaces of test functions, in which the inductive limit for $m \to \infty$ yields a DFS-space, are the following: $F = E^{\{M_p\}}$, sheaves of ultradifferentiable functions of the Roumieu type, cf. Komatsu [2, where in (2.13) h is replaced by m] and $F = A$, the sheaf of real-analytic functions. They are the test functions for ultradistributions of the Roumieu type or hyperfunctions, respectively. Here, in *type* II spaces of generalized functions, for fixed K the sequence of norms (2) is increasing for growing m.

In spaces of both types the property dual to (1) is

$$\|f\|_{K_2,m} \leq \|f\|_{K_1,m}, \quad K_1 \subset K_2, \quad f \in F_c'(K_1^o). \tag{3}$$

Throughout this paper we shall assume that the sequence (M_p) defining the ultradifferentiable functions satisfies the usual properties given in Komatsu [2, conditions M.1, M.2 and M.3], and moreover we shall not be concerned with questions whether K is regular, or whether (2) is also defined for sections with a support K. For we shall only use (2) if K is a neighbourhood of the support of f.

Finally we need the following property of test functions, which is true for $F = E$ or $E*$ (as in [2] $E*$ denotes either $E^{(M_p)}$ or $E^{\{M_p\}}$), but not for $F = A$. The proof is easily obtained from Komatsu [2, lemma 5.1 and prop. 2.7].

Lemma 1 Let K be a compact subset of Ω. Then for every compact neighbourhood S of K in Ω and for every compact neighbourhood K* of K in S^o there is a test function $\chi = \chi(K,K*) \in F_c(\Omega)$ which is identically one in a neighbourhood of K and vanishes outside K*. Moreover, for every norm m there are constants $M = M_m(K,K*)$, such that for all $\phi \in F(\Omega)$

$$\|\chi\phi\|_{S,m-1} \leq M\|\phi\|_{K*,m} \quad (type\ I)$$
$$\|\chi\phi\|_{S,m+1} \leq M\|\phi\|_{K*,m} \quad (type\ II) \tag{4}$$

(If $F = E$, there is no need to set m-1 in $type$ I, as m instead of m-1 would have been correct, too).

3. SHEAF MORPHISMS

Let $P : F_c'(\Omega) \to F_c'(\Omega)$ be an arbitrary cosheaf morphism (or local operator). Such a map can uniquely be extended to a sheafmorphism $P : F'(\Omega) \to F'(\Omega)$ in Ω. For F_c' equal to E' or $E*'$ we shall show that P is continuous outside discrete points in the following sense.

Theorem 1 For all the points $x \in \Omega$ and for all the norms m there are an open neighbourhood V of x and a norm k such that for all the compact sets $K \subset\subset V \setminus \{x\}$ and all the compact neighbourhoods K* of K in $V \setminus \{x\}$ there is a constant C with

$$\|Pf\|_{K*,k} \leq C\|f\|_{K*,m} \quad (type\ I)$$
$$\|Pf\|_{K*,m} \leq C\|f\|_{K*,k} \quad (type\ II) \qquad f \in F_K'(V \setminus \{x\}). \tag{5}$$

Before proving this theorem, let us briefly mention its consequences. First we remark that a similar theorem can be shown for local operators in the cosheaves E_c^* ($= \mathcal{D}*$) (in (5) *type* I and *type* II should be interchanged then, and in that case there is no need for the neighbourhoods K^*, nor does the constant C need to depend on K). As in Peetre's theorem [3], it follows that P is a suitable ultradifferential operator with ultradifferentiable coefficients.

If $F_c' = E'$ Theorem 1 implies that in each relatively compact subset P acts on m^{th}-order distributions as a differential operator with C^∞-coefficients outside discrete points (this set of points can depend on m, cf. the example given below). A similar result can be derived for ultradistributions, where in Roumieu type spaces (*type* II) the set of discrete points outside which P acts as an ultradifferential operator does not depend on the sections on which P is applied.

The following example shows that Theorem 1 is rather sharp and that the discrete points which are excluded can really be points of discontinuity. Thus, although in Peetre's original theorem there are no such points for sheaf morphisms between test functions, the idea to prove it by means of a theorem like Theorem 1 is essential.

<u>Example</u> (a discontinuous sheaf morphism in the sheaf $\mathcal{D}'$ of distributions). For each $m = 0,1,\ldots$ let $\{x_n^m \mid n = 1,2,\ldots\}$ be a set of discrete points in Ω (for ex. if $\Omega = (0,1) \subset \mathbb{R}$, take $x_n^m = (m+1)/(m+n+1)$; then the total collection of points is dense in Ω). In every stalk $\mathcal{D}'_{x_n^m}$ choose a (Hamel) base $\{f_\alpha^{m,n} \mid \alpha \in I^{m,n}\}$ of the space of germs in x_n^m of the mth-order distributions, complemented with $\{g_\beta^{m,n} \mid \beta \in J^{m,n}\}$ to a base of the whole stalk. Then, for any $f \in \mathcal{D}'(\Omega)$ there is a unique, finite sum

$$f_{x_n^m} = \sum_\alpha a_\alpha^{m,n}(f) f_\alpha^{m,n} + \sum_\beta b_\beta^{m,n}(f) g_\beta^{m,n},$$

in which the coefficients $a_\alpha^{m,n}$ and $b_\beta^{m,n}$ depend linearly on f. Furthermore, if for $\omega \subset\subset \Omega$ $f|_\omega$ is a kth-order distribution, then for each pair (m,n) with $m \geq k$ and such that $x_n^m \in \omega$ all the coefficients $b_\beta^{(m,n)}(f)$ vanish. Finally, for each pair (m,n) choose a particular index $\beta_0 = \beta_0^{m,n} \in J^{m,n}$ and define a linear map $P : \mathcal{D}'(\Omega) \to \mathcal{D}'(\Omega)$ by

$$Pf = \sum_{m,n} b_{\beta_0}^{m,n}(f) \delta_{x_n^m}, \quad f \in \mathcal{D}'(\Omega).$$

This is a good definition, because in every relatively compact subset ω of Ω and for every m there are only finitely many points $x_n^m \in \omega$, while $b_{\beta_0}^{m,n}(f) = 0$ for sufficiently large m. Thus, Pf is a sum of δ-functions concentrated in discrete points of Ω, hence $Pf \in \mathcal{D}'(\Omega)$. Clearly P is linear. Restricting f to $\Omega_1 \subset \Omega$ and applying on it the obvious restriction of P to Ω_1 (namely summing only over (m,n) for which $x_n^m \in \Omega_1$), then the result is the same if one restricts to the points $x_n^m \in \Omega_1$ after applying P on f. Hence P commutes with restrictions and thus is a sheaf morphism in Ω.

Proof of Theorem 1 If the theorenm were not true, there would exist a point $x_0 \in \Omega$ and a norm m_0 such that

$$\forall k, \forall V, \exists K, K^* \subset\subset V \setminus \{x_0\}, K \subset\subset K^*, \forall C, \exists f \in F_K'(V \setminus \{x_0\})$$

with

$$\|Pf\|_{K^*,k} > C, \quad \|f\|_{K^*,m_0} = 1 \quad (type\ I)$$

$$\|Pf\|_{K^*,m_0} > C, \quad \|f\|_{K^*,k} = 1 \quad (type\ II)$$
(6)

Then one could construct a sequence $(f_k) \subset F_c'(\Omega \setminus \{x_0\})$ as follows: let $V_1 = \Omega$ and if $k \geq 2$ for $j = 1,\ldots,k-1$ let, furthermore, relatively compact open neighbourhoods V_{j+1} of x_0 in Ω, compacts sets $K_j \subset V_j \setminus \{x_0\}$ with compact neighbourhoods K_j^* in $V_j \setminus \{x_0\}$, constants C_j and sections $f_j \in F_c'(\Omega \setminus \{x_0\})$ with the support in K_j have been determined already, then according to (6) to the norm k and to the neighbourhood V_k there belongs a compact set $K_k \subset V_k \setminus \{x_0\}$ and a compact neighbourhood K_k^* of K_k in $V_k \setminus \{x_0\}$ and, moreover, if F is a type I space, to the constant $C_k = 4^k M_k(K_k,K_k^*)$, where M is determined in (4) of Lemma 1, or in type II spaces take $C_k = 4^k M_{m_0}(K_k,K_k^*)$, there belongs a section $f_k \in F_{K_k}'(\Omega \setminus \{x_0\})$ such that

$$\|Pf_k\|_{K_k^*,k} > C_k, \quad \|f_k\|_{K_k^*,m_0} = 1 \quad (type\ I)$$

$$\|Pf_k\|_{K_k^*,m_0} > C_k, \quad \|f_k\|_{K_k^*,k} = 1 \quad (type\ II)$$
(7)

Finally, the construction of the sequence (f_k) can be continued by choosing a relatively compact open neighbourhood V_{k+1} of x_0 in Ω with $K_j^* \cap V_{k+1} = \emptyset$ for $j = 1,\ldots,k$.

Formula (7) means that for every k there is a test function $\phi_k \in F(\Omega)$, such that

$$|<Pf_k, \phi_k>| \geq C_k \|\phi_k\|_{K_k^*,k} \quad (type\ I)$$

$$|<Pf_k, \phi_k>| \geq C_k \|\phi_k\|_{K_k^*,m_0} \quad (type\ II)$$
(8)

Define the locally finite sum f in $\Omega \setminus \{x_0\}$ by $f = \sum_k 2^{-k} f_k$. Then, not only $f \in F'(\Omega \setminus \{x_0\})$, but even $f \in F_S'(\Omega)$, where S is a compact subset of Ω

containing all the sets K_k^* as well as a neighbourhood of the closure in Ω of $\bigcup_k K_k$, because by (3)

$$\|f\|_{S,m_0} \leq \sum_k 2^{-k} \|f_k\|_{S,m_0} \leq \sum_k 2^{-k} \|f_k\|_{K_k^*,m_0} = 1 \qquad (\textit{type I})$$

$$\forall j : \|f\|_{S,j} \leq \sum_k 2^{-k} \|f_k\|_{S,j} \leq \sum_{k<j} 2^{-k} \|f_k\|_{S,j} + \sum_{k \geq j} 2^{-k} \|f_k\|_{K_k^*,k} < \infty$$
$$(\textit{type II})$$

Now also $Pf \in \bar{F}'_S(\Omega)$, so that in *type* I spaces for some norm j, or in *type* II spaces for all norms j, $\|Pf\|_{S,j}$ must be finite. Thus, there must be a number C, or for every j there is a C, respectively, such that

$$\frac{|<Pf,\phi>|}{\|\phi\|_{S,j}} \leq C, \quad \forall \phi \in F(\Omega).$$

However, with χ_k as in Lemma 1 where $K = K_k$, by (8) we have:
(*type* I) for every $k \geq j+1$

$$\frac{|<Pf,\chi_k\phi_k>|}{\|\chi_k\phi_k\|_{S,j}} \geq \frac{2^{-k}|<Pf_k,\chi_k\phi_k>|}{\|\chi_k\phi_k\|_{S,k-1}} \geq \frac{2^{-k}|<Pf_k,\phi_k>|}{M_k(K_k,K_k^*)\|\phi_k\|_{K_k^*,k}} \geq 2^k$$

(*type* II) with $j = m_0+1$, for every k

$$\frac{|<Pf,\chi_k\phi_k>|}{\|\chi_k\phi_k\|_{S,m_0+1}} \geq \frac{2^{-k}|<Pf_k,\phi_k>|}{M_{m_0}(K_k,K_k^*)\|\phi_k\|_{K_k^*,m_0}} \geq 2^k. \quad \square$$

3. HYPERFUNCTIONS

For $F = A$ Lemma 1 does not hold and formula (7) for *type* II spaces is not impossible for hyperfunctions $f_k \in B_c(\Omega \setminus \{x_0\})$. Actually, since the sheaf B of hyperfunctions is flabby, any locally finite sum $\sum_k P f_k$ in $\Omega \setminus \{x_0\}$ can be extended to a section in $B(\Omega)$. And indeed, there is nothing like Theorem 1 for hyperfunctions, as the following example shows that there are sheaf morphisms in B which are discontinuous on sets containing more than countably many points.

We shall give an example of a sheaf morphism P in $B(\Omega)$ which itself has a compact support $\bar{\omega}$ in Ω. Such a sheaf morphism cannot be continuous in any point x of $\partial\omega$. For if it were, the stalk at x of the transposed of P (P considered as a cosheaf morphism), would be a continuous map: $A_x \to A_x$, but such a map would be zero, because sections in A vanishing on one side of $\partial\omega$ at x, vanish everywhere.

Example (discontinuous sheaf morphism with a compact support in the sheaf of hyperfunctions). Let P_1 be a continuous sheaf morphism in Ω (an analytic differential operator of infinite order in the sense of hyperfunctions). Let $\{f_\alpha\}$ be a base for the space $B(\omega)$ and let $\bar{f}_1$ be an extension of f_1 to $B(\Omega)$ vanishing in $\Omega \setminus \bar{\omega}$. Although the f_α's are linearly independent in ω, there might be a relation for the restrictions to an open subset. Therefore, the other extensions $\bar{f}_\alpha$ of f_α should be taken more carefully. Assume that for all $\beta < \alpha$ we have already extensions $\bar{f}_\beta$ of f_β vanishing in $\Omega \setminus \bar{\omega}$, such that in any open subset Ω_s of Ω, where there is a relation $\sum_{\beta<\alpha} b_\beta^s f_\beta|_{\Omega_s \cap \omega} = 0$ for a finite number of non-vanishing coefficients b_β^s, also $\sum_{\beta<\alpha} b_\beta^s \bar{f}_\beta|_{\Omega_s} = 0$. Let $\Omega_{\alpha,s}$ be an open subset such that there is a finite sum $f_\alpha|_{\Omega_{\alpha,s} \cap \omega} = \sum_{\beta<\alpha} c_\beta^s f_\beta|_{\Omega_{\alpha,s} \cap \omega}$. Define a section $\tilde{f}_\alpha$ in $\bigcup_s \Omega_{\alpha,s} \cup \omega$ by

$$\tilde{f}_\alpha|_\omega = f_\alpha, \quad \tilde{f}_\alpha|_{\Omega_{\alpha,s}} = \sum_{\beta<\alpha} c_\beta^s \bar{f}_\beta|_{\Omega_{\alpha,s}}.$$

This is a good definition, because in $\Omega_{\alpha,s} \cap \Omega_{\alpha,t}$ we have

$$\sum_{\beta<\alpha}(c_\beta^s - c_\beta^t) f_\beta|_{\Omega_{\alpha,s} \cap \Omega_{\alpha,t} \cap \omega} = 0,$$

so that the same relation holds for the sections $\bar{f}_\beta|_{\Omega_{\alpha,s} \cap \Omega_{\alpha,t}}$ and, hence,

$$\sum_{\beta<\alpha} c_\beta^s \bar{f}_\beta|_{\Omega_{\alpha,s} \cap \Omega_{\alpha,t}} = \sum_{\beta<\alpha} c_\beta^t \bar{f}_\beta|_{\Omega_{\alpha,s} \cap \Omega_{\alpha,t}}.$$

Let now $\bar{f}_\alpha$ be an extension of $\tilde{f}_\alpha$ to $B(\Omega)$ vanishing in $\Omega \setminus \bar{\omega}$. Then, for any vanishing finite sum $\sum_{\beta \leq \alpha} b_\beta^s f_\beta|_{\Omega_{\alpha,s} \cap \omega} = 0$ in an open $\Omega_{\alpha,s}$ in Ω also $\sum_{\beta \leq \alpha} b_\beta^s \bar{f}_\beta|_{\Omega_{\alpha,s}} = 0$.

Since B is flabby, for every open $\Omega_s \subset \Omega$ and for every $g \in B(\Omega_s \cap \omega)$, there are finitely many nonvanishing coefficients $a_\alpha^s(g)$ (not necessarily unique) with $g = \sum_\alpha a_\alpha^s(g) f_\alpha|_{\Omega_s \cap \omega}$. We define a map $P|_{\Omega_s}: B(\Omega_s) \to B(\Omega_s)$ by

$$P|_{\Omega_s} f = \sum_\alpha a_\alpha^s(g) \bar{f}_\alpha|_{\Omega_s}, \quad f \in B(\Omega_s),$$

where $g = P_1 f|_{\Omega_s \cap \omega}$. This is a good definition, for if also $g = \sum_\alpha \tilde{a}_\alpha^s(g) f_\alpha|_{\Omega_s \cap \omega}$, then the relation $\sum_\alpha (a_\alpha^s(g) - \tilde{a}_\alpha^s(g)) f_\alpha|_{\Omega_s \cap \omega} = 0$ for the sections $f_\alpha|_{\Omega_s \cap \omega}$ holds for $\bar{f}_\alpha|_{\Omega_s}$, too, so that

$$\sum_\alpha a_\alpha^s(g) \bar{f}_\alpha|_{\Omega_s} = \sum_\alpha \tilde{a}_\alpha^s(g) \bar{f}_\alpha|_{\Omega_s}.$$

Finally, if $\Omega_t \subset \Omega_s$ for $f \in B(\Omega_s)$ we have on the one hand

$$(P|_{\Omega_s} f)|_{\Omega_t} = \sum_\alpha a_\alpha^s(g) \bar{f}_\alpha|_{\Omega_t}$$

and on the other hand

$$P|_{\Omega_t}(f|_{\Omega_t}) = \sum_\alpha a_\alpha^t(g|_{\Omega_t \cap \omega}) \bar{f}_\alpha|_{\Omega_t},$$

while $P_1 f|_{\Omega_t \cap \omega} = g|_{\Omega_t \cap \omega} = \sum_\alpha a_\alpha^s(g) f_\alpha|_{\Omega_t \cap \omega}$. Hence, for all α we can take

$$a_\alpha^t(g|_{\Omega_t \cap \omega}) = a_\alpha^s(g).$$

Thus the collection $\{P|_{\Omega_s} \mid \Omega_s \subset \Omega\}$ commutes with restrictions and therefore determines a sheaf morphism in Ω.

In a similar way it can be shown that the sheaf of sheaf morphisms in the flabby sheaf B is itself flabby. Hence B is fine, but contrarily to (ultra)distributions in a non-constructive and discontinuous way.

Remark   There still remains to investigate whether Peetre's theorem is true in the sheaf A of real-analytic functions.

REFERENCES

1. G. E. Bredon, "Sheaf theory", McGraw-Hill, New York, (1967).
2. H. Komatsu, Ultradistributions I, Structure theorems and a characterization, J. Fac. Sci. Univ. Tokyo, Sec. IA 20, 25-105 (1973).
3. J. Peetre, Rectification á l'article "Une caractérisation abstraite des opérateurs différentiels", Math. Scand. 8, 116-120 (1960).

INFINITE DIMENSIONAL FOCK SPACES AND AN ASSOCIATED GENERALIZED

LAPLACIAN OPERATOR

John Schmeelk

Department of Mathematical Sciences
Virginia Commonwealth University
Richmond, Virginia 23284-0001 U.S.A.

1. INTRODUCTION

Quantum field theory has a long history of cooperation between physicists and mathematicians. The approach in this paper utilizes the theory developed in generalized functions. This technique was pioneered by Dr. Paul Dirac. His results of the 1920's still remain a very elegant treatment of the theory. A recent publication [9] written by Dirac in 1966 gives a concise assessment of the subject. Many other contributors such as Bergmann, Bogoliubov, Cholewinski, Colombeau, Friedrichs, Kastler, Kristensen, Mejblo, Poulsen, Rzewuski, Schiff, Shapiro, and Wightman incorporate generalized functions into their development.

Especially during the 1950's, it became apparent that the Hilbert space setting may prove to be too tight to allow for a much needed flexible theory. The notion of a generalized Fock space began to attract researchers. We will begin with the past development of a Fock space and then consider a mathematical space, $\Gamma^{p,sB}$, which serves as our generalized Fock space.

Classical quantum field theory formulates its general theory in the setting of entire functions. A classical entire function, F, with power series representation,

$$f(Z) = \sum_{\substack{m_i \\ 1 \le i \le n}} \alpha_{m_1, m_2, \ldots, m_n} Z_1^{m_1}, \ldots, Z_n^{m_n}, \tag{1.1}$$

or a functional power series of the form,

$$V[\alpha] = \sum_n \frac{1}{\sqrt{n!}} (v_n, \alpha^n) \tag{1.2}$$

where

$$(v_n, \alpha^n) = \int \cdots \int v_n(\vec{p}_1, \ldots, \vec{p}_n) \alpha(\vec{p}_1) \cdots \alpha(\vec{p}_n) d\vec{p}_1, \ldots, d\vec{p}_n, \qquad (1.3)$$

are discussed by Bergmann and Rzewuski [2, 40]. The complex valued functions, $\alpha(\vec{p}_i)$, $(1 \leq i \leq n)$, in expression (1.3) are defined on points, $\vec{p}$, belonging to Euclidean space, E_3. The n-point functions, $v_n(\vec{p}_1, \ldots, \vec{p}_n)$, are symmetric and defined on Euclidean space, E_{3n}.

However as the general theory advances, it was noticed that free field operators should perhaps be operator valued distributions [8]. The entire functions developed in this paper will have functional representations having tempered distributions as their domain. The range will be the reals. The wave functions developed by the physicists usually represent a probability distribution. A probability distribution when integrated over the entire space must equal one. Rapid descent test functions, $S(R^{3n})$, are excellent candidates for this job since when normalized their integral over the entire space will equal one.

The notion of p-particles changing to p+1 particles is mathematically treated as a creation operator. Similarly the destruction of one particle is represented as an annihilation operator. These operators will have domains developed in this paper and termed infinite dimensional Fock spaces.

A form of a kernel theorem is proven in a previous paper [46]. Therefore, our space, Γ^{pB}, will be homeomorphic to a space having the classical annihilation and creation operators defined in it as they are presented in reference [7].

We have presented the cannonical commutator relationships in several spaces and suggest the selection of the appropriate space be determined by the physical application [46]. The generalized Laplacian operator motivated in [25] will be developed in our setting and its connection to the one discussed in [46] will be examined. We can then develop a form of an infinite dimensional Schrödinger equation.

2. THE SCALE OF FRÉCHET SPACES, $\Gamma^{p^B} = \bigcup_{s \geq 1} \Gamma^{p,sB}$

For each $s \geq 1$, the space, $\Gamma^{p,sB}$ $(p > 1, B = \{B_i\}_{i=0}^{\infty}, B_i > B_j, j > i)$, is called an infinite dimensional Fock space. The p and B_i, $i \geq 0$ are all real numbers. These spaces are topological spaces of real-valued functionals on $S'(R^{3n}; R)$, the space of real-valued tempered distributions. The set of functionals which are members of $\Gamma^{p,sB}$ are all $C^{\infty}(S'(R^{3n}; R))$. We also require if $\Phi \in \Gamma^{p,sB}$, then

$$\Phi(x) = \sum_{q=0}^{\infty} a_q x^q = \sum_{q=0}^{\infty} a_q [x,\ldots,x] \qquad (2.1)$$

where $a_0 \in \mathbf{R}$ and a_q, $q \geq 1$ are q-multilinear symmetric continuous functionals on $S'(\mathbf{R}^{3n}) \times \ldots \times S'(\mathbf{R}^{3n})$ to $\mathbf{R}$.

We identify for each $\Phi \in \Gamma^{p,sB}$ the associated state vector,

$$\Phi \leftrightarrow \begin{pmatrix} a_0 \\ a_1 \\ \vdots \\ a_q \\ \vdots \end{pmatrix}. \qquad (2.2)$$

Each multilinear functional, a_q, $q \geq 1$, has an infinite dimensional domain space. This observation and the association (2.2) permits us to speak of an infinite dimensional Fock space.

The members, Φ, belonging to our infinite dimensional Fock space are entire functionals and the topological structure will be homeomorphic to a scale of spaces whose members can be represented in the form (2.2). However within this new form the individual entries will be members of $S(\mathbf{R}^{3n})$, the space of rapid descent test functions. This will generalize the Fock or Fischer space described in [6].

We first generalize the creation and annihilation operators which motivates a generalized commutator. In order to understand the topological properties of our creation and annihilation operators, we equip our infinite dimensional Fock space with the following sequence of norms:

$$|||\Phi|||_{sB_m} = \sup_q \frac{\|a_q\|_m \, q!^{1/p}}{(sB_m)^q} < \infty, \quad m = 0,1,\ldots \qquad (2.3)$$

where

$$\|a_q\|_m = \sup_{\|x\|_{-m} \leq 1} |a_q x^q| \quad m = 0,1,\ldots, \ x \in S'(\mathbf{R}^{3n}) \qquad (2.4)$$

and

$$\|x\|_{-m} = \sup_{\|\phi\|_m \leq 1} |<x,\phi>| \quad m = 0,1,\ldots, \ \phi \in S'(\mathbf{R}^{3n}) \qquad (2.5)$$

and

$$\|\phi\|_m = \sup_{\substack{\sigma_1 + \ldots + \sigma_n \leq m \\ (\tau_1,\ldots,\tau_n) \in \mathbf{R}^{3n}}} (1+|\tau_1|^2)^m \ldots (1+|\tau_n|^2)^m \left| \phi_{(\tau_1 \ldots \tau_n)}^{(\sigma_1,\ldots,\sigma_n)} \right| \qquad (2.6)$$

where

$$\phi^{(0,\ldots,\sigma_i,\ldots,0)}(\tau_1\ldots\tau_n) = \frac{\partial^{\sigma_i}}{\partial \tau_i^{\sigma_i}} \phi \qquad 1 \leq i \leq n. \qquad (2.7)$$

The functions, ϕ, are test functions of rapid descent and the functionals, x, are tempered distributions [8, 53]. The set of entire functionals belonging to $\Gamma^{p,sB}$ equipped with the natural topology induced by the sequences of norms, (2.3), is easily seen to be a Fréchet space. We then consider $1 \leq s \leq s'$ where clearly $\Gamma^{p,sB} \subset \Gamma^{p,s'B}$. Also the canonical injection, $J_{s's} : \Gamma^{p,sB} \mapsto \Gamma^{p,s'B}$, is continuous.

The kernel representation for the multilinear symmetric functionals, a_q, $q \geq 1$, will have a square summable property which the multilinear symmetric functionals also enjoy in the following sense.

<u>Proposition 2.8</u>   The sequence of multilinear symmetric functionals, $\{a_q\}_{q=1}^\infty$, $a_0 \in \mathbb{R}$, described in expression (2.2) is square summable in each norm,

$$\sum_{q=0}^\infty \|a_q\|_m^2 < \infty, \qquad m = 0,1,2,\ldots\,.$$

<u>Proof</u>   If $\Phi \in \Gamma^{p,sB}$, then $\Phi(x) = \sum_{q=0}^\infty a_q x^q$. Select any norm, $\|\cdot\|_{sB_m}$, and consider

$$\|\Phi\|_{sB_m} = \sup_q \frac{\|a_q\|_m q!^{1/p}}{(sB_m)^q} < C_m < \infty.$$

We therefore have

$$\|a_q\|_m < \frac{C_m (sB_m)^q}{q!^{1/p}}$$

for every q.

From these statements and returning to the square summable notion, we obtain

$$\sum_{q=0}^\infty \|a_q\|_m^2 = \sum_{q=0}^\infty \frac{\|a_q\|_m^2 q!^{1/p} (sB_m)^q}{(sB_m)^q q!^{1/p}} = \sum_{q=0}^\infty \frac{\|a_q\|_m q!^{1/p} \|a_q\|_m (sB_m)^q}{(sB_m)^q q!^{1/p}}$$

$$\leq \|\Phi\|_{sB_m} \sum_{q=0}^\infty \frac{\|a_q\|_m (sB_m)^q}{q!^{1/p}} \leq \|\Phi\|_{sB_m} C_m \left(\frac{(sB_m)^q}{q!^{1/p}}\right)^2 < \infty. \quad \square$$

3. ANNIHILATION AND CREATION OPERATORS

The annihilation and creation operators will first be defined on the functional representation, (2.2). The algebraic and topological properties for these operators will be investigated and a theorem will be proved that under appropriate hypothesis the annihilation operator can be applied to a member of the infinite dimensional Fock space an infinite number of times. However the composition of the annihilation operator will require the utilization of the scale of Fréchet spaces introduced in section 2.

Definition 3.1 For $\Phi \in \Gamma^{p,sB}$ and $h \in S'(\mathbb{R}^{3n})$, we define the annihilation operator, D_h, as follows:

$$D_h : \Gamma^{p,sB} \mapsto \Gamma^{p,sB}$$

$$D_h : \begin{pmatrix} a_0 \\ a_1 \\ \vdots \\ a_q \\ \vdots \end{pmatrix} \mapsto \begin{pmatrix} a_1[h] \\ 2a_2[h,.] \\ \vdots \\ q+1\, a_{q+1}[h,\ldots] \end{pmatrix}. \quad (3.2)$$

Similarly the operator, D_h^ν, ν a positive integer, is defined as follows:

$$D_h^\nu : \Gamma^{p,sB} \mapsto \Gamma^{p,sB}$$

$$D_h^\nu : \begin{pmatrix} a_0 \\ a_1 \\ \vdots \\ a_q \\ \vdots \end{pmatrix} \mapsto \begin{pmatrix} \nu!\, a_\nu[h^\nu] \\ (1+\nu)!\, a_{\nu+1}[h^\nu,.] \\ \vdots \\ \frac{(q+\nu)!}{q!}\, a_{q+\nu}[h^\nu,\ldots] \end{pmatrix}. \quad (3.3)$$

We recall that the multilinear functionals, a_q, $q \geq 1$, are symmetric and so the location of the tempered distribution, h, is invariant. For convenience, we place the tempered distribution, h, an appropriate number of times in the first ν arguments. We also note that the operator, D_h^ν, can be viewed as a differential operator acting on Φ and evaluated in the "direction", h. This is then a generalization of a derivative which can be evaluated in an infinite number of directions.

Definition 3.4 For $\Phi \in \Gamma^{p,sB}$ and $\phi \in S(\mathbb{R}^{3n})$, we define the creation operator, Q_ϕ^ν, as follows:

$$Q_\phi^\nu : \Gamma^{p,sB} \mapsto \Gamma^{p,sB}$$

$$Q_\phi^\nu : \begin{pmatrix} a_0 \\ a_1 \\ \cdot \\ \cdot \\ a_q \\ \cdot \\ \cdot \end{pmatrix} \mapsto \begin{pmatrix} 0 \\ 0 \\ \underbrace{<\cdot,\phi>\cdot<\cdot,\phi>\cdot\ldots\cdot<\cdot,\phi>}_{\nu \text{ copies}} a_0 \\ \cdot \\ \text{sym } <\cdot,\phi>\cdot\ldots\cdot<\cdot,\phi>\cdot a_{q-\nu}[\cdot,\ldots\cdot] \\ \cdot \\ \cdot \end{pmatrix} \qquad (3.5)$$

where

$$\text{sym} < \cdot,\phi > \cdot\ldots\cdot < \cdot,\phi > \cdot a_{q-\nu}[\cdot,\ldots\cdot] =$$

$$\frac{1}{(q+\nu)!} \sum_\sigma < \cdot_{\sigma(1)}, \phi > \cdot \ldots < \cdot_{\sigma(\nu)}, \phi > a_{q-\nu}[\cdot_{\sigma(\nu+1)}, \ldots, \cdot_{\sigma(q)}]$$

and σ varies over all permutations of $\{1,2,\ldots,q\}$. The sym operator is introduced so that the qth entry remains symmetric. Again the creation operator is applied to a state vector an infinite number of times in an appropriate sense.

4. INFINITE DIMENSIONAL LAPLACIAN OPERATOR

We briefly review cylinder functionals developed by K. O. Friedrichs and H. N. Shapiro [25]. Cylinder functionals have p-variables where each variable takes its value from a set of functions containing the classical piecewise constant functions. The p variable functions can be written as

$$\phi(t) \leftrightarrow \{\phi_1(t), \phi_2(t), \ldots, \phi_p(t)\}, \qquad (4.1)$$

where each $t \in \mathbf{R}^3$. A cylinder functional can be written as

$$f_p(\phi(t)) \leftrightarrow f_p(\phi_1(t), \phi_2(t), \ldots, \phi_p(t)), \qquad (4.2)$$

where the subscript p denotes the number of variables.

More specifically, when the quantum theory of fields is introduced in Chapter II [25] each t varies in a "cell" contained within $\mathbf{R}^3$. We consider $\phi_\gamma(t) = 0$, $1 \leq \gamma \leq p$, when t does not belong to any n-cell. We then define [25]

$$\delta f_p(\phi(t)) = 0.$$

If $t \in \gamma^{th}$ cell [25], we set

$$\frac{\delta}{\delta\phi(t)dt} f_p(\phi(t)) = \frac{1}{\Delta_\gamma} \frac{\partial}{\partial\phi_\gamma} f_p[\phi_1(t), \ldots, \phi_p(t)].$$

The Δ_γ is the "volume" of the γth cell. Schiff [42] requires the "volume of the cell" to tend to zero. This allows Schiff the capability of enjoying the presence of the Dirac delta functional.

Friedrichs and Shapiro [25] select the following quadratic functional,

$$f_2[\phi] = \iint b(x',x'')\phi(x')\phi(x'')dx'dx'',$$

and its generalized Laplacian becomes

$$Lf_2[\phi] = 2\int b(x,x)dx.$$

We observe that if we select a $\phi(t,t') \in S(\mathbf{R}^2)$, the space of $\mathbf{R}^2$ rapid descent test functions, and define h as

$$< h, \phi > \triangleq 2\int \phi(x,x)dx \qquad (4.3)$$

then h is a tempered distribution.

Proposition 4.4 The functional, h, defined in expression (4.3) is a tempered distribution.

Proof The linearity of h is obvious and we verify the continuity by checking the boundedness of h. We select any rapid descent test function, $\phi(t,t')$, and compute the following:

$$|< h,\phi(t,t') >| = 2\left|\int \phi(t,t)dt\right| \leq 2\left|\int \frac{\phi(t,t)(1+t^2)^4}{(1+t^2)^4}dt\right|$$

$$\leq 2\Pi \sup_{\substack{|\sigma|\leq 2 \\ t\in \mathbf{R}}} |(1+t^2)^4\phi^{(\sigma)}(t,t)| \leq 2\Pi \sup_{\substack{|\sigma|\leq 2 \\ (t_1,t_2)\in \mathbf{R}\times\mathbf{R}}} |(1+t_1^2)^2(1+t_2^2)^2\phi^{(\sigma)}(t_1,t_2)|$$

$$= 2\Pi \|\phi(t_1,t_2)\|_2. \quad \square$$

If we select a special $\phi \in \Gamma^{p,sB}$ where its Fock representation (2.2) is given as

$$\Phi \leftrightarrow \begin{pmatrix} a_0 \\ a_1 \\ a_2 \\ 0 \\ 0 \\ 0 \\ \cdot \\ \cdot \\ \cdot \end{pmatrix} \qquad (4.5)$$

and then apply two annihilation operators, it will yield

$$D_h^2 \Phi \leftrightarrow \begin{pmatrix} 2a_2[h,h] \\ 0 \\ 0 \\ \vdots \\ 0 \end{pmatrix}. \qquad (4.6)$$

Selecting $h = \delta(t-t')$ and implementing an integral operator on (4.6) results in

$$\Delta\Phi \leftrightarrow \begin{pmatrix} 2\int a_2[\delta(t-t'),\delta(t-t')]dt \\ 0 \\ \vdots \end{pmatrix}. \qquad (4.7)$$

Noting that $a_2[\delta(t-t'),\delta(t-t')] = \phi(t,t)$ is a member of $S(\mathbf{R}^2)$ [46], we have an analogue to expression (4.3). Our Laplacian is mathematically developed in [44] but here we have its Fock representation. If $\Phi \in \Gamma^{p,sB}$ where Φ has Fock representation,

$$\begin{pmatrix} a_0 \\ a_1 \\ a_2 \\ \vdots \\ a_q \\ \vdots \end{pmatrix}$$

then its generalized Laplacian becomes

$$\begin{pmatrix} 2\int a_2[\delta(t-t'),\delta(t-t')]dt \\ 2\cdot 3\int a_3[\delta(t-t'),\delta(t-t'),\cdot]dt \\ \vdots \\ (q+1)(q+2)\int a_{q+2}[\delta(t-t'),\delta(t-t'),\ldots]dt \\ \vdots \end{pmatrix}$$

We observe each a_q is a multilinear symmetric functional so for convenience we insert the pair of translated Dirac delta's in the first two arguments. We can now develop an infinite dimensional Schrödinger equation.

REFERENCES

1. V. Bergmann, Remarks on a Hilbert Space of Analytic Functions, Proceedings N.A.S., 48, 199-204, (1962).
2. V. Bergmann, On a Hilbert Space of Analytic Functionas and an Associated Integral Transform, Communications on Pure and Applied Mathematics, 19, 187-214. (1961).
3. V. Bergmann, A Family of Related Function Spaces Application to Distribution Theory Part II, Communications on Pure and Applied Mathematics, 20, 1'101, (1967).
4. V. M. Bogdan, Existence of Solutions to Differential Equations of Relativistic Mechanics Involving Lorentzian Time Delays, (to appear Journal of Mathematical Analysis and Applications).
5. N. N. Bogoliubov, D. V. Shirkov, "Introduction to the Theory of Quantized Fields", New York, John Wiley & Sons, (1976).
6. F. Cholewinski, Generalized Fock Spaces and Associated Operators, SIAM Journal of Mathematical Analysis, 15, No. 1, 177-202, (1984).
7. J. F. Colombeau, "New Generalized Functions, and Multiplication of Distributions", North Holland Mathematical Studies #90, New York, Elsevier, (1984).
8. F. Constantinescu, "Distributions and their Application in Physics", International Series in Natural Philosophy #100, New York, Pergamon Press, (1980).
9. P. A. M. Dirac, "Lectures in Quantum Field Theory", Belfer Graduate School Monograph Series, Yeshiva University, (1966).
10. P. Duchateau, New Proofs and Generalizations of Theorems of Existence and Uniqueness for the Goursat Problem, Applicable Analysis, 2, 61-78, (1972).
11. P. Duchateau, The Cauchy-Goursat problem, Mem. Amer. Math. Soc., No. 118, (1972).
12. P. Duchateau, A Holmgren Type Theorem for Pseudo Differential Operators in Gevrey Classes, J. Diff. Equ., 13, 319-328, (1973).
13. T. Dwyer, Partial Differential Equations in Fischer-Fock Spaces for the Hilbert-Schmidt Holomorphy Type, Bull. Amer. Math. Soc., 77, 725-730, (1971).
14. T. Dwyer, Holomorphic Fock Representations and Partial Differential Equations on Countably Hilbert Spaces, Bull. Amer. Math. Soc., 79, 1045-1050, (1973).

15. T. Dwyer, "Partial Differential Equations in Holomorphic Fock Spaces", Functional Analysis and Applications, L. Nachbin, ed., Lecture Notes in Math. No. 384, Springer-Verlag, Berlin - Heidelberg - New York, (1974).

16. T. Dwyer, "Holomorphic Representation of Tempered Distributions and Weighted Fock Spaces", Analyse Fonctionelle et Applications, L. Nachbin, ed., Actualite's Sci. et Industr. No. 1367, Hermann, Paris, 95-118, (1975).

17. T. Dwyer, "Differential Equations of Infinite Order in Vector-Valued Holomorphic Fock Spaces", Infinite-Dimensional Holomorphy and Applications, M. Matos, ed., North-Holland Mathematisc Studies, North-Holland Publ. Co., Amstedram, 167-200, (1977).

18. T. Dwyer, Equations Différentielles d'Ordre Infini dans des Espaces Localement Convexes, C. R. Acad. Sci. Paris, Ser. A, 280, 1439-1442, (1975).

19. T. Dwyer, Duallité des Espaces des Fonctions Entié en Dimension Infinie, Ann. Inst. Fourier (Grenoble), 26, 151-195, (1976).

20. T. Dwyer, Differential Operator of Infinite Order in Locally Convex Spaces, II, Rendiconti di matematica, 10, Serie VI, 149-179, (1977).

21. T. Dwyer, Differential Operators of Infinite Order in Locally Convex Spaces, II, Rendiconti di matematica, 10, no. 2, 278-293, (1977).

22. T. Dwyer, "Analytic Evolution Equations in Banach Spaces", Proc. 1977 Dublin Conference on Vector Space Measures, S. Cinnen, ed., Lecture Notes in Math., Springer-Verlag, Berlin - Heidelberg - New York, 48-61, (1978).

23. T. Dwyer, Infinite Dimensional Analytic Systems, Proc. 1977 Conference on Decision and Control, IEEE, 285-290, (1977).

24. A. Friedman, "Generalized Functions and Partial Differential Equations", New Jersey, Prentice Hall, Inc., (1963).

25. K. O. Friedrichs, H. N. Shapiro, "Integration of Functionals", New York University Notes, (1957).

26. A. Frolicher, W. Bucher, "Calculus in Vector Spaces without Norm", Lecture Notes in Mathematics 30, New York, Springer-Verlag, (1966).

27. I. M. Gelfand, and G. E. Shilov, "GeneralizeddFunctions", Volume II, Translated by Morris D. Friedman, Amiel Feinstem, Christian P. Peltzer, New York, Academic Press, Inc., (1968).

28. D. Kastler, "Introduction a L'Electrodynamique Quantique", Paris, Dunod, (1961).

29. G. Köthe, Dualität in der Funktionentheorie, Journal für die Reine and Angewandte Math., 191, 30-49, (1953).
30. G. Köthe, and O. Toeplitz, Lineare Räume mit Unendlichvielen Koordinaten und Ringe unendlicher Matrizen, Journal für die Reine and Angewandte Math., 171, 193-226, (1934).
31. G. Köthe, "Topological Vector Spaces I", New York, Springer-Verlag Inc., (1969).
32. P. Kristensen, I. Mejlbo, E. T. Poulsen, Tempered Distribution in Infinitely Many Dimensions I, Canonical Field Operators, Communications Mathematical Phyisics, I, 175-214, (1965).
33. P. Kristensen, I. Mejlbo, E. T. Poulsen, Tempered Distribution in Infinitely Many Dimensions II, Displacement Operators, Math. Scand., 14, 129-150, (1964).
34. P. Kristensen, Tempered Distributions in Infinitely Many Dimensions III, Linear Transformations of Field Operators, Commun. Math. Physics, 6, 29-48, (1967).
35. T. P. G. Liverman, "Generalized Functions and Direct Operational Methods", New Jersey, Prentice-Hall Inc., (1964).
36. G. Marinescu, "Espaces Vectorials Pseudo-topologiques et Theoriee des Distributions", Berlin, VEB Deutscher Verlag der Wissenschaften, (1963).
37. C. M. Roumieu, Sur Quelques Extensions de la Notion de Distributions, Ann. Scient. E. Norm. Sup., 77, 41-121, (1960).
38. J. Rzewuski, On a Triplet Including the Hilbert Space of Entire Functionals, Bull. de L'Academie Polonaise des Sciences, Ser. des Sciences Math., Astro. and Physics, Vol. 17, No. 7, 459-466, (1969).
39. J. Rzewuski, On a Hilbert Space of Functional Power Series, Bull. de L'Academie Polonaise des Sciences, Ser. des Sciences Math., Astr. et Physics, 18, No. 11, 677-685, (1970).
40. J. Rzewuski, On Entire Functionals in Quantum Field Theory, Reports on Mathematical Physics, 1, No. 1, 1-27, (1970).
41. J. Rzewuski, Some Estimates for Generating Functionals with An Application to Quantum Field Theory, Bull. de L'Academie Polonaise des Sciences, Ser. des Sciences Math., Astro. and Physics, 19, No. 3, 235-249, (1971).
42. L. Schiff, "Quantum Mechanics", New York, McGraw Hill, (1968).
43. J. Schmeelk, Application of Test Surfunctions, Applicable Analysis, 17, No. 3, 169-185, (1984).
44. J. Schmeelk, An Infinite Dimensional Laplacian Operator, Journal of Differential Equations, 36, No. 1, 74-88, (1980).

45. J. Schmeelk, Infinite Dimensional Parametric Distributions, <u>Applicable Analysis</u>, (to appear).
46. J. Schmeelk, Infinite Dimensional Fock Spaces and Associated Creation and Annihilation Operators, <u>Journal of Mathematical Analysis and Applications</u>, (to appear).
47. J. Schmeelk, Heisenberg's Uncertainty Principle Motivating Some Functional Analysis, Proceedings of the 1981 ASEE National Meeting, U.S.C., Los Angeles, California, June 21-25, 898-900, (1981).
48. J. Schmeelk, Heisenberg's Uncertainty Principle and Functional Analysis - Part II, Proceedings of the 1982 ASEE National Meeting, Texas A & M, College Station, Texas, June 20-23, 735-740, (1982).
49. J. Schmeelk, Test Surfunctions and Surdistributions, Ph. D. thesis, George Waschington University, February 1976.
50. L. Schwartz, "Theorie des Distributions", Paris, Hermann, (1966).
51. A. S. Wightman, K. O. Friedrich, Differential Equations of Mathematical Physics, <u>An Air Force Office of Scientific Research Scientific Report</u>, American University, October 1, (1966).
52. A. S. Wightman, R. F. Streater, "PCT, Spin and Statistics and all That", New York, W.A. Benjamin Inc., (1964).
53. A. H. Zemanian, "Distribution Theory and Transform Analysis", New York, McGraw-Hill Book Co., (1965).
54. A. H. Zemanian, "Realizability Theory for Continuous Linear Systems", New York, Academic Press Inc., (1972).

THE n-DIMENSIONAL STIELTJES TRANSFORMATION

Arpad Takači

Institute of Mathematics
University of Novi Sad
21000 Novi Sad, Dr I. Đuričića 4, Yugoslavia

ABSTRACT

In a previous paper ([8]), we analysed the relation between Silva's order of growth (introduced in [3]) with equivalence at infinity of distributions ([7]) and applied them to the distributional Stieltjes transformation (see [5]).

In this paper we shall define the n-dimensional versions of these notions, and, in particular, we shall prove some Abelian theorems for the n-dimensional Stieltjes transformation.

1. THE S-INTEGRAL

Throughout the paper n will denote the dimension of the Euclidean space $\mathbb{R}^n$, while $\mathbb{N}_0^n$ denotes its subspace of n-tuples of non-negative integers. Let

$$\mathbb{R}_1^n := \mathbb{R}_+^n := \{x = (x_1,\ldots,x_n) \in \mathbb{R}^n \mid x_i > 0\},$$

$$\mathbb{R}_2^n := \{x = (x_1,\ldots,x_n) \in \mathbb{R}^n \mid x_1 > 0, x_2 > 0, x_3 > 0,\ldots,x_n < 0\},$$

and in general $\mathbb{R}_k^n$, $k = 1,2,\ldots,2^n$ will denote the k-th orthant. We put $D_i = \partial/\partial x_i$, $i = 1,2,\ldots,n$, and $D^m = D_1^{m_1} \ldots D_n^{m_n}$ for $m = (m_1,\ldots,m_n)$; as usual, we use the same letter for the derivation in the classical and in the distributional sense, whenever no confusion in possible. Further more, we put $x^r := x_1^{r_1} \ldots x_n^{r_n}$ for $x = (x_1,\ldots,x_n) \in \mathbb{R}_k^n$ for some $k \in \{1,2,\ldots,2^n\}$ and $r = (r_1,\ldots,r_n) \in \mathbb{R}^n$.

We shall rewrite now the definition of the distributional limit at infinity from [3].

<u>Definition 1</u> If T is a distribution on $\mathbb{R}^n$, then the limit

$$\lim_{\substack{|x|\to\infty \\ x\in\mathbb{R}^n_+}} T_n = \lambda, \quad \lambda \in \mathbb{C},$$

means that there exist $m = (m_1,\ldots,m_n) \in \mathbb{N}^n_0$, $m_i \geq 1$, for each $i = 1,\ldots,n$, and a continuous function F on $\mathbb{R}^n$ such that

$$T = D^m F \text{ and}$$

$$\lim_{\substack{|x|\to\infty \\ x\in\mathbb{R}^n_+}} \frac{F(x)}{x^m} = \frac{\lambda}{m_1! \ldots m_n!}.$$

The notation "$|x| \to \infty$, $x \in \mathbb{R}^n_+$" (or, more generally, $|x| \to \infty$, $x \in \mathbb{R}^n_k$, $k \in \{1,2,\ldots,2^n\}$) means that for each $i \in \{1,\ldots,n\}$ $|x_i| \to \infty$ and x remains in the orthant $\mathbb{R}^n_+$ (respectively, in the orthant $\mathbb{R}^n_k$).

Of course, one can define in a similar manner the limit

$$\lim_{\substack{|x|\to\infty \\ x\in\mathbb{R}^n_k}} T \text{ for } k = 1,2,\ldots,2^n.$$

Clearly, this concept of limit is a variant of Lojasiewicz's limit (see [2]) for "infinity". As can be expected, the existence of the usual limit, in the case when T is in fact a locally integrable function in the orthant $\mathbb{R}^n_+$ (resp. $\mathbb{R}^n_k$, $k = 1,2,\ldots,2^n$), implies the existence of this "distributional" limit, while the opposite should not be true.

The importance of this limit can be seen in the following definition of the "S-integral" of a distribution ("S" stands for Silva).

<u>Definition 2</u> ([3]) Let T be a distribution on $\mathbb{R}^n$ and let U be another distribution with the property

$$D^m U = T, \quad m = (1,1,\ldots,1)$$

(it is well known that such a U always exists). Then, T is called an "S-integrable" on $\mathbb{R}^n$, if the limits

$$\lim_{\substack{|x|\to\infty \\ x\in\mathbb{R}^n_k}} U$$

exist for each $k = 1,2,\ldots,2^n$, and by definition

$$S - \int_{\mathbb{R}^n} T := \sum_{\varepsilon_{k_1},\ldots,\varepsilon_{k_n} \in \{-1,1\}} (-1)^{\varepsilon_{k_1}+\ldots+\varepsilon_{k_n}} \lim_{\substack{x_1 \to \varepsilon_{k_1} \cdot \infty, \ldots, x_n \to \varepsilon_{k_n} \cdot \infty \\ |x| \to \infty}} U$$

where the summation is being done over all the n-typles of numbers -1 and 1.

As can be checked easily, this "S-integral" does not depend on the choice of U, it is a linear functional on the set of "S-integrable distributions" and it is equal to the usual integral of an $L^1(\mathbb{R}^n)$ function. However, it makes sense for some elements from $\mathcal{D}'$, in particular for all the distributions with a compact support.

2. STIELTJES TRANSFORMATION OF DISTRIBUTIONS

The S-integral from Section 2 allows us to introduce the Stieltjes transformation on a subset of $\mathcal{D}'(\mathbb{R}^n)$. First, we shall recal the definition of the classical Stieltjes transform of a locally integrable function f on $\mathbb{R}^n$, which is supposed to be zero on $\mathbb{R}^n - \overline{\mathbb{R}}_+^n$:

$$\sigma_r(f)(s) := \int_{\overline{\mathbb{R}}_+^n} \frac{f(x)dx}{(x+s)^{r+\overline{1}}} = \int_0^\infty \ldots \int_0^\infty \frac{f(x)dx_1 \ldots dx_n}{(x_1+s_1)^{r_1+1} \ldots (x_n+s_n)^{r_n+1}} \quad (1)$$

for $r = (r_1,\ldots,r_n) \in \mathbb{R}^n$, $r_i > -1$, $i = 1,\ldots,n$. The transform in (1) is a function of the n-dimensional complex variable $s = (s_1,\ldots,s_n)$, provided that $s_i \notin (-\infty,0]$ for each $i = 1,\ldots,n$, and that the integral exists. (We used the convenient notation $\overline{1} = (1,\ldots,1)$.) However from now on we suppose that $s \in \mathbb{R}_+^n$.

For later considerations we note that (1) exists, if the functions

$$(1 + |x|)^{-\alpha_i} f(x)$$

are bounded for each $i = 1,\ldots,n$, where $\alpha_i < r_i$, $i = 1,\ldots,n$.

We have come to

Definition 3 If T is a distribution with a support in $\overline{\mathbb{R}}_+^n$, then its Stieltjes transform is defined by the S-integral

$$\sigma_r(T)(s) := S - \int_{\mathbb{R}^n} \frac{T(x)dx}{(x+s)^{r+\overline{1}}}$$

provided it exists, and $r = (r_1,\ldots,r_n)$, $r_i > -1$, $i = 1,\ldots,n$.

The Stieltjes transform of distributions was defined and analysed in several papers; one can consult the expository paper [5]. Definition 3 is more general (even in the case n = 1) from the one used in [1]. This can be checked by using the formula of partial integration for the S-integral of distributions (see [4]). We note that we can use the same symbol (σ_r) for the classical ((1)), and distributional Stieltjes transformation ((2)), since the two coincide if T is a continuous function on $\mathbb{R}^n$, which is zero outside $\bar{\mathbb{R}}_+^n$.

In order to find a sufficient condition on T for the existence of $\sigma_r(T)(s)$, we use the notions of the boundedness and Landau symbols of a distribution, both introduced by S. e Silva in [3].

Definition 4 A distribution T is <u>bounded</u> on $\mathbb{R}^n$, if there exist $m \in \mathbb{N}_0^n$ such that

$$T = D^m F$$

and for every regular matrix A of order n the function $x^{-m} F(Ax)$ is bounded on $\mathbb{R}^n$.

Definition 5 Let ϕ be an infinitely differentiable function on $\mathbb{R}_+^n \cap V$, where V is some neighbourhood of infinity. Then, distribution T is $O(\phi)$ <u>(big "oh" of ϕ)</u> in $\mathbb{R}_+^n$, if there exists a bounded distribution T_0 on $\mathbb{R}^n$ and a ball $B(0,\varepsilon)$, $\varepsilon > 0$, such that

$$T = \phi \cdot T_0 \text{ on } (\mathbb{R}_+^n \setminus L(0,\varepsilon)) \cap V.$$

(One defines the small "oh" of ϕ by supposing T_0 to "tend to zero" in the sense of Definition 1.1.)

Observe that Definition 5 reduces to Definition 4 if $\phi(x) \equiv 1$. From now on, ϕ will be a power function, i.e. of the form x^α. We have

Lemma 1 If $T = O(x^\alpha)$ in $\mathbb{R}_+^n$, then

$$D^\beta T = O(x^{\alpha-\beta}) \text{ in } \mathbb{R}_+^n, \text{ where } \alpha = (\alpha_1,\ldots,\alpha_n),$$

$\beta = (\beta_1,\ldots,\beta_n)$ and $\alpha-\beta = (\alpha_1-\beta_1,\ldots,\alpha_n-\beta_n)$.

Proof We start with the case $\alpha = (0,\ldots,0)$, $\beta = (1,0,\ldots,0)$. Taking m and F from Definition 4, we get

$$D_1 T = x_1^{-1}(x_1 D^{m'} F) = x_1^{-1}(D^{m'}(x \cdot F) - D^{m'} F),$$

where $m' = (m_1-1, m_2, \ldots, m_n)$. Hence, $D_1 T = O(x_1^{-1})$. If $\phi(x) = x^\alpha$ for $x \in \mathbb{R}_+^n \setminus B(0,\varepsilon)$, $(\varepsilon > 0)$, then taking T_0 and m from Definition 4, we get

$$D_1 T = \alpha_1 x^{\alpha'} T_0 + x_1 D_1 T_0,$$

which, in view of the previous case $(\alpha = (0,\ldots,0))$, implies the statement for $\beta = (1,0,\ldots,0)$. For β arbitrary the proof follows by induction in a similar way.

3. ASYMPTOTIC BOUND OF STIELTJES TRANSFORMS

Let $\alpha = (\alpha_1,\ldots,\alpha_n)$ and $r = (r_1,\ldots,r_n)$ be fixed.

Theorem 1 If the distribution T with a support in $\overline{\mathbb{R}}_+^n$ satisfies the condition

$$T = O(x^\alpha) \text{ as since } |x| \to \infty, \; x \in \mathbb{R}_+^n \text{ for } \alpha < r$$

(i.e. $\alpha_i < r_i$ for each $i = 1,\ldots,n$), then it has the Stieltjes transform (Definition 3).

Proof In view of Definition 3.3, we can write

$$T = B + x^\alpha T_0$$

where the supports of distributions B and T_0 are respectively contained in the sets

$$S_1 = \overline{L_n(0,R) \cap \mathbb{R}_+^n} \text{ and } S_2 = \overline{\mathbb{R}_+^n \setminus (L_n(0,R) \cap \mathbb{R}_+^n)},$$

where $L_n(0,R)$ denotes the central ball in $\mathbb{R}^n$ with a radius $R > 0$.

We shall prove first that $\sigma_r(B)(s)$ exists. Let $B_1 = (1/(x+s)^{r+1})B$; then B_1 also has a compact support, and hence its antiderivative U_1 is a constant in x (in general different) in a neighbourhood of infinity, staying in an orthant. Hence, $\sigma_r(B)(s)$ exists.

Let us observe now $\sigma_r(x^\alpha T_0)(s)$, where T_0 is a bounded distribution. By definition

$$\sigma_r(x^\alpha T_0)(s) = S - \int_{\mathbb{R}^n} \frac{(x^\alpha T_0)\,dx}{(x+s)^{r+\overline{1}}} = S - \int_{\mathbb{R}^n} \frac{(x^\alpha D^m F(x))\,dx}{(x+s)^{r+\overline{1}}},$$

where $F(x) = 0$ for $x \notin \overline{\mathbb{R}}_+^n$ and $x^{-m} F(Ax)$ is bounded for each regular matrix A. The last integral can be written as

$$(-1)^{|m|} \int_{\mathbb{R}^n} F(x) \cdot D^m \left(\frac{x^\alpha}{(x+s)^{r+\bar{1}}} \right) dx, \quad |m| = m_1 + \ldots + m_n$$

Calculating the derivative $D^m \cdot (x^\alpha/(x+s)^{r+\bar{1}})$ and using the conditions $\alpha_i < r_i$, $i = 1, \ldots, n$, we prove the existence of the integral in (3), and hence, that of $\sigma_r(T)(s)$. □

It is well known that $(n = 1)$

$$\sigma_r(x^\alpha)(s) = B(r-\alpha, \alpha+1) s^{\alpha-r}, \quad s > 0, \tag{4}$$

provided that $-1 < \alpha < r$. In view of (4), we have

$$\sigma_r(x_+^\alpha)(s) = \prod_{i=1}^{n} B(r_i - \alpha_i, \alpha_i + 1) s^{\alpha-r}, \tag{5}$$

if $\alpha = (\alpha_1, \ldots, \alpha_n)$, $r = (r_1, \ldots, r_n)$ and $-1 < \alpha_i < r_i$ for each $i = 1, \ldots, n$, where x_+^α denotes the regular distribution

$$\langle x_+^\alpha, \phi(x) \rangle := \int_{\mathbb{R}^n} x^\alpha \phi(x) dx, \quad \phi \in \mathcal{D}(\mathbb{R}^n). \tag{6}$$

(If $\alpha_i \leq -1$ for some $i \in \{1, \ldots, n\}$, then an appropriate regularization is needed.)

Our goal is to prove

Theorem 2 If T satisfies the conditions of Theorem 1 and if (additionally) $-1 < \alpha_i < r_i$ for each $i = 1, 2, \ldots, n$, then

$$\sigma_r(T)(s) = O(s^{\alpha-r}) \text{ as } s \to \infty.$$

The proof relies on the following two lemmas.

Lemma 2 If T has a compact support contained in $\bar{\mathbb{R}}_+^n$, then

$$\sigma_r(T)(s) = O(s^{-r-\bar{1}}) \text{ as } s \to \infty.$$

Lemma 3 Let f be a continuous function with a support in $\bar{\mathbb{R}}_+^n$ satisfying the condition

$$f(x) = O(x^\alpha), \text{ as } x \to \infty \text{ in the ordinary sense.}$$

Then

$$\sigma_r(f)(s) = O(s^{\alpha-r}) \tag{7}$$

if $-1 < \alpha_i < r_i$, $i = 1,\ldots,n$ and $\alpha = (\alpha_1,\ldots,\alpha_n)$, $r = (r_1,\ldots,r_n)$.

The proof of Lemma 2 is similar to that of Lemma 3.1 from [1] and it is omitted. In fact, one can prove more; for simplicity, we take n = 1. Namely, a distribution B with a compact support has a quasiasymptotic behaviour of the order $-m \in (-\mathbb{N})$, and in [6] it was proved that its Stieltjes transform then necessarily has the asymptotic behaviour

$$\sigma_r(B)(s) \sim C_m \cdot s^{-m-1} \text{ as } s \to +\infty \text{ for some } C_m \neq 0.$$

Proof of Lemma 3 We shall consider the case n = 2. The case n = 1 was considered in [8], while for n > 2 the proof is essentially the same.

So, let $|f(x)| \leq C \cdot x^\alpha$ if $x_1 \geq 1$, $x_2 \geq 1$, $x = (x_1, x_2)$ for some $C > 0$ and $-1 < \alpha_i < r_i$, $i = 1,2$ and $\alpha = (\alpha_1, \alpha_2)$, $r = (r_1, r_2)$. We have

$$\sigma_r(f)(s) = \int_0^\infty \int_0^\infty \frac{f(x_1,x_2)dx_1 dx_2}{(x_1+s_1)^{r_1+1}(x_2+s_2)^{r_2+1}}$$

$$= \int_0^1 \int_0^1 + \iint_{\substack{0 \leq x_1 \leq 1 \\ 1 \leq x_2 \leq \infty}} + \iint_{\substack{0 \leq x_2 \leq 1 \\ 1 \leq x_1 \leq \infty}} + \int_1^\infty \int_1^\infty = \sum_{\nu=1}^{4} I_\nu(s).$$

The estimate $|I_1(s)| \leq C_1 s^{-r-\bar{1}}$ follows from Lemma 2. Further more, we have

$$|I_2(s)| \leq C_1 \int_0^1 \frac{x_1^{\alpha_1} dx_1}{(x_1+s_1)^{r_1+1}} s_2^{\alpha_2-r_2}, \text{ which easily gives}$$

$$|I_2(s)| \leq C_2 s_1^{-r_1-1} s_2^{\alpha_2-r_2}; \text{ analogously } |I_3(s)| \leq C_3 s_1^{\alpha_1-r_1} s_2^{-1-r_2}.$$

By using the assumption and formula (5) we obtain

$$|I_4(s)| \leq C \int_1^\infty \int_1^\infty \frac{x_1^{\alpha_1} x_2^{\alpha_2} dx_1 dx_2}{(x_1+s_1)^{r_1+1}(x_2+s_2)^{r_2+1}} \leq C_4 \cdot s^{\alpha-r}.$$

Summing up the estimates for $I_\nu(s)$, $\nu = 1,\ldots,4$, we finish the proof of Lemma 2. □

Proof of Theorem 2 Let $T = 0(x^\alpha)$ in $\mathbb{R}_+^n$ since $|x| \to \infty$. Then

$$T = x^\alpha D^m F$$

on $\mathbb{R}_+^n \setminus B(0,\varepsilon)$ for some $m \in \mathbb{N}_0^n$, a continuous function F on $\mathbb{R}^n$ with the property from Definition 4 and some $\varepsilon > 0$. We put

$$T_1 = T - x^\alpha D^m G,$$

where $G(x) = F(x)$ for $x \in \mathbb{R}_+^n$, $|x| > \varepsilon$, and zero otherwise. Now T_1 has a compact support and by applying Lemma 2 we obtain

$$|\sigma_r(T_1)(s)| \leq C\, s^{-(r+\bar{1})} \quad \text{as } |s| \to \infty, \ s \in \mathbb{R}_+^n \text{ for some } C > 0. \tag{8}$$

Further more, partial integration gives

$$\sigma_r(x^\alpha D^m G)(s) = S - \int_{\mathbb{R}^n} \frac{x^\alpha D^m G(x)}{(x+s)^{r+\bar{1}}} = S - \int_{\mathbb{R}^n} \frac{\sum_{0 \leq |\nu| \leq |m|} C_\nu x^{\alpha-\nu} s^\nu}{(x+s)^{r+m+\bar{1}}} G(x)\, dx$$

where the summation is taken over by the n-typles $\nu = (\nu_1, \ldots, \nu_n)$, such that $0 \leq \nu_i \leq m_i$, $i = 1, \ldots, n$, and C_ν are some constants. By using (5) we obtain

$$|\sigma_r(x^\alpha D^m G(x))(s)| \leq C \cdot s^{\alpha-r} \tag{9}$$

for $s \in \mathbb{R}_+^n$ and $|\ | \geq \varepsilon' > 0$. Hence, (8) and (9) imply the statement. □

In a similar way we can prove

Theorem 3 If T is a distribution with a support in $\mathbb{R}_+^n$ and $T = o(x^\alpha)$ since $|x| \to \infty$, $x \in \mathbb{R}_+^n$, then

$$\sigma_r(T)(s) = o(s^{\alpha-r}) \quad \text{as } s \to \infty$$

for $\alpha = (\alpha_1, \ldots, \alpha_n)$, $r = (r_1, \ldots, r_n)$ and $-1 < \alpha_i < r_i$, $i = 1, \ldots, n$.

In theorems 2 and 3 conditions $-1 < \alpha_i$, $i = 1, \ldots, n$, are necessary, since the Stieltjes transformation of T, $\sigma_r(T)(s)$ does not exist otherwise. The quasiasymptotic behaviour of distributions, however, does not demand such a constraint. As an easily be guessed, the reason for this lies in the completely different nature of the quasiasymptotic, namely it is a global property of a distribution and since $\alpha \leq -1$ is neither comparable with the equivalence at infinity nor with the big or small "oh" from Definition 5.

REFERENCES

1. J. Lavoine, O. P. Misra, Théoremes abéliens pour la transformation de Stieltjes des distributions, C.R. Acad. Sci. Paris, 279, Ser. A, 99-102, (1974)
2. L. Lojasiewicz, Sur la valeur et la limit d'une distribution dans une point, St. Math., 16, 1, 1-36, (1957).
3. J. S. e Silva, Integrals and orders of growth of distributions, Proc. Int. Summer Institute, Lisboa, (1964).

4. J. S. e Silva, Les series de multiples des physiciens et la theorie des ultradistributions, Math. Ann., 174, 109-142, (1967).

5. B. Stanković, Abelian and Tauberian theorem for the distributional Stieltjes transformation, Uspehi mat. Nauk, T. 40, V. 4 (244), 91-103, (1985), (in Russian).

6. A. Takači, A note on the distributional Stieltjes transformation, Math. Proc. Cambridge, Phil. Soc., 94, 523-527, (1983).

7. A. Takači, On the equivalence at infinity of distributions, Zb. PMF N. Sad, 15, No. 1, 175-187, (1985).

8. A. Takači, On Silva's order of growth of distributions, Zb. PMF N. Sad, (to appear).

COLOMBEAU'S GENERALIZED FUNCTIONS AND NON-STANDARD ANALYSIS

T. D. Todorov

Institute of Nuclear Research and Nuclear Energy
Bulgarian Academy of Sciences
Sofia 1784, Bulgaria

ABSTRACT

Using some methods of Non-Standard Analysis we modify one of Colombeau's classes of generalized functions. As a result we define a class $\hat{E}$ of so-called *metafunctions* which possesses all the good properties of Colombeau's generalized functions, i.e. (i) $\hat{E}$ is an associative and commutative algebra over the system of so-called *complex meta-numbers* $\hat{\mathbb{C}}$; (ii) Every meta-function has partial derivatives of any odrer (which are meta-functions again); (iii) Every meta-function is integrable on any compact set of $\mathbb{R}^n$ and the integral is a number from $\hat{\mathbb{C}}$; (iv) $\hat{E}$ contains all the tempered distributions S', i.e. $S' \subset \hat{E}$ isomorphically with respect to all the linear operations (including the differentiation). Thus, within the class $\hat{E}$ the problem of multiplication of the tempered distributions is satisfactorily solved (every two distributions in S' have a well-defined product in $\hat{E}$). The crucial point is that $\hat{\mathbb{C}}$ is a field in contrast to the system of Colombeau's *generalized numbers* $\bar{\mathbb{C}}$ which is a ring only ($\bar{\mathbb{C}}$ is the counterpart of $\hat{\mathbb{C}}$ in Colombeau's theory). In this way we simplify and improve slightly the properties of the integral and the notion of "values of the meta-functions", as well as the properties of the whole class $\hat{E}$ itself if compared with the original Colombeau theory. And, what is maybe more important, we clarify the connection between Non-Standard Analysis and Colombeau's theory of *new generalized functions* in the framework of which the problem of the multiplication of distributions was recently solved.

INTRODUCTION

Several years ago J. F. Colombeau proposed a new class of generalized functions G [2 - 4], in which a solution was found of the problem of multiplication of Schwartz distributions [10]. G turns out to be an associative and commutative algebra with partial differentiation, integration, etc. and the space of distributions D' is included in G in a canonical way, i.e. $D' \subset G$, with respect to all *good operations* in D': the addition, multiplication by a complex number and partial differentiation. So, every two distributions can be correctly multiplied within G. In particular, we have the following pleasant result:

$$(\delta(x) \cdot x^n) x^{-n} = \delta(x) \cdot (x^n \cdot x^{-n})$$

which, as we know, does not hold in D' [11]. The new theory of generalized functions was immediately applied to Quantum Field Theory [3] and the Theory of Non-Linear Differential Equations [5] which opens the door for several other investigations.

A typical feature of the Colombeau theory is the usage of larger systems of generalized numbers $\bar{\mathbb{R}}$ and $\bar{\mathbb{C}}$ which are enlargements of $\mathbb{R}$ and $\mathbb{C}$ respectively, i.e.

$$\mathbb{R} \subset \bar{\mathbb{R}}, \quad \mathbb{C} \subset \bar{\mathbb{C}}, \quad \bar{\mathbb{C}} = \bar{\mathbb{R}} + i\bar{\mathbb{R}}$$

For example, the integral:

$$a = \int_K f(x) dx$$

(if it exists) is defined in $\bar{\mathbb{C}}$, i.e. $a \in \bar{\mathbb{C}}$, for $f \in G$ and $K \subseteq \mathbb{R}^n$. In particular, for $f = T \cdot \tau$, when $T \in D'$ and $\tau \in D$, the number a coincides with the usual complex number $\langle T, \tau \rangle$, i.e.

$$\int_{\mathbb{R}^n} T(x) \tau(x) dx = \langle T, \tau \rangle, \quad \tau \in D,$$

so, in Colombeau's theory, the distributions are identified with their kernels in G (see also [13]). Notice that the above equality has been used by physicists as a heuristic notation even in the framework of the Schwartz theory of distributions.

Finally, every generalized function $f \in G$ has values in $\bar{\mathbb{C}}$ for any real point $x \in \mathbb{R}^n$. In particular,

$$\delta(x) = \begin{cases} 0, & x \neq 0, \\ \ell, & x = 0, x \in \mathbb{R}^n, \end{cases}$$

where ℓ is a real generalized number, i.e. $\ell \in \bar{\mathbb{R}}$ (ℓ is uniquely determined as an element of $\bar{\mathbb{R}} \setminus \mathbb{R}$).

We feel that ℓ must, in a sense, be an infinite (infinitely large) number. Recall that a generalized number a is called *infinitely small* or *infinitesimal* if $|a| < r$ for all positive real numbers r; a is called *finite* if $|a| < r$ for some $r \in \mathbb{R}_+$ and a is called *infinitely large* or *infinite* if $|a| >$ for all $r \in \mathbb{R}_+$.

Unfortunately, we cannot apply directly the above definition to the systems $\bar{\mathbb{R}}$ and $\bar{\mathbb{C}}$ used in Colombeau's theory and there are two reasons for this: (i) $\bar{\mathbb{R}}$ and $\bar{\mathbb{C}}$ are rings, not fields [3, pp. 147]; $\bar{\mathbb{R}}$ is partially but not totally ordered. So that the formula:

$$|a| = \max\{-a, a\}$$

for the *magnitude of* a has no sense for any $a \in \bar{\mathbb{R}}$ and there exist generalized numbers from $\bar{\mathbb{R}}$ and $\bar{\mathbb{C}}$ which are neither infinitesimal nor finite, nor infinite.

The Purpose   Preserving all the good properties of the Colombeau class of generalized functions, we shall slightly modify the Colombeau theory such that the new class of generalized functions and the new systems of generalized numbers - denoted by $\hat{E}, \hat{\mathbb{R}}$ or $\hat{\mathbb{C}}$ respectively - will satisfy: (i) the integral and the values of the functions from $\hat{E}$ will be in $\hat{\mathbb{C}}$; (ii) $\hat{\mathbb{R}}$ and $\hat{\mathbb{C}}$ will be fields and $\hat{\mathbb{R}}$ will be totally-ordered. The inclusions $\mathbb{R} \subset \hat{\mathbb{R}}$ and $\mathbb{C} \subset \hat{\mathbb{C}}$ will be in the sense of *totally ordered fields* and *fields* respectively; (iii) $\hat{\mathbb{R}}$ and $\hat{\mathbb{C}}$ are non-Archimedean, i.e. contain non-zero infinitesimals and infinite numbers.

It is natural to use the methods of Non-Standard Analysis since, as we know, Non-Standard Analysis is a modern version of the *theory of infinitesimals* [1], [6 - 9], [12].

Notations   By

$$\mathcal{D} \equiv \mathcal{D}(\mathbb{R}^n), \quad S \equiv S(\mathbb{R}^n), \quad E \equiv C^\infty(\mathbb{R}^n), \quad C^0 \equiv C^0(\mathbb{R}^n)$$

$$L^1_{loc} \equiv L^1_{loc}(\mathbb{R}^n), \quad S' \equiv S'(\mathbb{R}^n), \quad \mathcal{D}' \equiv \mathcal{D}'(\mathbb{R}^n), \quad G \equiv G(\mathbb{R}^n)$$

will be denoted the classes of: test-functions with compact supports, the class of all C^∞-functions on $\mathbb{R}^n$ which (together with all their derivatives) decrease rapidly at infinity, C^∞-functions, continuous functions, locally-integrable functions, tempered distributions, Schwartz distributions and Colombeau's class of generalized functions defined on $\mathbb{R}^n$, respectively. Then, if

$$x = (x_1, x_2, \ldots, x_n) \in \mathbb{R}^n \text{ and } \alpha = (\alpha_1, \alpha_2, \ldots, \alpha_n),$$

$\alpha_i \in \mathbb{N}_0 = \mathbb{N} \cup \{0\}$

$$\|x\| = (x_1^2 + x_2^2 + \ldots + x_n^2)^{1/2},$$

$$|\alpha| = \alpha_1 + \alpha_2 + \ldots + \alpha_n,$$

and

$$x^\alpha \equiv x_1^{\alpha_1} \cdot x_2^{\alpha_2} \ldots x_n^{\alpha_n}.$$

All the classical integrals will be in the Lebesgue (or Riemann) sense.

1. THE CLASS OF NON-STANDARD FUNCTIONS *E

The set

$$I = S \times \mathbb{R}_+ \qquad (1.1)$$

will be called an *index set* and its elements will be denoted by $\phi \otimes \varepsilon$, i.e. "$\phi \otimes \varepsilon \in I$" means "$\phi \in S$ and $\varepsilon \in \mathbb{R}_+$".

Definition 1.2 (Fixing of the U) Let U be any (arbitrarily chosen and fixed) free ultrafilter in I having the property: there exist a double-sequence of positive real numbers:

$$\{a_{q,k} : q \in \mathbb{N}, k \in \mathbb{N}_0\}$$

and a double-sequence of subsets of S:

$$\{A_{q,k} : q \in \mathbb{N}, k \in \mathbb{N}_0\}$$

such that (for any $q, p \in \mathbb{N}$, $k, \ell \in \mathbb{N}_0$ and $\alpha, \beta \in \mathbb{N}_0^n$):

(i) $A_{q,k} \times (0, 1/q) \in U$

(ii) $A_{q,k} \cap A_{p,\ell} = A_{\max(q,p), \max(k,\ell)}$;

(iii) $\phi \in A_{1,k}$; implies $\int_{\mathbb{R}^n} \phi(x) dx = 1$;

(iv) $\phi \in A_{q,k}$ implies $\int_{\mathbb{R}^n} x^\alpha \phi(x) dx = 0$, $1 \leq |\alpha| \leq q$,

and

$$\int_{\mathbb{R}^n} |x^\alpha \partial^\beta \phi(x)| dx \leq a_{q,k}, \quad |\alpha| \leq q+1, |\beta| \leq k.$$

Notice that the sequence $\{A_{q,k}\}$ is a slight modification of the sequences $\{A_q\}$ of sets of test-functions used in the original works of Colombeau [3 - 5].

The triple $(\mathbb{C}, I, \mathcal{U})$, where $\mathbb{C}$ is the system of complex numbers (considered as a set of *individuals* [12]), determines a non-standard model of analysis according to the general superpower and superstructure construction [12]. This model contains (as internal sets) the non-standard versions: $*\mathbb{N}$, $*\mathbb{R}$, $*\mathbb{C}$, $*E$, $*C^0$, $*L_{loc}$, etc. of $\mathbb{N}$, $\mathbb{R}$, $\mathbb{C}$, $E \equiv C^\infty(\mathbb{R}^n)$, C^0, L_{loc}, etc. respectively, together with all the relations, operations, definitions and theorems obtained with the help of the Transfer Principles [12]. In particular, $*\mathbb{R}$ and $*\mathbb{C}$ turn out to be non-Archimedean fields (containing non-zero infinitesimals and infinite numbers), $*\mathbb{R}$ is a totally ordered field and we have:

$$\mathbb{R} \subset *\mathbb{R}, \quad \mathbb{C} \subset *\mathbb{C}, \quad *\mathbb{C} = *\mathbb{R} + i*\mathbb{R}. \tag{1.3}$$

We shall call the elements of $*\mathbb{R}$ and $*\mathbb{C}$ hyperreal and hypercomplex numbers respectively. $*E$ is a class of non-standard functions of the type:

$$f : *\mathbb{R}^n \to *\mathbb{C} \tag{1.4}$$

and $*E$ turns out to be an associative and commutative differential algebra over $*\mathbb{C}$ supplied, also, with integration, convolution, etc. $*E$ contains a copy $^\sigma E$ of E, i.e.

$$E \cong {}^\sigma E \subset *E \tag{1.5}$$

where $^\sigma E$ is the class of *standard* functions [12]. This isomorphism preserves all the operation in E and $*E$ (in particular, $^\sigma E$ is a *differential subalgebra* of $*E$; see definition (1.12), (iii)).

For readers who are not familiar with non-standard methods, we shall briefly present the definitions of $*\mathbb{R}$, $*\mathbb{C}$ and $*E$. For details we refer the reader to any book on Non-Standard Analysis: [1], [6 - 9], [12], etc.

<u>Definitions 1.6</u> (Ultrapowers)   Let

$$\mathbb{C}^I = \{A \mid A : I \to \mathbb{C}\} \tag{1.7}$$

and

$$E^I = \{F \mid F : I \to E\} \tag{1.8}$$

be the ultrapowers of $\mathbb{C}$ and E supplied with the corresponding pointwise operations taken from $\mathbb{C}$ and E respectively.

Definition 1.9 (Equivalence Relation)

(i) Let $A, B \in \mathbb{C}$. Then $A \sim B$, if

$$\{\phi \otimes \varepsilon \mid A_{\phi,\varepsilon} = B_{\phi,\varepsilon}\} \in \mathcal{U}; \tag{1.10}$$

(ii) Let $F, G \in E^I$. Then $F \sim G$, if

$$\{\phi \otimes \varepsilon \mid F_{\phi,\varepsilon}(x) = G_{\phi,\varepsilon}(x), x \in \mathbb{R}^n\} \in \mathcal{U}. \tag{1.11}$$

Definition 1.12 (*$\mathbb{R}$, *$\mathbb{C}$, *E)

(i) We define:

$$*\mathbb{C} = \mathbb{C}^I/\sim; \quad *E = E^I/\sim; \tag{1.13}$$

(ii) *$\mathbb{R}$ is the set of all $a \in *\mathbb{C}$ for which

$$\{\phi \otimes \varepsilon \mid A_{\phi,\varepsilon} \in \mathbb{R}\} \in \mathcal{U} \tag{1.14}$$

for some $A \in a$;

(iii) Inclusions (1.3) and (1.5) will be defined by the constant elements of $\mathbb{C}^I$ and E^I respectively, e.g. $^\sigma E$ is the set of all $*f \in *E$ such that $F_{\phi,\varepsilon}(x) \equiv f(x)$ for some $F \in *f$ and some $f \in E$. If $\Phi \subseteq E$, then $^\sigma \Phi = \{*f \mid f \in \Phi\}$.

All operations in *$\mathbb{R}$, *$\mathbb{C}$ and *E are introduced by the corresponding representatives (as in any factor-space). We shall give the definitions of the order relation in *$\mathbb{R}$ and the partial differentiation in *E only as examples.

Definition 1.15 (Order Relation)   Let $a, b \in *\mathbb{R}$. Then $a < b$ if

$$\{\phi \otimes \varepsilon \mid A_{\phi,\varepsilon} < B_{\phi,\varepsilon}\} \in \mathcal{U} \tag{1.16}$$

for some $A \in a$ and $B \in b$.

Definition 1.17 (differentiation)   Let $f \in *E$ and let

$$\partial \equiv \partial_x^\alpha \equiv \frac{\partial^{|\alpha|}}{\partial x_1^{\alpha_1} \partial x_2^{\alpha_2} \cdots \partial x_n^{\alpha_n}} \tag{1.18}$$

be some partial derivative (see the notations in the Introduction). Then $\partial f \in *E$ will be defined by $\partial F \in \partial f$ for any $F \in f$ (where ∂F is defined pointwisely, i.e.

$$(\partial F)_{\phi,\varepsilon}(x) = \partial_x^\alpha F_{\phi,\varepsilon}(x), \quad x \in \mathbb{R}^n, \ \phi \otimes \varepsilon \in I.) \tag{1.19}$$

Example 1.20 (Infinitesimals)   The hyperreal number $s \in {}^*\mathbb{R}$ defined by $A \in s$ where

$$A_{\phi,\varepsilon} = \varepsilon, \quad \phi \circledast \varepsilon \in I, \tag{1.21}$$

is an example of a positive (non-zero) infinitesimal, i.e. $0 < s < r$ for all $r \in \mathbb{R}_+$. The hyperreal numbers $s^2, s^3, \ldots, \exp(-s^{-1}), (\ln s)^{-1}$, etc. are, of course, also infinitesimals. The hyperreals $s^{-1}, s^{-2}, \ldots, \exp(s-1), \ln s$, are infinitely large numbers. Finally, it can be proved [6] that every finite number $a \in {}^*\mathbb{C}$ (or $a \in {}^*\mathbb{R}$) can be uniquely presented as $a = c+h$ for some $c \in \mathbb{C}$ (or $x \in \mathbb{R}$) and some infinitesimal h. The complex (real) number $c = {}^0 a \equiv \text{sta}$ is called the *standard part of* a.

Remark 1.22 (Uniqueness of $*E$)   It is known that every standard structure, in particular the class $E \equiv C^\infty(\mathbb{R}^n)$, has many (non-isomorphic in general) non-standard models, denoted usually as the same by $*E$. In our construction, although the index set I (1.1) is *fixed*, we still have many $*E$. First, we have different sequences $\{A_{q,k}\}$, $\{a_{q,k}\}$ having the properties listed in (1.2). And second, even these sequences had been somehow chosen and fixed, still we would have many ultrafilters U having the property (1.2)-(i). So we have, in fact, different non-isomorphic, in general, classes of non-standard functions $*E$. But, it is not worth worrying too much about the uniqueness of $*E$. All these models $*E$ of E are, in a certain sense (which is specific, for Non-Standard Analysis), similar to each other being non-standard models of the same standard class E (in particular, E and any of these $*E$ are indistinguishable under the first-order formal language). This is the reason why we consider all $*E$ as more or less equivalent. For more details concerning this question, we refer the reader to any book on Non-Standard Analysis. What is more important in our case is that the particular choice of the index set I and the special properties of the ultrafilter U allow us (as we shall see in the next section) to embed the space of tempered distributions and the ring of polynoms in $*E$ in a way which is, in a sense, canonical; something which cannot be done in any non-standard model $*E$ of E (obtained for arbitrarily chosen I and U).

2. TEMPERED DISTRIBUTIONS IN $*E$

Now, we shall include the class of tempered distributions S' in $*E$ in a way which, in a sense, is canonical.

Definition 2.1 (Colombeau's Regularization)   If $T \in S'$, then $t \in E^I$

determined by the formula:

$$t_{\phi,\varepsilon}(x) = < T(\xi), \frac{1}{\varepsilon^n}\phi\left(\frac{\xi-x}{\varepsilon}\right) >, \quad x \in \mathbb{R}^n, \quad \phi \otimes \varepsilon \in I, \qquad (2.2)$$

will be called *Colombeau's regularization of* T.

Remark 2.3 (Canonical)   The above formula is, of course, well-known in Distribution Theory [10]. Notice that ϕ in (2.2) is a parameter (but not fixed) and $\phi \otimes \varepsilon$ plays the role of the *index* of the *sequence* $t_{\phi,\varepsilon}(\cdot)$ (see Todorov [14]). This is the reason why we look upon inclusion (2.6) defined below as a canonical one.

Theworem 2.4 ($S' \subset {}^*E$)   The mapping:

$$S' \ni T \to f_T \in {}^*E \qquad (2.5)$$

defined by $t \in f_T$, where t is the Colombeau regularization of T (2.1), is injective and it preserves the *good operations* in S': addition, multiplication by a complex number and partial differentiation. We shall write this as:

$$S' \subset {}^*E \qquad (2.6)$$

and sometimes we shall identify in notations a given distribution T with its image f_T (putting $T \equiv f_T$).

Proof   The non-trivial part of the proof is the injective property of the mapping, since the preservation of the operations follows directly from formula (2.2). Suppose that $f_T = 0$ in ${}^*E$, which means

$$\{\phi \otimes \varepsilon | t_{\phi,\varepsilon}(x) = 0, x \in \mathbb{R}^n\} \in U$$

and which, on its part, implies

$$V_1 \equiv \{\phi \otimes \varepsilon | \int_{\mathbb{R}^n} t_{\phi,\varepsilon}(x)\tau(x)dx = 0\} \in U$$

for any (arbitrarily chosen and fixed) test-function $\tau \in S$. On the other hand, for any $r \in \mathbb{R}_+$ we have

$$V_2 = \{\phi \otimes \varepsilon \big| |< T,\tau > - \int_{\mathbb{R}^n} t_{\phi,\varepsilon}(x)\tau(x)dx| < r\} \supseteq A_{q,0} \times (0,\varepsilon_0)$$

for some (sufficiently large) $q \in \mathbb{N}$ and some (sufficiently small) $\varepsilon_0 \in \mathbb{R}_+$. This implies $V_2 \in U$ which implies $V_1 \cap V_2 \in U$, i.e.

$$\{\phi \otimes \varepsilon \big| |< T,\tau >| < r\} \in U.$$

So, we obtain $< T, \tau > = 0$, for any $\tau \in S$, i.e. $T = 0$ in S'. The proof is complete. □

Remark 2.7 (Multiplication of Distrinutions) Inclusion (2.6), in a sense, solves the problem of multiplication of tempered distributions since every two distributions from S' can be correctly multiplied within the algebra $*E$.

Notations 2.8 Let $0^\infty \equiv 0^\infty(\mathbb{R}^n)$ be the class of tempered C^∞-functions (all C^∞-functions which are increasing at infinity, together with their derivatives not faster than some polynom). Then, if $\Phi \subseteq 0^\infty$, then by $^d\Phi$ we shall denote the set of all the tempered regular distributions with kernels from Φ, i.e. which are of the type:

$$\int_{\mathbb{R}^n} F(x)\tau(x)dx, \quad \tau \in S, \tag{2.9}$$

where $F \in \Phi$. Notice that $^d\Phi \subset S'$. In particular, we have

$$^d 0^\infty \subset S' \subset *E \tag{2.10}$$

by injection (2.5).

Corollary 2.11 $^d 0^\infty \subset *E$ as differential linear subspaces.

Proof A direct consequence of Theorem (2.4). □

Theorem 2.12 (Polynoms) Let P be the ring of all polynomials P of the type $P : \mathbb{R}^n \to \mathbb{C}$. Then,

$$^\sigma P = {^d}P. \tag{2.13}$$

(see (1.5) and (iii)-part of definition (1.12)).

Proof Let $P \in P$ and let us set

$$\Delta_{\phi,\varepsilon}(x) = \int_{\mathbb{R}^n} P(x + \varepsilon\xi)\phi(\xi)d\xi - P(x), \quad x \in \mathbb{R}^n, \quad \phi \circledast \varepsilon \in I. \tag{2.14}$$

We have to show that $\Delta \sim 0$ (1.9), i.e. that

$$V = \{\phi \circledast \varepsilon \mid \Delta_{\phi,\varepsilon}(x) = 0, x \in \mathbb{R}^n\} \in U.$$

Indeed, using Taylor's formula for $P(x + \varepsilon\xi)$ (and keeping in mind the (iv)-part of definition (1.2)), we obtain:

$$V \supseteq A_{q,0} \times \mathbb{R}_+$$

for all $q \in \mathbb{N}$ large enough. This implies $V \in U$. The proof is complete. □

Corollary 2.15 If we set ${}^\sigma P = {}^d P \equiv P$, then

$$P \subset S' \subset {}^*E \tag{2.16}$$

where the inclusion of P in $*E$ is in the sense of *differential rings* (*differential algebras over* $\mathbb{C}$).

3. THE CLASS OF META-FUNCTIONS

The class $*E$ defined in Sec. 1 has one serious disadvantage of having two different copies of the class O^∞ of the tempered C^∞-functions: the first ${}^\sigma O^\infty$ obtained by the constant elements of E^I and the second ${}^d O^\infty$ obtained by the mapping (2.5), i.e. as a subspace of S'. It turns out that these two copies do not coincide, i.e. ${}^\sigma O^\infty \neq {}^d O^\infty$, with the exception of the ring (algebra) of polynoms P (2.12). Now, we shal *repair* this disadvantage by using some kind of factorization of the systems $*\mathbb{R}$, $*\mathbb{C}$ and $*E$ defined in Sec. 1.

Definition 3.1 (Moderate Elements) Let s be the positive infinitesimal defined in (1.20). Then, we set

$$\mathbb{C}_M = \{c \in {}^*\mathbb{C} \mid |c| < s^{-p} \text{ for some } p \in \mathbb{N}\}, \tag{3.2}$$

$$E_M = \{f \in {}^*E \mid \text{ for any } \partial \text{ there exist } p \in \mathbb{N} \text{ and } P \in \mathcal{P} \text{ such that } |\partial f(x)|$$
$$< s^{-p} * P(x), \ x \in {}^*\mathbb{R}^n\} \tag{3.3}$$

where $*P$ is the non-standard extension of P. The elements of $\mathbb{C}_M$ and E_M will be called *moderate hypercomplex numbers* (or just *moderate numbers*) and *moderate $*E$-functions*, respectively. (The partial derivative ∂ in (3.3) is is, of course, in $*E$ (1.17)).

It is easy to check that $\mathbb{C}_M$ is a subring of $*\mathbb{C}$ and E_M is a differential subalgebra (over $\mathbb{C}_M$) of $*E$. The next result is important for our construction, but we shall omit the proof.

Theorem 3.4 All tempered distributions (considered as $*E$-functions (2.6)) are moderate, i.e.

$$S' \subset E_M. \tag{3.5}$$

Definitions 3.6 (Ideals) Let s be as in Definition 3.1. Then,

$$\mathbb{C}_0 = \{c \in {}^*\mathbb{C} \mid |c| < s^p \text{ for all } p \in \mathbb{N}\} \quad (3.7)$$

$$E_0 = \{f \in {}^*E \mid \text{for any } \partial \text{ and any } p \in \mathbb{N} \text{ there exists } P \in \mathcal{P},$$

$$\text{such that } |\partial f(x)| < s^p * P(x), \ x \in {}^*\mathbb{R}^n\}. \quad (3.8)$$

Obviously $\mathbb{C}_0$ and E_0 are ideals of $\mathbb{C}_M$ and E_M respectively, and $\mathbb{C}_0$ is maximal.

The next result is *the key* to this paper but we shall omit the proof again.

Lemma 3.9 Let $T \in S' \subset {}^*E$ be a tempered distribution (considered as ${}^*E$-function). Then $T \in E_0$, if and only if $T = 0$ in S'.

Definition 3.10 ($\hat{\mathbb{R}}, \hat{\mathbb{C}}, \hat{E}$)

(i) The elements of the factor-spaces

$$\hat{\mathbb{C}} = \mathbb{C}_M / \mathbb{C}_0 \quad (3.11)$$

and

$$\hat{E} = E_M / E_0 \quad (3.12)$$

will be called *meta-complex numbers* (or just *meta-numbers*) and *meta-functions* respectively. $\hat{\mathbb{R}}$ will be the set of $\alpha \in \hat{\mathbb{C}}$ for which there exists $a \in \alpha$, such that $a \in {}^*\mathbb{R}$.

(ii) All the operations in $\hat{\mathbb{C}}$ (in particular, in $\hat{\mathbb{R}}$) and in $\hat{E}$ will be defined by the corresponding representatives (as in any factor-space). In particular, the order relation in $\hat{\mathbb{R}}$ will be defined as follows: Let $\alpha, \beta \in \hat{\mathbb{R}}$. Then $\alpha < \beta$, if $a < b$ (in ${}^*\hat{\mathbb{R}}$) for some $a \in \alpha$ and some $b \in \beta$.

Lemma 3.13 Let ${}^\sigma O^\infty$ and ${}^d O^\infty$ be the subsets of ${}^*E$ defined in (1.12), (iii), and (2.8) respectively. Then:

$${}^\sigma O^\infty = {}^d O^\infty \quad (\text{mod } E_0). \quad (3.14)$$

We shall omit the proof again.

Theorem 3.15

(i) $\hat{\mathbb{R}}$ and $\hat{\mathbb{C}}$ are non-Archimedean fields (containing copies of $\mathbb{R}$ and $\mathbb{C}$ respectively, as well as non-zero infinitesimals). $\hat{\mathbb{R}}$ is a totally-ordered field;

(ii) $\hat{E}$ is an associative and commutative differential algebra over $\hat{\mathbb{C}}$

(supplied also with integration);

(iii) The inclusions:

$$O^\infty \subset S' \subset \hat{E} \tag{3.16}$$

are valid, defined by

$$O^\infty = {}^\sigma O^\infty / E_0 = {}^d O^\infty / E_0 \tag{3.17}$$

and

$$S' = (S' \text{ in } {}^*E)/E_0 \tag{3.18}$$

respectively. The inclusion $O^\infty \subset \hat{E}$ is in the sense of *differential algebras* (*supplied also with integration*) and $S' \in \hat{E}$ is in the sense of *differential linear spaces*.

(iv) Every meta-function $\hat{f} \in \hat{E}$ is a mapping of the type (the values of $\hat{f}$):

$$\hat{f} : \hat{\mathbb{R}}^n \to \hat{\mathbb{C}} \tag{3.19}$$

in the sense that there exists an injective correspondence from $\hat{E}$ into the set of all mappings of type (3.19), such that the algebraic operations in $\hat{E}$ go into the corresponding pointwise operations.

<u>Remark 3.20</u> (Connection with the Colombeau Original Class G) Let us define the free filter F_c in I by:

$$F_c = \{V \subseteq I \mid A_{q,k} \times (0,1/q) \subseteq V \text{ for some } q \in \mathbb{N} \text{ and some } k \in \mathbb{N}_0\} \tag{3.21}$$

We shall call F_c *Colombeau's filter*. Let us replace U by F_c everywhere in our construction. Then, instead of $\hat{\mathbb{R}}$, $\hat{\mathbb{C}}$ and $\hat{E}$ we shall obtain $\bar{\mathbb{R}}$, $\bar{\mathbb{C}}$ and G respectively, which are number systems and the class of generalized functions of *Colombeau's type* [3 - 5], (see also [14], Section 15). In this case however, we shall lose the field properties of $\hat{\mathbb{R}}$ and $\hat{\mathbb{C}}$ as well as the property of $\hat{\mathbb{R}}$ to be *totally-ordered* with all the consequences for the properties of the corresponding class of generalized functions.

ACKNOWLEDGMENTS

The author would like to thank Professor Abdus Salam, the International Atomic Energy Agency and UNESCO for hospitality at the International Centre for Theoretical Physics, Trieste, Italy.

REFERENCES

1. S. Albeverio, J. E. Fenstad, R. Hoegh-Krohn and T. Lindstrom, "Nonstandard Methods in Stochastic Analysis and Mathematical Physics", Academic Press, New York, (1986).
2. J. F. Colombeau, "Differential Calculus and Holomorphy, Real Complex Analysis in Locally Convex Spaces", North Holland, (1982).
3. J. F. Colombeau, "New Generalized Functions and Multiplication of Distributions", North Holland, (1984).
4. J. F. Colombeau, "Elementary Introduction to Now Generalized Functions", North Holland, (1985).
5. E. E. Rosinger, "Generalized Solutions of Nonlinear Partial Differential Equations", North Holland, Mathematics Studies 144, (1987).
6. M. Davis, "Applied Nonstandard Analysis", John Wiley, New York, (1977).
7. H. J. Keisler, "Elementary Calculus", Prindle, Weber and Schmidt, Boston, (1976).
8. H. J. Keisler, "Foundation of infinitesimal calculus", Prindle, Weber and Schmidt, Boston, (1976).
9. A. Robinson, "Nonstandard analysis", North Holland, Amsterdam, (1966) (2nd revised edition 1974).
10. L. Schwartz, "Theory des distributions", Hermann, Paris, (1950).
11. L. Schwartz, Sur l'impossibilite de la multiplication des distributions, Comptes Rendus Acad. Sci. Paris, 239, 847-848, (1954).
12. K. D. Stroyan and W. A. J. Luxemburg, "Introduction to the theory of infinitesimals", Academic Press, New York, (1976).
13. T. D. Todorov, The products $\delta^2(x), \delta(x) \cdot x^{-n}, \theta(x) \cdot x^{-n}$, etc. in the class of the asymptotic functions, Bulg. J. Phys., Vol. 12, 5, 465-480, (1985).
14. T. D. Todorov, Sequential approach to Colombeau's theory of generalized functions, ICTP, Trieste, Preprint IC/87/26, (1987).

ONE PRODUCT OF DISTRIBUTIONS

Miloš Tomić

Mašinski fakultet Sarajevo

71000 Sarajevo, Yugoslavia

ABSTRACT

In this paper we prove the inequality by which the Fourier transform of the product $|x|^r s(x)$ for $s \in S(\mathbb{R}_n)$, $x \in \mathbb{R}_n$, $r > 0$, is estimated. The product $|x|^r \cdot \hat{f}$ is also defined, where f is a real locally integrable function on $\mathbb{R}_n$ and $\hat{f}$ it is the Fourier transforms and its proved that this product is a tempered distribution.

1. INTRODUCTION

By S (see [3], [6], [8]) we shall denote the space of infinitely differentiable functions $s(x)$ on $\mathbb{R}_n$ for which

$$K(\alpha, k, s) = \sup_x (1 + |x|^2)^{\alpha/2} |s^{(k)}(x)| < \infty$$

for every $\alpha \geq 0$ and every $k = (k_1, \ldots, k_n)$, k_j are non-negative integers and

$$|x| = \left(\sum_{j=1}^{n} x_j^2 \right)^{1/2}.$$

S' denotes the space of continuous linear functionals on S (tempered distributions).

<u>Definition 1.1</u> $M_\alpha = M_\alpha(\mathbb{R}_n)$, $\alpha > 0$, is the space of all locally integrable functions f such that

$$\|f\|_{M_\alpha} = \int_{R_n} \frac{|f(x)|}{(1+|x|^2)^{\alpha/2}} dx < \infty. \tag{1.1}$$

The Fourier transform $\hat{f} = F(f)$ of a distribution $f \in S'$ is defined by $<\hat{f},s> = <f,\hat{s}>$, $s \in S$. The inverse Fourier trnasform $F^{-1}(f)$ of f is defined by $<F^{-1}(f),s> = <f,F^{-1}(s)>$.

In [5] the case of a function of one variable was considered and using the notion of a bounded distribution, the product $\lambda \hat{f}$ was defined. We shall define in the same way this product for $f \in M_\alpha(R_n)$ and the existence of this product we shall prove by Theorems 1 and 2 for $\lambda = |x|^r$, $r > 0$, $x = (x_1,\ldots,x_n)$. We can prove the existence of the product in a new way. Besides that, Theorem 2.1 serves as a proof that the product $|x|^r \cdot \hat{f}$ is a distribution from S'; Theorem 2.1 has an independent interest because it gives the estimation of a function $[|y|^r s(y)]\hat{\ }(x)$, $s \in S$.

Definition 1.2 0_{M_α} is the space of all the functions $\lambda(x)$, $x \in R_n$, so that for any $s \in S$, $\lambda s \in L^1(R_n)$ and

$$\sup_x (1+|x|^2)^{\alpha/2} |(\lambda(u)s(u))\hat{\ }(x)| < \infty. \tag{1.2}$$

Definition 1.3 Let $f \in M_\alpha$ and $\lambda \in 0_{M_\alpha}$; then the product $\lambda \cdot \hat{f}$ is defined by

$$<\lambda \cdot \hat{f}, s> = \int_{R_n} f(x)(\lambda(u)s(u))\hat{\ }(x)dx, \quad s \in S.$$

Clearly, it is a linear functional on S.

Obviously, $0_M \subset 0_{M_\alpha}$, where 0_M denotes the set of multipliers for S. If $\lambda \in 0_M$, then $\lambda \cdot \hat{f} = \lambda \hat{f}$. We shall prove that $|x|^r \in 0_{M_\alpha}$ and that $|x|^r \cdot \hat{f} \in S'$ for any $f \in M_\alpha$.

We shall use the inequality

$$\frac{1}{1+|x-y|^2} \leq C_1 \frac{1+|x|^2}{1+|y|^2} \tag{1.4}$$

which follows from the inequality (see [6], (VII, 5; 7))

$$1+|\eta|^2 \leq C_1(1+|\xi|^2)(1+|\xi+\eta|^2),$$

where C_1 is a constant.

2. RESULTS

Theorem 2.1 Let $0 < \alpha < r$. Then, for every function $s \in S(\mathbb{R}_n)$ and every transform $A \in \{F, F^{-1}\}$ the inequality

$$\sup_x (1 + |x|^2)^{\alpha/2} |A[|y|^r s(y)](x)| < \infty. \tag{2.1}$$

For every number $r > 0$ and every function $s \in S(\mathbb{R}_n)$, the function $A[|y|^r s(y)](x)$ belongs to the space $L_p(\mathbb{R}_n)$ for every number $p \in [1, \infty]$.

Remark 2.1 The first part of Theorem 2.1 follows from [4], Lemma 1, but its proof which is given here is quite different and original and enables us to prove the second part of the theorem and Theorem 2.2.

Theorem 2.2 For $f \in M_\alpha(\mathbb{R}_n)$, $0 < \alpha < r$, the product $|x|^r \cdot \hat{f}$ is a distribution from S'.

Proof of Theorem 2.1 We shall use the development of the function

$$h(t) = (1-t)^{r/2} = 1 + \sum_{m=1}^{\infty} A_{m,r} t^m, \quad (|t| < 1, r > 0),$$

where (see [7], V. 3.2(32))

$$\sum_{m=1}^{\infty} |A_{m,r}| < \infty.$$

Putting in the previous equality $t = (1 + |x|^2)^{-1}$, we obtain

$$|x|^r s(x) = (1 + |x|^2)^{r/2} s(x) + \sum_{m=1}^{\infty} A_{m,r} \frac{s(x)}{(1 + |x|^2)^{m-r/2}}. \tag{2.2}$$

We shall divide the sum from (2.2) into two sums. Let $N = N(r) = [r/2]$ be the biggest natural number contained in $r/2$ (in the case of $0 < r < 2$ the sum is not divided). So, $N(r) \leq r/2$ and $m - r/2 \leq 0$ for $m \leq N(r)$; $m - r/2 > 0$ for $m > N(r)$.

In order to simplify the notation, we shall consider (write) $A = F = \hat{}$ but we can write $A = F^{-1}$ too. From (2.2) it follows

$$[|x|^r s(x)]\hat{}(y) = [(1 + |x|^2)^{r/2} s(x)]\hat{}(y)$$
$$+ \sum_{m=1}^{N} A_{m,r} [(1 + |x|^2)^{r/2 - m} s(x)]\hat{}(y) + \sum_{m=N+1}^{\infty} A_{m,r} \left[\frac{s(x)}{(1+|x|^2)^{m-r/2}}\right]\hat{}(y).$$

$$\tag{2.3}$$

Since $[(1 + |x|^2)^{r/2} s(x)]\hat{\ }(y) \in S(\mathbb{R}_n)$, then from (2.3) we get

$$(1 + |y|^2)^{\alpha/2} |[|x|^r s(x)]\hat{\ }(y)| \leq C_1(\alpha,r,s) +$$

$$+ \sum_{m=N+1}^{\infty} |A_{m,r}| (1 + |y|^2)^{\alpha/2} \left| \left[\frac{s(x)}{(1 + |x|^2)^{m-r/2}} \right]\hat{\ }(y) \right|. \qquad (2.4)$$

By the equality (see [6], (VII; 8, 4))

$$A(u,v) = A(u) * A(v), \quad A \in \{F, F^{-1}\}, \ u \in S, \ v \in S' \qquad (2.5)$$

for the sum from (2.4), we have

$$\Sigma = \sum_{m=N+1}^{\infty} |A_{m,r}| (1 + |y|^2)^{\alpha/2} |(G_{2m-r} * \hat{s})(y)|, \qquad (2.6)$$

where G_β is the kernel of Bessel

$$G_\beta(x) = [(1 + |y|^2)^{-\beta/2}]\hat{\ }(x), \ x \in \mathbb{R}_n$$

and it is given by the equality (see [7], V. 3.1 (2.6))

$$G_\beta(x) = \frac{(2\pi)^{n/2}}{(4\pi)^{\beta/2} \Gamma(\beta/2)} \int_0^\infty e^{-\pi|x|^2/\delta} e^{-\delta/4\pi} \delta^{(-n+\beta)/2} \frac{d\delta}{\delta}. \qquad (2.7)$$

Using the inequality (1.4), we get

$$|(G_{2m-r} * \hat{s})(y)| \leq (2\pi)^{-n/2} \int_{\mathbb{R}_n} G_{2m-r}(x) |\hat{s}(y-x)| dx$$

$$\leq \int_{\mathbb{R}_n} G_{2m-r}(x) \frac{K(\alpha,0,\hat{s})}{(1 + |y-x|^2)^{\alpha/2}} dx$$

$$\leq C_2 \frac{K(\alpha,0,\hat{s})}{(1+|y|^2)^{\alpha/2}} \int_{\mathbb{R}_n} (1 + |x|^2)^{\alpha/2} G_{2m-r}(x) dx$$

and

$$(1 + |y|^2)^{\alpha/2} |(G_{2m-r} * \hat{s})(y)| \leq C_2 K(\alpha,0,\hat{s}) \int_{\mathbb{R}_n} (1 + |x|^2)^{\alpha/2} G_{2m-r}(x) dx. \qquad (2.8)$$

For the estimation of the integral in (2.8), we use (2.7) and the theorem of Fubini. We have

$$I = \int_{\mathbb{R}_n} (1 + |x|^2)^{\alpha/2} G_{2m-r}(x) dx =$$

$$= \frac{(2\pi)^{n/2}}{(4\pi)^{m-r/2}\Gamma(m - r/2)} \int_0^\infty e^{-\delta/4\pi} \delta^{-\frac{n}{2}+m-\frac{r}{2}} \frac{d\delta}{\delta} \int_{\mathbb{R}_n} (1 + |x|^2)^{\alpha/2} e^{-\pi|x|^2/\delta} dx. \quad (2.9)$$

Moreover, we have

$$\int_{\mathbb{R}_n} (1 + |x|^2)^{\alpha/2} e^{-\pi|x|^2/\delta} dx = P(n) \int_0^\infty (1 + t^2)^{\alpha/2} e^{-\pi t^2/\delta} t^{n-1} dt \quad (2.10)$$

where $P(n)$ is the constant.

For the integral from (2.10), we have

$$\int_0^\infty (1 + t^2)^{\alpha/2} t^{n-1} e^{-\pi t^2/\delta} dt = \int_0^1 + \int_1^\infty,$$

$$\int_0^1 (1 + t^2)^{\alpha/2} t^{n-1} e^{-\pi t^2/\delta} dt \leq \int_0^1 (1 + t^2)^{\alpha/2} t^{n-1} dt = C_3(\alpha, n),$$

$$\int_1^\infty (1 + t^2)^{\alpha/2} t^{n-1} e^{-\pi t^2/\delta} dt \leq C_4 \int_1^\infty t^{\alpha+n-1} e^{-\pi t^2/\delta} dt$$

$$\leq C_4 \int_0^\infty t^{\alpha+n-1} e^{-\pi t^2/\delta} dt.$$

By the substitution $\sqrt{\pi} t = \sqrt{\delta} v$, we obtain

$$\int_0^\infty t^{\alpha+n-1} e^{-\pi t^2/\delta} dt = \frac{\delta^{(\alpha+n)/2}}{\pi^{(\alpha+n)/2}} \int_0^\infty v^{\alpha+n-1} e^{-v^2} dv \leq C_5(\alpha, n) \delta^{(\alpha+n)/2}.$$

Thus,

$$\int_0^\infty (1 + t^2)^{\alpha/2} t^{n-1} e^{-\pi t^2/\delta} dt \leq C_3(\alpha, n) + C_5(\alpha, n) \delta^{(\alpha+n)/2}$$

$$\leq C_6(\alpha, n)(1 + \delta^{(\alpha+n)/2}). \quad (2.11)$$

Using (2.10) and (2.11), it follows from (2.9) that

$$I \leq \frac{C_7(\alpha, n)}{(4\pi)^{m-r/2}\Gamma(m - r/2)} \int_0^\infty e^{-\delta/4\pi} \delta^{-\frac{n}{2}+m-\frac{r}{2}} (1 + \delta^{(\alpha+n)/2}) \frac{d\delta}{\delta}. \quad (2.12)$$

Since (see [7], V. 3.1)

$$\frac{\Gamma(\beta)}{t^\beta} = \int_0^\infty e^{-t\delta} \delta^\beta \frac{d\delta}{\delta},$$

345

it follows from (2.12) that

$$I \leq \frac{C_8(\alpha,n)}{\Gamma(m-r/2)} \{\Gamma(m-r/2-n/2) + \Gamma(m-r/2-\alpha/2)\}$$

$$\leq 2C_8(\alpha,n) \frac{\Gamma(m-r/2+\alpha/2)}{\Gamma(m-r/2)} \cdot \quad (2.13)$$

In view of (2.8), (2.9) and (2.13) from (2.6), it follows that

$$\Sigma \leq C_9(\alpha,n)K(\alpha,0,\hat{s}) \sum_{m=N+1}^{\infty} |A_{m,r}| \frac{\Gamma(m-r/2+\alpha/2)}{\Gamma(m-r/2)} \cdot \quad (2.14)$$

We shall prove that the series from (2.14) converges for $\alpha < r$. Since $A_{m,r} = h^{(m)}(0)/m!$, we get

$$A_{m,r} = \frac{r(r-2)(r-4) \ldots [r-2(m-1)]}{m! \, 2^m} \cdot$$

For function $\Gamma(x)$ the equality $\Gamma(1+x) = x\Gamma(x)$ (see [1], V. 3.1 or [2], 1.2) holds. In view of this equality for

$$C_m = |A_{m,r}| = \frac{\Gamma(m-r/2+\alpha/2)}{\Gamma(m-r/2)},$$

we have

$$\frac{C_m}{C_{m+1}} = \frac{m-r/2}{m-r/2+\alpha/2} \frac{2m-2}{2m-r}$$

whence

$$m\left(\frac{C_m}{C_{m+1}} - 1\right) \to 1 + \frac{r}{2} - \frac{\alpha}{2}, \quad m \to \infty.$$

We conclude that the series from (2.14) converges for $\alpha < r$ and diverges for $\alpha > r$. The series diverges also for $\alpha = r$. Thus,

$$\Sigma \leq C_{10}(n,r,\alpha)K(\alpha,0,\hat{s}), \quad (2.15)$$

which completes the proof of the inequality (2.1).

From (2.2) it follows that

$$[|x|^r s(x)]^{\wedge}(y) = \hat{s}_r(y) + \sum_{m=1}^{\infty} A_{m,r}(G * \hat{s}_r)(y), \quad (2.16)$$

where $s_r(x) = (1+|x|^2)^{r/2} s(x)$.

Since (see [7], V. 3.1)

$$G_{2m} \geq 0, \quad \int_{\mathbb{R}_n} G_{2m}(y) \, dy = (2\pi)^{n/2},$$

we obtain

$$\int_{\mathbb{R}_n} |(G_{2m} * \hat{s}_r)(y)| \, dy \leq (2\pi)^{-n/2} \int_{\mathbb{R}_n} G_{2m}(x) \, dx \int_{\mathbb{R}_n} |\hat{s}_r(y-x)| \, dy = \|\hat{s}_r\|_{L_1}.$$
(2.17)

From (2.16) in view of (2.17), we obtain

$$\|[|x|^r s(x)]\hat{\,}(y)\|_{L_1} \leq (1 + \sum_{m=1}^{\infty} |A_{m,r}|) \|\hat{s}_r\|_{L_1}.$$
(2.18)

The inequality (2.18) means that $[|x|^r s(x)]\hat{\,}(y) \in L_1(\mathbb{R}_n)$. Since $|x|^r s(x) \in L_p(\mathbb{R}_n)$ for every $p \in [1,2]$, then $[|x|^r s(x)]\hat{\,}(y) \in L_q(\mathbb{R}_n)$ for every $q \in [2,\infty]$. This means that $[|x|^r s(x)]\hat{\,}(y) \in L_p(\mathbb{R}_n)$ for every $p \in [1,\infty]$. The theorem has been proved. □

Proof of Theorem 2.2 Assume that $f \in M_\alpha(\mathbb{R}_n)$, $0 < \alpha < r$, and $s_j \to 0$ in S. We have to prove that $< |x|^r \cdot \hat{f}, s_j > \to 0$, $(j \to \infty)$.

By definition 1.2, we have

$$< |y|^r \cdot \hat{f}, s_j > = \int f(y) [|x|^r s_j(x)]\hat{\,}(y) \, dy$$

$$= \int \frac{f(y)}{(1 + |y|^2)^{\alpha/2}} (1 + |y|^2)^{\alpha/2} [|x|^r s_j(x)]\hat{\,}(y) \, dy,$$

whence

$$|< |y|^r \cdot \hat{f}, s_j >| \leq \|f\|_{M_\alpha} \sup_y (1 + |y|^2)^{\alpha/2} |[|x|^r s_j(x)]\hat{\,}(y)|.$$

From (2.19) in view of (2.3), (2.6) and (2.15), it follows that $< |y|^r \cdot \hat{f}, s_j > \to 0$ if $s_j \to 0$ in the sense of S.
The theorem has been proved. □

Remark 2.2 Theorems 2.1 and 2.2 can be used for determination of classes of saturation of singular integrals in spaces L_p and M_α. The properties of the product which we have defined above will be given in a separate paper.

REFERENCES

1. E. Janke, F. Emde, F. Losch, "Special functions", Moscow (1977), (in Russian).

2. D. S. Mitrinović, "Introduction in special functions", Beograd, (1972) (in Serbocroatian).
3. S. M. Nikolskii, "Approximation of functions in several variables and theorems of inclusion", Moscow, (1969), (in Russian).
4. N. Ortner, Faltung hypersingulärer Integraloperatoren, Math. Ann. 248, 19-46, (1980).
5. T. Ostrogorski, Global and local saturation theorems in some spaces of temperate functions, Publ. de l'inst. math. t 20(40), 199-213, (1979).
6. L. Schwartz, "Théorie des distributions", t. 2, Paris, (1951).
7. E. M. Stein, "Singular integrals and differentiability properties of functions", Moscow, (1973), (in Russian).
8. V. S. Vladimirov, "Generalized functions in mathematical physics", Moscow, (1976), (in Russian).

ABEL SUMMABILITY FOR A DISTRIBUTION SAMPLING THEOREM

Gilbert G. Walter

University of Wisconsin-Milwaukee, U.S.A.

ABSTRACT

Let $F(w)$ be an L^2 function with compact support on $[-\sigma,\sigma]$; let $T = \pi/\sigma$ and $f(t)$ be the Fourier transform of $F(w)$. Then the well-known sampling theorem says

$$f(t) = \sum_{n=-\infty}^{\infty} f(nT) \frac{\sin \sigma(t - nT)}{\sigma(t - nT)},$$

where convergence is uniform in $\mathbb{R}^1$. If $F(w)$ is now a distribution with compact support on $[-\sigma,\sigma]$ the Fourier transform is still a function but the series does not converge necessarily. However it is shown, under mild conditions of $F(w)$, that the series is Abel summable, i.e.

$$f(t) = \lim_{r \to 1^-} \sum_{n=-\infty}^{\infty} r^{|n|} f(nT) \frac{\sin \sigma(t - nT)}{\sigma(t - nT)}$$

where the convergence is uniform on bounded sets in $\mathbb{R}^1$.

1. INTRODUCTION

Let $F(w)$ be an $L^2(\mathbb{R}^1)$ function with compact support in $[-\sigma,\sigma]$; let $f(t)$ be its Fourier transform and let $T = \pi/\sigma$. For such functions the sampling theorem says

$$f(t) = \sum_{n=-\infty}^{\infty} f(nT) \frac{\sin \sigma(t - nT)}{\sigma(t - nT)} \qquad (1.1)$$

This has become part of the folk knowledge of mathematics but its importance was first recognized by Shannon [6], who exploited it in communications theory. Its convergence is easily seen to be uniform on the entire real line.

However in communications theory many signals do not have finite energy and hence are not in $L^2(\mathbb{R}^1)$. For such signals the series (1.1) does not necessarily converge. A general class of such signals is given by the Fourier transforms of distributions with compact support. The signals $f(t)$ will then be entire functions of polynomial growth.

Versions of the sampling theorem appropriate for such signals have been studied by a number of authors beginning with Campbell [2] and including Pfaffelhuber [5], Hoskins and De Sousa Pinto [3], Lee [4], and Walter [7]. All suffered from the same shortcoming, viz., that the form (1.1) had to be modified considerably to obtain a convergence theorem.

In this work we shall try to retain the elegance of (1.1) but require only that the convergence of the series be interpreted in the sense of Abel summability. We shall show that this is the case for all distributions which have support in $[-\sigma,\sigma]$ and are strongly integrable.

2. INTEGRATION OF DISTRIBUTIONS

There are several ways to define the integral and hence the Fourier transform of a distribution of compact support. However not all are appropriate for a sampling theorem since both a Fourier transform and a Fourier series are needed. We shall consider three definitions of integration of a distribution with support in $[-\pi,\pi]$.

__Definition 2.1__ A distribution $F(w)$ is integrable in the sense of Schwartz over $[-\pi,\pi]$ if its support is in this interval. Its Fourier transform is

$$f(t) = \frac{1}{2\pi} < F(w), e^{iwt} >. \qquad (2.1)$$

The other two definitions involve the value of distribution at a point and may be found in [1].

__Definition 2.2__ A distribution $F(w)$ is weakly (strongly) integrable over $[-\pi,\pi]$ if its anti derivative $F^{(-1)}$ is such that $F^{(-1)}(w+\pi) - F^{(-1)}(w-\pi)$ has a value at 0 ($F^{(-1)}$ has a value at π and $-\pi$).

Clearly if $F(w)$ is strongly integrable over $[-\pi,\pi]$ it is also weakly integrable. The Definition 2.2 does not require that F have compact support. If F does have compact support on $[-\pi,\pi]$ it may not be weakly integrable as shown by the example.

Example 2.1 Let $F(w) = \delta(w - \pi)$; then F has its support on $[-\pi,\pi]$, but $F^{(-1)}(w + \pi) - F^{(-1)}(w - \pi) = H(w) - H(w - 2\pi)$ which does not have a value at 0. Hence F is Schwartz integrable but neither weakly nor strongly integrable.

Example 2.2 Let $F(w) = \delta(w - \pi) - \delta(w + \pi)$; then again $F(w)$ is not weakly integrable. Its Fourier transform is

$$f(t) = \frac{1}{2\pi} < F(w), e^{iwt} > = \frac{i}{\pi} \sin\pi t$$

but its Fourier coefficients are

$$c_n = \frac{1}{2\pi} < F(w), e^{iwn} > = 0, \quad n = 0, \pm 1, \ldots .$$

Hence there is no way in which f(t) can be represented by its sampling expansion.

Example 2.3 Let $F(w) = \delta(w - \pi) + \delta(w + \pi)$; now $F(w)$ is weakly integrable over $[-\pi,\pi]$ but is not strongly integrable. However $F(w)e^{-iwt}$ is not weakly integrable since $H(w)e^{-i\pi t} - H(w - 2\pi)e^{-i\pi t} - H(w)e^{+i\pi t} + H(w + 2\pi)e^{+i\pi t}$ does not have a value at 0 when t is not an integer.

Since we must take the Fourier transform of F(w) for each value of t, we choose strong integrability as the appropriate definition. Clearly if $F^{(-1)}(w)$ has a value at $\pm\pi$, so does $(F(w)e^{-iwt})^{(-1)}$.

The strong integral $\int_{-\infty}^{\infty} F(w)e^{-iwt} dw$ is defined as the difference between the values of $(F(w)e^{-iwt})^{(-1)}$ at π and $-\pi$. This can be shown to be consistent with the Schwartz definition of the Fourier transform (2.1) for distributions with support in $[-\pi,\pi]$. The latter is sometimes more useful for calculations.

Proposition 2.1 Let F(w) be a distribution with support on $[-\pi,\pi]$. Then there is a distribution G(w) with support on $[-\pi,\pi]$ such that

$$F(w) = DG(w) + c\delta(w) \tag{2.2}$$

If F is also strongly integrable on $[-\pi,\pi]$ then so is G and

$$\int_{-\pi}^{\pi} F = \langle F, 1 \rangle = c. \tag{2.3}$$

Proof Since F has support in $[-\pi,\pi]$, any antiderivative $F^{(-1)}$ is constant in $(-\infty,-\pi)$ and (π,∞). We may take the constant in $(-\infty,-\pi)$ to be zero and let $c = F^{(-1)}(w)$ for $w > \pi$. Take $G(w) = F^{(-1)}(w) - cH(w)$, which satisfies (2.2). Then $G(w)$ also has a value at $\pm\pi$ since both $F^{(-1)}$ and H do. Since, however, $G(w) = 0$ for $|w| > \pi$, the value must be zero there.

Now we use (2.2) to calculate

$$\langle F, 1 \rangle = \langle DG, 1 \rangle + c \langle \delta, 1 \rangle = -\langle G, D1 \rangle + c = c$$

and

$$\int_{-\pi}^{\pi} F = F^{(-1)}(\pi) - F^{(-1)}(-\pi) = cH(\pi) - cH(-\pi) = c. \quad \square$$

Corollary 2.2 Let $F(w)$ have compact support on $[-\pi,\pi]$; then there is an L^2 function G with support on $[-\pi,\pi]$, an integer p, and constants $c_0, c_1, \ldots, c_{p-1}$ such that

$$F = D^p G + \sum_{j=0}^{p-1} c_j \delta^{(j)} \tag{2.4}$$

and

$$f(t) = \frac{1}{2\pi} \langle F, e^{iwt} \rangle = \frac{(it)^p}{2\pi} \langle G, e^{iwt} \rangle + \sum_{j=0}^{p-1} c_j (it)^j.$$

The proof involves iterating (2.2) p times until the resulting distribution G is a locally integrable function in $L^2[-\pi,\pi]$. $\square$

We must also consider periodic distributions since we will be calculating the Fourier series. Any distribution F with support on $[-\pi,\pi]$ has a periodic extension F^* given by

$$F^*(w) = \sum_{k=-\infty}^{\infty} F(w + 2k\pi)$$

and $F^* = F$ in the open interval $(-\pi,\pi)$.

For F which are strongly integrable over and have support in $[-\pi,\pi]$, the periodic extension is easily shown to be given by

$$F^* = DG^* + c\delta^* \tag{2.5}$$

where G is the distribution with compact support given in (2.2).

The Fourier coefficient of F^* are

$$c_n = \frac{1}{2\pi} < F^*, e^{inw} >_{2\pi} = \frac{in}{2\pi} < G^*, e^{inw} >_{2\pi} + \frac{c}{2\pi} = \frac{1}{2\pi} < F, e^{inw} > \quad (2.6)$$

It should be observed that (2.6) is not true in general as can be seen by example 2.1. Here $<,>_{2\pi}$ denotes the value of a periodic distribution in $\mathcal{D}_{2\pi}$.

3. ABEL SUMMABILITY

The kernel of Abel summability $P_r(w)$ is given by

$$2\pi P_r(w) = \frac{1 - r^2}{1 - 2r\cos w + r^2} = \sum_{n=-\infty}^{\infty} r^{|n|} e^{inw} \qquad 0 < r < 1. \quad (3.1)$$

It is used to show Abel summability of Fourier series of functions or distributions. The following result is well known [9].

Let $F^*(w)$ be a bounded periodic function on $\mathbb{R}^1$ continuous on (a,b); then

$$(P_r * F^*) \to F^* \quad \text{as } r \to 1-, \quad (3.2)$$

bounded by and uniformly on interior subintervals of (a,b).

We are now ready for our main result.

Theorem Let $f(t)$ be a function on $\mathbb{R}^1$ which is the Fourier transform of a distribution $F(w)$ with compact support on $[-\sigma,\sigma]$ which is strogly integrable over this interval, let $T = \pi\sigma$, then

$$f(t) = \sum_{n=-\infty}^{\infty} f(nT) \frac{\sin \sigma(t - nT)}{\sigma(t - nT)} \quad (3.3)$$

where convergence is in the sense of Abel summability, uniform on bounded sets.

Proof We may assume without loss of generality that $\sigma = \pi$, since other values merely involve a change of scale. We must show that

$$\sum_{n=-\infty}^{\infty} r^{|n|} f(n) \frac{\sin \pi(t - n)}{\pi(t - n)} \to f(t) \quad (3.4)$$

uniformly on bounded sets as $r \to 1^-$.

First we observe that the series in (3.4) converges uniformly since $f(n) = O(n^p)$ for some integer p. Also $f(t)$ is an entire function of polynomial growth by the Paley-Wiener-Schwartz Theorem. The inverse Fourier transform of the left side of (3.4) restricted to $(-\pi,\pi)$ is

$$\sum_{n=-\infty}^{\infty} r^{|n|} f(n) e^{iwn} = \sum_{n=-\infty}^{\infty} \frac{r^{|n|}}{2\pi} < F, e^{iwn} > e^{iwn} \qquad (3.5)$$

since $f(n)$ is just the Fourier coefficient of F. But

$$\sum_{n=-\infty}^{\infty} r^{|n|} e^{i(w-\lambda)n}$$

converges in the sense of E, i.e. it converges uniformly on $\mathbb{R}^1$ and each of its derivatives also converges uniformly. Hence (3.5) may be written as

$$\frac{1}{2\pi} < F(\lambda), \sum_{n=-\infty}^{\infty} r^{|n|} e^{i(\lambda-w)n} > = < F(\lambda), P_r(\lambda - w) >$$

and we need merely show that

$$\frac{1}{2\pi} \int_{-\pi}^{\pi} e^{-iwt} < F(\lambda), P_r(\lambda - w) > dw \to \frac{1}{2\pi} \int_{-\pi}^{\pi} e^{-iwt} F(w) \, dw \qquad (3.6)$$

as $r \to 1^-$. We may write the left side of (3.6) as

$$\frac{1}{2\pi} < F(\lambda), \int_{-\pi}^{\pi} P_r(\lambda - w) e^{iwt} \, dw >.$$

But

$$\phi_r(\lambda, t) = \int_{-\pi}^{\pi} P_r(\lambda - w) e^{iwt} \, dw \to \phi^*(\lambda, t) \qquad (3.7)$$

uniformly in interior intervals of $(-\pi,\pi)$ and boundedly, where $\phi^*(w,t)$ is the periodic extension of e^{iwt}. It is discontinuous at $\pm\pi$ except when t is an integer.

If we differentiate (3.7) p times the resulting function converges in the sense of distributions. However this is insufficient for convergence of (3.6) unless $F \in \mathcal{D}$. Rather we must use

Lemma 3.1 Let

$$K_r(w) = \frac{(-1)^k w^k P_r^{(k)}(w)}{k!} \; ;$$

then $\{K_r\}$ is a quasi-positive delta-family on $\left(\frac{-3\pi}{2}, \frac{3\pi}{2}\right)$ as $r \to 1^-$.

Clearly $K_r(w) \to \delta(w)$ as $r \to 1^-$ since $P_r(w)$ does. A quasi-positive delta sequence is one such that

(i) $\displaystyle\int_{-3\pi/2}^{3\pi/2} K_r \to 1,$

(ii) $\displaystyle\int_{-3\pi/2}^{3\pi/2} |K_r| \leq M,$ a constant,

(iii) $K_r(w) \to 0$ uniformly as $r \to 1$ for $\delta \leq |w| < 3\pi/2$.

That K_r satisfies these properties is a slight modification of the proof given in [8].

For such delta families we have, when $\phi \in L'$

$$(K_r * \phi)(w) \to \phi(w)$$

uniformly on closed intervals interior to intervals of continuity of $\phi(w)$.

We now use the representation (2.4) for strongly integrable F, i.e.

$$F = D^p G + \sum_{j=0}^{p-1} c_j \delta^{(j)}. \tag{3.8}$$

where $D^{p-1}G$ has a value at $\pm\pi$, which we have shown, must be zero. By Lojasiewicz's theorem, $(D^{p-1}G)(w) = D^r H_\pi(w)$ where $H_\pi(w)$ is continuous and $(H_\pi(w)/(w-\pi)^r) \to 0$ as $w \to \pi$ and similarly for $-\pi$. By increasing p if necessary we may assume $p-1 = r$ and hence $H_\pi(w) - G(w)$ is a polynomial of degree $< p-1$. Similarly $H_{-\pi}(w) - G(w)$ is such a polynomial. Since $G(w)$ has support on $[-\pi,\pi]$, $H_\pi(w)$ must itself be this polynomial for $w > \pi$, and hence must be zero there. Thus we may assume that

$$\frac{G(w)}{(w \pm \pi)^{p-1}} \to 0 \text{ as } w \to \pm\pi$$

and may further assume that G is continuous except at $w = 0$. We now substitute (3.8) into (3.6) to obtain a term of the form

$$\frac{1}{2\pi} \int_{-\pi}^{\pi} e^{-iwt} < D^p G, P_r(\lambda - w) > dw$$

$$= \frac{1}{2\pi} \int_{-\pi}^{\pi} e^{-iwt} (-1)^p \int_{-\pi}^{\pi} G(\lambda) P_r^{(p)}(\lambda - w) d\lambda \, dw$$

$$= \frac{1}{2\pi} \int_{-\pi}^{\pi} G(\lambda)(-1)^p \int_{-\pi}^{\pi} e^{-iwt} P_r^{(p)}(\lambda - w) dw \, d\lambda$$

$$= \frac{1}{2\pi} \int_{-\pi}^{\pi} G(\lambda)(-1)^p \{P_r^{(p-1)}(\lambda - \pi)(-2i \sin\pi t) +$$

$$+ \int_{-\pi}^{\pi} (it)e^{-iwt}P_r^{(p-1)}(\lambda - w)dw\}d\lambda$$

$$= \frac{1}{2\pi}\int_{-\pi}^{\pi} \frac{G(\lambda)}{(\lambda - \pi)^{p-1}} \frac{(-1)^{p-1}(\lambda - \pi)^{p-1}P_r^{(p-1)}(\lambda - \pi)}{p!} \, 2i\sin\pi t \, d\lambda$$

$$+ \frac{1}{2}\int_{-\pi}^{\pi} G(\lambda)(-1)^p\{\int_{-\pi}^{\pi} (it)e^{-iwt}P_r^{(p-1)}(\lambda - w)dw\}d\lambda. \tag{3.9}$$

The first integral converges to zero uniformly for t in bounded sets since $G(\lambda)/(\lambda - \pi)^{p-1}$ is continuous at π by Lemma 3.1. We now repeat the integration by parts to obtain

$$\frac{1}{2\pi}\int_{-\pi}^{\pi} e^{-iwt} < D^p G, \, P_r(\lambda - w) > dw$$

$$= \frac{1}{2}\int_{-\pi}^{\pi} G(\lambda) \int_{-\pi}^{\pi} (-it)^p e^{-iwt} P_r(\lambda - w) dw \, d\lambda + o(1)$$

$$= \frac{1}{2\pi}(-it)^p \int_{-\pi}^{\pi} e^{-iwt}(G*P_r)(w)dw + o(1) \to \frac{1}{2\pi}(-it)^p \int_{-\pi}^{\pi} e^{-iwt} G(w)dw \tag{3.10}$$

where the convergence is again uniform for t in bounded sets. Turning now to the other terms in (3.8), we have

$$\frac{1}{2\pi}\int_{-\pi}^{\pi} e^{-iwt} < \sum_{j=0}^{p-1} c_j \delta^{(j)}(\lambda), \, P_r(\lambda - w) > dw$$

$$= \frac{1}{2}\int_{-\pi}^{\pi} e^{-iwt} \sum_{j=0}^{p-1} c_j(-1)^j P_r^{(j)}(w) \, dw$$

$$= \frac{1}{2}\sum_{j=0}^{p-1} c_j(-1)^j \int_{-\pi}^{\pi} e^{-iwt} P_r^{(j)}(w) \, dw \tag{3.11}$$

But

$$(-1)^j \int_{-\pi}^{\pi} e^{-iwt} P_r^{(j)}(w) dw = o(1) + \int_{-\pi}^{\pi} (-it)^j e^{-iwt} P_r(w) \, dw \to (-it)^j \tag{3.12}$$

uniformly for bounded t. By combining (3.10), (3.11) and (3.12) into the left side of (3.6) we have

$$\frac{1}{2\pi}\int e^{-iwt} < F(\lambda), P_r(\lambda - w) > dw$$

$$\to \frac{1}{2\pi}(-it)^p \int e^{-itw} G(w)dw + \frac{1}{2\pi}\sum_{j=0}^{p-1} c_j(-it)^j = \frac{1}{2\pi}\int e^{-iwt} F(w)dw. \tag{3.13}$$

Hence (3.4) holds and by a change of scale (3.3). □

REFERENCES

1. P. Antosik, T. Mikusiński, and R. Sikorski, "Theory of Distributions", PWN-Warsaw, (1973).
2. L. L. Campbell, Sampling theorem for the fourier transform of a distribution with bounded support, SIAM J. Applied Math., 16 626-636 (1968).
3. R. F. Hoskins and J. De Sousa Pinto, Sampling expansions for functions band-limited in the distributional sense, SIAM J. Appl. Math., 44, 605-610 (1984).
4. A. J. Lee, Characterization of band-limited functions and processes, Info. and Control, 31, 258-271 (1976).
5. E. Pfaffelhuber, Sampling series for band-limited generalized functions, IEEE Trans. on Info. Theory, IT-17, 650-654 (1971).
6. C. E. Shannon, Communications in the presence of noise, Proc. IRE, 37, 10-21 (1949).
7. G. Walter, Sampling band-limited functions of polynomial growth, SIAM J. Math. Anal., (1988).
8. G. Walter, Fourier Series and Analytic Representations of Distributions, SIAM Review, 12, 272-276 (1970).
9. A. Zygmund, "Trigonometric Series", I & II, Cambridge Press, Cambridge (1959).

ON THE VALUE OF A DISTRIBUTION AT A POINT

Ryszard Wawak

Institute of Mathematics
Polish Academy of Sciences
Sniadeckich 8, 00-950 Warsaw, Poland

ABSTRACT

We show connections between the notion of the value of a distribution at a point in the Łojasiewicz sense and the integrability of its Fourier transform. We consider the one-dimensional case.

We will define first the notion of the weak integrability of a distribution on $\mathbb{R}$. To this end we introduce the following sequences of seminorms (see [4]):

$$\ell_k(\phi) = \sum_{0 \leq \alpha \leq k} (\sup|\phi^{(\alpha)}(x)| + \int |\phi^{(\alpha+1)}(x)|dx),$$

$$w_k(\phi) = \sum_{0 \leq \alpha \leq k} (\sup|<x>^\alpha \phi^{(\alpha)}(x)| + \int |(<x>^\alpha \phi^{(\alpha)}(x))'| dx)$$

for $\phi \in D$, where $<x> = (1 + |x|^2)^{1/2}$.

We define a space $D_{\overline{L}}$:

$$D_{\overline{L}} = \{\phi \in C^\infty : \ell_k(\phi) < +\infty,\ k = 0,1,\ldots\}$$

with the following topology: $\phi_v \to 0$ in $D_{\overline{L}}$ if:

(i) $\ell_k(\phi_v) \leq C_k < +\infty$, $k = 0,1,\ldots$, $v = 1,2,\ldots$

(ii) $\forall \alpha \in N_0\ \phi_v^{(\alpha)} \to 0$ almost uniformly,

and a space $D_{\overline{W}}$, replacing ℓ_k by w_k in this definition. Notice that $D_{\overline{L}}$ is a space of such functions that all their derivatives have bounded variation on $\mathbb{R}$. (For the definition of variation see e.g. [3]).

Remark 1 D is dense in $D_{\overline{L}}$ ($D_{\overline{W}}$) and the constant functions belong to $D_{\overline{L}}$ ($D_{\overline{W}}$). The Schwartz space S is contained in $D_{\overline{L}}$ ($D_{\overline{W}}$), so the dual space $D_{\overline{L}}'$ ($D_{\overline{W}}'$) is a space of tempered distributions.

Proposition 1 If $\phi, \psi \in D_{\overline{L}}$ ($D_{\overline{W}}$), then $\phi\psi \in D_{\overline{L}}$ ($D_{\overline{W}}$).

Definition 1 We say that a sequence $\eta_v \in C_0^\infty$ is L-convergent (strongly convergent) to 1 if:

(i) $\ell_k(\eta_v) \leq C_k < +\infty$, $k = 0, 1, \ldots$, $v = 1, 2, \ldots$

  $(w_k(\eta_v) \leq C_k < +\infty,\ k = 0,1,\ldots,\ v = 1,2,\ldots)$

(ii) $\forall K \subset R\ \exists M \in N\ \forall v \geq M\ \eta_v(x) = 1$ for $x \in K$.

Definition 2 A distribution T is an improper integrable (weak integrable), if the limit $\lim_v T[\eta_v]$ exists and is finite for every sequence η_v L-convergent (strongly convergent) to 1. We write: $\int T = \lim_v T[\eta_v]$.

Theorem 1 (see [4] Theorem 2.2) For $T \in D'$ the following conditions are equivalent:

(i) T is an improper integrable distribution

(ii) $T \in D_{\overline{L}}'$

(iii) $T = \sum\limits_{\alpha \leq r} f_\alpha^{(\alpha)}$, where f_α are improper integrable continuous functions (in the sense $\lim\limits_{a \to -\infty, b \to +\infty} \int_a^b f(x)\,dx$, $f \in L_{loc}^1$)

(iv) $T * \phi$ is an improper integrable continuous function for $\phi \in D$.

Proposition 2 If f is an improper integrable function, then $f\omega$ is an improper integrable function for every function ω which has bounded variation on R.

Theorem 2 For $T \in D'$ the following conditions are equivalent:

(i) T is a weak integrable distribution

(ii) $T \in D_{\overline{W}}'$

(iii) $T = \sum_{\alpha \leq r} <x>^\alpha g_\alpha^{(\alpha)}$, where g_α are improper integrable contunuous functions.

Proof The implications (i) $\Rightarrow$ (ii) and (iii) $\Rightarrow$ (i) easily follow from Propositions 1 and 2. We will show the implication (ii) $\Rightarrow$ (iii):

Let $\ell(x) = \int_0^x dt/(1+t^2)^{1/2}$ and ℓ^{-1} denote the function inverse to ℓ. It is not difficult to see that $\ell^{(\alpha)}(x)$ can be written in the form $\omega_\alpha(x)/<x>^\alpha$ for $\alpha = 1, 2, \ldots$, where ω_α are C^∞ functions with bounded variation on R. Obviously $\omega_1 \equiv 1$. Hence, for $\phi \in C^\infty$

$$<x>^\alpha (\phi \cdot \ell)^{(\alpha)}(x) = \phi^{(\alpha)} \cdot \ell(x) + \phi^{(\alpha-1)} \cdot \ell(x) \bar{\omega}_2(x) + \ldots$$

$$\ldots + \phi' \cdot \ell(x) \bar{\omega}_\alpha(x) \quad \text{for} \quad \alpha = 0, 1, \ldots,$$

where $\bar{\omega}_2, \ldots, \bar{\omega}_\alpha$ are some C^∞ functions with bounded variation on $\mathbb{R}$.

An elementary inductive proof leads to the conclusion that $\phi_y \in D_{\bar{L}}(R_y)$ iff $(\phi \cdot \ell)_x \in D_{\bar{W}}(R_x)$, and consequently $T_x \in D_{\bar{W}}'(R_x)$, iff $(T \cdot \ell^{-1})_y \ell^{-1'}(y) \in D_{\bar{L}}'(R_y)$. Using Theorem 1 we obtain:

$$T_x \in D_{\bar{W}}'(R_x) \text{ iff } (T \cdot \ell^{-1})_y \ell^{-1'}(y) = \sum_{\alpha \leq r} f_\alpha^{(\alpha)}(y)$$

for some improper integrable continuous functions f_α. So:

$$T_x = \sum_{\alpha \leq r} \ell'(x) f_\alpha^{(\alpha)} \cdot \ell(x).$$

One can prove inductively that the last sum can be written in the form:

$$\sum_{\beta \leq r} <x>^\beta (\ell'(x) h_\beta \cdot \ell(x))^{(\beta)},$$

where h_β are sums of the products of the functions f_α times some of $\omega_1 \cdot \ell^{-1}, \ldots, \omega_{r+1} \cdot \ell^{-1}$. So, using Propositions 1 and 2, we obtain: h_β are improper integrable continuous functions and, as a consequence, $g_\beta(x) = \ell'(x) h_\beta \cdot \ell(x)$ are such functions.

This finishes the proof. □

Remark 2 Another definition of integrability of distributions has been introduced in a similar way in [1]. The space $\overset{\circ}{B}_1{}'$ of integrable distributions is connected there with the following sequence of seminorms:

361

$$p_k(\phi) = \sum_{\alpha \leq k} \sup |<x>^\alpha \phi^{(\alpha)}(x)|.$$

We see that p_k is "a part" of w_k. One can prove the following properties:

$$\overset{\circ}{B}_i{}' \subsetneq D_{\overline{W}}{}', \quad D_{\overline{L}}{}' \subsetneq D_{\overline{W}}{}', \quad \overset{\circ}{B}_i{}' \not\subset D_{\overline{L}}{}', \quad \overset{\circ}{B}_i{}' \not\supset D_{\overline{L}}{}'.$$

We will recall now the Łojasiewicz definition of the value of a distribution at a point (see [2]):

Definition 3 We say that a distribution u has the value C at a point x_0, if the limit $\lim_{v \to +\infty} u_x[v\phi(v(x-x_0))]$ exists and equals $C\int\phi$ for every function $\phi \in D$. We will write $u(x_0) = C$.

Remark 3 If $u \in S'$, we can equivalently write $\sigma \in S$ instead of $\phi \in D$ in Definition 3.

For $u \in S'$ let $\hat{u}$ and $\check{u}$ denote the Fourier and the inverse Fourier transforms of u, i.e.:

$$\hat{\sigma}(\xi) = (2\pi)^{-1/2} \int_{-\infty}^{+\infty} \sigma(x) e^{-ix\xi} dx \quad \text{for} \quad \sigma \in S.$$

Remark 4 If $\hat{T}$ is a weak integrable distribution, then the distribution T has the value $(2\pi)^{-1/2} \int \hat{T}$ at the origin.

Proof For every fixed function $\sigma \in S$ the sequence $\sigma_v(x) = \sigma(x/v)$ converges to $\sigma(0)$ in $D_{\overline{W}}$, so $\hat{T}[\sigma_v] \to \int \hat{T} \sigma(0)$ and at the same time $T[\check{\sigma}_v] = \hat{T}[\sigma_v] = (2\pi)^{1/2} \hat{T}_x[(2\pi)^{-1/2} v\check{\sigma}(xv)]$. Since $\int (2\pi)^{-1/2} v\check{\sigma}(xv) dx = \sigma(0)$, then $\hat{T}$ has the value $(2\pi)^{-1/2} \int \hat{T}$ at the point 0.

Theorem 3 Let $u \in S'$. Assume that supp $\hat{u}$ is contained in a half-line. Then, u has a value at the origin iff $\hat{u}$ is a weak integrable distribution. If one of these conditions hold, then

$$u(0) = (2\pi)^{-1/2} \int \hat{u}.$$

Proof We need only to prove the implication "$\Rightarrow$". Obviously, we can assume that $u(0) = 0$.

It is enough to prove that the distribution $(\hat{u} \cdot \ell^{-1})_y \ell^{-1'}(y)$ is an improper integrable distribution in the sense of Definition 2 (function ℓ has been defined in the proof of Theorem 2). To this end we show, using Theorem 1, that $((\hat{u} \cdot \ell^{-1}) \ell^{-1'}) * \psi$ is an improper integrable continuous function for $\psi \in D$, i.e. for any sequence $\chi(r_v^-, r_v^+)$ of the characteristic functions of the intervals $< r_v^-, r_v^+ >$, $r_v^- \to -\infty$, $r_v^+ \to +\infty$, the limit $\lim ((\hat{u} \cdot \ell^{-1}) \ell^{-1'}) * \psi[X(r_v^-, r_v^+)]$ exists and is finite. Since the support of $\hat{u}^v$ is contained in a halfline, the same is true for $(\hat{u} \cdot \ell^{-1}) \ell^{-1'}$, and we can take the sequences $X(-r_v, r_v)$, $r_v \to +\infty$ instead of $X(r_v^-, r_v^+)$ in our considerations. We calculate:

$$((\hat{u} \cdot \ell^{-1}) \ell^{-1'}) * \psi[X(-r_v, r_v)] = (\hat{u} \cdot \ell^{-1}) \ell^{-1'} [X(-r_v, r_v) * \tilde{\psi}]$$

$$= \hat{u}[(X(-r_v, r_v) * \tilde{\psi}) \cdot \ell], \text{ where } \tilde{\psi}(x) = \psi(-x).$$

The assumptions and the Banach-Steinhaus theorem imply that $\hat{u}[\phi_v] \to 0$, for every sequence $\phi_v \in D$ satisfying the following conditions:

$$\text{supp } \phi_v \subset <-a_v, a_v > \text{ and } \sup |\phi_v^{(\alpha)}| \leq A_\alpha / a_v^\alpha \text{ for } \alpha = 0,1,\ldots, \quad (*)$$
$$v = 1, 2, \ldots, \quad a_v \to +\infty.$$

A simple calculation shows that the sequence $(X(-r_v, r_v) * \tilde{\psi}) \cdot \ell$ satisfies the conditions (*). Hence, $\hat{u}[(X(-r_v, r_v) * \tilde{\psi}) \cdot \ell] \to 0$, and this finishes the proof. □

<u>Corollary 1</u> Let $u \in D'$ and $\phi \in D$. Then, $(u\phi)^\wedge$ is a weak integrable distribution iff u and $(u\phi) * Vp(1/x)$ have values at the origin. If one of these conditions hold, then $u(0)\phi(0) = (2\pi)^{-1/2} \int (u\phi)^\wedge$.

<u>Corollary 2</u> For $u \in S'$ the following conditions are equivalent:

(i) $\hat{u}$ is a weak integrable distrribution

(ii) u and $u * \sigma Vp \frac{1}{x}$ have values at the origin for $\sigma \in S$

(iii) u and $u * \sigma Vp \frac{1}{x}$ have values at the origin for some $\sigma \in S$, $\sigma(0) \neq 0$.

If one of these conditions hold, then $u(0) = (2\pi)^{-1/2} \int \hat{u}$.

<u>Corollary 3</u> If $u \in S'$ and u is a C^1 function in a neighbourhood of the point 0, then $\hat{u}$ is a weak integrable distribution.

The following example shows that the statement from Remark 4 cannot be converted:

<u>Example</u> $\tilde{T}(x) = \text{sgn } x \, (\ln|x|)^{-1}$ for $0 < |x| < e^{-1}$ and $u(x) = 0$ for other $x \in \mathbb{R}$.

REFERENCES

1. P. Dierolf, J. Voigt, Calculation of the bidual for some function spaces, Integrable distributions, <u>Math</u>. <u>Ann</u>., 253, 63-87 (1980).
2. S. Łojasiewicz, Sur la valeur et la limite d'une distribution dans un point, <u>Studia Math</u>., 16, 1-36 (1957).
3. S. Łojasiewicz, "Wstęp do teorii funkcji rzeczywistych", PWN, Warszawa (1973).
4. R. Wawak, Improper integrals of distributions, <u>Studia Math</u>., (to appear).

SECTION III. CONVERGENCE STRUCTURES

ON INTERCHANGE OF LIMITS

Piotr Antosik

Institute of Mathematics
Polish Academy of Sciences
Katowice, Poland

1. INTRODUCTION

Using matrix methods we prove theorems on interchange of limits for matrices (double sequences) whose elements are in an abelian group equiped with a convergence. Proofs of theorems on interchange of limits, uniform convergence, equicontinuity, uniform countable additivity, uniform boundedness can be reduced to the problem of convergence to zero of diagonals of certain matrices, so-called K-matrices (see, [1], [2]). A matrix $\{x_{ij}\}$ whose elements x_{ij} for $i,j \in \mathbb{N}$ belong to an abelian group equiped with a convergence is said to be a K-matrix if for every increasing sequence $\{m_i\}$ of positive integers there are a subsequence $\{n_i\}$ of $\{m_i\}$ and a sequence $\{x_i\}$ such that the following conditions hold:

(i) $\sum_{k=1}^{j} x_{n_i n_k} \to x_i$ as $j \to \infty$ for $i \in \mathbb{N}$;

(ii) $x_i \to 0$

and

(iii) $x_{n_i n_j} \to 0$ as $i \to \infty$ for $j \in \mathbb{N}$.

In [1] it is shown that main diagonals of K-matrices whose elements are in a topological group converge to zero, i.e., if X is a topological group, $x_{ij} \in X$ for $i,j \in N$ and $\{x_{ij}\}$ is a K-matrix, then $x_{ii} \to 0$. In this paper we prove the same for matrices whose elements are in an abelian group equiped with a convergence satisfying some conditions. The conditions are expressed in terms of convergent sequences without using topological concepts.

2. THE MAIN LEMMA

We start this section with recalling the basic FLUSH-properties of a convergence in an abelian group. By a convergence in a set X we shall mean a set G of ordered pairs $(\{x_n\}, x)$ where $\{x_n\}$ is a sequence in X and $x \in X$. To denote that $(\{x_n\}, x) \in G$ we shall write $x_n \to x$ in (X,G) or, simply, $x_n \to x$. In the sequal we shall refer to the following properties of convergences in X.

F. If $x_n \to x$ and $\{u_n\}$ is a subsequence of $\{x_n\}$, then $u_n \to x$;

U. If for every subsequence $\{u_n\}$ of a given sequence $\{x_n\}$ there is a subsequence $\{v_n\}$ of $\{u_n\}$ such that $v_n \to x$ for a given x, then $x_n \to x$;

S. If $x_n = x$ for $n \in N$, then $x_n \to x$;

H. If $x_n \to x$ and $x_n \to y$, then $x = y$.

In the case when X is a abelian group we shall refer to the folloving property

L. If $x_n \to x$ and $y_n \to y$, then $x_n - y_n \to x - y$.

The next property of convergence is concerned with infinite matrices of elements in an abelian group.

Y. If $x_{ij} \in X$ for $i,j \in \mathbb{N}$, $x_{ij} \to 0$ as $j \to \infty$ for $i \in \mathbb{N}$ and $x_{ij} \to 0$ as $i \to \infty$ for $j \in \mathbb{N}$, then there is a subsequence $\{m_i\}$ of $\{i\}$ such that

$$\sum_{j \in A_i} x_{m_i m_j} \to 0,$$

whenever A_i is a finite subset of N and $i \notin A_i$ for $i \in N$.

A set (abelian group, linear space) equiped with a convergence is said to be a convergence space (convergence group, convergence linear space). Saying, for instance, that X is a FLYUS-convergence group we mean that X is equiped with a convergence satisfying conditions F, L, Y, U and S.

Obviously, the convergence in a quasi-normed group satisfies conditions FLUS. If $\|x_{ij}\| \to 0$ as $j \to \infty$ for $i \in \mathbb{N}$ and $\|x_{ij}\| \to 0$ as $i \to \infty$ for $j \in \mathbb{N}$, then, by induction, one can select a subsequence $\{m_i\}$ of $\{i\}$ such that $\|x_{m_i m_j}\| \leq 2^{-i-j}$ for $i,j \in \mathbb{N}$ and $i \neq j$. Hence it follows that the convergence in a quasi-normed group is a FLYUS-convergence.

Assume that L is an Archimedean Riesz space (see, [3]) and assume that $\{x_{ij}\}$ is a matrix of elements in L whose rows and columns are relatively uniformly star convergent to zero. If L has the σ-property, then there is $u \in L$ and a subsequence $\{m_i\}$ of $\{i\}$ such that $|x_{m_i m_j}| < 2^{-i-j} u$ for $i,j \in \mathbb{N}$ and $i \neq j$. Hence it follows that the relatively uniform star convergence in an Archimedean-Riesz space with the σ-property is a Hausdorff FLYSU-convergence.

Proposition 1 If X is a FLYS-convergence group, $x_{ij} \in X$ for $i,j \in \mathbb{N}$, $x_{ij} \to x_i$ as $j \to \infty$ for $i \in \mathbb{N}$ and $x_i \to x$, then for every increasing sequence $\{(m_i, n_i)\}$ in $\mathbb{N} \times \mathbb{N}$ there is a subsequence $\{(p_i, q_i)\}$ of $\{(m_i, n_i)\}$ such that $x_{p_i q_{r_i}} \to x$ whenever $r_i > i$ for $i \in \mathbb{N}$.

Proof Let $\{(m_i, n_i)\}$ be an increasing sequence in $\mathbb{N} \times \mathbb{N}$. Then, by F,L and S, $x_{m_i m_j} - x_{m_i} \to 0$ as $j \to \infty$ for $i \in \mathbb{N}$. Let $\{y_{ij}\}$ be a matrix such that

$$y_{ij} = 0 \text{ if } i \geq j \text{ and } y_{ij} = x_{m_i n_j} - x_{m_i} \text{ if } i < j$$

for $i,j \in \mathbb{N}$. By Y, there is a subsequence $\{s_i\}$ of $\{i\}$ such that

$$\sum_{j \in A_i} y_{s_i s_j} \to 0$$

whenever A_i are finite subsets of $\mathbb{N}$ and $i \notin A_i$ for $i \in \mathbb{N}$. We put $p_i = m_{s_i}$ and $q_i = n_{s_i}$ for $i \in \mathbb{N}$ and claim that for every sequence $\{r_i\}$ in $\mathbb{N}$ such that $r_i > i$ for $i \in \mathbb{N}$

$$x_{p_i q_{r_i}} \to x.$$

In fact, let $\{r_i\}$ be a sequence in $\mathbb{N}$ and let A_i be an onepoint set consisting only r_i for $i \in \mathbb{N}$. Then

$$\sum_{j \in A_i} y_{s_i s_j} = x_{p_i q_{r_i}} - x_{p_i}$$

and

$$x_{p_i q_{r_i}} - x_{p_i} \to 0.$$

By F, $x_{p_i} \to x$. Hence, by L, $x_{p_i q_{r_i}} \to x$ which was to be proved.

The Main Lemma If $\{x_{ij}\}$ is a K-matrix in FLYUS-convergence group, then $x_{ii} \to 0$.

Proof Assume that X is a FLYUS-convergence group and $\{x_{ij}\}$ is a K-matrix in X. Let $\{m_i\}$ be a subsequence of $\{i\}$ and let $u_{ij} = x_{m_i m_j}$ for $i,j \in \mathbb{N}$. Since $\{x_{ij}\}$ is a K-matrix, there is a square submatrix $\{v_{ij}\}$ of $\{u_{ij}\}$ whose rows and columns converge to zero. By property Y, there is a square submatrix $\{t_{ij}\}$ of $\{v_{ij}\}$ such that

$$\sum_{j \in A_i} t_{ij} \to 0 \tag{1}$$

whenever A_i are finite subsets of $\mathbb{N}$ and $i \notin A_i$ from $i \in \mathbb{N}$. Again, since $\{x_{ij}\}$ is a K-matrix, there is a square submatrix $\{w_{ij}\}$ of $\{t_{ij}\}$ such that

$$\sum_{j=1}^{\infty} w_{ij} \to 0 \tag{2}$$

By Proposition 1, there is a subsequence $\{(m_i, m_i)\}$ of $\{(i,i)\}$ such that

$$\sum_{k=1}^{m_{i+1}} w_{m_i k} \to 0.$$

We note that

$$w_{m_i m_i} = \sum_{j=1}^{m_{i+1}} w_{m_i j} - \sum_{j \in A_i} w_{m_i j}$$

where $A_i = \{j: 1 \leq j \leq m_{i+1}\} \setminus \{i\}$. Hence, by (1), (2) and L, we get $w_{m_{i+1} m_{i+1}} \to 0$. In this way we have shown that every subsequence of $\{x_{ii}\}$ has a subsequence which converges to zero. Consequently, by U, $x_{ii} \to 0$ which was to be proved. □

3. APPLICATIONS OF THE LEMMA

We prove theorems on existence and equality of itterated and double limits of matrices. Assume that X is a convergence space and $x_{ij} \in X$ for $i,j \in \mathbb{N}$. We shall write

$$\lim_{i \to \infty} \lim_{j \to \infty} x_{ij} = x$$

if there is a sequence $\{x_i\}$ such that $x_{ij} \to x_i$ as $j \to \infty$ for $i \in \mathbb{N}$ and $x_i \to x$. We write

Proof   Assume that $\{(m_i, n_i)\}$ is an increasing sequence in $\mathbb{N} \times \mathbb{N}$. We note that under conditions of the theorem, $x_{m_i n_j} \to 0$ as $j \to \infty$ for $i \in \mathbb{N}$. Therefore, by Proposition 1, there is a subsequence $\{(p_i, q_i)\}$ of $\{(n_i, n_i)\}$ such that for every $k \in \mathbb{N}$ we have $x_{p_i q_{i+k}} \to 0$. It is easy to check that under conditions of Theorem 1, the following

$$\{x_{p_{i+1} q_{j+1}} - x_{p_i q_{j+1}}\}$$

is a K-matrix. Hence, by the Lemma an L, $x_{p_{i+1} q_{i+1}} \to 0$. Consequently, $x_{ij} \to 0$ as $i, j \to \infty$. Hence, by Proposition 2, we get Theorem 1. □

Theorem 2   If for every subsequence $\{m_j\}$ of $\{i\}$ the itterated limit

$$\lim_{i \to \infty} \lim_{j \to \infty} \sum_{k=0}^{j} x_{im_k}$$

exists, then we have

$$\lim_{i,j \to \infty} \sum_{k=1}^{j} x_{ik} = \lim_{j \to \infty} \lim_{i \to \infty} \sum_{k=1}^{j} x_{ik} = \lim_{i \to \infty} \lim_{j \to \infty} \sum_{k=1}^{j} x_{ik}.$$

Proof   We note that under conditions of Theorem 2, we have

$$\lim_{i \to \infty} \lim_{j \to \infty} \sum_{k=1}^{j} x_{ik} = x$$

for some $x \in X$. Assume that $\{(m_i, n_i)\}$ is a increasing sequence in $\mathbb{N} \times \mathbb{N}$. By proposition 1, there is a subsequence $\{(p_i, q_i)\}$ of $\{(m_i, n_i)\}$ such that for every $r \in \mathbb{N}$ we have

$$\sum_{k=1}^{q_{i+r}} x_{p_i k} \to x.$$

It is easy to check that under conditions of Theorem 2, the matrix

$$\left\{ \sum_{k=1}^{q_{j+1}} x_{p_{i+1} k} - \sum_{k=1}^{q_{j+2}} x_{p_{i+1} k} + \sum_{k=1}^{q_{j+2}} x_{p_i k} - \sum_{k=1}^{q_{j+2}} x_{p_i k} \right\}$$

is a K-matrix. Hence, be the Lemma,

$$\sum_{k=1}^{q_{i+1}} x_{p_{i+1} k} \to x.$$

This and Proposition 2 imply Theorem 2. □

$$\lim_{i,j \to \infty} x_{ij} = x$$

if for every increasing sequence $\{(m_i, n_i)\}$ in $\mathbb{N} \ \mathbb{N}$ we have $x_{m_i n_i} \to x$. If U holds and for every subsequence $\{(m_i, n_i)\}$ there is a subsequence $\{(p_i, q_i)\}$ of $\{(m_i, n_i)\}$ such that

$$x_{p_i q_i} \to x$$

then (1) holds.

In the following proposition and three theorems we assume that X is a FLYUS-convergence, $x_{ij} \in X$ for $i, j \in \mathbb{N}$ and we assume that rows and columns of the matrix $\{x_{ij}\}$ converge.

Proposition 2 If

$$\lim_{i,j \to \infty} x_{ij} = x,$$

then

$$\lim_{j \to \infty} \lim_{i \to \infty} x_{ij} = \lim_{i \to \infty} \lim_{j \to \infty} x_{ij} = x.$$

Proof Assume that $x_{ij} \to x_i$ as $j \to \infty$ for $i \in \mathbb{N}$. Let $\{m_i\}$ be a subsequence of $\{i\}$. Then $x_{m_i m_j} - x_{m_i} \to 0$ as $j \to \infty$ for $i \in \mathbb{N}$. By Proposition 1, there is a subsequence $\{(p_i, p_i)\}$ of $\{(m_i, m_i)\}$ such that $x_{p_i p_{i+1}} - x_{p_i} \to 0$. Since $x_{p_i p_{i+1}} \to x$ we see that $x_{p_i} \to x$. Hence, by L and U, $x_i \in x$ or, equivalently,

$$\lim_{i \to \infty} \lim_{j \to \infty} x_{ij} = x.$$

Similarly one can show that

$$\lim_{j \to \infty} \lim_{i \to \infty} x_{ij} = x. \quad \square$$

Theorem 1 If for every subsequence $\{m_i\}$ of $\{i\}$ there is a subsequence $\{n_i\}$ of $\{m_i\}$ such that the itterated limit

$$\lim_{i \to \infty} \lim_{j \to \infty} \sum_{k=1}^{j} x_{i n_k}$$

exists, then

$$\lim_{i,j \to \infty} x_{ij} = \lim_{j \to \infty} \lim_{i \to \infty} x_{ij} = \lim_{i \to \infty} \lim_{j \to \infty} x_{ij} = 0.$$

Theorem 3 If for every $j \in \mathbb{N}$ the limit

$$\lim_{i \to \infty} \sum_{k=1}^{i} x_{kj}$$

exists and for every subsequence $\{m_i\}$ of $\{i\}$ the itterated limit

$$\lim_{i \to \infty} \lim_{j \to \infty} \sum_{k=1}^{i} \sum_{\ell=1}^{j} x_{km_\ell}$$

exists, then we have

$$\lim_{i,j \to \infty} \sum_{k=1}^{i} \sum_{\ell=1}^{j} x_{k\ell} = \lim_{i \to \infty} \lim_{j \to \infty} \sum_{k=1}^{i} \sum_{\ell=1}^{j} x_{k\ell} = \lim_{j \to \infty} \lim_{i \to \infty} \sum_{k=1}^{i} \sum_{\ell=1}^{j} x_{k\ell}.$$

Proof We note that under conditions of Theorem 3, we have

$$\lim_{i \to \infty} \lim_{j \to \infty} \sum_{k=1}^{i} \sum_{\ell=1}^{j} x_{k\ell} = x$$

for some $x \in X$. Assume that $\{(m_i, n_i)\}$ is an increasing sequence in $\mathbb{N} \times \mathbb{N}$. By Proposition 1, there is a subsequence $\{(p_i, q_i)\}$ of $\{(m_i, n_i)\}$ such that for every $r \in \mathbb{N}$ we have

$$\sum_{k=1}^{p_i} \sum_{\ell=1}^{q_{i+r}} x_{k\ell} \to x.$$

It is easy to check that under conditions of Theorem 3, the matix

$$\left\{ \sum_{k=1}^{p_{i+1}} \sum_{\ell=1}^{q_{j+1}} x_{k\ell} - \sum_{k=1}^{p_{i+1}} \sum_{\ell=1}^{p_{j+2}} x_{k\ell} + \sum_{k=1}^{p_i} \sum_{\ell=1}^{q_{j+2}} x_{k\ell} - \sum_{k=1}^{p_i} \sum_{\ell=1}^{q_{j+2}} x_{k\ell} \right\}$$

is a K-matrix. Hence, by the Lemma, we get

$$\sum_{k=1}^{p_{i+1}} \sum_{\ell=1}^{q_{i+1}} x_{k\ell} \to x.$$

This and Proposition 2 imply Theorem 3. □

REFERENCES

1. P. Antosik, A lemma on matrices and its applications, <u>Contemporary Mathematics</u>, Volume 52, 89-95, (1986).

2. P. Antosik, Ch. Swartz, "Matrix Methods in Analysis", Springer-Verlag, V. 1113, (1985).
3. W. A. J. Luxemburg, A. C. Zaanen, "Riesz spaces", V. 1., North-Holland Publishing Company, Amsterdam – London, (1971).

COUNTABILITY, COMPLETENESS AND THE CLOSED GRAPH THEOREM

R. Beattie[1] and H. -P. Butzmann[2]

[1] Dept. of Mathematics and Computer Science
Mount Allison University, Sackville, N.B., Canada, EOA 3C0

[2] Fakultät für Mathematik und Informatik
Universität Mannheim, 6800 Mannheim, BRD

The webs of M. De Wilde [4] have made an enormous contribution to the closed graph theorems in locally convex spaces(ℓcs). Although webs have a very intricate layered construction, two properties in particular have contributed to the closed graph theorem. First of all, webs possess a strong countability condition in the range space which suitably matches the Baire property of Fréchet spaces in the domain space; as a result the zero neighbourhood filter is mapped to a p-Cauchy filter, a filter attempting to settle down. Secondly webs provide a completeness condition which allow p-Cauchy filters to converge.

In [1] and [2] webs were examined in the context of convergence spaces. The webs gave rise to a convergence vector space (cvs), the web-space, and the countability and completeness properties of the web were reflected in similar properties of the web-space. The web-space turned out to play a central role in De Wilde type closed graph theorems.

In this paper we show that neither the web nor the web-space is required for this type of closed graph theorem. Its validity depends precisely on the countability and completeness properties of the web-space and not on the intrinsic structure of the web itself.

We assume throughout that cvs possess point-separating linear functionals and that the filters are balanced, i.e., for every convergent filter there is a coarser convergent filter with a basis of balanced sets.

1. COUNTABILITY

Definition 1.1   Recall a cvs E is called <u>first</u> <u>countable</u> if for every $G \to 0$ in E, there is an $F \subset G$ such that $F \to 0$ and F has a countable basis.

E is called <u>strongly first countable</u> if it has a countable local basis at 0, i.e., there is a countable collection C of subsets of E such that for every $G \to 0$ in E there is an $F \subset G$ with $F \to 0$ and F has a basis of elements from C.

Clearly strongly first countable cvs are first countable but not conversely. (See [3].)

<u>Definition 1.2</u>   Let F be a cvs. A filter G in F is <u>p-Cauchy</u> if $\exists A \to 0$ in E such that
$$\forall A \in A, \; \exists G_0 \in G, \; \forall G \in G, \; G_0 \subset G + A.$$
Note that if F is a locally convex topological vector space one can always use the neighbourhood filter V of 0 for A.

<u>Definition 1.3</u>   Let E,F be cvs and $f : E \to F$ a linear mapping. Then f is <u>nearly continuous</u> if $\forall G \to 0$ in E, $\exists F \subset G$ such that $F \to 0$ in E and $f(F)$ is p-Cauchy.

It is easy to see that if E is a lcs with 0 neighbourhood filter U, then f is nearly continuous iff $f(U)$ is p-Cauchy.

Similarly, if both E,F are lcs with 0 neighbourhood filters U and V respectively, then f is nearly continuous iff $f^{-1}(V) \subset U$.

<u>Lemma 1.4</u>   Let E be a Fréchet space with 0 neighbourhood filter U, F a strongly first countable cvs, $\underline{f : E \to F}$ a linear mapping. Then there is a filter $G_0 \to 0$ in F such that $f^{-1}(G_0) \subset U$.

<u>Proof</u>   Let C be a countable local basis of 0 in F. Suppose the claim is false. Consider the family
$$\tilde{C} = \{C \in C \mid \overline{f^{-1}(C - C)} \not\subset U\}.$$
We show $\tilde{C}$ is a local covering in F at 0. Suppose $G \to 0$ in F. Then $\exists G' \subset G$ such that $G' \to 0$ <u>in F and G' has</u> a basis of elements in C. Since $G' - G' \to 0$, by assumption $\overline{f^{-1}(G' - G')} \not\subset U$ and so there is some $G \in G'$ with $\overline{f^{-1}(G - G)} \not\subset U$. Since G' has a basis in C, there is a $C \in C \cap G'$ with $C \subseteq G$ and hence $\overline{f^{-1}(C - C)} \not\subset U$. Thus $C \in G' \cap \tilde{C} \subset G \cap \tilde{C}$ as required.

Now assume $y \in F$; then $\left\{\frac{1}{i} y\right\} \to 0$ in F and so $\exists C \in \tilde{C} \cap \left[\left\{\frac{1}{i} y\right\}\right]$, i.e. $\exists i_0$ such that $\left\{\frac{1}{i} y \mid i \geq i_0\right\} \subset C$. It follows that $y \in i_0 C$ and so

376

$$F = \cup\{iC \mid i \in \mathbb{N}, C \in \check{C}\}.$$

Thus

$$E = f^{-1}(F) = \cup\{if^{-1}(C) \mid i \in \mathbb{N}, C \in \check{C}\}.$$

Since E is second category, $\exists i_0 \in \mathbb{N}$, $C_0 \in \check{C}$, $x_0 \in E$ and $U_0 \in U$ such that

$$x_0 + U_0 \subset \overline{i_0 f^{-1}(C_0)}$$

and thus $\left(\frac{1}{i_0}\right)x_0 + \left(\frac{1}{i_0}\right)U_0 \subset \overline{f^{-1}(C_0)}$.

Let $\tilde{x} = \left(\frac{1}{i_0}\right)x_0$, $\tilde{U} = \left(\frac{1}{i_0}\right)U_0$; then $\tilde{x} + \tilde{U} \subset \overline{f^{-1}(C_0)}$.

Hence $\tilde{x} + \tilde{U} - (\tilde{x} + \tilde{U}) \subset \overline{f^{-1}(C_0)} - \overline{f^{-1}(C_0)} \subset \overline{f^{-1}(C_0) - f^{-1}(C_0)} \subset \overline{f^{-1}(C_0 - C_0)}$.

It follows that $\overline{f^{-1}(C_0 - C_0)} \in U$ contradicting the choice of C_0. □

Lemma 1.5 Let E be a lcs, F a cvs and $f : E \to F$ linear. Suppose for some filter $G \to 0$ in F, $f^{-1}(G) \subset U$. Then $f(U)$ is p-Cauchy and so f is nearly continuous.

Proof We show $\forall G \in G$, $\exists U \in U$, $\forall V \in U$, $f(U) \subset f(V) + G$.
So fix $G \in G$ and consider $U = f^{-1}(G) \in U$. Then $U \subset f^{-1}(G) + V$ for all $V \in U$ and so $f(U) \subset f(f^{-1}(G)) + f(V) \subset G + f(V)$. □

Lemmas 1.4 and 1.5 combine to show that if E is a Fréchet space with 0 neighbourhood filter U, F is a strongly first countable cvs and $f : E \to F$ is a linear mapping, then $f(U)$ is p-Cauchy.

In the following section we study conditions which permit p-Cauchy filters to converge.

2. COMPLETENESS AND THE CLOSED GRAPH THEOREM

Definition 2.1 Let E be a first countable cvs. A filter F on E is said to have <u>a rapid basis</u> if a countable basis (F_n) of F can be found such that

$$F_{n+1} + F_{n+1} \subset F_n \quad \text{for all n.}$$

It is easy to show that if F has a rapid basis, then F is equable, i.e. $N \cdot F = F$ where N is the 0 neighbourhood filter in $\mathbb{R}$.

The reason for the name rapid basis is the following:

377

Lemma 2.2 Let (F_n) be a rapid basis for F and let (x_n) be a sequence in E with $x_n \in F_n$ for all n. Then the series $s_n = \sum_{k=1}^{n} x_k$ is Cauchy in E if $F \to 0$ in E.

Proof Fix n_0 and let $n > m \geq n_0$. Then

$$s_n - s_m = \sum_{k=m+1}^{n} x_k = x_{m+1} + x_{m+2} + \ldots + x_{n-1} + x_n$$

$$\in F_{m+1} + F_{m+2} + \ldots + F_{n-1} + F_n$$

$$\subset F_m$$

$$\subset F_{n_0}.$$

Hence the series is Cauchy in E. □

Definition 2.3 A cvs E is called <u>ultracomplete</u> if $\forall G \to 0$, $\exists F \subset G$, $F \to 0$, F has a rapid basis (F_n) and the filter $\left[\sum_{k=n}^{\infty} F_k \mid n \in \mathbb{N}\right]$ converges in E.

Tacitly assumed in the definition is that the filter elements make sense, i.e., for every rapid basis (F_n) and every sequence (x_n), $x_n \in F_n$, the series $\sum_{n=1}^{\infty} x_n$ converges in E. For equable cvs, this latter condition is equivalent to completeness. Thus, for our setting, ultracompleteness implies completeness.

Theorem 2.4 Let E be a Fréchet space, F a strongly first countable, ultracomplete cvs $f : E \to F$ a linear mapping with a closed graph. Then f is continuous.

Proof By Lemma 1.5, f is nearly continuous so that $f(U)$ is p-Cauchy. Thus $\exists G \to 0$, $\forall G \in G$, $\exists U \in U$, $\forall V \in U$, $f(U) \subseteq f(V) + G$. Without loss of generality, assume G has a rapid basis and $\left[\sum_{k=n}^{\infty} G_k \mid n \in \mathbb{N}\right]$ converges in F. We can find $U_n \in U$ so that $f(U_n) \subseteq f(V) + G_n$ for all $V \in U$ and so $U_n \subseteq V + f^{-1}(G_n)$ for all $V \in U$. Suppose (wlog again) (U_n) is a neighbourhood basis of 0. Then it follows that

$$U_n \subseteq f^{-1}(G_n) + U_{n+1}.$$

We show that $f(U_n) \subseteq \sum_{k=n}^{\infty} G_k$ for all $n \in \mathbb{N}$. So let $z \in U_n$. Then

378

$$z = w_n + u_{n+1}, \qquad w_n \in f^{-1}(G_n), \qquad u_{n+1} \in U_{n+1}$$

$$= w_n + w_{n+1} + u_{n+2}$$

$$\vdots$$

$$= w_n + w_{n+1} + \ldots + w_{n+r} + u_{n+r+1}, \qquad w_i \in f^{-1}(G_i), \quad u_i \in U_i.$$

Clearly $\sum_{k=n}^{m} w_k$ converges to z and $f\left\{\sum_{k=n}^{m} w_k\right\} = \sum_{k=n}^{m} f(w_k)$ converges also. Since the graph of f is closed,

$$f(z) = \lim_{m \to \infty} \sum_{k=n}^{\infty} f(w_k) = \sum_{k=n}^{\infty} f(w_k) \in \sum_{k=n}^{\infty} G_k.$$

Hence $f(U_n) \subset \sum_{k=n}^{\infty} G_k$ and

$$f(U) \supseteq \left[\sum_{k=n}^{\infty} G_k \mid n \in \mathbb{N}\right].$$

Since $\left[\sum_{k=n}^{\infty} G_k \mid n \in \mathbb{N}\right]$ converges, so does $f(U)$. □

The following corollary increases the scope of Theorem 2.4.

Corollary 2.5 Let E be a Fréchet space, F a cvs which admits a finer cvs structure which is both strongly first countable and ultracomplete. Any linear mapping $f : E \to F$ with a closed graph is continuous.

Proof If F_λ denotes the new structure on F, then by Theorem 2.4, $f : E \to F_\lambda$ is continuous. □

3. APPLICATIONS TO WEB-SPACES

Definition 3.1 A <u>web</u> $\mathcal{W}$ on a vector space F is a countable family of balanced subsets of F indexed by the finite sequences of integers and arranged in layers. The first layer is a sequence (W_i) whose union absorbs each point of F. For each set W_i, there is a sequence W_{ij} of subsets of $(1/2)W_i$ whose union as j varies absorbs W_i. The collection W_{ij} as both i and j vary form the second layer. In the same way, for each set W_{ij} there is a sequence (W_{ijk}) of subsets of $(1/2)W_{ij}$ whose union as k varies absorbs W_{ij}. The collection W_{ijk} as i, j and k vary make up the third layer. And so on for subsequent layers.

By a strand S of $\mathcal{W}$ is meant any sequence of sets $W_i, W_{ij}, W_{ijk}\ldots$, one from each layer, each set (from the second on) being one of those determined by its predecessor. When considering a single strand S, we adopt the simpler notation ([6]) $S = (W_k)$ to indicate that for each k, W_k is a set in the kth layer of the web.

The strands are filter bases and give rise to a convergence space. With a slight assumption on the web (see [2]) this becomes a convergence vector space which we call the web-space. A cvs F is said to be webbed if the strands of the web-space F_W converge in F, i.e., if the identity id : $F_W \to F$ is continuous.

A web is called <u>strict</u> [5] (tight [6]) if for each strand $S = (W_k)$ and every sequence (x_k) with $x_k \in W_k$ for all k, the series $\sum_{k=1}^{\infty} x_k$ converges in F and $\sum_{r=k+1}^{\infty} x_r \in W_k$ for all k.

By the very definition of a web we have the following:

<u>Proposition 3.2</u>   Let F be a cvs webbed by W. Then the web-space F_W has the following properties:

(i)   Each strand $S = (W_k)$ is a rapid basis, i.e. $W_{k+1} + W_{k+1} \subset W_k$ for all k.

(ii) F_W is strongly first countable.

<u>Proposition 3.3</u>   Let F be a cvs webbed by W. If the web W is strict, then the web-space is ultracomplete.

A close look at the condition of strictness shows that this is apparently a stronger condition than the ultracompleteness of F_W. For F_W to be ultracomplete $\left[\sum_{k=n}^{\infty} W_k \mid n \in \mathbb{N}\right]$ must converge in F_W for each strand $S = (W_k)$. For W to be strict, $\left[\sum_{k=n}^{\infty} W_k \mid n \in \mathbb{N}\right]$ must equal S for $S = (W_k)$.

<u>Theorem 3.4</u>   (cf. [2, Theorem 2.6])   Let E be a Fréchet space, F a cvs with a strict web, f : E $\to$ F a linear mapping with a closed graph. Then f is continuous.

<u>Proof</u>   The web-space F_W is strongly first countable and ultracomplete and so f : E $\to$ F_W is even continuous. □

Some questions arise in connection with the previous results. First of all, do ultracomplete cvs share any of the remarkable permanence properties of cvs with strict webs. Secondly, is the full force of ultracompleteness necessary in Corollary 2.5 for the continuity of f : E $\to$ F? What is actually obtained is a much sharper result, namely the continuity of f : E $\to$ F_λ for a finer cvs F_λ.

REFERENCES

1. R. Beattie, Convergence spaces with webs, <u>Math</u>. <u>Nachr</u>. 116, 159-164 (1984).
2. R. Beattie, A convenient category for the closed graph theorem, Categorical Topology, Proc. Conference Toledo, Ohio 1983, Heldermann, Berlin, 29-45 (1984).
3. R. Beattie and H. -P. Butzmann, Strongly first countable convergence spaces, Convergence Structures 1984, Proc. Conference on Convergence, Bechyne, Czechoslovakia, Akademie-Verlag, Berlin, 39-46 (1985).
4. M. De Wilde, Closed Graph Theorems and Webbed Spaces, <u>Research Notes in Mathematics</u> 19, Pitman, London (1978).
5. H. Jarchow, "Locally Convex Spaces", Teubner, Stuttgart (1981).
6. W. Robertson, On the closed graph theorem and spaces with webs, <u>Proc</u>. <u>London</u> <u>Math</u>. <u>Soc</u>. (3) 24, 692-738 (1972).

INDUCTIVE LIMITS OF RIESZ SPACES

Wolfgang Filter

Mathematik ETH-Zentrum
CH-8092 Zürich
Switzerland

ABSTRACT

The inductive limit of a family of Riesz spaces is introduced and investigated.

1. INTRODUCTION

In this paper we shall present the basic properties of inductive limits of families of Riesz spaces. Such inductive limits appear in various places in literature: The extended order dual of a Riesz space (introduced by Luxemburg and Masterson [6]) is one example, the space of generalized functions in measure theory (considered by Constantinescu [3]) is a special case of the first example, and also the space $C_\infty(X)$ of extended real-valued continuous functions f with a dense set $\{|f| = \infty\}$ on some Stonian space X is an inductive limit; see also Section 4.

The basic properties of inductive limits are listed in Section 2. There are no really surprising results, in particular the (set-theoretical) inductive limit of a family (E_ι) of Riesz spaces is a Riesz space again provided the maps $\phi_{\lambda\iota} : E_\iota \to E_\lambda$ are Riesz homomorphisms (Corollary 2.3). Since most of the proofs are simply a matter of routine, we shall often omit them. In contrast, lateral completeness seems to be a more delicate problem; we treat it in Section 3.

For notations, we refer to the standard book [7].

2. BASIC PROPERTIES

We fix an upper directed ordered non-empty set I, a family $(E_\iota)_{\iota \in I}$ of Riesz spaces and, for all $\iota, \lambda \in I$, $\iota \leq \lambda$, Riesz homomorphisms $\phi_{\lambda\iota} : E_\iota \to E_\lambda$ such that $\phi_{\kappa\iota} = \phi_{\kappa\lambda} \circ \phi_{\lambda\iota}$ whenever $\iota \leq \lambda \leq \kappa$.

F denotes the disjoint union of the sets E_ι, and $\iota(x)$ the element of I with $x \in E_{\iota(x)}$, for each $x \in F$. An equivalence relation $\sim$ is defined on F by setting

$$x \sim y :\Leftrightarrow \exists \lambda \in I,\ \lambda \geq \iota(x),\ \lambda \geq \iota(y),\ \phi_{\lambda,\iota(x)}(x) = \phi_{\lambda,\iota(y)}(y).$$

Denoting by $\dot x$ the equivalence class of $x \in F$, the set

$$E := \{\dot x : x \in F\}$$

is the inductive limit of the sets E_ι with respect to the maps $\phi_{\lambda\iota}$ ([2], §1, no. 11).

Obviously, $x \sim \phi_{\lambda,\iota(x)}(x)$ if $\lambda \geq \iota(x)$ (in particular $x \sim \phi_{\iota\iota}(x)$ for $x \in E_\iota$), and $\phi_{\lambda,\iota(x)}(x) \leq \phi_{\lambda,\iota(y)}(y)$ whenever $\lambda \geq \kappa \geq \iota(x), \iota(y)$ and $\phi_{\kappa,\iota(x)}(x) \leq \phi_{\kappa,\iota(y)}(y)$.

<u>Proposition 2.1</u> For $x, y \in F$, $x' \in \dot x$ and $y' \in \dot y$, the following are equivalent:

a) $\exists \kappa \geq \iota(x), \iota(y),\ \phi_{\kappa,\iota(x)}(x) \leq \phi_{\kappa,\iota(y)}(y)$.

b) $\exists \lambda \geq \iota(x'), \iota(y'),\ \phi_{\lambda,\iota(x')}(x') \leq \phi_{\lambda,\iota(y')}(y')$.

<u>Proof</u>. We need only show a $\Rightarrow$ b. There are $\iota \geq \iota(x), \iota(x')$ such that $\phi_{\iota,\iota(x)}(x) = \phi_{\iota,\iota(x')}(x')$ and $\iota' \geq \iota(y), \iota(y')$ such that $\phi_{\iota',\iota(y)}(y) = \phi_{\iota',\iota(y')}(y')$. There exists $\lambda \geq \iota, \iota', \kappa$, and we get

$$\phi_{\iota,\iota(x')}(x') = \phi_{\lambda\iota}(\phi_{\iota,\iota(x')}(x')) = \phi_{\lambda\iota}(\phi_{\lambda,\iota(x)}(x)) =$$

$$= \phi_{\lambda,\iota(x)}(x) = \phi_{\lambda\kappa}(\phi_{\kappa,\iota(x)}(x)) \leq \phi_{\lambda\kappa}(\phi_{\kappa,\iota(y)}(y)) = \dots =$$

$$= \phi_{\lambda,\iota(y')}(y'). \quad \square$$

By 2.1, we can introduce unambiguously an order relation on E by setting

$$\dot x \leq \dot y :\Leftrightarrow \exists \kappa \geq \iota(x), \iota(y),\ \phi_{\kappa,\iota(x)}(x) \leq \phi_{\kappa,\iota(y)}(y).$$

In the same way, one can show that

$$\alpha \dot x + \beta \dot y := \overline{\phi_{\kappa,\iota(x)}(\alpha x) + \phi_{\kappa,\iota(y)}(\beta y)},\ \text{where}\ \kappa \geq \iota(x), \iota(y)\ \text{is arbitrary, introduces a well-defined vector space structure on } E.$$

Proposition 2.2 For $x, y \in F$, there exists $\dot{x} \wedge \dot{y}$ and is equal to $\overline{\phi_{\kappa,\iota(x)}(x) \wedge \phi_{\kappa,\iota(y)}(y)}$, where $\kappa \geq \iota(x), \iota(y)$ is arbitrary; an analogous statement holds for $\dot{x} \vee \dot{y}$.

Proof. Let $\kappa \geq \iota(x), \iota(y)$. $\phi_{\kappa\kappa}(\phi_{\kappa,\iota(x)}(x) \wedge \phi_{\kappa,\iota(y)}(y)) = \phi_{\kappa,\iota(x)}(x) \wedge \phi_{\kappa,\iota(y)}(y)$, hence $\overline{\phi_{\kappa,\iota(x)}(x) \wedge \phi_{\kappa,\iota(y)}(y)} \leq \dot{x}, \dot{y}$, by definition of $\leq$. On the other hand, if $\dot{z} \leq \dot{x}, \dot{y}$, then there is $\lambda \geq \iota(z), \kappa$ such that $\phi_{\lambda,\iota(z)}(z) \leq \phi_{\lambda,\iota(x)}(x), \phi_{\lambda,\iota(y)}(y)$. We get $\phi_{\lambda,\iota(z)}(z) \leq \phi_{\lambda\kappa}(\phi_{\kappa,\iota(x)}(x) \wedge \phi_{\kappa,\iota(y)}(y))$ which implies $\dot{z} \leq \overline{\phi_{\kappa,\iota(x)}(x) \wedge \phi_{\kappa,\iota(y)}(y)}$. □

Now, we can easily verify our first natural results:

Corollary 2.3 E endowed with the structures introduced above is a Riesz space.

Corollary 2.4 For each $\iota \in I$, the map
$$\phi_\iota : E_\iota \to E, \ x \to \dot{x}$$
is a Riesz homomorphism.

Corollary 2.5 A map ψ from E to a Riesz space G is a Riesz homomorphism iff all $\psi \circ \phi_\iota$ are Riesz homomorphisms.

In the most interesting cases, the maps $\phi_{\lambda\iota}$ are injective. <u>So from now on, we make this assumption.</u> Then, by the definition of $\sim$, each ϕ_ι is injective, and $\phi_{\iota\iota}$ is the identity map.

We identify E_ι with a Riesz subspace of E via the Riesz homomorphism ϕ_ι and with a Riesz subspace of E_λ (for $\iota \leq \lambda$) via $\phi_{\lambda\iota}$. Hence, under these identifications, $E = \bigcup_{\iota \in I} E_\iota$ and $E_\iota \subset E_\lambda$ for $\iota \leq \lambda$, and for $x \in E$ the assertion "$x \in E_\iota$" has to be read as: There is a unique $x_\iota \in E_\iota$ with $\dot{x}_\iota = x$.

Proposition 2.6 For $\iota \in I$, the following are equivalent:
a) E_ι is a solid subspace of E.
b) E_ι is a solid subspace of E_λ, for each $\lambda \geq \iota$.

Proof. a $\Rightarrow$ b is obvious. For b $\Rightarrow$ a, take $x \in E$, $y \in E_\iota$ with $0 \leq x \leq y$. There exists $\lambda \geq \iota$ with $x \in E_\lambda$, and hence $0 \leq x \leq y$ in E_λ which implies $x \in E_\iota$. □

Proposition 2.7   For $\iota \in I$, the following assertions hold:

a)   If E_ι is a band of E_λ for all $\lambda \geq \iota$, then E_ι is a band of E.

b)   If E_ι is a band of E, and if $\lambda \geq \iota$ such that E_λ is order dense in E_κ whenever $\kappa \geq \lambda$, then E_ι is a band of E_λ.

Proof.   a)   follows from the fact: If (x_α) is a family from E_ι with $x_\alpha \uparrow x$ in E, and if $\lambda \geq \iota$ with $x \in E_\lambda$, then $x_\alpha \uparrow x$ in E_λ.

b)   Let $0 \leq x_\alpha \uparrow x$ in E_λ, with $x_\alpha \in E_\iota$. If $z \geq x_\alpha$ for all α, and if $\kappa \geq \lambda$ with $z \in E_\kappa$, then by [5], 17A, $z \geq x$, which implies $x_\alpha \uparrow x$ in E. Hence $x \in E_\iota$. □

One easily verifies the following facts as well:

Proposition 2.8   For $\iota \in I$, the following are equivalent:

a) E_ι is order dense in E.

b) E_ι is order dense in E_λ, for each $\lambda \geq \iota$.

Proposition 2.9 E is Archimedean iff each E_ι is Archimedean.

Proposition 2.10   Assume E_ι is a solid subspace of E_λ whenever $\lambda \geq \iota$. Then E has the countable-sub-property iff each E_ι has it.

Using [7], 24.9, it is not difficult to prove

Proposition 2.11   Assume E_ι is a solid subspace of E_λ whenever $\lambda \geq \iota$. Then E has the projection property (resp. the principal projection property) iff each E_ι has the same property.

Also, the Dedekind completeness shows the expected behaviour:

Proposition 2.12   Assume E_ι is a solid subspace of E_λ whenever $\lambda \geq \iota$. Then E is Dedekind complete (resp. Dedekind σ-complete) iff all E_ι are.

Proof.   Suppose all E_ι are Dedekind complete, and let $0 \leq x_\alpha \uparrow \leq x$ in E. There exists $\iota \in I$ with $x \in E_\iota$ and, for each α, an index $\kappa(\alpha) \geq \iota$ with $x_\alpha \in E_{\kappa(\alpha)}$. Hence $0 \leq x_\alpha \uparrow \leq x$ in E_ι, which implies the existence of z with $x_\alpha \uparrow z$ in E_ι. By 2.6, $x_\alpha \uparrow z$ in E. □

The following Theorem tells us that the processes of constructing Dedekind completions and inductive limits can be interchanged if the $\phi_{\lambda\iota}$ are order continuous:

Theorem 2.13 Assume E_ι is a solid subspace of E_λ and $\phi_{\lambda\iota}$ is order continuous whenever $\lambda \geq \iota$. Suppose further that all E_ι are Archimedean, and denote by $\bar{E}_\iota$ the Dedekind completion of E_ι. Then:

a) For $\iota \leq \lambda$, $\phi_{\lambda\iota}$ extends to a uniquely determined injective order continuous Riesz homomorphism $\bar{\phi}_{\lambda\iota} : \bar{E}_\iota \to \bar{E}_\lambda$.

b) $\bar{\phi}_{\kappa\iota} = \bar{\phi}_{\kappa\lambda} \circ \bar{\phi}_{\lambda\iota}$ whenever $\iota \leq \lambda \leq \kappa$.

c) $\bar{\phi}_{\lambda\iota}(\bar{E}_\iota)$ is a solid subspace of $\bar{E}_\lambda$, for $\iota \leq \lambda$.

Denoting by $\bar{E}$ the inductive limit of $(\bar{E}_\iota)_{\iota \in I}$ with respect to the maps $\bar{\phi}_{\lambda\iota}$, we have

d) $\bar{E}$ is the Dedekind completion of E.

Proof. a) Denote by $\psi_{\lambda\iota}$ the injective order continuous Riesz homomorphism $E_\iota \to \bar{E}_\lambda$, $x \to \phi_{\lambda\iota}(x)$. For each $0 \leq x \in \bar{E}_\iota$ there is $z \in E_\iota$ with $x \leq z$, which implies $\sup\{\psi_{\lambda\iota}(y) : y \in E_\iota, 0 \leq y \leq x\} < \infty$. [5], 17B, 17C give the assertion.

b) follows from the order denseness of E_ι in $\bar{E}_\iota$ and the order continuity of $\bar{\phi}_{\kappa\iota}$ and $\bar{\phi}_{\kappa\lambda} \circ \bar{\phi}_{\lambda\iota}$.

c) Let $x \in \bar{E}_\iota$, $y \in \bar{E}_\lambda$ such that $0 \leq y \leq \bar{\phi}_{\lambda\iota}(x)$. Since $y = \bigvee_{\substack{z \in E_\lambda \\ 0 \leq z \leq y}} z$, and all these z are members of $\phi_{\lambda\iota}(E_\iota)$, we get

$$y \leq \bigvee_{\substack{w \in E_\iota \\ 0 \leq \phi_{\lambda\iota}(w) \leq y}} \phi_{\lambda\iota}(w) \leq \bigvee_{\substack{v \in E_\iota \\ 0 \leq \bar{\phi}_{\lambda\iota}(v) \leq y}} \bar{\phi}_{\lambda\iota}(v) \leq y,$$

which gives, using [5], 17D, the assertion.

d) By 2.12, $\bar{E}$ is Dedekind complete, and it is easily checked that E is a Riesz subspace of $\bar{E}$ such that, for $0 < z \in \bar{E}$, there exist $x, y \in E$ with $0 < x \leq z \leq y$. □

Proposition 2.14 Assume E_ι is a solid subspace of E_λ whenever $\lambda \geq \iota$. Then, E is uniformly complete iff all E_ι are.

Proof. We shall only show the "if"-part. So let $0 \leq u \in E$, and let $(x_n)_{n \in \mathbb{N}}$ be a u-uniform Cauchy sequence in E. There is $p \in \mathbb{N}$ such that $|x_n - x_p| \leq u$ whenever $n \geq p$. Furthermore, there exists $\iota \in I$ such that $u, x_1, \ldots, x_p \in E_\iota$. Hence $x_n = (x_n - x_p) + x_p \in E_\iota$, for each $n \geq p$. Thus $(x_n)_{n \in \mathbb{N}}$ is a u-uniform Cauchy sequence in E_ι, which consequently u-converges in E_ι and hence also in E. □

Finally, we want to investigate the consequences if the E_ι are equipped with locally convex-solid topologies τ_ι, and E is equipped with the inductive topology τ with respect to the maps ϕ_ι. For the notation used here, see [1].

Proposition 2.15 Assume E_1 is a solid subspace of E_λ whenever $\lambda \geq 1$. Then, with the just mentioned meaning of τ_1 and τ, we have:

a) τ is locally convex-solid.

b) If all τ_1 satisfy the Lebesgue property (resp. the σ-Lebesgue property, resp. the pre-Lebesgue property), then the same holds for τ.

Now suppose $I = \mathbb{N}$, under the natural order.

c) If τ_{n+1} induces τ_n on E_n, for each $n \in \mathbb{N}$, then the inverse implication of b) is valid.

d) If τ_{n+1} induces τ_n on E_n, for each $n \in \mathbb{N}$, and E_n is closed in (E_{n+1}, τ_{n+1}), then τ is a Levi topology provided the same is true for all τ_n.

Proof. a) is [8], ch. 2, 4.16. b) is easy to see, and c), d) follow by applying [9], 6.4 and 6.5. □

2.15d) cannot be generalized to uncountable I, as the following example shows:

Let X be an uncountable set, and for each countable $A \subset X$ let $E_A := \mathbb{R}^A$ be endowed with the product toplogy τ_A. Ordering the set of countable subsets of X by inclusion, $E := \{f \in \mathbb{R}^X : \{f \neq 0\}$ is countable$\}$ is the inductive limit of the E_A with respect to the natural embedding maps, and τ is the restriction of the product topology on $\mathbb{R}^X$ to E. All τ_A are Levi, but the τ-bounded set of characteristic functions of the countable subsets of X has no supremum in E.

Problem: What can be said about the Fatou topologies?

3. LATERAL COMPLETENESS

The example at the end of 2 can also serve to show that lateral completeness of the spaces E_1 does not imply lateral completeness of E.

In contrast, we want to prove now that for a certain class of spaces of mappings the inductive limit is laterally complete. To formulate our Theorem as general as possible (in particular it will apply to all examples in Section 4), we have to make a number of assumptions.

So let X be a set, and let U be a subset of the power set of X such that $U, V \in U \Rightarrow U \cap V \in U$. We set

$$U \leq V :\iff U \supset V.$$

Hence $U \vee V = U \cap V$, with respect to $\leq$. We assume that $(U, \leq)$ is a complete lattice with the greatest element U_{max} and the smallest element U_{min} such that $(\bigwedge_{\iota \in I} U_\iota) \vee U = \bigwedge_{\iota \in I}(U_\iota \vee U)$ for each non-empty family $(U_\iota)_{\iota \in I}$ from $\mathcal{U}$.

Further, we suppose that to each $U \in \mathcal{U}$ there is assigned a subset U_+ of U such that

(+) $\quad U \subset V \Rightarrow U_+ \subset V_+$.

Let H be a Riesz space. For $U \in \mathcal{U}$ and $f, g \in H^U$ we set

$$f \leq g : \leftrightarrow (x \in U_+ \Rightarrow f(x) \leq g(x))$$

and suppose that E_U is a subset of H^U such that $\leq$ induces an order relation on E_U under which E_U is a Riesz space. Moreover, we require $g|_U \in E_U$ whenever $U \subset V$ and $g \in E_V$.

We make, furthermore, the following assumptions:

(1) If $(U_\iota)_{\iota \in I}$ is a non-empty family from $\mathcal{U}$ and $x \in \bigwedge_{\iota \in I} U_\iota$, then there exists a finite $J \subset I$ with $x \in \bigwedge_{\iota \in J} U_\iota$.

(2) If $U, V \in \mathcal{U}$ with $U \vee V = U_{max}$, and if $0 \leq f \in E_U$, $0 \leq g \in E_V$, then there is exactly one $h \in E_{U \wedge V}$ with $h|_{E_U} = f$, $h|_{E_V} = g$; we set $\widetilde{f+g}_{U \wedge V} := h$.

(3) For each $U \in \mathcal{U}$ and each family $(U_\iota)_{\iota \in I}$ from $\mathcal{U}$ with $U = \bigwedge_{\iota \in I} U_\iota$:

(3a) $f, g \in E_U$, $f \geq g$ pointwise on $\bigcup_{\iota \in I}(U_\iota)_+ \Rightarrow f \geq g$, provided $(U_\iota)_{\iota \in I}$ is finite and lower directed or provided $U_\iota \vee U_\lambda = U_{max}$ whenever $\iota \neq \lambda$.

(3b) $f \in H^U$, $f|_{U_\iota} \in E_{U_\iota}$ for all $\iota \in I \Rightarrow f \in E_U$, provided $U_\iota \vee U_\lambda = U_{max}$ whenever $\iota \neq \lambda$.

For $U, V \in \mathcal{U}$, $V \subset U$, we set $V_U^\perp := \bigwedge \{W : W \in \mathcal{U}, W \subset U, W \vee V = U_{max}\}$.
For $U \in \mathcal{U}$ and $f \subset E_U$ we set $N(f, U) := \bigwedge \{V : V \in \mathcal{U}, V \subset U, f|_V = 0\}$ and $C(f, U) := (N(f, U))_U^\perp$.

Let $\mathcal{V}$ be a subset of $\mathcal{U}$ with $U_{min} \in \mathcal{V}$, $(U, V \in \mathcal{V} \Rightarrow U \cap V \in \mathcal{V})$, and

(4) $U \in \mathcal{U}$, $V \in \mathcal{V}$, $U \subset V \Rightarrow U \wedge U_V^\perp \in \mathcal{V}$

(5) $U, V \in \mathcal{V}$, $0 \leq f \in E_U$, $0 \leq g \in E_V$, $f|_{U \vee V} \wedge g|_{U \vee V} = 0 \Rightarrow$
$\Rightarrow C(f, U) \vee C(g, V) = U_{max}$.

Finally, for $U, V \in \mathcal{V}$, $U \subset V$, assume that

$$\phi_{UV} : E_V \to E_U, \quad f \to f|_U$$

is an injective Riesz homomorphism, and let E be the inductive limit of $(E_V)_{V \in \mathcal{V}}$ with respect to the maps ϕ_{UV}.

Theorem 3.1 Under the assumptions made above, E is laterally complete.

Proof. Let $(f_\iota)_{\iota \in I}$ be an orthogonal family of positive elements of E. For each $\iota \in I$, let $U_\iota \in \mathcal{V}$ with $f_\iota \in E_{U_\iota}$, and set $C_\iota := C(f_\iota, U_\iota)$, $N_\iota := N(f_\iota, U_\iota)$, $g_\iota := f_\iota|_{C_\iota} \in E_{C_\iota}$. By (5), $C_\iota \vee C_\lambda = U_{max}$ for $\iota \neq \lambda$. For $J \subset I$, J finite, we set $C_J := \bigwedge_{\iota \in J} C_\iota$ and, using (2), $g_J := \widetilde{\sum_{\iota \in J} g_\iota}_{C_J}$.

389

Furthermore, we define $U := \bigwedge_{\iota \in I} C_\iota$

To show that for $x \in U$ and finite J, L with $x \subset C_J \cap C_L$ we have $g_J(x) = g_L(x)$, we set $L_1 := L \cup J \setminus J$ and $J_1 := L \cup J \setminus L$. Then $\widetilde{g_{L_1} + g_J}\,_{C_{L \cup J}} = g_{L \cup J} = \widetilde{g_{J_1} + g_L}\,_{C_{L \cup J}}$ by (2), which implies the assertion.

This last remark and (1) allow us to define, for $x \in U$,

$$g(x) := g_J(x) \quad \text{where } J \subset I, \; J \text{ finite, and } x \in C_J.$$

Then, $g|_{C_\iota} = g_\iota$ for all $\iota \in I$, and, by (3b), $g \in E_U$.

We set

$$f := \widetilde{g + 0}_{U \wedge U_{min}^\perp}.$$

By (4), $V := U \wedge U_{min}^\perp \in V$, and $f \in E_V$.

Applying (3a) twice, we get $f \geq 0$.

Now, let $\iota \in I$. By (4), $W := V \wedge (C_\iota \wedge N_\iota) \in V$. We have $f|_{V \vee C_\iota} = f_\iota|_{V \vee C_\iota}$, and by (+), $f|_{V \vee N_\iota} \geq 0 = f_\iota|_{V \vee N_\iota}$, where the last equality follows from $f_\iota|_{N_\iota} = 0$ (which is proved by using (1) and (3a)). Since $N_\iota \vee C_\iota = U_{max}$, (3a) yields $f|_W \geq f_\iota|_W$, hence $f \geq f_\iota$.

To complete the proof, let $h \in E$ such that $h \geq f_\iota$ for all $\iota \in I$. There is $S \in V$ with $h \in E_S$. Then $T := V \vee S \in V$, and $T = (\bigwedge_{\iota \in I} (C_\iota \vee S)) \wedge (U_{min}^\perp \vee S)$. Again using (+) and (3a), we get $h|_T \geq f|_T$, hence $h \geq f$. □

4. EXAMPLES

1.) Let F be an Archimedean Riesz space.

For $A \subset F$, we denote by $S(A)$ the solid subspace of F generated by A. We set

$G := \{G \subset F : G \text{ is a solid subspace of } F\}$

$H := \{G \in G : G \text{ is order dense in } F\}$.

Then we have in G: $\bigvee_{\iota \in I} G_\iota = \bigcap_{\iota \in I} G_\iota$, $\bigwedge_{\iota \in I} G_\iota = S(\bigcup_{\iota \in I} G_\iota)$, hence also $(\bigwedge_{\iota \in I} G_\iota) \vee G = \bigwedge_{\iota \in I} (G_\iota \vee G)$. $\{0\}$ is the greatest element of G, and F is the smallest. Furthermore $G_H^\perp = G^d \cap H$, for $G \subset H$.

We set $G_+ := \{x \in G : x \geq 0\}$ for $G \in G$, and consider the family $(G_n^\sim)_{G \in G}$, where $G_n^\sim$ denotes the set of order continuous linear forms on G.

For $\xi \in G_n^\sim$, we have $N(\xi, G) = \{x \in G : |\xi|(|x|) = 0\}$ and $C(\xi, G) = N(\xi, G)^d \cap G$.

One easily verifies (1) - (5), where for (4) and (5) one can use [1], 1.12 and 3.10.

For $G, H \in H$, $G \subset H$, the map

$$\phi_{GH} : H_n^\sim \to G_n^\sim, \quad \xi \mapsto \xi|_G$$

is an injective Riesz homomorphism, and via ϕ_{GH}, $\tilde{H}_n$ is a solid subspace of $\tilde{G}_n$.

The inductive limit of $(G_n)_{G \in H}$ with respect to the maps ϕ_{GH} is called the extended order dual of F ([6]). It is universally complete (cf. 2.12 and 3.1).

2.) Let X be a topological space.

We set

$U := \{U \subset X : U \text{ open}\}$

$V := \{V \in U : V \text{ dense in } X\}$.

Then $\bigvee_{i \in I} U_i = \overline{\bigcap_{i \in I} U_i}$, $\bigwedge_{i \in I} U_i = \bigcup_{i \in I} U_i$, $(\bigwedge_{i \in I} U_i) \vee U = \bigwedge_{i \in I} (U_i \vee U)$, $U_{\min} = $

$= X$, $U_{\max} = \emptyset$, and $V_U^{\perp} = U \setminus \bar{V}^{\text{in } U}$ for $V \subset U$.

We set $U_+ := U$, for all $U \in U$.

Let H be a locally solid Riesz space. Then, for each $U \in U$, $E_U := $
$= \{f \in H^U : f \text{ continuous}\}$ is a Riesz space, and $(f \wedge g)(x) = f(x) \wedge g(x)$ for $f, g \in E_U$ and $x \in U$.

For $f \in U$, we have $N(f,U) = U \setminus \text{supp}(f)$ and $C(f,U) = \overline{\text{supp}(f)}^{\circ}$.

The maps $\phi_{UV} : E_V \to E_U$, $f \to f|_U$ are injective Riesz homomorphisms $(U, V \in V, U \subset V)$.

By 3.1, the inductive limit E of $(E_U)_{U \in V}$ with respect to the maps ϕ_{UV} is laterally complete; but in general it is not Dedekind σ-complete, as the example $X =]0,1[$ shows.

We remark that in the special case where X is a Stonian space and $H = \mathbb{R}$, E can be identified with $C_{\infty}(X)$ ([4], 3.12).

3.) Let X be a locally compact space, and let U, V as in 2.).

We set

$B(U) := \{B \subset U : B \text{ is a relatively compact Borel set of } U\}$ for $U \in U$

and

$B := \{B(U) : U \in U\}$.

Ordering B by

$B(U) \leq (V) : \iff B(U) \supset B(V)$,

we see that the map $U \to B$, $U \to B(U)$ is an order isomorphism; hence all order assertions can be transferred from U to B.

For $U \in U$, we set $B(U)_+ := B(U)$ and

$M(U) := \{\mu \in \mathbb{R}^{B(U)} : \mu \text{ is a normal Radon measure on } U\}$.

(A Radon measure is called normal if it is interior regular with respect to the open sets.)

For $U \in U$ and $\mu \in M(U)$, we have $N(\mu, B(U)) = B(U \setminus \text{supp}(\mu))$ and $C(\mu, B(U)) = B(\overline{\text{supp}(\mu)}^{\circ})$.

(1) - (4) can easily be verified. (5) is derived by applying Hahn's decomposition and the normality of the measures.

It is obvious that for $U, V \in \mathcal{V}, U \subset V$, the map

$$\phi_{UV} : M(V) \to M(U), \quad \mu \to \mu|_U$$

is a Riesz homomorphism, which is injective by the normality of the measures.

Hence, the inductive limit $M_q(X)$ of $(M(U))_{U \in \mathcal{V}}$ with respect to the maps ϕ_{UV} is universally complete. By [4], 3.16, $M_q(X)$ is the extended order dual of the space E, where E is as in 2.), so one can conclude its universal completeness also from 1.).

REFERENCES

1. C. D. Aliprantis, O. Burkinshaw, "Locally solid Riesz spaces", Academic Press, New York - San Francisco - London (1978).
2. N. Bourbaki, "Théorie des ensembles", ch. III, Hermann et Cie., Paris (1956).
3. C. Constantinescu, "Duality in measure theory", LN 796, Springer Verlag, Berlin - Heidelberg - New York (1980).
4. W. Filter, Dual spaces of $C_\infty(X)$, Rend. Circ. Mat. Palermo 35, 135 - 158 (1986).
5. D. H. Fremlin, "Topological Riesz spaces and measure theory", Cambridge Univ. Press, London - New York (1974).
6. W. A. J. Luxemburg, J. J. Masterson, An extension of the concept of the order dual of a Riesz space, Can. J. Math. 19, 488 - 498 (1967).
7. W. A. J. Luxemburg, A. C. Zaanen, "Riesz spaces I", North-Holland Publ. Comp., Amsterdam - London (1971).
8. A. L. Peressini, "Ordered topological vector spaces", Harper and Row, New York - Evanston - London (1967).
9. H. H. Schaefer, "Topological vector spaces", Springer-Verlag, New York - Heidelberg - Berlin (1971).

CONVERGENCE COMPLETION OF PARTIALLY ORDERED GROUPS

Isidore Fleischer

University of Windsor, Canada

As is well known, the (MacNeille, [7]) conditional completion by non-void cuts of a partially ordered group (which is the unique conditionally complete lattice each element of which is a join and a meet of the group's elements) cannot in general (in fact whenever the group fails to be "Archimedean") be made into a partially ordered group. There is a largest subset of the completion to which the group composition can be extended so as to achieve a partially ordered group (see Fuchs [6] or below). In the totally ordered case this subset may be attained intrinsically as the completion of the original group in its order-topology (Cohen-Goffman [4]). There is no suitable order-topology even for lattice-ordered groups; one can use certain down-directed subsets of positive elements with meet zero to induce order-theoretic topological group structures whose completions may then be shown to be canonically contained in the MacNeille completion (Banaschewski [1]); and more general such down-directed subsets to induce order-theoretic non-group convergence structures whose completions are also so contained - indeed, the totality of these suffice to attain the largest group subextension of the MacNeille completion of a commutative lattice-ordered group (Ibid.). This procedure has been identified as completion with respect to a form of order-convergence by Papangelou [8], who is then able to give a much more efficient proof of the same result. To extend this to partially ordered (possibly non-commutative) groups, it is necessary to isolate the appropriate notion of order-convergence and to devise a proof independent of the more special properties this notion has in ℓ-groups. The result is to make every partially ordered group into a convergence group whose convergence completion is exactly the largest possible partially ordered group in the conditional (order) completion.

A paradigm for the MacNeille completion – indeed, its most important and motivating example – is the Dedekind completion of the rationals Q to the reals R. The latter is obtained as the "cuts" in Q – where a "cut" is usually defined to be a decomposition of Q into a hereditary V and a (consequently) co-hereditary complement U. Of course this entails that every element of V is < each element of U, a situation which will be symbolized as V < U; moreover, that they are complementary comes to: every (proper) upper/lower bound to all of V/U is in U/V; conversely, these suffice to make (U,V) a "cut". Alternatively, one could alter the definition of "cut" by making every $v \in V \leq$ each $u \in U$ – symbolized by $V \leq U$ – and drop the parenthesized "proper". This does not change the irrational cuts but will make the rational ones cease to be decompositions, U and V then intersecting in just the element of Q they represent. This obviates the familiar disadvantage of having the elements of Q correspond to two distinct "cuts" which must subsequently be identified in obtaining R; and it is the one suitable for generalizations to posets, where the halves of a cut no longer exhaust the set.

Thus a <u>cut</u> in a poset Q will be understood to be a pair (U,V) of its non-void subsets in the relation $U \geq V$ such that every upper/lower bound to V/U is in U/V. The set R of cuts is ordered by: $(U,V) \geq (U',V')$ just when $U \geq V'$ – equivalently $V \supseteq V'$ (also $U' \supseteq U$). This is readily seen to be a partial order; and R contains an order-isomorph of Q as the principal cuts: $U(q) = \{x \geq q\}$, $V(q) = \{y \leq q\}$. Since $q = \vee q_i$ in Q if and only if $U(q) = \cap U(q_i)$, existent sups/infs in Q remain such in R. Furthermore, R is a conditionally complete lattice: Indeed, the U's/V's are closed for non-void intersection – if $u \in \cap U_i$ then $u \geq$ every V_i i.e. the lower bounds V of $\cap U_i$ contains $\cup V_i$ and therefore the upper bounds of V are contained in those of $\cup V_i$, which is $\cap U_i$ – thus $(\cap U_i, V)$ constitutes a cut. Now if $u \geq S \subset R$ then $U = \cap \{U(s): s \in S\}$ is the upper half of $\vee S$; dually, if $v \leq S$ then $V = \cap \{V(s): s \in S\}$ is the lower half of $\wedge S$. These forms show that every cut is the sup and inf of the part of $Q \leq$ and $\geq$ it: i.e. Q is "order-dense" in R. It may be shown that R is the largest poset in which Q is order-dense. The further functorial properties of this construction are not as simple as one might hope: the interested reader could be sent to Arch. Math. <u>18</u> (1967), 369-377, MR <u>36</u> #5036 and J. Austral. Math. Soc. <u>21</u> (1976), 220-223. What is clear (which is all that will be needed in the sequel) is that every (order-) isomorphism between posets extends uniquely to a (lattice-) isomorphism between their MacNeille completions.

Now let Q be a pogroup, Although commutativity is not assumed, additive notation will (as usual in this subject) be used. The basic axiom is that (left or right) translation by a group element is an order endo(hence auto)morphism: $p \leq q$ entails $p+r \leq q+r$; hence also $-p+p-q \leq -p+q-q$ i.e. $-q \leq -p$ so that inversion is an antiautomorphism. By the functorial remark just made, left and right translation <u>by elements of Q</u> (and inversion) extend uniquely to lattice (anti)automorphisms of R: explicitly $q + (U,V) = (q+V, q+U)$, $-(U,V) = (-V,-U)$. Fixing (U,V) in the former and letting q vary yields an action on R on Q which sends it into R; this action is isotone ($q \geq p$ entails $q+U \geq p+V$) and even "complete" in that existent extrema in Q are preserved - $V' \leq U+q_i$ i.e. every $v' \leq$ each $u+q_i$, entails $v' \leq$ each $u + \wedge q_i$ i.e. $V' \leq U + \wedge q_i$. Since group translation must be complete, this could be hailed as a hopeful indication for the possibility of extending the addition in Q to a group composition in R. It is indeed possible to extend the action of R on Q to one of R on R; and so as to retain order-preservation and even sup- <u>or</u> inf-preservation (see the above cited references); but not (in general) both - the resulting composition does not make R into a group. Here is a definitive example:

Let the group be the (totally ordered) lexicographic rational plane: that is, the product $Q \times Q'$ of the rationals with themselves, ordered according to the first differing component (i.e. as would be the two letter words with rational letters in a dictionary) - the order thus imposes traversal of the plane up the vertical lines with constant rational abscissa, these lines being taken in the order of their abcscissas. This group has a cut with U the pairs having first component a non-negative rational $q \geq 0$ and V the pairs with first component strictly negative $q < 0$. The translation by any $(0,q')$ on $Q \times Q'$ extends so as to send this cut on itself: thus (U,V) acting on $Q \times Q'$ absorbs all $(0,q')$ - which could not happen in a group. Another way to see the impossibility of extending to a group composition: $-(U,V)$ was seen to be $(-V,-U)$ which is here the cut with lower half $q \leq 0$, upper half $q > 0$ - but adding (elementwise) the upper and lower halves of (U,V) and $(-V,-U)$ does not result in a cut at all: V-U and U-V are each bounded away from 0 by the $(0,q')$.

There are two ways of dealing with this dilemma: The more drastic is to refuse to consider groups in which this unpleasantness can occur. The groups Q whose R is a group can be given an internal characterization: In any such group, $\{nq: n > 0\}$ is bounded only if $q \leq 0$ - For $\vee nq$ exists in a group only if $\vee nq + q = \vee(n+1)q \leq \vee nq$, whence $q \leq 0$. Conversely, if $q \leq U-V$ i.e. $q+V \leq U$ i.e. the automorphism of q-translation decreases (V,U), then so also does nq i.e. $nq \leq U-V$, so that $\{nq\}$ is bounded: if this entails $q \leq 0$ then $\wedge U-V = 0 (= \vee V-U)$; this is turn entails that (U,V) acts (not just isotonely but) as an order isomorphism of Q into R ($q+V \leq p+U$ only if $-p+q \leq U-V$ hence $q \leq p$); and this will be shown below to ensure that the set of all cuts constitutes a group.

In rational vector spaces, this "Archimedean" characterizing condition comes to: $q \leq \frac{1}{n}p$ only for $q \leq 0$; in a directed space it suffices to take p positive and the condition reads: $\wedge \frac{1}{n} p = 0$, i.e. the transformation from the scalars into the space effected by every positive p is continuous at 0. This should indicate why functional analysts are quite comfortable with this restriction to Archimedean groups. (It might also be mentioned that the condition implies Zq bounded only for $q = 0$, the converse implication holding in lattice-ordered groups.)

The less drastic option is to stay with the most general group but to restrict to a part of its MacNeille completion which can be made into a group extension. It turns out that there is always a unique largest such part, which consists of those cuts for which $\wedge U-V = 0$ (and also $\wedge -V+U = 0$, although this will turn out to be a consequence by the completability proved below); it will be appropriate to call these cuts <u>invertible</u> - more generally, to so call the pairs $U \geq V$ (possibly not making up a cut) for which $\wedge U-V = 0$. This condition ensures that $\wedge U = \vee V$ in R - indeed if $v^* \leq U$, $u^* \geq V$ then $U-V \geq v^*-u^*$ hence $v^* \leq u^*$: thus (U,V) is contained in a unique cut (U^*,V^*) of R; and this will still have $\wedge U^*-V^* = 0$. These are the only elements of R which could belong to a pogroup containing Q, since $\wedge U-V = 0$ is equivalent to: $p+U \geq q+V$ only for $p \geq q$. Now the pairs $U \geq V$ may be added by taking the complex sum of corresponding halves (since $U+U' \geq V+V'$) and the sum of invertible pairs is invertible (since $p+U+U' \geq q+V+V'$ only if every $p+u \geq$ each $q+v$.) For those pairs with $\wedge U = \vee V$ in Q, the addition agrees with that in Q; and it is uniquely determined as inducing the addition in Q and distributing across $\vee$ and $\wedge$ in R (as is required in a pogroup). Finally, the antiautomorphism of inversion sends each pair $U \geq V$ to the pair $-V \geq -U$. whose sum with an invertible $U \geq V$ will be invertible with $\wedge$ and $\vee = 0$. Thus the extrema in R of the invertible pairs do indeed constitute a largest possible subgroup containing Q. (This subgroup even acts as automorphisms on all of R: if $U' \geq V'$ and $p \geq V'+V$, $q \leq U'+U$, i.e. $p-V \geq V'$, $q-U \leq U'$ then $\wedge U' = \vee V'$ in R entails $p-V \geq q-U$ therefore $p \geq q$ by invertibility of (U,V) so that $\wedge U'+U = \vee V'+V$ in R. The reader should have no trouble determining that for the lexicographic rational plane considered above, this largest subgroup is the lexicographic product $Q \times R$ - that is, just the gaps on each vertical line can be filled without losing the group property.)

In a conditionally complete lattice it is possible to define, for an eventually bounded net x_n, lim sup x_n as $\wedge_n \vee_{m \geq n} x_m$; and then <u>order-convergence</u> for such a net as lim sup x_n = lim inf x_n, this common value being the (unique) <u>limit</u> on the net. It should be pointed our that this convergence is in general not topological: for example, in the conditionally complete lattice of measurable functions under $\leq$ almost everywhere (hence identified under equality a.e.), order-convergence of sequences comes to convergence a.e.; on the other hand, every sequence converging in measure has each of its subsequences converging in measure and moreover some subsequence converging a.e. – thus in any topology which makes a.e. convergent sequences converge, the is measure sequences must also converge: but there are sequences which converge in measure without converging a.e., hence the a.e. order-convergence is not topological.

It has been proposed to define order-convergence in an arbitrary poset as the relativized convergence induced by its MacNeille completion [2, p. 60]. This may be defined intrinsically, i.e. within the poset: Note that the same nets in Q are eventually bounded in R as in Q (since Q bounds the elements of R) and the lim sup of a bounded net in Q is the inf in R of the elements which eventually dominate it (this being the inf of the sups in R of its final subsets); insofar as it belongs to Q, this is the inf in Q. Thus a net <u>order-converges</u> in Q just when the inf of its eventually dominating elements is the sup of its eventually dominated ones, this common value being the (unique) limit.* (It obviously suffices to have the equality for <u>any</u> set of eventually dominating/dominated elements).

In a pogroup this convergence is algebraically compatible: Adding elements each of which eventually dominates its net yields an element which eventually dominates the sum net (on the product directed index set) and the inf of all these sums is the sum of the infs (by the automorphic character of group translation); and inversion, being algebraically an induces a convergence homeomorphism. Therefore a convergent net is (both left and right) Cauchy – a <u>right Cauchy net</u> being understood as one for which $x_m - x_n$ converges to 0 as a net on the square directed index set (the "right" refers to the generation of the uniformity by right translation e.g. in a topological group, using the neighbourhood filter at 0, $x-y \in N$ is equivalent to $x \in N+y$; or to $y \in -N+x$); since this net is inverted on interchanging the indices, it suffices to have lim sup $x_m - x_n$ – i.e. the inf of the w eventually $\geq x_m - x_n$ – equal to 0. More generally this last holds for every net order-converging to an invertible element: since every u-v is such a w, $\wedge U - V = 0$ ensures right Cauchy.

Conversely, every right Cauchy net order-converges to an element of R; since $w + x_n \geq x_m$ eventually in m and n, $w + \wedge x_n \geq \vee x_m$ eventually in R, hence $w + \lim \inf x_n \geq \lim \sup x_n$, which entails convergence since R-action

---

*In a totally ordered set an element is lim sup/inf if and only if every strictly larger/smaller element eventually dominates/is dominated by the net-hence order-convergence coincides with convergence in the order topology, defined as having the open intervals as base.

preserves $\wedge w = 0$. Finally to see that only invertibles can be limits of right Cauchy nets**, suppose $U-V \geq p$ i.e. $U \geq p+V$ for the limit cut (U,V), hence $\vee x_n \geq p + \wedge x_n$ in R for every final subset of the x_n converging to (U,V). Now if w and w' are eventually $\geq x_m - x_n$ i.e. $w, w' + \wedge x_n \geq \vee x_n$ eventually in R, then $w - p + w' + \wedge x_n \geq w + \wedge x_n \geq \vee x_n$ eventually in R - thus $W - p + W \subset W$ whence by right Cauchy, $0 = \wedge W \leq \wedge(W - p + W) = -p$ or $p \leq 0$.

REFERENCES

1. B. Banaschewski, Über die Vervollständigung geordneter Gruppen, M. Nachr., 16, 51-71, (1957).

2. G. Birkhoff, Lattice Theory, revised ed. 2., A. M. S. Colloq. Publ. vol. 25, (1948).

3. N. Bourbaki, "Topologie genéralé", Ch. III Groupes topologiques, Paris,

4. L. W. Cohen, C. Goffman, The topology of ordered Abelian groups, Trans. A. M. S., 67, 310-319, (1949).

5. C. J. Everett, Sequence completion of lattice moduls, Duke Math. J., 11, 109-119, (1944).

6. L. Fuchs, "Partially Ordered Algebraic Systems", Pergamon Press, (1963).

7. H. M. MacNeille, Partially ordered sets, Trans. A. M. S., 42, 416-460, (1937).

8. F. Papangelou, Order convergence and topological completion of commutative lattice-groups, Math. Annalen, 155, 81-107, (1964).

** The simpler argument of [8, 6.1] may be used when it is known that $\wedge 2W = 0$ e.g. if W is down-directed.

SOME RESULTS FROM NONLINEAR ANALYSIS IN LIMIT VECTOR SPACES

Olga Hadžić

Institute of Mathematics
University of Novi Sad
21000 Novi Sad, Dr I. Đuričića 4, Yugoslavia

ABSTRACT

In paper [10] Bieri Hanspeter obtained some results from nonlinear analysis in limit vector spaces. Using a variant of the KKM lemma in limit vector spaces, we shall prove in this paper some fixed point theorems in limit vector spaces. A generalization of the Ky Fan minimax principle in limit vector spaces is also obtained.

1. INTRODUCTION

In his Doctoral Thesis [3] Simon Courant obtained some interesting results about limit vector spaces, which are very useful for Analysis on these spaces. The interest for limit vector spaces has been increased in connection with differential calculus in non-normed vector spaces and investigations of some spaces of mappings. Some results about the Hahn-Banach theorem in limit vector spaces (or convergence vector spaces) have been obtained by Bieri Hanspeter [10], Ronald Beatti [1] (a survey is given in paper by T. S. McDermott [15]). Bieri Hanspeter proved in [10] some fixed point and minimax theorems in limit vector spaces.

In this paper we shall obtain, using a variant of the Knaster-Kuratowski-Mazurkiewicz lemma, some results from nonlinear analysis in limit vector spaces.

Theorem 1 is a generalization of the well known Ky Fan result from [6] to limit vector spaces. Using theorem 1 we prove a generalization of theorem 1 from the paper by P. Deguire and A. Granas [4] which has many applications in nonlinear analysis.

The KKM lemma is one of the most interesting results in nonlinear analysis. There are a great number of papers connected with this well known lemma ([4], [6], [7], [9], [12], [14], [16], [17], [18], [19], [20]). This lemma can be used in minimax theorems, variational inequalities, fixed point theorems and other branches of mathematics. Some information about the KKM lemma and Hahn-Banach theorem can be found in papers by W. Takahashi [20] and S. Simons [19].

Now, we shall give some preliminaries.

First, we shall recall some definitions from [10]. If F_1 and F_2 are two filters on a set then:

$$F_1 \cup F_2 = \{F \cup G;\ F \in F_1, G \in F_2\}$$

$$F_1 \cap F_2 = \{F \cap G;\ F \in F_1, G \in F_2\}$$

and by $F = \{F_i;\ i \in I\}$ we shall denote the filter induced by the filter bases $\{F_i;\ i \in I\}$. The main filter, such as $\{\{x\}\}$ or $\{G\}$, will be denoted by $[x], [G]$.

All vector spaces in this paper will be assumed to be over the field of real numbers and by V we shall denote the neigbourhood filter over $\mathbb{R}$ (in the natural topology). The notation $F \downarrow x$ means that filter F converges to x ($x \in M$ and M is a limit space) and we have that the following relations are satisfied:

- $[x] \downarrow x$;
- $F_1 \downarrow x, F_2 \downarrow x \Rightarrow F_1 \cup F_2 \downarrow x$;
- $F_1 \downarrow x, F_2 \leq F_1 \Rightarrow F_2 \downarrow x$.

If E is a <u>limit vector space</u> then [10]:

- $F_1 \downarrow 0, F_2 \downarrow 0 \Rightarrow F_1 + F_2 \downarrow 0$;
- $F \downarrow 0 \Rightarrow V \cdot F \downarrow 0$;
- $F \downarrow 0, \lambda \in \mathbb{R} \Rightarrow \lambda F \downarrow 0$;
- $x \in E \Rightarrow Vx \downarrow 0$;
- $F \downarrow x \Leftrightarrow F - x \downarrow 0$.

A subset A of a limit vector space E is compact if every ultra-filter on A is convergent and [10] if A is compact, every open covering of A has a finite subcovering of A.

It is well known that every Hausdorff limit vector space of finite dimension is isomorphic to $\mathbb{R}^n$ [3].

2. KKM MAPPING IN LIMIT VECTOR SPACES

In the fixed point theory and nonlinear analysis in topological vector spaces, the KKM lemma takes an important place (see books [5] and [8]).

The following definition is well known [8]:

Definition Let E be a vector space, X a nonempty subset of E and $\Gamma : X \to 2^E$ (the family of nonempty subsets of E). The mapping Γ is said to be a KKM mapping if for every finite subset $\{x_1, x_2, \ldots, x_k\}$ of X we have:

$$co\{x_1, x_2, \ldots, x_k\} \subseteq \bigcup_{i=1}^{k} \Gamma x_i \quad \text{(where co is convex hull)}.$$

The following lemma is known as the Knaster-Kuratowski-Mazurkiewicz lemma [13] (for short, the KKM lemma).

Lemma Let S be the set of vertices of a simplex in $\mathbb{R}^n$ and $\Gamma : S \to 2^{\mathbb{R}^n}$ a KKM mapping with compact values. Then $\bigcap_{x \in S} \Gamma x \neq \emptyset$.

The next result, obtained by Ky Fan, is in fact a generalization of the KKM lemma [6].

Theorem A Let X be a nonempty set of a topological vector space E and $\Gamma : X \to 2^E$ a KKM mapping such that Γx is closed for every $x \in X$ and there exists $x_0 \in X$ such that Γx_0 is compact. Then:

$$\bigcap_{x \in X} \Gamma x \neq \emptyset.$$

Some generalizations of Theorem A are given in Fan's paper [7]. In [12] Won Kyu Kim obtained the following result.

Theorem B Let $X = \{x_1, x_2, \ldots, x_n\}$ be the set of vertices of a simplex S^{n-1} in $E = \mathbb{R}^n$ and $\Gamma : X \to 2^E$ an open valued KKM mapping. Then:

$$\bigcap_{i=1}^{n} \Gamma x_i \neq \emptyset.$$

We shall generalize Theorem A to limit vector spaces.

Theorem 1 Let X be a nonempty subset of a Hausdorff limit vector space E and $\Gamma : X \to 2^E$ a KKM mapping with closed values such that for some $x_0 \in X$, Γx_0 is compact. Then:

$$\bigcap_{x \in X} \Gamma x \neq \emptyset. \tag{1}$$

Proof It is enough to prove that the family $\{\Gamma x\}_{x \in X}$ has the finite intersection property. Indeed, if the family $\{\Gamma x\}_{x \in X}$ has the finite intersection property, we shall show that the assumption $\bigcap_{x \in X} \Gamma x = \emptyset$ leads to a contradiction. Suppose that $\bigcap_{x \in X} \Gamma x = \emptyset$. From the relation $\bigcap_{x \in X} \Gamma x = \emptyset$ it follows that:

$$[\bigcap_{x \in X} \Gamma x]^c_{\Gamma x_0} = \bigcup_{x \in X} [\Gamma x]^c_{\Gamma x_0} = \Gamma x_0 \quad ([A]^c_{\Gamma x_0} \text{ is the complement of } A \text{ in } \Gamma x_0).$$

Since Γx is closed for every $x \in X$, it follows that $\{[\Gamma x]^c_{\Gamma x_0}\}_{x \in X}$ is an open covering of the set Γx_0. From the compactness of the set Γx_0 it follows that there exists a finite covering $\{[\Gamma x_i]^c_{\Gamma x_0}\}_{i=1}^n$ of the set Γx_0. From the relation:

$$\Gamma x_0 = \bigcup_{i=1}^n [\Gamma x_i]^c_{\Gamma x_0}$$

it follows that $\bigcap_{i=1}^n \Gamma x_i = \emptyset$. But this contradicts the assumption that the family $\{\Gamma x\}_{x \in X}$ has the finite intersection property.

Hence, it remains to be proved that the family $\{\Gamma x\}_{x \in X}$ has the finite intersection property. Let $\{x_1, x_2, \ldots, x_r\}$ be an arbitrary subset of X. We shall prove that $\bigcap_{i=1}^r \Gamma x_i \neq \emptyset$. Let $v_1 = (1,0,0,\ldots,0)$, $v_2 = (0,1,\ldots,0), \ldots, v_r = (0,0,\ldots,1)$ and, as in [10], let $h : co\{v_1, v_2, \ldots, v_r\} \to co\{x_1, x_2, \ldots, x_r\}$ be a linear mapping defined by: $h(v_i) = x_i$ ($i \in \{1,2,\ldots,r\}$). Then if $(t_1, t_2, \ldots, t_r) \in co\{v_1, v_2, \ldots, v_r\}$ (i.e. $(t_1, t_2, \ldots, t_r) = \sum_{i=1}^r t_i v_i$, $\sum_{i=1}^r t_i = 1$, $t_i \geq 0$, $i \in \{1,2,\ldots,r\}$) we have:

$$h(t_1, t_2, \ldots, t_r) = \sum_{i=1}^r t_i x_i.$$

Since Γ is a KKM mapping, we have that for every $I \subseteq \{1,2,\ldots,r\}$:

$$co\{x_i; i \in I\} \subseteq \bigcup_{i \in I} \Gamma x_i$$

which implies that:

$$co\{v_i; i \in I\} = h^{-1}(co\{x_i; i \in I\}) \subseteq h^{-1}(\bigcup_{i \in I} \Gamma x_i) \subseteq \bigcup_{i \in I} h^{-1}(\Gamma x_i)$$

$$= \bigcup_{i \in I} \Gamma' v_i$$

where $\Gamma' v_i = h^{-1}(\Gamma x_i)$ ($i \in \{1,2,\ldots,r\}$). Limit structure in E induces on the set $h(co\{v_1, v_2, \ldots, v_r\})$ the natural topology and h is continuous. Since Γx_i is closed, we obtain that $\Gamma' v_i$ is closed in $co\{v_1, v_2, \ldots, v_r\}$. This implies that $\Gamma' v_i$ ($i \in \{1,2,\ldots,r\}$) is compact. From the KKM lemma it follows that $\bigcap_{i=1}^r \Gamma' v_i \neq \emptyset$, since Γ' is a KKM mapping. Hence, $\bigcap_{i=1}^r h^{-1}(\Gamma x_i) =$

$= \bigcap_{i=1}^{r} \Gamma' v_i \neq \emptyset$, which implies that $\bigcap_{i=1}^{r} \Gamma x_i \neq \emptyset$. From this we obtain that (1) holds. □

   Remark   Using theorem B we can prove easily an "open version" theorem for a nonempty intersection (Theorem 3 in this paper).

3. APPLICATIONS

There are many possibilities for applications of Theorem 1, similarly as in the case of topological vector spaces (see, for example papers by S. Simons [19] and W. Takahashi [20]).

Here, we shall give some generalizations of the results from [4] and [12], as an illustration.

Let $(P, \leq)$ be an ordered set. A mapping $\phi : X \to P$ of a limit space X into P is said to be <u>lower semicontinuous on</u> X if for every $\lambda \in P$ the set:

$$\{x;\ x \in X,\ \phi(x) > \lambda\}$$

is open. If X is a convex subset of a limit vector space, $\phi : X \to P$ is <u>quasi-concave on</u> X, if for every $\lambda \in P$ the set:

$$\{x;\ x \in X,\ \phi(x) > \lambda\}$$

is convex.

The following theorem is a generalization of Theorem 1 from [4]. (A proof of Theorem 1 for $P = \mathbb{R}$ from [4], in which the Brouwer fixed point in used, is given in [16]).

   Theorem 2   Let X be a nonempty, compact and convex subset of a Hausdorff limit vector space E, $(P, \leq)$ an ordered set and $\phi, \psi : X \times X \to P$ so that the following conditions are satisfied:

   (i) $\phi(x,y) \leq \psi(x,y)$, for every $(x,y) \in X \times X$;

   (ii) $x \mapsto \psi(x,y)$ is quasi-concave on X, for every $y \in X$;

   (iii) $y \mapsto \phi(x,y)$ is lower semicontinuous on X, for every $x \in X$.

Then, for every $\lambda \in P$ one of the following two conditions is satisfied.

   (1)   There exists $y_0 \in X$ such that:

$$\{x;\ x \in X,\ \phi(x,y_0) > \lambda\} = \emptyset,$$

   (2)   There exists $w \in X$ such that $\psi(w,w) > \lambda$.

Proof Let $\lambda \in P$ and define $A : X \to 2^X$ and $\Gamma : X \to 2^X$ in the following way:

$$Ax = \{z; z \in X, \phi(x,z) > \lambda\}, \quad \Gamma x = X \setminus Ax, \quad x \in X.$$

Since the set A is open, it follows that Γx is closed. From $\Gamma x \subseteq X$ and Satz 2.3 [10] we obtain that Γx is compact, for every $x \in X$. The rest of the proof is as in [4], and we shall repeat it for completeness sake.

We have the following two possibilities.

(a) The mapping Γ is not a KKM mapping. This means that there exists a finite subset $\{x_1, x_2, \ldots, x_n\} \subseteq X$ and $w = \sum_{i=1}^{n} \lambda_i x_i \in \text{co}\{x_1, x_2, \ldots, x_n\}$, such that $w \notin \bigcup_{i=1}^{n} \Gamma x_i$. This implies that $w \in \bigcap_{i=1}^{n} Ax_i$, which means that $\lambda < \phi(x_i, w) \leq \psi(x_i, w)$ for every $i \in \{1, 2, \ldots, n\}$. Since the set $\{x; x \in X, \psi(x,w) > \lambda\}$ is convex, we obtain that:

$$\sum_{i=1}^{n} \lambda_i x_i \in \{x; x \in X, \psi(x,w) > \lambda\}.$$

This means that $\psi(\sum_{i=1}^{n} \lambda_i x_i, w) > \lambda$ and so $\psi(w,w) > \lambda$. Hence, in this case we proved (2).

(b) Suppose that Γ is a KKM. Since Γx is compact for every $x \in X$, all the conditions of Theorem 1 are satisfied and so:

$$\bigcap_{x \in X} \Gamma x \neq \emptyset.$$

This implies that $X \setminus \bigcup_{x \in X} Ax \neq \emptyset$ and let $y_0 \in X \setminus \bigcup_{x \in X} Ax$. Then $y_0 \notin Ax$, for every $x \in X$ which means that $\{x; x \in X, \phi(x, y_0) > \lambda\} = \emptyset$. This proves (1). □

The Ky Fan minimax principle can be generalized to limit vector spaces by using Theorem 2.

Corollary Let X be a nonempty, convex, compact subset of a Hausdorff limit vector space, $\phi : X \times X \to \mathbb{R}$ and let the following two conditions be satisfied:

(i) $x \to \phi(x,y)$ is quasi-concave on X, for every $y \in X$,

(ii) $y \to \phi(x,y)$ is lower semicontinuous on X for every $x \in X$.

Then:

(a) For every $\lambda \in \mathbb{R}$ one of the following two conditions is satisfied:

1^0 There exists $y_0 \in X$ so that $\phi(x, y_0) < \lambda$, for every $x \in X$.
2^0 There exists $w \in X$ so that $\phi(w,w) > \lambda$.

(b) $\inf_{y \in X} \sup_{x \in X} \phi(x,y) \leq \sup_{x \in X} \phi(x,x)$.

Proof (a) Follows from Theorem 2 if we take that $P = \mathbb{R}$ and $\phi = \psi$. Further, let $\lambda = \sup_{x \in X} \phi(x,x)$ (we can suppose that $\lambda < \infty$). Then (b) follows from (a). □

There are many applications of Ky Fan minimax principle in the following fields (see [4] for references):

- Systems of inequalities;
- Minimization of convex functionals;
- Variational inequalities;
- Fixed point theory;
- Game theory.

Hence, using the Corollary we can obtain some generalizations of the results from nonlinear analysis, similarly as in the case of topological vector spaces.

Using Theorem 1 from [12] it is easy to prove the following generalization of Theorem 3 from the paper of Won Kyu Kim [12], using the fact that every Hausdorff limit space of finite dimensions isomorphic to $\mathbb{R}^n$.

Theorem 3 Let X be an arbitrary set of a Hausdorff limit vector space E and $F : X \to 2^E$ an open valued KKM mapping. Then, the family $\{F(x)\}_{x \in X}$ has the finite intersection property.

As in [12] from Theorem 3 we can deduce the following fixed point theorem.

Theorem 4 Let X be a nonempty, closed, convex subset in a Hausdorff limit vector space E and $F : X \to 2^X$ such that $F(x)$ is closed for every $x \in X$. If $F^{-1}(y)$ is convex for each $y \in X$ and $X = \bigcup_{i=1}^{n} F(x_i)$ for some $\{x_1, x_2, \ldots, x_n\} \subseteq X$, then there exists $w \in X$ such that $w \in F(w)$.

Proof As in [12], let $G : X \to 2^X$ be defined by $G(x) = [F(x)]_X^c$. Then, $G(x)$ is open and from $X = \bigcup_{i=1}^{n} F(x_i)$ it follows that:

$$\bigcap_{i=1}^{n} G(x_i) = \emptyset.$$

From Theorem 3 it follows that G is not a KKM mapping. If $w = \sum_{i=1}^{r} \lambda_i y_i$ is such an element from co $\{y_1, y_2, \ldots, y_r\}$ ($y_i \in X$, $i \in \{1,2,\ldots,r\}$) that $w \notin \bigcup_{i=1}^{r} G(y_i)$, then $w \in \bigcap_{i=1}^{r} F(y_i)$. Then the convexity of $F^{-1}(w)$ implies that $w \in F(w)$. □

Using Theorem 2 we shall prove the following fixed point theorem, which generalizes Corollary 6 from [4]

405

Theorem 5 Let X be a nonempty, convex and compact subset of a Hausdorff limit vector space E and $A : X \to 2^X$, so that the following conditions are satisfied:

(i) $A^{-1}y$ is convex for every $y \in X$;

(ii) There exists a mapping $B : X \to 2^X$ so that:

(a) $Bx \subseteq Ax$, for every $x \in X$;

(b) $B^{-1}y \neq \emptyset$, for every $y \in X$;

(c) Bx is open, for every $x \in X$.

Then, there exists an element $w \in X$ such that $w \in Aw$.

Proof As in [4] the mappings $\phi, \psi : X \times X \to \mathbb{R}$ are defined by:

$$\phi(x,y) = \begin{cases} 1, & y \in Bx \\ 0, & y \notin Bx \end{cases}, \quad \psi(x,y) = \begin{cases} 1, & y \in Ax \\ 0, & y \notin Ax. \end{cases}$$

We shall show that all the conditions of Theorem 2 are satisfied and that there exists $w \in X$, so that $\psi(w,w) = 1$. The lower semicontinuity of the mapping $y \to \phi(x,y)$ on X for every $x \in X$ follows from (c) and the quasi-concavity of the mapping $x \to \psi(x,y)$ on X, for every $y \in X$ follows from (i). Further, (i) from Theorem 2 follows from (ii)(a). Hence, one of the conditions 1^0 or 2^0 of Theorem 2 is satisfied. But (ii)(b) implies that 1^0 is not satisfied. Hence, if $\lambda \in (0,1)$ and $w \in X$ is such that $\psi(w,w) > \lambda$, then $\psi(w,w) = 1$. This means that $w \in Aw$. □

REFERENCES

1. R. Beattie, The Hahn-Banach problem in convergence vector spaces, Math. Nachr., 93, 319-330, (1979).

2. E. Binz, Ein Differenzierbarkeitsbegriff in limitierten Vektorräumen, Comment. Math. Helv., 41, 2, 137-156, (1966/67).

3. S. Courant, Beiträge zur Theorie der limitierten Vektorräume, Comment. Math. Helv., 249-268, (1969).

4. P. Deguire and A. Granas, Sur une certaine alternative nonlinéaire en analyse convexe, Studia Math., TLXXXIII, 127-138, (1986).

5. J. Dugundji and A. Granas, "Fixed Point Theory", Vol. 1, Monografie Mat., 61, PWN, Warszawa, 209 pp., (1982).

6. K. Fan, A generalization of Tychonoff's fixed point theorem, Math. Ann., 142, 305-310, (1961).

7. K. Fan, Some properties of convex sets related to fixed point theorems, Math. Ann., 266, 519-537, (1984).

8. O. Hadžić, "Fixed Point Theory in Topological Vector Spaces",

University of Novi Sad, Institute of Mathematics, 337 pp., (1984).

9. O. Hadžić, Some remarks on a theorem on best approximations, <u>Anal. Num. Theor. Approx.</u>, Tom. 15, No. 1, 27-35, (1986).

10. B. Hanspeter, Der Satz von Hahn-Banach und Fixpunktsätze in limitierten Vektorräumen, <u>Comment. Math. Helv.</u>, 339-404, (1970).

11. H. H. Keller, Räume stetiger multilinearer Abbildungen als Limesräume, <u>Math. Ann.</u>, 159, 259-270, (1965).

12. W. Kyu Kim, Some applications of the Kakutani fixed point theorem, <u>J. Math. Anal. Appl.</u>, 121, 119-122, (1987).

13. B. Knaster, C. Kuratowski and S. Mazurkiewicz, Ein Beweis des Fixpunktsatzes für n-dimensionale Simplexe, <u>Fund. Math.</u> 14, 132-137, (1929).

14. M. Lasonde, On the use of KKM multifunctions in fixed point theory and related topics, <u>J. Math. Anal. Appl.</u>, 97, 151-201, (1983).

15. T. S. McDermott, The Hahn-Banach theorem in convergence vector spaces, Conference on Convergence Spaces, 1976, Proc. Univ. Nevada, Reno, 148-154, (1976).

16. M. H. Shih, K. K. Tan, The Ky Fan minimax principle, sets with convex sections and variational inequalities, Diff. Geom.- Calculus of Variations and Their Applications, Marcel Dekker, Inc., 471-481, (1984).

17. M. H. Shih, K. K. Fan, A further generalization of Ky Fan's minimax inequality and its applications, <u>Studia Math.</u> T. LXXVIII, 279-278, (1984).

18. M. H. Shih, K. Keong Tan, Minimax inequalities and applications, <u>Contemporary Mathematics</u>, Vol. 54, 45-63, (1986).

19. S. Simons, Two-function minimax theorems and variational inequalities for functions on compact and noncompact sets, with some comments on fixed-point theorems, <u>Proc. Pure. Math.</u>, Vol. 45, Part 2, 377-392, (1986).

20. W. Takahashi, Fixed point, minimax and Hahn-Banach theorems, <u>Proc. Symp. Pure Math.</u>, Vol. 45, Part 2, 419-427, (1986).

COMPLETIONS OF CAUCHY VECTOR SPACES

D. C. Kent* and G. D. Richardson**

* Department of Pure and Applied Mathematics
  Washington State University, Pullman, WA 99164 USA
** Departments of Mathematics and Statistics
  University of Central Florida, Orlando, FL 32816 USA

ABSTRACT

T_2 and T_3 completions of Cauchy vector spaces are studied. Every T_2 Cauchy vector space is shown to have a strict T_2 Cauchy vector space completion. A Cauchy vector space has a T_3 Cauchy vector space completion exactly when the underlying Cauchy space has a T_3 Cauchy space completion.

1. INTRODUCTION

Let $\dot{x}$ denote the filter on set S containing the subset $\{x\}$ and let us recall the axioms for a Cauchy structure defined by Keller [6]. A collection of filters C on set S is called a Cauchy structure if (1) $\dot{x} \in C$ for each $x \in S$, (2) $F \in C$ when $G \subseteq F$ and $G \in C$, and (3) $F \cap G \in C$ when $F, G \in C$ and F and G are not disjoint filters. The pair (S,C) is called a Cauchy space. Keller [6] proved that these three axioms characterize the allowable set of Cauchy filters for the uniform convergence structures of Cook and Fischer [3]. Reed showed that T_2 completions of Cauchy spaces lead to T_2 completions of uniform convergence spaces ([10], Theorem 15). Hence, it seems that Cauchy spaces are a convenient setting for studying completion theory.

The triple $(S,\cdot,C)$ is called a Cauchy group when $(S,\cdot)$ is a group and (S,C) is a Cauchy space such that the operation is Cauchy continuous. Fric and Kent [4] showed that every T_2 Cauchy group has a T_2 Cauchy group completion possessing the universal mapping property. Further, inherent Cauchy lattices and Cauchy ℓ-groups have proved to be fruitful in finding completions of lattices and lattice ordered groups. The reader is referred to Ball ([1], [2]) and Kent [7] for work in this direction.

Our aim is to extended this study to the vector space setting. Let $(S,+,\cdot)$ denote a vector space over the real or complex field K and let (S,C) be a Cauchy space such that addition and scalar multiplication are Cauchy continuous. Then $(S,+,\cdot,C)$ is called a <u>Cauchy vector space</u> (over K) and C is said to be an <u>admissible Cauchy vector structure</u> for $(S,+,\cdot)$. Let CHY denote the category of all Cauchy vector spaces over K, whose morphisms are the class of all continuous linear maps between objects. The class of all objects in CHY is denoted by $|CHY|$.

2. PRELIMINARIES

Let $V(\lambda)$ denote the neighbourhood filter of λ in K and denote $V(0)$ simply by V. Suppose that $(S,+,\cdot)$ is a vector space over K and let C be a collection of filters on S, satisfying the following axioms:

(1) $\dot{x} \in C$ for each $x \in S$
(2) $G \in C$ when $F \subseteq G$ and $F \in C$
(3) $F + G \in C$ when $F, G \in C$ (2.1)
(4) $\lambda F \in C$ when $F \in C$
(5) $VF \in C$ when $F \in C$

Note that an admissible Cauchy structure for $(S,+,\cdot)$ must necessarily satisfy these axioms. Conversely, assume that C satisfies these axioms. If F, G each belong to C and are not disjoint filters, then $(F + G) - G \subseteq F \cap G$ and thus $F \cap G \in C$. It follows that C is a Cauchy structure and C is easily shown to be an admissible Cauchy vector structure for $(S,+,\cdot)$. Hence, the axioms listed in (2.1) characterize the admissible Cauchy vector structures for $(S,+,\cdot)$.

Quite often it is convenient to denote $(S,+,\cdot,C) \in |CHY|$ by (S,C) and vector space $(S,+,\cdot)$ simply by S. The axioms listed in (2.1) may be used to verify the next result. Let $\mathbb{N}$ denote the set of all natural numbers.

<u>Proposition 2.1</u> Let S be a vector space and let B be a collection of filters on S such that $\dot{x} \in B$ for each $x \in S$. Then $C = \{G \mid \sum_{1}^{n} V(\lambda_i) F_i \subseteq G, \lambda_i \in K, F_i \in B, n \in \mathbb{N}\}$ is the smallest admissible Cauchy vector structure for S containing B.

Recall that a Cauchy space (S,C) is <u>regular</u> if cl $F \in C$ (closure of F) when $F \in C$. Further, $(S,+,\cdot,C) \in |CHY|$ is <u>regular</u> (T_2, T_3) when the Cauchy space (S,C) is regular (T_2, T_3).

Proposition 2.2   Let $(S,C) \in |CHY|$. There exists a smallest admissible regular Cauchy vector structure rC containing C. Moreover, rC coincides with the smallest regular Cauchy structure containing C.

Proof   Let us use transfinite induction on the nonnegative ordinals to define rC. Define $r_0 C = C$ and if α is non-limit ordinal, let $r_\alpha C$ be the smallest Cauchy structure containing $\{cl_p^n F | F \in C, n \in \mathbb{N}\}$, where p denotes the convergence structure for S induced by $r_{\alpha-1} C$. Since $cl_p^n \sum_1^k V(\lambda_i) F_i \subseteq \sum_1^k V(\lambda_i) cl_p^n F_i$, it follows by Proposition 2.1 that $r_\alpha C = \{G | cl_p^n F \subseteq G, F \in C, n \in \mathbb{N}\}$ is an admissible Cauchy vector structure for S. If α is a limit ordinal, define $r_\alpha C = \bigcup_{\beta < \alpha} r_\beta C$. The construction coincides with the regularity series for a Cauchy space (see [8]) and thus the desired conclusion follows.

The space (S, rC) defined in the result above is called the <u>regular modification</u> of (S,C). An object in CHY is said to be <u>complete</u> when the associated Cauchy space is complete; that is, each element in the Cauchy structure converges. Further, $(T, \mathcal{D}, f)$ is called a <u>completion</u> of (S,C) if $(T, \mathcal{D})$ is a complete object in CHY and $f : (S,C) \to (T, \mathcal{D})$ is a Cauchy isomorphism onto a dense subspace of $(T, \mathcal{D})$. Note that by construction, (S, rC) is complete when (S,C) is complete. Moreover, $F : CHY \to CHY$ defined by $F(S,C) = (S, rC)$, $F(f) = f$, defines a functor.

3. COMPLETIONS

Let (S,C) denote a T_2 object in CHY and recall that $F \sim G$ iff $F \cap G \in C$ defines an equivalence relation on C. Denote by S^* the corresponding set of all equivalence classes and let $j(x) = [\dot{x}]$, $x \in S$. Define addition and scalar multiplication in S^* by $[F] + [G] = [F + G]$ and $\lambda [F] = [\lambda F]$; then S^* becomes a vector space over K. For convenience, let us reduce the notation $F + \dot{x}$, $F \cap \dot{x}$, and $V\dot{x}$ to simply $F + x$, $F \cap x$, and Vx, respectively. Denote by $C^* = \{H | jF \cap [F] \subseteq H, F \in C\}$ and $C_* = \{G | \sum_1^n V(\lambda_i)(jF_i \cap [F_i]) \subseteq G, \lambda_i \in K, F_i \in C, \text{ and } n \in \mathbb{N}\}$. It follows from Proposition 2.1 that C_* is the smallest admissible Cauchy vector structure containing C^*. Wyler [11] has shown that (S^*, C^*, j) is a T_2 Cauchy space completion of (S,C) possessing the universal mapping property for Cauchy continuous functions from (S,C) into T_2 complete Cauchy spaces.

Given a T_2 object (S,C) in CHY, define $\mathcal{D} = \{H | [F] + \bigcap_1^m (jF_i - [F_i]) + \sum_1^n V[G_i] \subseteq H, F, F_i \in C, \text{ and } n, m \in \mathbb{N}\}$. It can be shown that the axioms listed in (2.1) are satisfied and thus $(S^*, \mathcal{D})$ is a Cauchy vector space. Note that if $F \in C$, then $[F] + (jF - [F]) \cap (j(\dot{0}) - [\dot{0}]) \subseteq jF \cap [F]$ and thus jF converges to $[F]$ and $jF - [F]$ converges to $[\dot{0}]$ in $(S^*, \mathcal{D})$. It follows that $\mathcal{D} = C_*$ and $(S^*, \mathcal{D})$ is a complete Cauchy vector space. Indeed, $(S^*, \mathcal{D})$ is the convergence vector space introduced by S. Gähler, W. Gähler, and G. Kneis [5]. These authors showed in Theorem 10 [5] that a T_2 convergence vector space has a T_2 convergence vector space completion iff when $F - F \to 0$, $VF \to 0$. It is important to keep in mind that the Cauchy filters for a convergence vector space are precisely those F satisfying $F - F \to 0$, whereas the Cauchy filters of a Cauchy vector space (S, C) are exactly those numbers of C. Given a T_2 object (S, C) in CHY, (S^*, C_*) is a complete Cauchy vector space but the induced convergence vector space is not necessarily complete.

A Cauchy vector space completion (T, E, k) of (S, C) is said to be <u>strict</u> if when $H \to t$ in T, there exists an F on S such that $kF \to t$ in T and $clkF \subseteq H$.

Theorem 3.1 Let (S, C) be a T_2 object in CHY. Then (S^*, C_*, j) is a strict T_2 Cauchy vector space completion of (S, C) and, moreover, it possesses the universal mapping property.

Proof The proof relies heavily on the work done by S. Gähler, W. Gähler, and G. Kneis [5] in the convergence vector space setting. Recall that by construction, (S^*, C_*) is a complete Cauchy vector space and also that $C_* = \mathcal{D}$. Assume that $[H] \to [\dot{0}]$ in (S^*, C_*). Then $\bigcap_1^m (jF_i - [F_i]) + \sum_1^n V[G_i] \subseteq [H]$ for some $F_i, G_i \in C$ and $m, n \in \mathbb{N}$. The proof given in part 1 of Theorem 8 [5] shows that there exists an $F_0 \in C$ such that $jF_0 - [F_0] \subseteq [H]$ and thus $jF_0 \subseteq [F_0 \dotplus H]$. Since (S, C) is T_2, the relationship above implies that $jF_0 = [\dot{x}]$ for some $x \in S$. Hence $H = (x + H) - x \to 0$ in (S, C) and thus $[H] = [\dot{0}]$; therefore, (S^*, C_*) is T_2.

Since $jG \to [G]$ in (S^*, C_*) for each $G \in C$, $j : (S, C) \to (S^*, C_*)$ is Cauchy continuous and $j(S)$ is a dense subset of (S^*, C_*). Assume that F is any filter on S such that $jF \to [\dot{0}]$ in (S^*, C_*). Then $\bigcap_1^m (jF_i - [F_i]) + \sum_1^n V[G_i] \subseteq jF$ for some $F_i, G_i \in C$ and $m, n \in \mathbb{N}$. The proof given in part 2 of Theorem 9 [5] shows that there exists a $G \in C$ with $jG \subseteq jF$ and thus $F \in C$. Next, suppose that H is any filter on S such that $jH \to [G]$ in (S^*, C_*) for some $G \in C$. Since $jG \to [G]$, it follows that $j(H - G) \to [\dot{0}]$ in (S^*, C_*) and by the preceding argument, $H - G \in C$. Hence $(H - G) + G \subseteq H \in C$ and thus $j : (S, C) \to (S^*, C_*)$ is a Cauchy embedding. The universal mapping property is easily verified.

Finally, note that $clj(\sum_1^n V(\lambda_i)F_i) = cl \sum_1^n V(\lambda_i)jF_i \subseteq \sum_1^n cl(V(\lambda_i)jF_i) \subseteq$

$\sum_{1}^{n} V(\lambda_i) \text{clj} F_i \subseteq \sum_{1}^{n} V(\lambda_i)(jF_i \cap [F_i])$ and thus it follows that (S^*, C_*) is a strict T_2 Cauchy vector space completion of (S,C). □

An example is given by S. Gähler, W. Gähler, and G. Kneis ([5], p. 191) of a T_2 convergence vector space having a Cauchy filter F such that VF fails to converge to zero and thus the space does not possess a T_2 convergence vector space completion. This can not occur in CHY, for if $(S,C) \in |CHY|$, $F \in C$, then $VF \in C$. Since zero is an adherent point of VF, VF converges in (S,C). Hence a completion theory for CHY seems to be more natural than that in the convergence vector space setting. Next, let us consider T_3 completions in CHY.

<u>Theorem 3.2</u> Let $(S,+,\cdot,C)$ denote a T_3 Cauchy vector space. A necessary and sufficient condition for $(S,+,\cdot,C)$ to have a T_3 Cauchy vector space completion is for (S,C) to possess a T_3 Cauchy space completion. Moreover, when the latter exists, (S^*, rC_*, j) is a T_3 Cauchy vector space completion of (S,C) possessing the universal mapping property.

<u>Proof</u> Assume that (S,C) has a T_3 Cauchy space completion. It is shown in Corollary 1.3 [9] that in this case, (S^*, rC^*, j) is a T_3 Cauchy space completion of (S,C) possessing the universal property for Cauchy continuous maps from (S,C) into T_3 complete Cauchy spaces. Note that $C^* \subseteq C_*$ and thus $rC^* \subseteq rC_*$. Conversely, let $H \in C_*$. Then $\sum_{1}^{n} V(\lambda_i)(jF_i \cap [F_i]) \subseteq H$ for some $\lambda_i \in K$, $F_i \in C$, and $n \in \mathbb{N}$. It was shown in the proof of Theorem 3.1 that $\text{clj}(\sum_{1}^{n} V(\lambda_i) F_i) \subseteq \sum_{1}^{n} V(\lambda_i)(jF_i \cap [F_i]) \subseteq H$ and thus $H \in rC^*$. Hence $C_* \subseteq rC^*$, or $rC_* \subseteq rC^*$ and thus $rC_* = rC^*$. However, $(S^*, C_*) \in |CHY|$ and thus by Proposition 2.2, (S^*, rC_*, j) is a T_3 Cauchy vector space completion of (S,C) which possesses the universal mapping property. □

It is hoped that this work will inspire other authors to consider the merits of Cauchy vector spaces. Other properties of Cauchy vector spaces need to be studied and connections with convergence vector spaces need to be investigated. Convergence vector spaces have proved to be an important setting for extending certain theorems in functional analysis.

REFERENCES

1. R. Ball, Convergence and Cauchy Structures on Lattice Ordered Groups, <u>Trans. Amer. Math. Soc.</u>, 259, 357-392, (1980).

2. R. Ball, Distributive Cauchy Lattices, Algebra Universalis, 18, 134-174, (1984).

3. C. H. Cook and H. R. Fisher, Uniform Convergence Structures, Math. Ann., 173, 290-306, (1967).

4. R. Fric and D. C. Kent, A Completion Functor for Cauchy Groups, Internat. J. Math. and Math. Sci., 4, 55-65, (1981).

5. S. Gähler, W. Gähler, and G. Kneis, Completion of Pseudo-Topological Vector Spaces, Math. Nachr., 75, 185-206, (1976).

6. H. H. Keller, Die Limes-Uniformisierbarkeit der Limesräume, Math. Ann., 176, 334-341, (1968).

7. D. C. Kent, Completions of ℓ-Cauchy Groups, in "Ordered Groups", Marcel Dekker, Inc., New York and Basel, 93-97, (1980).

8. D. C. Kent, The Regularity Series of a Cauchy Space, Internat. J. Math. and Math. Sci., 7, 1-13, (1984).

9. D. C. Kent and G. D. Richardson, Cauchy Spaces with Regular Completions, Pac. J. Math., 111, 105-116, (1984).

10. E. E. Reed, Completions of Uniform Convergence Spaces, Math. Ann. 194, 83-108, (1971).

11. O. Wyler, Ein Komplettierungsfunktor für Uniforme Limesräume, Math. Nachr., 46, 1-12, (1970).

REGULAR INDUCTIVE LIMITS

Jan Kucera

Washington State University
Pullman, Washington, USA

ABSTRACT

Given a sequence $E_1 \subset E_2 \subset \ldots$ of locally convex spaces with continuous inclusions, the locally convex inductive limit $E = \text{indlim } E_n$ is called regular if every set bounded in E is also bounded in some E_n. Several necessary and sufficient conditions for the regularity of E are derived.

A number of spaces of generalized functions are either inductive limits, or duals of inductive limits, of simpler spaces. Hence it is desirable to know all bounded sets in an inductive limit.

Let $E_1 \subset E_2 \subset \ldots$ be a sequence of locally convex spaces with continuous inclusions, and $E = \text{indlim } E_n$ their locally convex, not necessarily strict, inductive limit. In accordance with [4] we call E to be regular, resp. α-regular, if every set bounded in E is bounded, resp. contained, in some E_n.

Dieudonné and Schwartz proved that if each space E_n is closed in E_{n+1} and the inductive limit is strict then E is regular, [1, Prop. 4-4]. In [4] Floret presented a sufficient condition for regularity of LB-spaces, and in [2] Gothendieck has a different sufficient condition for regularity of LF-spaces. In [9] Bosch and Kucera used the Grothendieck work to show that if each space E_n is webbed then fast completeness of E implies its regularity. The regularity of (general) locally convex inductive limits has been studied in [6, 7, 8].

<u>Lemma 1</u>   Let $E = \text{indlim } E_n$ and $B \subset E$ be bounded. Then B is contained,

resp. bounded, in some E_k iff for any $n \in \mathbb{N}$ there exist $m \in \mathbb{N}$ such that $B \cap \overline{E}_n^E$ is contained, resp. bounded, in E_m.

Here $\overline{E}_n^E$ denotes the closure of E_n in the space E.

Proof The only if part is evident. For the if part, denote by τ the topology of E and put $F = \text{indlim}(\overline{E}_n^E, \tau)$. Then all inclusions $E_n \to (\overline{E}_n^E, \tau) \to F$, $n \in \mathbb{N}$, are continuous and the inclusion $E \to F$ is continuous, too.

If a set B is bounded in E, it is bounded in F and, by the Dieudonné-Schwartz theorem, it is bounded in some $(\overline{E}_n^E, \tau)$. Hence $B = B \cap \overline{E}_n^E$ is contained, resp. bounded, in E_m. □

As a consequence of Lemma 1, we get:

Theorem 1 E is α-regular, resp. regular, iff for every set B bounded in E and every $n \in \mathbb{N}$, there is $m \in \mathbb{N}$ such that $B \cap \overline{E}_n^E$ is contained, resp. bounded, in E_m.

Example The assumption of B being bounded in E in Theorem 1 cannot be removed. Take the set R of all real numbers and equip $E_n = R^n \times \{0\}^{\mathbb{N}}$ with the locally convex topology. Then E is the space of all finite sequences with the finest locally convex topology. In such topology every bounded set is contained in a finite dimensional subspace. Hence $B = [-1,1]^{\mathbb{N}} \cap E$ is not bounded in E and for any $n \in \mathbb{N}$, the set $B \cap \overline{E}_n^E = B \cap E_n$ is bounded in E_n.

Lemma 2 Let E'_{nb}, resp. E'_b, be the strong dual of E_n, $n \in \mathbb{N}$, resp. E. Then $\text{projlim } E'_{nb} = E'_b$ iff $\text{projlim } E'_{nb}$ is barreled.

Proof Assume $\text{projlim } E'_{nb} = E'_b$. Take a barrel B in $\text{projlim } E'_{nb}$. The polar B^0 is weakly bounded in E. Since the weak topology of E is compatible with the original topology of E, B^0 is bounded in E. By the Bipolar Theorem, $B = B^{00}$ is a 0-neighbourhood in E'_b as well as in $\text{projlim } E'_{nb}$.

Assume $\text{projlim } E'_{nb}$ to be barreld. The topology of $\text{projlim } E'_{nb}$ is weaker than the bounded topology of E'. If V is an absolutely convex and closed 0-neighbourhood in E'_b, then V is a barrel in $\text{projlim } E'_{nb}$, hence a 0-neighbourhood, and both topologies are equal. □

Denote, for brevity, by (P_i), $i = 1, 2$, the following properties:

(P_1) For each set B bounded in E, there exists a set A bounded in some E_n such that $B \subset \overline{A}^E$.

(P_2) For each set B bounded in some E_n, there exists $m \in \mathbb{N}$ such that $\overline{B}^E$ is bounded in E_m.

It was proved that in [8] that the property (P_1) holds iff projlim E'_{nb} = E'_b.

Theorem 2   The following properties are equivalent:

(a)   E regular

(b)   projlim E'_{nb} barreled & (P_2)

(c)   projlim E'_{nb} = E'_b & (P_2)

(d) (P_1) & (P_2)

Proof   By Lemma 2 and [8] the last three properties are equivalent. It suffices to prove (a) <=> (d). The if implication is evident. Assume (d). Let a set B be bounded in E. By (P_1), we have $B \subset \bar{A}^E$, where A is bounded in some E_n. By (P_2), the set $\bar{A}^E$ is bounded in some E_m. Hence $B \subset \bar{A}^E$ is also bounded in E_m. □

Theorem 3

(a)   If projlim E'_{nb} = E'_b and each space E_n is semireflexive, then E is regular.

(b)   If each space E'_{nb} is Frechet, then (P_2) <=> E regular.

Proof   (a) It is sufficient to show that the semireflexivity of all spaces E_n, $n \in N$, implies (P_2). Let B be bounded in some E_n. Denote by A the absolutely convex closed hull of B in E_n. Then A is bounded in E_n. Since E_n is semireflexive, A is weakly compact in E_n. The identity map: $E_{n\sigma} \to E_\sigma$ (spaces E_n and E with their weak topologies) in continuous and A is weakly compact in E, hence, weakly closed in E. As a convex set, A is also closed in E and $\bar{B}^E \subset \bar{A}^E$ = A is bounded in E_n.

(b)   follows from Theorem 2(b) since the projective limit of Fréchet spaces E'_{nb} is a Fréchet, hence barreled, space. □

Example   Let $L_q = \{\phi : R^n \to C; \|\phi\|_q^2 = \sum_{|\alpha+\beta| \leq q} \int_{R^n} |x^\alpha D^\beta \phi|^2 dx < \infty\}$, $q \in \mathbb{N}$.

If we assume that the derivatives $D^\beta \phi$ are Sobolev (weak) derivatives, then each space L_q is Hilbert. We denote, for brevity, $W(x) = (1 + |x|^2)^{1/2}$, $x \in \mathbb{R}^n$. For each $p \in \mathbb{N}$ the mapping $W^p : f \to W^p f : S' \to S'$ is bijective. Hence we can provide the image $W^p L_q$ with a topology which makes W^p a topological isomorphism. If we denote O_q = indlim $W^p L_q$, $q \in \mathbb{N}$, then projlim O_q = O_M, the space of multipliers on temperate distributions.

It is easy to verify that indlim O_q satisfies the condition (b) of Theorem 2. This gives us a description of bounded sets in O_M as follows: A set $B \subset O_M$ is bounded iff for every multi-index $\beta \in \mathbb{N}^n$ there is $p \in \mathbb{N}$ such that the set $\{W^{-p}D^{\beta}f;\ f \in B\}$ is bounded in $L^{\infty}(\mathbb{R}^n)$.

REFERENCES

1. J. Dieudonné, L. Schwartz, La dualité danse les spaces (F) et (LF), Ann. l'Inst. Fourier, Tome 1, 61-101, (1949).
2. A. Grothendieck, Produit tensoriels topologiques et espace s nucléaires, Memoirs AMS, 16, (1955).
3. J. Kucera, K. McKennon, Bounded sets in inductive limits, Proc. AMS, Vol. 69, No. 1, 62-64, (1978).
4. K. Floret, On bounded sets in inductive limits of normed spaces, Proc. AMS, Vol. 75, No. 2, 1979, 221-225.
5. K. Floret, "Some aspects of the theory of locally convex inductive limits" in Functional Analysis, North-Holland, (1980).
6. J. Kucera, K. McKennon, Dieudonné-Schwartz theorem on bounded sets in inductive limits, Proc. AMS., Vol. 78, No. 3, 366-368, (1980).
7. J. Kucera, K. McKennon, Dieudonné-Schwartz theorem on bounded sets in inductive limits II, Proc AMS, Vol. 86, No. 3, 392-394, (1982).
8. C. Bosch, J. Kucera, K. McKennon, Dual characterization of the Dieudonné-Schwartz theorem on bounded sets, Int. J. & Math. Sci., Vol. 6, No. 1, 189-192, (1983).
9. J. Kucera, C. Bosch, Bounded sets in fast complete inductive limits, Int. J. Math. & Math. Sci., Vol. 7, No. 3, 615-617, (1984).

WEAK CONVERGENCE IN A K-SPACE

Endre Pap

Institute of Mathematics
University of Novi Sad
dr I. Đuričića 4, 21000 Novi Sad, Yugoslavia

The notions of K-convergence and K-space have proved to be very useful in functional analysis. In monographs [2] and [3] there are many results on these subjects and we shall use the notations and notions from them.

We shall investigate in this paper the weak convergence in a locally convex K-space.

First, we need some definitions.

<u>Definition</u> A sequence $\{x_n\}$ of elements in a convergence group X (see for example [3]) is said to be a <u>K-sequence</u> in X if for each subsequence $\{y_n\}$ of $\{x_n\}$ there exist a subsequence $\{z_n\}$ of $\{y_n\}$ and an element $z \in X$ such that $\sum_{k=1}^{n} z_k \to z$. X is a <u>K-space</u> if each sequence $\{x_n\}$ from X such that $x_n \to 0$ is a K-sequence.

Specially, if X is a locally convex space, then a sequence $\{x_n\}$ from X is called a <u>weak K-sequence</u> if it is a K-sequence with respect to the weak convergence in X.

We shall start with the well known Banach-Saks theorem: If $\{x_n\}$ is a sequence from a Hilbert space H such that it weakly converges to an element x_0 from H, then there exists a subsequence $\{x_{n_i}\}$ of $\{x_n\}$ such that the sequence $\{y_n\}$, where

$$y_n = \frac{1}{n} \sum_{i=1}^{n} x_{n_i},$$

is norm convergent to x_0. We shall obtain a version of the Banach-Saks theorem for locally convex space. Namely, we have

Theorem 1 Let $\{x_n\}$ be a sequence from a locally convex vector space X such that the sequence $\{x_n - x_0\}$ is a weakly K-sequence for some $x_0 \in X$. Then the sequence $\{x_n\}$ has a subsequence $\{x_{n_i}\}$ such that the sequence $\{y_n\}$, where

$$y_n = \frac{1}{n} \sum_{i=1}^{n} x_{n_i},$$

is (strongly) convergent to x_0.

Proof: Let $\{x_n - x_0\}$ be a weakly K-sequence for some $x_0 \in X$. Then there exists a subsequence $\{x_{n_i}\}$ of $\{x_n\}$ such that for some $x \in X$

$$\lim_{n \to \infty} f\left(\sum_{i=1}^{n} (x_{n_i} - x_0) \right) = f(x) \quad (f \in X^*).$$

Hence, the sequence

$$\left\{ \sum_{i=1}^{n} (x_{n_i} - x_0) \right\}$$

is weakly bounded. Since, by the Mackey theorem, each weakly bounded subset of a locally convex space is bounded, it follows that the preceding sequence is also bounded. Then, taking an arbitrary seminorm p from the family of seminorms which generates the topology of this space, we have

$$\lim_{n \to \infty} p\left(\frac{1}{n} \sum_{i=1}^{n} x_{n_i} - x_0 \right) = \lim_{n \to \infty} p\left(\frac{1}{n} \sum_{i=1}^{n} (x_{n_i} - x_0) \right) = 0.$$

Hence, the sequence $\{y_n\}$, $y_n = \frac{1}{n} \sum_{i=1}^{n} x_{n_i}$, converges to x_0. □

Remark 1 In the special case, when X is a normed vector space the preceding result is obtained in [4] (Theorem 1). For normed spaces, the preceding theorem could also be obtained as an easy consequence of Corollary 3.8 from [2]. Namely, in a normed space, a sequence is a weak K-sequence iff it is a norm K-sequence. In connection with this fact see, Problem 9 in these proceedings.

We have the following consequence, which is important in the optimization theory.

Corollary 1 Let g be a convex continuous functional on a locally convex space X. If $\{x_n\}$ is a sequence from X such that the sequence $\{x_n - x_0\}$ is a weak K-sequence for some $x_0 \in X$, then

$$\lim \inf g(x_a) \geq g(x_0).$$

Proof: We take a subsequence $\{y_n\}$ of $\{x_n\}$ such that $\liminf g(x_n) = \lim_{n\to\infty} g(y_n)$. Now, using Theorem 1, we obtain a subsequence $\{z_n\}$ of $\{y_n\}$ such that the sequence

$$\left\{\frac{1}{n}\sum_{k=1}^{n}z_k\right\}$$

is convergent to x_0. We have

$$\lim_{n\to\infty}\frac{1}{n}\sum_{k=1}^{n}g(z_k) = \lim_{n\to\infty}g(z_n) \geq \lim_{n\to\infty}g\left(\frac{1}{n}\sum_{k=1}^{n}z_k\right) = g(x_0). \quad \square$$

We shall need in the proof of Theorem 2 the following result from [1].

<u>Lemma</u> Assume that $\{x_{ij}\}$ $(i,j \in \mathbb{N})$ is a matrix of elements x_{ij} in a topological group X. If for each increasing sequence $\{m_i\}$ of positive integers there exists a subsequence $\{n_i\}$ of $\{m_i\}$ such that

(i) $\lim_{i\to\infty} x_{n_i n_j} = 0$ $(j \in N)$

and

(ii) $\lim_{i\to\infty} \sum_{j=1}^{\infty} x_{n_i n_j} = 0$,

then $\lim_{i\to\infty} x_{ii} = 0$.

Remark 2 It is unknown if the preceding Lemma holds also for a FLUSH convergence group [1]) (for the notation see for example [3]). Now we have

Theorem 2 Let X be a convergence vector space which is a K-space and let Y be a normed vector space. If $T : X \to Y$ is a linear operator such that $x_n \to 0$ implies $T(x_n) \to 0$ weakly, then T is (sequentially) continuous.

Proof: We take a sequence $\{x_n\}$ from X such that $x_n \to 0$. We assume that the normed vector space Y is separable. Namely, we can always replace Y by the closed linear subspace generated by the sequence $\{T(x_n)\}$. As a consequence of the Hahn-Banach theorem, we obtain that for each $T(x_n)$ there exists $f_n \in Y^*$ such that $\|f_n\| = 1$ and

$$f_n(T(x_n)) = \|T(x_n)\|. \tag{1}$$

---

1) For the partial answer see the paper of P. Antosik in this Proceedings.

By the Banach-Alaoglu theorem there exists a subsequence $\{f_{k_i}\}$ of $\{f_n\}$ such that $\{f_{k_i}\}$ is weak $*$ convergent to an element $f \in Y^*$.

Let

$$x_{ij} = (f_{k_i} - f)(T(x_{k_j})) \quad (i,j \in \mathbb{N}).$$

Then, the matrix $\{x_{ij}\}$ $(i,j \in N)$ satisfies the conditions of the Lemma. Namely, condition (i) is obvious, since for each subsequence $\{x_i\}$ of any increasing sequence $\{m_i\}$ of positive integers m_i holds

$$\lim_{i \to \infty} (f_{k_{n_i}} - f)(T(x_{k_{n_j}})) = 0 \quad (j \in \mathbb{N}).$$

Since X is a K-space, there exists a subsequence $\{n_i\}$ of $\{m_i\}$ such that

$$\lim_{n \to \infty} \sum_{j=1}^{n} x_{k_{n_j}} = x$$

for some $x \in X$. Then, we have

$$(f_{k_{n_i}} - f)\left[T\left(\sum_{j=1}^{n} x_{k_{n_j}}\right)\right] \to (f_{k_{n_i}} - f)(T(x)) \text{ for } n \to \infty.$$

Hence,

$$\lim_{i \to \infty} \sum_{j=1}^{\infty} (f_{k_{n_i}} - f)(T(x_{k_{n_j}})) = 0,$$

i.e. condition (ii) from Lemma. Now, the Lemma implies

$$\lim_{i \to \infty} x_{ii} = \lim_{i \to \infty} (f_{k_i} - f)(T(x_{k_i})) = 0.$$

Hence, by the weak continuity of T, we obtain

$$\lim_{i \to \infty} f_{k_i}(T(x_{k_i})) = 0.$$

By (1) this implies $\lim_{n \to \infty} \|T(x_{k_i})\| = 0$. Since the preceding procedure is true if we take an arbitrary subsequence of $\{x_n\}$ instead of $\{x_n\}$, we obtain, by the Urysohn property of the set of real numbers, that $\lim_{n \to \infty} \|T(x_n)\| = 0$, i.e. T is continuous.

A lineaer operator $T : X \to Y$ is weakly (sequentially) continuous if for each sequence $\{x_n\}$ from X such that $\{x_n\}$ from X such that $x_n \to x$ weakly for some $x \in X$, then $T(x_n) \to T(x)$ weakly.

<u>Corollary 2</u> (Theorem 3, [4]) Let X be a normed K-space and let Y be a normed space. If $T : X \to Y$ is a linear weakly continuous operator, then T is continuous.

<u>Remark 3</u> Let S_w be the family of all weakly summable sequences in a locally convex vector space X, i.e. $\{x_n\} \in S_w$ iff $\sum_{k=1}^{n} x_k \to x$ weakly as $n \to \infty$ for some $x \in X$.

Theorem 3.1 from [6] implies

<u>Theorem 3</u> Let f be a map from a locally convex weakly K-space X into a locally convex space Y such that $f(0) = 0$. The function f is weakly zero-continuous ($x_n \to 0$ weakly implies $f(x_n) \to 0$ weakly) iff $f(x_n) \to 0$ weakly for any $\{x_n\} \in S_w$.

REFERENCES

1. P. Antosik, A lemma on matrices and its applications, <u>Contemporary Mathematics</u> 52, 89 - 95 (1986).
2. P. Antosik, Ch. Swartz, "Matrix Methods in Analysis", Lecture Notes in Mathematics, Vol. 1113, Springer-Verlag, 1985.
3. E. Pap, "Funkcionalna analiza", Institute of Mathematics, Novi Sad, 1982.
4. E. Pap, Some theorems of functional analysis with K-convergence, <u>Zbornik radova PMF u Novom Sadu</u> 15,2, 43 - 49 (1985).
5. E. Pap, Comparison of K-convergences, Proceedings of III Italian Natural Meeting on Topology (to appear).
6. E. Pap, Zero-continuous functions on K-, N- and complete spaces, <u>Zbornik radova PMF u Novom Sadu</u>, 16,2, 9-15, (1986).

THE BANACH-STEINHAUS THEOREM FOR ORDERED SPACES

Charles Swartz

Department of Mathematical Sciences
New Mexico State University
Las Cruces, New Mexico 88003, U.S.A.

ABSTRACT

Let X and Y be vector lattices and $T_i : X \to Y$ a sequence of linear operators which are sequentially continuous with respect to relative uniform convergence. If $\{T_j x\}$ is relatively uniformly convergent to Tx for each $x \in X$, under appropriate assumptions on the spaces, we show that the linear operator T is also continuous and that the $\{T_i\}$ are order equicontinuous in a certain sense. We also establish an order version of the Uniform Boundedness Principle.

1. INTRODUCTION

The Banach-Steinhaus Theorem which assets that the pointwise limit of a sequence of continuous linear operators between Banach spaces is also a continuous linear operator has proven to be one of the most fundamental and useful results in functional analysis. In this note we consider the problem of obtaining a version of the Banach-Steinhaus Theorem for linear mappings between ordered spaces. We first present a simple example which shows that a straightforward version of the Banach-Steinhaus Theorem does not hold for linear functionals which are sequentially continuous with respect to order convergence. By considering linear operators which are continuous with respect to relative uniform convergence and by employing the analogue of the matrix methods used in [1] for ordered spaces, we then establish, under appropriate assumptions on the range space, two versions of the Banach-Steinhaus Theorem for such operators. We also establish an order version of the Uniform Boundedness Principle.

Throughout this note we let X and Y be vector lattices with positive cones X_+ and Y_+, respectively. We adhere to the terminology of [4]. A decreasing sequence $\{\varepsilon_k\}$ in X_+ is said to order converge to 0 if $\inf \varepsilon_k = 0$; we write $\varepsilon_k \downarrow 0$. A sequence $\{x_k\} \subseteq X$ is said to be <u>order convergent</u> to x if there is a decreasing sequence $\varepsilon_k \downarrow 0$ such that $|x_k - x| \leq \varepsilon_k$ for all k; we write 0-$\lim x_k = x$.

If $\sigma(\tau)$ is a convergence in X(Y) and $T : X \to Y$ is a linear map, then T is (sequentially) continuous with respect to σ and τ if $\{x_k\}$ σ-convergent to 0 implies that $\{Tx_k\}$ is τ-convergent to 0: we say that T is <u>(σ,τ)-continuous</u>.

We now give an example of a sequence of linear functionals which are sequentially continuous with respect to order convergence but which converge pointwise to a linear functional which is not sequentially order continuous. That is, a straightforward analogue of the Banach-Steinhaus Theorem does not hold for order continuous linear functionals.

<u>Example 1</u> Let c be the space of all convergent real-valued sequences, ordered coordinatewise. Define $S_i : c \to \mathbb{R}$ by $\langle S_i, \{t_j\} \rangle = t_i$ and $L : c \to \mathbb{R}$ by $\langle L, \{t_j\} \rangle = \lim_j t_j$. Each S_i is sequentially continuous with respect to order convergence in c and $\{S_i\}$ is pointwise convergent to L. But, L is not sequentially order continuous (let $x_j = \{(1 - 1/k)^j\}_{k=1}^{\infty}$; then $x_j \downarrow 0$ but $\langle L, x_j \rangle = \lim_k (1 - 1/k)^j = 1$ for each j).

It should be noted that there is a Banach-Steinhaus type result due to Nakano for the case when the range space is $\mathbb{R}$ and the linear functions are order bounded ([4], IX.1.1); however, there do not seem to be any theorems of Banach-Steinhaus type in the literature when the range space is not $\mathbb{R}$. We will now estabilsh two Banach-Steinhaus type results when the operators are continuous with respect to the other most common convergence employed in vector lattices, relative uniform convergence.

We first recall the definition of relative uniform convergence.

<u>Definition 2</u> Let $u \in X_+$. The sequence $\{x_k\}$ is said to u-converge to $x \in X$ (converge u - uniformly to x ([2])) if there exists a scalar sequence $t_k \downarrow 0$ such that $|x_k - x| \leq t_k u$ for all k; the element u is called a <u>convergence regulator</u> for $\{x_k\}$ ([4], III.11). We write u - $\lim x_k = x$. The sequence $\{x_k\}$ is said to be <u>relatively uniformly convergent</u> to x if there exists $u \in X_+$ such that u - $\lim x_k = x$; we write r - $\lim x_k = x$. The

sequence is u*-convergent (r*-convergent) to $x \in X$ if every subsequence of $\{x_k\}$ has a subsequence which is u-convergent (relatively uniformly convergent) to x; we write $u^* - \lim x_k = x$ ($r^* - \lim x_k = x$).

We next make the definition of K convergent sequences in a vector lattice. These sequences have proven to be very useful in obtaining versions of the Banach-Steinhaus Theorem and the Uniform Boundedness Principle for normed linear spaces which are not necessarily complete ([1]). These sequences are used in one of our versions of the Banach-Steinhaus Theorem.

Definition 3 Let $u \in X_+$. The sequence $\{x_k\}$ is said to be $u - K -$ convergent to 0 ($r - K$ -convergent to 0) if every subsequence of $\{x_k\}$ has a subsequence $\{x_{n_k}\}$ such that the series $\sum_{k=1}^{\infty} x_{n_k}$ is u - convergent (r - convergent) to some $x \in X$.

Note that if $\{x_k\}$ is u - K convergent to 0 (r - K convergent to 0) then $\{x_k\}$ is u* - convergent to 0 (r* - convergent to 0). The converse of this does not hold in general as the following example shows.

Example 4 Let m_0 be the space of real-valued sequences $\{t_k\}$ which have finite range and order m_0 coordinatewise. Let $u = (1,1,\ldots)$ and let e_j be the sequence with a 1 in the jth coordinate and 0 elsewhere. Then $x_j = (\frac{1}{j})e_j \leq (\frac{1}{j})u$ for each j so $u - \lim x_j = 0$ but no subseries $\sum_{j=k}^{\infty} x_{k_j}$ can converge to a sequence belonging to m_0.

If X is Dedekind σ-complete (briefly σ-complete), the converse does hold. This is very analogous to the situation in normed spaces ([1], 3.2).

Proposition 5 Let X be σ-complete. If $u - \lim x_k = 0$ ($r - \lim x_k = 0$), then $\{x_k\}$ is u - K convergent to 0 (r - K convergent to 0).

Proof Let $|x_k| \leq t_k u$, where $t_k \downarrow 0$. Given any increasing sequence of positive integers $\{m_k\}$, choose a subsequence $\{n_k\}$ of $\{m_k\}$ such that $\sum_{k=1}^{\infty} t_{n_k} < \infty$. Then $|x_{n_k}| \leq t_{n_k} u$ and the σ-completeness of X implies that the series $\sum_{k=1}^{\infty} x_{n_k}$ is absolutely order convergent to some $x \in X$ ([5], IV.9). But $|x - \sum_{k=1}^{p} x_{n_k}| = |\sum_{k=p+1}^{\infty} x_{n_k}| \leq \sum_{k=p+1}^{\infty} |x_{n_k}| \leq \left[\sum_{k=p+1}^{\infty} t_{n_k}\right] u$ and $\left(\sum_{k=p+1}^{\infty} t_{n_k}\right) \downarrow 0$ implies that the series is actually u-convergent to x. □

2. BANACH-STEINHAUS THEOREM

In this section we obtain versions of the Banach-Steinhaus Theorem for operators which are continuous with respect to u - uniform convergence or relative uniform convergence. The results are fairly straightforward analogues of the classical Banach-Steinhaus Theorem.

We begin by establishing an analogue for ordered spaces of the matrix theorem given in [1], 2.1. This results was used in [1] to establish the Antosik Diagonal Theorem ([1], 2.2), and this Diagonal Theorem was then used to treat the Banach-Steinhaus Theorem as well as many other topics in functional analysis and measure theory ([1]). We use this matrix theorem in deriving our basic Banach-Steinhaus results.

Lemma 1   Let the matrix $[x_{ij}]$, $x_{ij} \in X$, be such that its rows and columns are u-convergent to 0. Let $[\varepsilon_{ij}]$ be a matrix in $\mathbb{R}$ with $\varepsilon_{ij} > 0$. Then there exists an increasing sequence of positive integers $\{m_j\}$ such that $|x_{m_i m_j}| \leq \varepsilon_{ij} u$ for $i \neq j$.

Proof   Put $m_1 = 1$. Since $u - \lim_j x_{1j} x_{ij} = 0$ and $u - \lim_i x_{i1} = 0$, there exists $m_2 > m_1$ such that $|x_{1m_2}| \leq \varepsilon_{12} u$ and $|x_{m_2 1}| \leq \varepsilon_{21} u$. Similarly, there exists $m_3 > m_2$ such that $|x_{m_1 m_3}| \leq \varepsilon_{13} u$, $|x_{m_2 m_3}| \leq \varepsilon_{23} u$, $|x_{m_3 m_2}| \leq \varepsilon_{32} u$ and $|x_{m_3 m_1}| \leq \varepsilon_{31} u$. Now continue by induction. □

We now use this matrix result to derive a lemma which will be used to obtain our Banach-Steinhaus result with respect to u-uniform convergence. In what follows below, we let $u \in X_+$ and $v \in Y_+$.

Lemma 2   Let $T_i : X \to Y$ be (u,v)-continuous and Y σ-complete. If $u - \lim_j x_j = 0$, $\{x_j\}$ is $u - K$ convergent to 0 and $v - \lim T_i x = 0$ for each $x \in X$, then $v^* - \lim_i T_i x_i = 0$.

Proof   The rows and columns of the matrix $[T_i x_j]$ are v-convergent to 0 so if we apply Lemma 2.1 to this matrix and the scalar matrix $[\varepsilon_{ij}] = [2^{-i-j}]$, there exists an increasing sequence of positive integers $\{m_i\}$ such that $|T_{m_i} x_{m_j}| \leq 2^{-i-j} v$ for $i \neq j$. Since $\{x_j\}$ is $u - K$ convergent, there is a subsequence $\{n_j\}$ of $\{m_j\}$ such that $\sum_{j=1}^{\infty} x_{n_j}$ is u-convergent to some element $x \in X$. Note that we still have $|T_{n_i} x_{n_j}| \leq 2^{-i-j} v$ for $i \neq j$, and also the series $\sum_{j=1}^{\infty} T_{n_i} x_{n_j}$ is absolutely order convergent to $T_{n_i}(x)$ by the σ-completeness of Y and the (u,v)-continuity of T_{n_i}.
We thus have

428

$$|T_{n_i}x_{n_i}| \leq |\sum_{\substack{j=1}}^{\infty} T_{n_i}x_{n_j}| + |\sum_{\substack{j=1 \\ j \neq i}}^{\infty} T_{n_i}x_{n_j}| \leq |T_{n_i}(x)| + \sum_{\substack{j=1 \\ j \neq i}}^{\infty} |T_{n_i}x_{n_j}|$$

$$\leq |T_{n_i}(x)| + \sum_{j=1}^{\infty} 2^{-i-j}v = |T_{n_i}(x)| + 2^{-i}v. \tag{1}$$

Since $v - \lim T_{n_i}(x) = 0$, (1) implies that $v - \lim T_{n_i}x_{n_i} = 0$. Since we can apply the same argument to any subsequence of $\{T_i x_i\}$, this shows that $\{T_i x_i\}$ is v^*-convergent to 0. □

Remark 3 Note that if X is σ-complete, then the assumption that $\{x_j\}$ is $u - K$ convergent to 0 can be dropped (Proposition 1.5).

Our first Banach-Steinhaus result is now easily obtained from Lemma 2.2. Recall, that if σ is a notion of convergence in X, then X is said to have the <u>diagonal property</u> with respect to σ if whenever $[x_{ij}]$ is a matrix in X such that $\sigma - \lim_j x_{ij} = 0$ for each i, then there exists an increasing sequence of positive integers $\{m_i\}$ such that $\sigma - \lim x_{im_i} = 0$ ([2], [3], 1.5.5). Theorem VI. 5.3 of [4] gives sufficient conditions for X to have the diagonal property with respect to u-uniform convergence or relative uniform convergence (see also [2], §71).

Theorem 4 Let $T_i : X \to Y$ be (u,v)-continuous and Y σ-complete and have the diagonal property with respect to v-convergence. If $v - \lim T_i x = Tx$ exists for each $x \in X$ and if $u - \lim x_j = 0$ and $\{x_j\}$ is $u - K$ convergent to 0, then

$$v^* - \lim T_i x_i = 0 \tag{2}$$

and

$$v^* - \lim Tx_j = 0. \tag{3}$$

<u>Proof</u> (2) Consider the matrix $[T_i x_j]$; the rows of this matrix are v-convergent to 0 so by the diagonal property, there is an increasing sequence of positive integers $\{m_i\}$ such that $v - \lim T_i x_{m_i} = 0$. Apply Lemma 2.2 to the sequences $\{T_{m_i} - T_i\}$ and $\{x_{m_j}\}$ to obtain that $v^* - \lim\{T_{m_i}x_{m_i} - T_i x_{m_i}\} = 0$. Hence, $v^* - \lim T_{m_i}x_{m_i} = 0$, and since the same argument can be applied to any subsequence of $\{T_i x_i\}$, it follows that $v^* - \lim T_i x_i = 0$.

(3) Consider the matrix $[T_j x_i - T x_i]$; the rows of this matrix are v-argument to 0 so by the diagonal property, there is an increasing sequence of positive integers $\{n_i\}$ such that $v - \lim(T_{n_i}x_i - Tx_i) = 0$.

By (2), $v^* - \lim T_{n_i} x_i = 0$ so $v^* - \lim T x_i = 0$. □

Remark 5 If X is σ-complete, the hypothesis that $\{x_j\}$ is u - K convergent to 0 can be dropped. In this case, condition (3) is just the condition that the limit operator T is (u,v*)-continuous so that Theorem 4 can be viewed as a Banach-Steinhaus type of result for (u,v)-continuous linear operators.

If E and F are normed spaces, recall that a sequence of continuous linear operators $T_i : E \to F$ is equicontinuous iff $\lim T_i x_i = 0$ for every sequence $\{x_i\}$ with $\lim x_i = 0$. Thus, if X is σ-complete, condition (2) can be viewed as an order-equicontinuity result for the sequence of operators $\{T_i\}$ with respect to u-convergence in X and v*-convergence in Y.

Of course, there is the annoying and undesirable situation in the theorem that the T_i are assumed to be (u,v)-continuous whereas the limit operator is only (u,v*)-continuous. Of course, when v and v* convergence in the range space coincide, this difficulty does not exist. This is the case when $Y = \mathbb{R}$, and in view of the lack of Banach-Steinhaus type results for general range spaces perhaps even this shortcoming should not be viewed too negatively.

We now consider the problem of obtaining a Banach-Steinhaus result for relative uniform convergence. For this, we consider the following property for a vector lattice X:

(C) any countable set of relatively uniformly convergent sequences has a common convergence regulator.

This property is abstracted from the conclusion of Theorem VI.5.2 of [4] and this theorem gives sufficient conditions for the vector lattice X to have property (C); see also property (σ) of [2], §70.

We establish the analogue of Lemma 2.2 for relative uniform convergence and then the Banach-Steinhaus result for relative uniform convergence will be an immediate consequence.

Lemma 6 Let X be σ-complete and let Y be σ-complete and have property (C). Let $T_i : X \to Y$ be linear and (r,r)-continuous. If $r - \lim T_i x = 0$ for each $x \in X$ and $r - \lim_j x_j = 0$, then $r^* - \lim T_i x_i = 0$.

Proof Let u be a convergence regulator for $\{x_j\}$ with $|x_j| \le t_j u$, $t_j \downarrow 0$. Pick a subsequence $\{t_{m_j}\}$ such that $\sum_{j=1}^{\infty} t_{m_j} < \infty$. Then by the σ-completeness of X, the subseries $\sum_{j=1}^{\infty} x_{m_j}$ is absolutely order convergent to some $x \in X$ ([4], IV.9). Since

$$\left|\sum_{j=1}^{\infty} x_{m_j} - x\right| = \left|\sum_{j=n+1}^{\infty} x_{m_j}\right| \le \left(\sum_{j=n+1}^{\infty} t_{m_j}\right) u,$$ the series $\sum_{j=1}^{\infty} x_{m_j}$ is actually u-convergent to x, and since the same argument can be applied to any subseries, the series is subseries u-convergent. Since we are only going to produce a subsequence of $\{T_i x_i\}$ which is r-convergent to 0, we assume for notational convenience that $m_j = j$.

Now $r - \lim_i T_i x_j = 0$ for each j and $r - \lim_j T_i x_j = 0$ for each i so there is a common convergence regulator v for each of these sequences (property (C)). Applying Lemma 2.1 to the matrix $[T_i x_j]$ and the scalar matrix $[2^{-i-j}]$ implies that there exists an increasing sequence of positive integer $\{n_j\}$ such that $|T_{n_i} x_{n_j}| \le 2^{-i-j} v$ for $i \ne j$. Note that for each i the series $\sum_{j=1}^{\infty} T_{n_i} x_{n_j}$ is absolutely order convergent in Y by the σ-completeness.

As noted above, the series $\sum_{j=1}^{\infty} x_{n_j}$ is u-convergent to say $x \in X$, and each T_i is (r,r)-continuous so $T_i x = \sum_{j=1}^{\infty} T_i x_{n_j}$, where the series is r-convergent. We then have

$$|T_{n_i} x_{n_i}| \le \left|\sum_{j=1}^{\infty} T_{n_i} x_{n_j}\right| + \left|\sum_{\substack{j=1 \\ j \ne i}}^{\infty} T_{n_i} x_{n_j}\right| \le |T_{n_i} x| + \sum_{\substack{j=1 \\ j \ne i}}^{\infty} |T_{n_i} x_{n_j}|$$

(4)

$$\le |T_{n_i} x| + \sum_{j=1}^{\infty} 2^{-i-j} v = |T_{n_i} x| + 2^{-i} v.$$

Since $r - \lim T_{n_i} x = 0$, (4) implies that $r - \lim T_{n_i} x_{n_i} = 0$, and the proof is complete. □

Theorem 7 Let X be σ-complete and let Y be σ-complete, have property (C) and the diagonal property with respect to relative uniform convergence. Let $T_i : X \to Y$ be (r,r)-continuous. If $r - \lim T_i x = Tx$ exists for each $x \in X$ and $r - \lim x_j = 0$, then

$$r^* - \lim T_i x_i = 0 \quad \text{(order equicontinuity of } \{T_i\}) \tag{5}$$

and

$$r^* - \lim Tx_j = 0 \quad ((r,r^*)\text{-continuity of T}). \tag{6}$$

Proof Y has the diagonal property with respect to r-convergence and the rows of the matrix $[T_i x_j]$ are r-convergent to 0 so there is an increasing sequence of positive integers $\{m_i\}$ such that $r - \lim T_i x_{m_i} = 0$. Now apply Lemma 2.6 to the sequences $\{T_{m_i} - T_i\}$ and $\{x_{m_i}\}$ to obtain $r^* - \lim T_{m_i} x_{m_i} = 0$. Since the same argument can be applied to every subsequence of $\{T_i x_i\}$, we have (5).

For (6), consider the matrix $[T_j x_i - Tx_i]$; the rows are r-convergent to 0 so by the diagonal property, there is a increasing sequence of positive integers $\{m_i\}$ such that $r - \lim(T_{m_i} x_i - Tx_i) = 0$. By (5), $r^* - \lim T_{m_i} x_i = 0$ so $r^* - \lim Tx_i = 0$ and (6) holds. □

Sufficient condition for Y to have property (C) and the diagonal property are given in [4] VI.5.2 and VI.5.3.

Again this result has the annoying feature that the operators T_i are assumed to be (r,r)-continuous while the limit operator T is concluded to be (r,r*)-continuous. However, for many vector lattices r and r* convergence coincide ([3], 4.2.1). Again, due to the lack of Banach-Steinhaus results for general range spaces, even this shortcoming should not be viewed too netively.

Lemma 6 can also be employed to obtain an order version of the classical Uniform Boundedness Principle from functional analysis. One version of the Uniform Boundedness Principle states if $T_i : X \to Y$ is a sequence of continuous linear operators from a Banach space X into a normed space Y which is pointwise bounded on X, then $\{T_i\}$ is equicontinuous, i.e., $T_i x_i \to 0$ whenever $x_i \to 0$. An order analogue of this statement is

Theorem 8 Let X and Y be as in Lemma 6 and let $T_i : X \to Y$ be (r,r)-continuous. If $\{T_i x\}$ is order bounded in Y for each $x \in X$, then $r^* - \lim T_i x_i = 0$ whenever $r - \lim x_i = 0$.

Proof Let $r - \lim x_i = 0$. Since relative uniform convergence is stable ([4], VI.4), there exists a scalar sequence $t_i \uparrow \infty$ such that $r - \lim t_i x_i = 0$. For each x, $r - \lim(1/t_i)T_i x = 0$ by the pointwise order bounded condition. By Lemma 6, $r^* - \lim(1/t_i)T_i(t_i x_i) = r^* - \lim T_i x_i = 0$. □

The author would like to thank Joe Kist for his assistance.

REFERENCES

1. P. Antosik and C. Swartz, "Matrix Methods in Analysis", Springer Verlag Lecture Notes in Mathematics # 1113, Heidelburg (1985).
2. W. A. J. Luxemburg and A. C. Zaanen, "Riesz Spaces I", North Holland, Amsterdam (1971).
3. A. L. Peressini, "Ordered Topological Vector Spaces", Harper and Row, New York (1967).
4. B. Z. Vulikh, "Introduction to the Theory of Partially Ordered Spaces", Wolters-Noordhoff, Groningen (1967).

SECTION IV. OPEN PROBLEMS

OPEN PROBLEMS

Proposed by prof. P. Antosik and J. Burzyk:

Assume that $f \in \mathcal{D}'$, $g \in C$, $g(x_0) \neq 0$ and $(f * \delta_n)(g, \tilde{\delta}_n) \xrightarrow{\mathcal{D}'} 0$ for every delta-sequences δ_n and $\tilde{\delta}_n$.

Problem 1 Is there a neighborhood V of x_0 such that $f = 0$ on V?

Proposed by prof. T. K. Boehme:

For $T \in \mathcal{D}'$ and $\phi \in \mathcal{D}$ one has that the pointwise product $\phi T \in \mathcal{D}'$. For hyperfunctions on the real line pointwise products with real analytic functions are always possible.

For regular operators pointwise products with polynomials are possible.

Problem 2 Can a larger class than polynomials be obtained such that pointwise products with regular operators are always possible?

Proposed by J. Burzyk:

Problem 3 Let x be a regular operator with bounded support. Is the operator $1/x$ regular?

Problem 4 We say that a sequence $\{x_n\}$ of regular operators is convergent to the operator x if there exists a delta sequence $\{\delta_n\}$ and continuous functions f_n, such that

$$x_n - x = \frac{f_n}{\delta_n} \text{ and } f_n \to 0 \text{ almost uniformly.}$$

Let us suppose that an operator x has support $\{0\}$. Is there a sequence $\{\alpha_n\}$ of complex numbers such that the series $\sum_{n=1}^{\infty} \alpha_n \delta^{(n)}$ is convergent to the operator x?

Proposed by prof. J. F. Colombeau:

Problem 5 Does the subalgebra of $G(\mathbb{R})$ spanned by the distributions contain nonclassical constants? i.e. let $G \in G(\mathbb{R})$ be a finite sum of finite products of distributions; does $G' = 0$ imply G that is a classical constant?

Problem 6 Let Ω be a connected open set in $\mathbb{C}^n$ (possibly n = 1). Let $F \in G(\Omega)$ be such that $\bar{\partial}F = 0$; let us assume there is a nonvoid open subset $\omega \in \Omega$ such that the restriction of F to ω is a distribution (thus a holomorphic function). Is F a holomorphic function on Ω? (the answer is yes if we assume that the restriction $F|_\omega$ can be holomorphically extended to Ω).

Problem 7 Investigate convergence structures on $G(\Omega)$ and use them (a uniform – but nonvector space-topology has already been considered). (For other problems see the paper of Colombeau in this Proceedings).

Proposed by prof. M. Oberguggenberger:

Problem 8 (Multiplication of distributions)
Let $S, T \in \mathcal{D}'(\mathbb{R}^n)$. Define the product of S and T by

$$ST = \lim_{\varepsilon \to 0} (S * \phi^\varepsilon)(T * \phi^\varepsilon)$$

provided the limit exists in $\mathcal{D}'(\mathbb{R}^n)$ for all delta-nets $\{\phi^\varepsilon\}_{\varepsilon > 0}$ (and is independent of the net chosen) with $\phi^\varepsilon \geq 0$, $\int \phi^\varepsilon(x)dx = 1$ for all ε, and which satisfy one of the following conditions:

(1) support $(\phi^\varepsilon) \to \{0\}$ as $\varepsilon \to 0$, and

$$\int |x^\alpha \partial^\alpha \phi^\varepsilon(x)| \leq M_\alpha \quad (\text{for all } \alpha \in \mathbb{N}_0^n, \varepsilon > 0)$$

(2) support $(\phi^\varepsilon) \subset \{|x| \leq \varepsilon\}$, and

$$\varepsilon^{|\alpha|} \int |\partial^\alpha \phi^\varepsilon(x)| dx \leq M_\alpha \quad (\text{for all } \alpha \in \mathbb{N}_0^n, \varepsilon > 0)$$

(3) $\phi^\varepsilon(x) = \varepsilon^{-n} \phi(x/\varepsilon)$ for some $\phi \in \mathcal{D}(\mathbb{R}^n)$

(4) $\phi^\varepsilon(x) = \varepsilon^{-n} \phi(x/\varepsilon)$ for some $\phi \in \mathcal{D}(\mathbb{R}^n)$ with $\phi(x) = \phi(|x|)$.

Further, in dimension n = 1, consider the assertion

(5) $\lim_{\varepsilon \to 0} \hat{S}_\varepsilon \hat{T}_\varepsilon$ exists in $\mathcal{D}'(\mathbb{R})$

where $\hat{S}(z) \in H(\mathbb{C} \setminus \mathbb{R})$ is such that $\hat{S}_\varepsilon(x) = \hat{S}(x+i\varepsilon) - \hat{S}(x-i\varepsilon)$ tends to $S(x)$

in $\mathcal{D}'(\mathbb{R})$; similar for T. Concerning the existance of the product of S and T in the sense of (j), j = 1,...,5, the implications (1) ⇒ (2) ⇒ (3) ⇒ (4) are obvious; the implication (1) ⇒ (5) is known to hold. No distinguishing examples are known.

Question: Determine which of these products are equivalent.

Remarks (a) Kamiński asked in [Studia math. 74 (1982),83-96] whether (2) and (3) are equivalent. (b) It has been shown recently by Jelinek [Comment. Math. Univ. Carolinae 27 (1986), 377-394] that the product of S and T in the Colombeau algebra $G(\mathbb{R}^n)$ admits an associated distribution if and only if the product (3) of S and T exists.

Proposed by prof. E. Pap:

Problem 9 Is a K-weakly sequence a K-sequence in a locally convex vector space?
(For the explanations see the paper of E. Pap in this Proceedings).

Proposed by prof. T. D. Todorov:

Problem 10 Rigged Hilbert Space for Quantum Mechanics and Quantum Field Theory.

Proposed by prof. Ch. Swartz:

Problem 11 Let X be a convergence group and Σ a σ-algebra of subsets of a set S. If $\mu_i : \Sigma \to X$ is countable additive for each $i \in \mathbb{N}$ and $\lim_{i \to \infty} \mu_i(E) = \mu(E)$ exists for each $E \in \Sigma$, when is μ necessarily countably additive?

FOLLOWING OPEN PROBLEMS WERE PRESENTED AT THE SYMPOSIUM ON GENERALIZED

FUNCTIONS AT SREBRNO (YUGOSLAVIA) June 16-21, 1971

Proposed by prof. T. K. Boehme:

Problem 1 Say that a space has the diagonal sequence property if

$$x_{n,m} \to x_m \quad \text{as } n \to \infty$$

each $m = 1, 2, ..$

and

$$x_m \to x \quad \text{as } m \to \infty$$

implies there exists a diagonal sequence $x_{n,m(n)}$ such that

$$x_{n,m(n)} \to x \text{ as } n \to \infty.$$

Every matric space has the diagonal sequence property.

Boehme pointed out (T. K. Boehme, The convolution Integral, SIAM Review 10 (1968), Corrolary 3.5, page 415) that the space of Mikusiński operators with non-negative support numbers has the diagonal sequence property when the usual convergence is used (Type I convergence).

Question: Is the sequential Topology on the space of operators with non-negative support numbers metrizable?

(Comment: yes - P. Antosik).

Problem 2 Let m_g be the mapping of $C(\mathbb{R}^N)$ into $C(\mathbb{R}^N)$

$$m_g : \{f(t)\} \to \{g(t)f(t)\}.$$

If g is continuous on $\mathbb{R}$ then m_g is a continuous mapping of $C(\mathbb{R})$ into $C(\mathbb{R})$. If $g \in C^\infty(\mathbb{R})$ then m_g has a unique continuous extension to a continuous linear map of $\mathcal{D}'_+(\mathbb{R})$ into $\mathcal{D}'_+(\mathbb{R})$ where $\mathcal{D}'_+$ is the space of right sided distributions.

Question: If g is an entire function, does the map m_g have a unique continuous extension to a map of M into M where M is the space of Mikusiński operators?

Problem 3 Let $\{\phi_n\}$ be an orthonormal system in $L^2[0,1]$. If $\phi_n(x) = e^{2\pi i n x}$ then the Fourier series of an $f \in L^1[0,1]$ has the Riemann localization property, i.e.,

if $f \in C^\infty(\alpha,\beta)$ $0 \le \alpha < \alpha_1 < \beta < \beta_1 \le 1$ then the Fourier series of f converges uniformly to f on $[\alpha_1, \beta_1]$.

This property fails f is a distribution. For example if $f = \delta$ is the delta function then $\delta(x) = 0$ on $(0,1)$, but the Fourier series

$$\delta(x) \sim \sum_{-\infty}^{\infty} e^{2\pi i n x}$$

does not converge uniformly on any interval.

Question: Is there an orthonormal sequence $\{\phi_n\}$ such that the Fourier expansion of distributions with respect to the ϕ_n's has the localization property?

Problem 4 Let P be a polynomial in k variables and let $P(D)$ be the corresponding partial differential operator with constant coefficients. Here

$$D^n = \frac{\partial^{|n|}}{\partial x_1^{n_1} \ldots \partial x_k^{n_k}} \qquad n = (n_1, n_2, \ldots, n_k) \\ |n| = n_1 + n_2 + \ldots + n_k.$$

Let s be the corresponding Mikusiński differentiation operator $s = (s_1, s_2, \ldots, s_k)$ and

$$s^n = s_1^{n_1} s_2^{n_2} \ldots s_k^{n_k}.$$

Question: If $P(D)$ is hyperbolic is the Mikusiński operator $1/P(s)$ necessarily a regular operator? Similarly, if $P(D)$ is parabolic is $1/P(s)$ necessarily a regular operator?

[An operator a is regular if for every $\varepsilon > 0$ there is a continuous non-negative function ϕ which vanishes outside of $|x| > \varepsilon$, $\int_{\mathbb{R}^N} \phi(x)dx = 1$, and an $f \in C(\mathbb{R}^N)$ such that $a = f/\phi$].

Proposed by V. A. Ditkin:

Problem 5 The continuous derivative $a'(\lambda)$ of the operator function $a(\lambda) = F(t;\lambda)/G(t;\lambda)$ is the class in the field of Mikusiński operators, the representative of which is the pair $(F_\lambda * G - G * G_\lambda, G * G)$ (see V. A. Ditkin and A. P. Prudnikov, Integral Transformations and Operational Calculus, Fizmatgiz, Moscow, 1961, p. 114). L. Berg proved with certain restrictions, that if $a'(\lambda) = 0$, $\alpha < \lambda < \beta$, the operator $a(\lambda)$ is independent on the parameter λ (see L. Berg, Asymptotische Auffassung der Operatorenrechnung, Studia Math. 21, 1962, p. 215-229).

Prove that it is the property valid in general case.

Proposed by A. P. Prudnikov:

Problem 6 Build a generatrix function and analogue of Rodrigues' formula for Ditkin's polynomials orthogonal on the interval $0 \leq x < \infty$ with respect to the weight function $p(x) = 2K_0(2\sqrt{x})$ where $K_0(2\sqrt{x})$ is the MacDonald function, as defined in V. A. Ditkin and A. P. Prudnikov, "Integral Transformations", Mathematical Analysis, 1966, (Progress in Mathematics, VINITI AN SSSR, Moscow, 1967, p. 19).

Proposed by B. Stanković:

We suppose that in the field K of Mikusiński's operators we have a topological structure given by the convergence of sequences type I (type II).

Problem 7 Find the conditions under which a subset of the field of operators K is a compact set.
(J. Burzyk solved this problem in his paper published in Studia Math. T. LXXV (1983), 313-333).

Problem 8 Find the conditions under which a mapping of a subset of the field of operators in itself has a fixed point.

Proposed by R. A. Struble:

Problem 9 Continuity of Operators.
Each operator x of the Mikusiński field may be considered a mapping which sends "denominators to numerators" in an equivalence class of fractions. (Studia Math. 37, 1971, 103-109). Using the ordinary representations of x as quotients of elements of the ring of continuous functions on $[0,\infty)$ with the topology of compact convergence, some of these mappings are continuous and some of them are not continuous. For example, the integration operator is continuous while the differentiation operator is not. It would be of interest to characterize in some simple fashion those operators which

are continuous. Of even more interest, perhaps, would be to characterize those operators x which are continuous mappings using the representations of x as quotients of elements of the ring of infinitely differentiable functions on $(0,\infty)$ with the topology of compact convergence of all derivatives. In this situation, both the integration and differentiation operators are continuous and, moreover, all distributions are continuous. The reciprocal of a nonzero ring element, however, is not a continuous mapping.

Problem 10   Continuous Linear Operator Transformations

E. Gesztelyi has defined (Publ. Math. Debrecen 14, 1967, 162-206) as continuous, a linear transformation F of the Mikusiński field if for every real interval $[\alpha,\beta]$ and every operator function $f(\lambda)$, continuous on $[\alpha,\beta]$, the operator function $F[f(\lambda)]$ is continuous on $[\alpha,\beta]$. All classical transformations (the ring generated by the algebraic derivative, the exponential shifts, the dilatations and the field elements), are continuous. Are there any nonclassical, continuous linear operator transformations?
(Comment: See A. Bleyer's paper in this Proceedings).

PARTICIPANTS

ANTOSIK, P.,   Institut Matematyczny PAN, Wieczorka 8, 40-013 Katowice, Poland

BEATTIE, R.,   Dept. of Mathem., Mount Allison University, Sackville, N. B., EOA 3C0, Canada

BERG, L.,   Wilhelm-Pieck Universität Rostock, Sektion Mathem., DDR, 2500 Rostock, Iniversitätsplatz 1

BLEYER, A.,   Techn. Univ. Budapest, Stoczek u. 3H, 1111 Budapest, Hungary

BOEHME, T.,   University of California, Santa Barbara, CA 93106, USA

BOZHINOV, N., Institute of Mathem., Bulgarian Academy of Sci., 1090 Sofia, P.O. Box 373, Bulgaria

BURZYK J.,   Institut Matematyczny PAN, Wieczorka 8, 40-013 Katowice, Poland

CARMICHAEL, R., Dept. of Mathem. and Computer Science Wake Forest Univ., P. O. Box 7311, Winston-Salem NC 27109, USA

COLLIER, M.,   Mary Washington College, Dept. of Mathem., Fredericksburg, Virginia, USA

COLOMBEAU, J-F., UER de Mathematiques Universite de Bordeaux I, 33405 Talence, France

DESPOTOVIĆ-NIKOLIĆ, D., Institute of Mathem., 21000 Novi Sad, Dr Ilije Đuričića 4, Yugoslavia

DIEROLF, P.,   FB IV - Mathematik der Universität Trier, Pf. 3825, D-5500 Trier, West Germany

DIMOVSKI, I., Institute of Mathem., Bulgarian Academy of Sci., 1090 Sofia, P.O. Box 373

DOLCHER, M.,   Universita di Trieste, Dept. di Matematica, Trieste, Via Tagliapietra 4, Italy

FILTER, W.,   Mathematik, ETH - Zentrum CH, 8092, Zürich, Switzerland

FISHER, B.,   Dept. of Mathem., The University Leicester, LE1 7RH, England

GÄHLER, W.,   Akademie der Wissenschaften der DDR, Karl-Weierstrass Inst. für Mathematik, 1086 Berlin, Mohrenstrasse 39, DDR

GEDEON, R.,   P.O. Box 88 Betlehem, Birzeit Univ., West Bank Israel

GESZTELY, E., Dept. og Mathem., Lajos Kossuth University, H-4010 Debrecen, P.f. 12, Hungary

GOYAL, A.N., Dept. of Mathem., University of Rajasthan, Jaipur-302004. India

de GRAAF, J., Eindhoven Univ. of Technology, Dept. of Math. and Comp. Sciences, Eindhoven, The Netherlands

HADŽIĆ, O., Institute of Mathematics, 21000 Novi Sad, Dr Ilije Đuričića 4, Yugoslavia

HIGGS, D., Pure Mathematics Sept., University of Waterloo, Ontario, Canada N2L 3G1

ISHIKAWA, S., Dept. of Mathem., Faculty of Science and Technology, Keio University, Hiyoshi 3-14-1, Kouhoku-ku Yokohama, 223 Japan

ITANO, M., Faculty of Integrated Arts & Sciences, Hiroshima University, Naka-ku, Hiroshima 730, Japan

KAMINSKI, A., Institute of Mathem., Polish Academy of Sciences, Wieczorka 8, 40-013 Katowice, Poland

KARTASHOVA, L., Dept. of Math. Rostov State Univ., 344010, Rostov-on Don, Mechnikova 79, USSR

KEITER, H., Institute of Physisk, Univ. Dortmund, Pf. 500 500 D-4600 Dortmund 50, West Germany

KELINGOS, J., Vanderbilt Univ., Box 6196 B, Nashville, TN 37325, USA

KIRYAKOVA, V., Inst. of Mathematics Bulgarian Academy of Sciences, Sofia 1090, Bulgaria

KOH, E., University of Regina, Dept. of Mathematics & Statistics, Regina, S4S OA2, Canada

KOMATSU, H., Dept. of Math., Fac. of Sci., Tokyo, 113 Japan

KUCERA, J., Dept. of Mathematics, Washington State University, Pullman, Washington 99164, USA

LANGENBRUCH, M., Mathematisches Institut der Westfällischen Wilhelms-Univer., Einsteinstr. 64, 4400 Münster, West Germany

LINDSTRÖM, M., Dept. of Mathematics, Abo Akademi, Fänriksgatan 3B 20500 Abo, Finland

LIU GIU, Eindhoven University of Technology, 5600 MB Eindhoven, The Netherlands

MARIĆ, V., Inst. of Applied Sciences, Technical Faculty, 21000 Novi Sad, V. Vlahovića 2, Yugoslavia

MICHAILOV, V., Steklov Inst. of Math., Moscow, Vavilona st. 24, USSR

MIELOSZYK, E., Gdansk Technical University, Majakowskiego 11/12, 80-952 Gdansk, Poland

MIKETINAC, M., University of Saskatchewan, dept. of Mathem., Saskatoon S7N OWO, Canada

MILOJEVIĆ, P., Dept. of Mathem., New Jersey Institute of Techn., Newark, N.J. 07102, USA

MISRA, Department of Mathematics, Indian Institute of Technology. Hauz Khas, New Delhi - 110 016, India

OBERGUGGENBERGER, M., Institut für Mathematik ind Geometrie, Univ. Innsbruck, Technikerstr. 13, A-6020 Innsbruck, Austria.

PAP, E., Institute of Mathematics, 21000 Novi Sad, Dr Ilije Đuričića 4, Yugoslavia

PELTOLA, V-P., Helsingin Teknillinen Oppilaitos, PL 166, SF -00181 Helsinki, Finland

PERSSON, J., Mathematiska Institutionen Lunds Universitet, Bpx 118, S-22100 Lunds, Sweden

PETZSCHE, J., Mathematisches Inst. der Universität Dortmund, Pf. 500 500, 4600 Dortmund, West Germany

PILIPOVIĆ, S., Institute of Mathematics, 21000 Novi Sad, Dr Ilije Đuričića 4, Yugoslavia

RICHARDSON, G., Dept. of Mathematics, University of Central Florida, Orlando, Fl 32816, USA

RIECKE, C.V., Cameron University, Lawton, OK 73505, USA

RITTER, U., Mathematik - Department, ETH - Zentrum, CH-8092 Zürich, Switzerland

ROEVER, J.W., University of Twente, Dept. of Mathematics, P.O. Box 217, 7500, AE, ENCHEDE, The Netherlands

SCHMEELK, J., Dept. of Mathem., Virginia Commonwealth University, 1015 West Main Street, Richmond, VA 23284, USA

SHANKER, H., Dept. of Civil Engineering, Indian Institute of Technology, Delhi, Hauz Khas, New Delhi 110016, India

SZÖKEFALVI-NAGY, B., Bolyai Inst., Aradi Vertanuk Tere 1, H-6720 Szeged, Hungary

SWARTZ, CH., Mathematics Department, New Mexico State Univ., Las Cruces, N. M. 88003, USA

STANKOVIĆ, B., Institute of Mathematics, 21000 Novi Sad, Dr Ilije Đuričića 4, Yugoslavia

SUHODOLC, Institut za matematiko Universe v Ljubljani, 61111 Ljubljana, Jadranska cesta 19, Yugoslavia

TAKAČI, A., Institute of Mathematics, 21000 Novi Sad, Dr Ilije Đuričića 4, Yugoslavia

TAKAČI, Đ., Institute of Mathematics, 21000 Novi Sad, Dr Ilije Đuričića 4, Yugoslavia

TILLMAN, H. G. Mathematisches Institut der Universität Münster, D-4400 Münster, Einsteinstr. 62, West germany

TODOROV, T.,   Institute for Nuclear Research and Nuclear Energy of the Bulgarian Academy of Sciences, Sofia 1784, Boul. Lenin 72, Bulgaria

TOMIĆ, M.,   Mašinski fakultet Sarajevo, Dž. Nehrua 10, 71000 Sarajevo, Yugoslavia

VLADIMIROV, V., Steklov Inst. of Math. Vavilona st. 42, Moscow, USSR

VAINIO, R.,   Dept. of Mathem., Abo Akademi, Fänriksgatan 3, 20500 Abo, Finland

VOGT, D.,   Bergische Universität-Gesamthochschule, Fachbereich 7 - Mathematik, Gaussstr. 20, D-5600 Wuppertal, West Germany

WALTER, G.,   Mathem. Dept., University of Wisconsin Milwaukee, Box 413, WI 53201, USA

WAWAK, R.,   Institute of Mathematics Polsih Academy of Sci., ul. Sniadeckich 8, 00-950 Warsawa, Poland

WIENER, J.,   Dept. of Mathem., Pan American University, Edinburg, TX 78539, USA

ZAYED, A.,   California Polytechnic State University, San Luis Obispo, CA 93407, USA

ZSIDO, L.,   Mathem. Inst. A, Pfaffenwaldring 57, 7000 Stuttgart, 7000 Stuttgart 80, West Germany

On the 27th of July 1987, one of the directors of Conference GFCA-87, prof. J. Mikusiński died at the age of 74. The mathematical community suddenly lost one of its eminent members. Professor Mikusiński will remain in our minds as one of the founders of the theory of generalized functions with his original operational approach which is called "the operational calculus" and also with his sequential approach to the theory of distributions and convergence structures.

Istvan Fenyö, Professor Emeritus at the Technical University of Budapest, Hungary, died on July 28, 1987, at the age of 70. He was one of the steady participants at Conferences on generalized functions.

Marija Skendžić, a full professor at PMF, University of Novi Sad, died on 2nd January 1988, at the age of 50. As the member of the Organizing commitee of the Conference she gave a considerable contribution in organizing the Conference in which she didn't take part because of her illness.

INDEX

Abel summable, 349
Abelian theorem, 317
Abelian type theorem, 80, 139
Abstract Wiener space, 179
  (Sobolev space on an abstract
  Wiener space, 180)
Adiabatic regularization, 52
A-function, 158
Algebra
  Φ-algebra, 30
  Ω-algebra, 31
  Ω/E-algebra, 31
  relational, 37
  Nevanlinna, 83
  Smirnov, 83
  Wiener, 90
  Eilenberg Moore, 29
Annihilation operator, 309
Anti derivative, 350
Approximate identity, 125
Archimedian-Riesz space, 369, 390
Associated elements, 14
Asymptotical expansion of distribution, 72
Asymptotic of distributions, 71
  S-asymptotic, 71
  quasiasymptotic, 71, 139
  generalized S-asymptotic, 81
Asymptotic sequence, 81

Banach-Saks theorem, 419
Banach-Steinhaus theorem, 363, 425
Barrel, 416
Beurling ultradistributions, 285
Boehmians, 3, 121, 125
  nonharmonic, 10
Boundary value, 131
  distributional, 137
  unrestricted, 137
Brouwer degree theory, 245
Burzyk-Paley-Wiener theorem, 121

Canonical partition function, 47
Cartwright class, 123
Cauchy group, 409
  kernel, 287
  structure, 409
  vector space,
Closed graph theorem, 378
Colombeau's algebra, 257
  filter, 338
  regularization, 333
Compatible sets, 188
Cone, 75
  acute, 75
  convex, 75, 132
Continuous iterations, 102
Continuous operator functions, 116
Convergence, 29, 368
  type I, 268, 440
  type I', 268
  type II, 440
  uniformly star, 369
  S-convergence, 29
Convergence regulator, 426
Convergence vector space, 375, 399
Convergence space
  Φ-convergence space
  F_0-convergence space, 38
Convergence structure of type I, 113
Convolution quotient, 171
Convolution product, 147
Creation operator, 309
Cut, 394
Cut-off function, 62
Cylindar functional, 310

Dedekind completnes, 386
Delta-sequence, 3
Defining function, 59
Diagonal property, 429
Different Heaviside functions, 14
Differential algebraic equation, 103

449

Differential linear space, 338
Dirac delta functional, 311
Distribution of exponential growth, 133
Distribution differential equation 279
Distributional limit, 370
Džrbasjan-Gelfand-Leontijev operators, 207

Endomorphism of the operator field, 113
Equivalence at infinity, 72
Erdélyi-Kober fractional integral, 207
Estimation
  in L, 271
  in F_0, 274

Fatou topology, 388
Fermi-like operator, 48
Fermi-liquid theory, 53
Finite functional, 94
Fixed point theorem, 399
Fluid dynamics, 17
FLUSH convergence group, 421
FLYS convergence group, 369
FLYUS convergence group, 368
Fock space, 306
Fourier hyper-functions, 68
Fourier transform, 80, 83, 87, 133
F.P. (finite part), 198
Fractional derivatives, 206
Function
  of infraexponential type, 61
  subharmonic, 62
Fundamental sequence, 4

Generalized fractional calculus, 205
Generalized function, 13
Generalized Laplacian, 311
Generalized Meier transformation, 219
Generalized number, 328
Generalized transcedental functions, 157
G-functions, 157
Green formula, 58
Group property, 102

Hamiltonian, 47
  Anderson-Hamiltonian, 49
Hardy H^p function, 131
Harmonic functions, 3
Heaviside, 13, 58
  function, 14, 128
Heavy Fermion system, 49
Heisenberg convolution algebra, 177

Hybridization term, 49
Hyper-Bessel operators of Dimovski, 208
Hyperbolic
  equations, 245
  systems, 257
Hypercomplex numbers, 331
Hyperfunctions, 3, 58, 297, 435
  Laplace, 60
  Fourier, 68
Hyperreal numbers, 331
Hyperspaces, 227

Improper integrable distribution, 360
Indicatrix function, 135
Individuals, 331
Inductive limit, 383
  α-regular, 415
Infinite, 329
Infinitesimal, 329
Integral distribution, 350
Interchange of limits, 367
Inverse Fourier transform, 342
Itterated limit, 370
Invertible cuts, 396

Jacobi matrix, 100

KKM lemma, 400
K-matrix, 367
$K'\{M_p\}$-spaces, 187
K-sequence, 419, 437
K-space

Laplace
  A-transform, 159
  equation, 3, 22
  hyperfunctions, 60
  transform of hyperfunctions, 57
Laterally complete, 389
Lattice ordered group, 393
LB-space, 415
Lebesgue property, 388
Left common multiple property, 171
Levi topology, 388
ℓ-group, 393
Linear acoustics, 20
Linear hyperbolic (n×n)-system, 257
Linear partial differential equation, 267
Linked part of a diagram, 53
Locally compact Abelian topological group, 176
Lojasiewicz's
  limit, 318
  theorem, 355, 362

Many-body perturbation expansions, 47
Matsubara-frequencies, 48
Meijer's
  G-function, 205
  transform, 219
Mikusiński's
  field, 121, 125, 176
  operators, 267, 438, 439, 440
Moderate numbers, 336
Monad, 29
  filter monad, 33
  powerset monad, 33
  ultrafilter monad, 33
  proper filter monad, 37
  generalized monad, 42
  relation monad, 42
  sequence monad, 42
Morphism, 410
M_p-compatible sets, 189

Nearly-continuous function, 376
Neumann series, 101
Neutrix, 148
  limit, 148
Non-Archimedian field, 331
Nonharmonic solutions of Laplace
  equation, 3
Nonhomogenous medium, 20
Nonlinear operator equation, 245
Normal Radon measure, 391

Object, 410
Obreschkoff method, 107
Operational calculus, 64, 235
Operator field, 113
  Ditkin-Berg, 113
Order-convergence, 396, 426
Ordinary differential equation, 99

Padé approximation, 107
Paley-Wiener-Schwartz theorem, 353
Partially ordered group, 393
p-Cauchy filter, 376
Periodic convolution quotient, 123
Periodic distributions, 353
Perturbation expansion, 54
Poisson kernel, 288
Poisson-Sonine type transformations, 208
Polynomial compatible sets, 189
Product of distributions, 436

Quantum field theory, 305, 437
Quasi-norm, 270
Quasi-positive delta sequence, 354
Quotient functor, 30

Rapid basis, 377
r^*-convergent sequence, 426
Regular Cauchy space, 410,
Regular convolution quotient, 126
Regular inductive limit, 415
Regularization of the numerical
  instability, 103
Regularizing sequence, 126
Regular modification, 411
Regular operator, 121, 435, 439
Regular Mikusiński operators, 126
Reguar spaces, 227
Regular support, 127
Riemann-Hilbert problem, 83
Riemann-Liouville fractional integral, 207
Riemann-Liouville integral, 219
Riemann localization property, 439
Riesz
  homomorphism, 383
  space, 383
Right Cauchy net, 397
R-module, 171
Roumieu ultradistributions, 293

Sampling theorem, 349
Saturation, 347
Semi-reflexive space, 417
Sequential quotients, 114
Sheaf morphism, 297
Sheaves of generalized functions, 297
Shock wave, 13, 17
Silva's order of growth, 317
Singular integral, 347
Slowly varying function, 139
Sokhotski formulas, 197
Solid subspace, 385
Space of multipliers, 417
Spaces of the type LA{}, LA'{}, 161
Stieltjes transformation, 139
Stonian space, 391
Strand, 379
Strict web, 380
Strongly correlated system, 49
Strongly first countable, 376
Structural theorem, 140
Subfunctor, 30
Substitution, 115
  W(p), 115
  refinement, 116

Tempered distributions, 139
Theory of characteristic, 100
Topological boundary, 131
Transcedental functions, 157
T-spaces, 176
Tube domain, 131
Twins, 53

451

u*-convergent sequence, 426
Ultracomplete, 378
ultradifferential operator, 286
Ultradistributions, 83
  of the Beurling type, 86
Union functor, 30
Uniform boundness principle, 425
Unit-sequence, 188
Universal mapping property, 412

Vector lattice, 425
Viscous media, 17

Weak integrable distributions, 360
Weak K-sequence, 419, 437
Weakly compact set, 419
Weakly summable sequence, 423
Web, 375
Web-space, 375
Wiener functional, 179
Wiener-Hopf equation, 83
  generalized, 83
Wiener space, 179
Wild constant, 22